Bone Marrow Transplantation

Bone Marrow Transplantation

Richard K. Burt, M.D.
Northwestern University
Chicago, Illinois, U.S.A.

H. Joachim Deeg, M.D.
Fred Hutchinson Cancer Center
Seattle, Washington, U.S.A.

Scott Thomas Lothian, R.Ph.
Northwestern Memorial Hospital
Chicago, Illinois, U.S.A.

George W. Santos, M.D.
Johns Hopkins University
Baltimore, Maryland, U.S.A.

AUSTIN, TEXAS
U.S.A.

VADEMECUM
Bone Marrow Transplantation
LANDES BIOSCIENCE
Austin

U.S. and Canada Copyright © 1996 Landes Bioscience
All rights reserved.
No part of this book may be reproduced or transmitted in any form or by any means, electronic or mechanical, including photocopy, recording, or any information storage and retrieval system, without permission in writing from the publisher.
Printed in the U.S.A.

Please address all inquiries to the Publishers:
R.G. Landes Company, 810 S. Church, Georgetown, Texas, U.S.A. 78626
Phone: 512/ 863 7762; FAX: 512/ 863 0081

U.S. and Canada ISBN: 1-57059-560-7

Library of Congress Cataloging-in-Publication Data
CIP information applied for but not received at time of publication.

Appendices

List of Tables

EDITORS

Richard K. Burt, M.D.
Northwestern University
Chicago, Illinois, U.S.A.
Chapters 4.1, 6.1, 6.2, 6.3, 6.4, 10.1

H. Joachim Deeg, M.D.
Fred Hutchinson Cancer Center
Seattle, Washington, U.S.A.
Chapters 6.1, 7.2, 13.2

Scott Thomas Lothian, R.Ph.
Northwestern Memorial Hospital
Chicago, Illinois, U.S.A.

George W. Santos, M.D.
Johns Hopkins University
Baltimore, Maryland, U.S.A.
Chapters 1.1, 6.2, 6.3, 6.4

CONTRIBUTORS

H. Richard Alexander, M.D.
Senior Investigator
NIH, NCI, DCT, Surgery Branch
Bethesda, Maryland, U.S.A.
Chapter 8.1

S. Martin-Algarra, M.D.
University of Nebraska Medical Center
Department of Internal Medicine
Section of Oncology/Hematology
Omaha, Nebraska, U.S.A.
Chapter 6.7

Emanuele Angelucci, M.D.
Hematology Department
Hospital of Pesaro
Pesaro, Italy
Chapter 6.12

Ellen Areman, M.S.
Director, Technical Cellular
   Engineering Laboratory
Georgetown University Hospital
Washington, D.C., U.S.A.
Chapter 3.3

James O. Armitage, M.D.
Chief, Hematology Oncology
University of Nebraska Medical Center
Department of Internal Medicine
Omaha, Nebraska, U.S.A.
Chapter 6.7

Bart Barlogie, M.D., Ph.D.
Professor of Medicine
Chief, Hematology Oncology
Arkansas Cancer Research Center
University of Arkansas
   for Medical Sciences
Little Rock, Arkansas, U.S.A.
Chapter 6.6

Charles Bennett, M.D.
Associate Professor of Medicine
Northwestern University
   Medical School
Senior Research Associate
Lakeside VA Medical Center
Chicago, Illinois, U.S.A.
Chapter 1.5

Sara L. Bergerson, M.S., R.D., C.N.S.D.
Clinical Nutrition Department
NIH, Clinical Center
Bethesda, Maryland, U.S.A.
Chapter 8.2

Philip J. Bierman, M.D.
University of Nebraska Medical Center
Department of Internal Medicine
Section of Oncology/Hematology
Omaha, Nebraska, U.S.A.
Chapter 6.7

Malcolm K. Brenner, M.D., Ph.D.
Saint Jude Hospital
Director, BMT Unit
Memphis, Tennessee, U.S.A.
Chapter 6.9

F. Peter Buckley, M.D.
Medical Director, Pain and Toxicity
Fred Hutchinson Cancer Center
Seattle, Washington, U.S.A.
Chapter 8.3

William H. Burns, M.D.
Professor of Medicine
Director, BMT Unit
Medical College of Wisconsin
Milwaukee, Wiconsin, U.S.A.
Chapter 10.3

Shalina Gupta-Burt, M.D.
Rush University
Chicago, Illinois, U.S.A.
Chapter 4.2

Richard Champlin, M.D.
Director, BMT
M.D. Anderson Cancer Center
Houston, Texas, U.S.A.
Chapters 1.4, 6.5

Nelson J. Chao, M.D.
Assistant Director, BMT Unit
Stanford University Hospital
Stanford, California, U.S.A.
Chapter 11.0

Cynthia DeLaat, M.D.
Assistant Professor of Pediatrics
Director of Late Effects Clinic
Children's Hospital Medical Center
Cincinnati, Ohio, U.S.A.
Chapter 6.10

Suzanne G. Demko, P.A.C.
Lead PA and Research Coordinator
BMT Unit
Johns Hopkins Hospital
Baltimore, Maryland, U.S.A.
Chapter 2.0

Lisa Filipovich, M.D.
Professor of Pediatrics
Director, Pediatric Immunology
University of Minnesota
Minneapolis, Minnesota, U.S.A.
Chapter 6.10

Mary E.D. Flowers, M.D.
The Fred Hutchinson Cancer
    Research Center
University of Washington
    School of Medicine
Seattle, Washington, U.S.A.
Chapter 14

Michele Cottler-Fox, M.D.
Medical Director
Stem Cell Laboratory
Marrow/Stem Cell Transplant Program
University of Maryland Cancer Center
Baltimore, Maryland, U.S.A.
Chapters 3.1, 3.2, 3.4

James Gajewski, M.D.
Section of BMT
Department of Hematology
MD Anderson Cancer Center
Houston, Texas, U.S.A.
Chapter 1.4

Colleen J. Gilbert, Pharm.D.
BMT Program
Duke University Medical Center
Durham, North Carolina, U.S.A.
Chapter 4.3

Philip Gold, M.D.
The Fred Hutchinson Cancer
    Research Center
University of Washington
    School of Medicine
Seattle, Washington, U.S.A.
Chapter 14

Richard E. Harris, M.D.
Director, BMT
University of Cincinnati
Children's Hospital
Cincinnati, Ohio, U.S.A.
Chapters 6.10, 6.11

Nancy Hensel, M.T.(ASCP),
    C.H.S.(ABHI)
National Institutes of Health
National Heart, Lung
    and Blood Institute
Bethesda, Maryland, U.S.A.
Chapter 5.0

Helen E. Heslop, M.D.
Saint Jude Hospital
Director, BMT Unit
Memphis, Tennessee, U.S.A.
Chapter 6.9

Lori A.S. Hollis, Pharm.D.
Johns Hopkins Oncology Center
Baltimore, Maryland, U.S.A.
Chapter 9

Sundar Jagannath, M.D.
Professor of Medicine
Division of Hematology/Oncology
Chief of BMT
Arkansas Cancer Research Center
University of Arkansas
   for Medical Sciences
Little Rock, Arkansas, U.S.A.
Chapter 6.6

Bruce Kaplan, M.D.
Division of Nephrology
Northwestern University
Chicago, Illinois, U.S.A.
Chapter 8.6

Issa Khouri, M.D.
M.D. Anderson Cancer Center
Houston, Texas, U.S.A.
Chapter 6.5

Robert A. Krance, M.D.
Saint Jude Hospital
Clinical Director, BMT Unit
Memphis, Tennessee, U.S.A.
Chapter 6.9

William Krivit, M.D., Ph.D.
Professor of Pediatrics
University of Minnesota
   Hospitals and Clinics
Minneapolis, Minnesota, U.S.A.
Chapter 6.11

Nancy Leslie, M.D.
Assistant Professor of Pediatrics
Head, Metabolic Services
Division of Human Genetics
Children's Hospital Medical Center
Cincinnati, Ohio, U.S.A.
Chapter 6.11

Michael C. Lill, M.D.
UCLA, BMT Unit
University of California
   School of Medicine
Los Angeles, California, U.S.A.
Chapter 6.8

Gwynn D. Long, M.D.
Stanford University Hospital
Stanford, California, U.S.A.
Chapter 13.1

Guido Lucarelli, M.D.
Hematology Department
Hospital of Pesaro
Pesaro, Italy
Chapter 6.12

Dimitrios A. Mavroudis, M.D.
Hematology-Oncology Fellow
BMT Unit, NHLBI, NHI
Bethesda, Maryland, U.S.A.
Chapter 12

Philip L. McCarthy Jr., M.D.
Director, Bone
   Marrow Transplant Program
Baylor College of Medicine
Methodist Hospital
Houston, Texas, U.S.A.
Chapters 1.2, 1.3

Peter A. McSweeney, M.D.
Fred Hutchinson Cancer Center
Seattle, Washington, U.S.A.
Chapter 7.1

Carole B. Miller, M.D.
Assistant Professor of Oncology
Johns Hopkins University
Baltimore, Maryland, U.S.A.
Chapter 9

Christopher L. Morris, M.D., Ph.D.
Marrow Transplant Program
Children's Hospital Medical Center
Cincinnati, Ohio, U.S.A.
Chapter 6.13

Salim Mujais, M.D.
Division of Nephrology
Northwestern University
Chicago, Illinois, U.S.A.
Chapter 8.6

Paul G. Okunieff, M.D.
Radiation Oncology Branch
National Institutes of Health
Bethesda, Maryland, U.S.A.
Chapter 4.2

William P. Petros, Pharm.D.
BMT Program
Duke University Medical Center
Durham, North Carolina, U.S.A.
Chapter 4.3

Philip A. Pizzo, M.D.
Infectious Diseases Section
Pediatric Branch
National Cancer Institute
Bethesda, Maryland, U.S.A.
Chapter 10.2

Chitra Rajagopal, M.D.
BMT Program
Georgetown University Medical Center
Washington, D.C., U.S.A.
Chapter 3.3

Philip Rowlings, M.D.
International BMT Registry
Milwaukee, Wisconsin, U.S.A.
Chapters 6.2, 6.3, 6.4

Dominic A. Solimando Jr., M.A.
Department of Pharmacy
Thomas Jefferson University Hospital
Philadelphia, Pennsylvania, U.S.A.
Chapter 16

F. Marc Stewart, M.D.
Chairman, Hematology/Oncology
University of Massachusetts
Worcester, Massachusetts, U.S.A.
Chapters 7.3, 7.4

Keith M. Sullivan, M.D.
Fred Hutchinson Cancer Center
and University of Washington
Seattle, Washington, U.S.A.
Chapter 14

Robert W. Taylor, M.D.
Director, Critical Care Unit
Saint John's Mercy Hospital
St. Louis, Missouri, U.S.A.
Chapters 8.4, 8.5

Mary C. Territo, M.D.
Director, UCLA BMT Unit
University of California
School of Medicine
Los Angeles, California, U.S.A.
Chapter 6.8

Ann E. Traynor, M.D.
Division of Hematology/Oncology
Northwestern University
Chicago, Illinois, U.S.A.
Chapter 6.6

Guido Tricot, M.D., Ph.D.
Professor of Medicine
Division of Hematology/Oncology
Arkansas Cancer Research Center
University of Arkansas
for Medical Sciences
Little Rock, Arkansas, U.S.A.
Chapter 6.6

Steven J. Trottier, M.D.
Clinical Assistant Professor
St. Louis University
School of Medicine
Department of Critical Care Medicine
St. John's Mercy Medical Center
St. Louis, Missouri, U.S.A.
Chapters 8.4, 8.5

David H. Vesole, M.D., Ph.D.
Assitant Professor of Medicine
Division of Hematology/Oncology
Arkansas Cancer Research Center
University of Arkansas
for Medical Sciences
Little Rock, Arkansas, U.S.A.
Chapter 6.6

Christopher E. Walsh, M.D., Ph.D.
Senior Investigator
NIH, CC, CPD, Hematology Service
Bethesda, Maryland, U.S.A.
Chapter 15

Thomas J. Walsh, M.D.
Senior Investigator
Infectious Diseases Section
Head, Mycology Unit
NIH, NCI, DCT, Pediatric Branch
Bethesda, Maryland, U.S.A.
Chapter 10.2

Ilana Westerman, B.A.
Northwestern University
Chicago, Illinois, U.S.A.
Chapter 1.5

Lori A. Williams, R.N.
Clinical Nurse Specialist
Oncology and BMT
Georgetown University Hospital
Washington, D.C., U.S.A.
Chapters 1.2, 1.3

Wyndham H. Wilson, Ph.D., M.D.
Special Assistant to the Director
NIH, NCI, DCT
Bethesda, Maryland, U.S.A.
Chapter 4.1

PREFACE

This book is a therapeutic manual designed as a quick practical guide and reference for houseofficers, fellows, pharmacists and nurses on the bone marrow transplant unit. Indications, complications, drug doses and approaches to clinical management problems are emphasized. Requirements for a bone marrow transplant unit are outlined. Chapters with procedures on patient evaluation, unrelated donors, marrow processing, engraftment/relapse and general patient care (including outpatient) are included along with many other pertinent concerns. Problems, such as GVHD, infection and toxicity are also discussed. Diseases requiring bone marrow transplants, such as multiple myeloma, lymphomas, pediatric malignancies and adult solid tumors have chapters devoted to them.

ACKNOWLEDGMENT

To those physicians who helped guide me by their example, thank you. To list a few: William Burns, M.D. (Medical College of Wisconsin); Shalina Gupta-Burt, M.D. (Rush); Martin Lidsky, M.D. (Baylor); Charles Link, M.D. (Human Gene Therapy Research Institute); Edward C. Lynch, M.D. (Baylor); Henry McFarland, M.D. (NIH); Arthur Nienhuis, M.D. (Saint Jude); Steven Rosen, M.D. (Northwestern); George Santos, M.D. (Johns Hopkins); H. I. Schweppe, M.D. (Baylor); and Ann Traynor, M.D. (Northwestern). Finally, thank you to Connie and Charles Liedske for their courage and determination.

Once upon a time there were three musketeers
Known throughout the land for their good cheers
Beauty was the first, so full of effervesent life
to the lucky man would she become his wife
Science was marveled as the eighth wonder
His logic and wisdom were as clear as thunder
Courage, a great horseman, would mount and ride
For no one could stand in his stride
Far and wide did they play
And many an evil dragon did they slay

Then one day, beauty lay fallen
Science and Courage were sullen
Do not be sad, for it was not long
In the future Beauty grew strong
Next Science fell to his side
And left alone was Courage to ride

With head and sword held high, he charged from the rear
But an Achilles' heel did he also have to fear
For he knew little of Cupid's arrow
Soon it found him too, so long and narrow

May all children grow up and then once again be given the
chance to re-live life through the eyes of their children and grand-
children.

Richard K. Burt, M.D.

1

1.0 General

1

History of Bone Marrow Transplantation

George W. Santos

I) INTRODUCTION

Although there were early sporadic attempts to use bone marrow for its therapeutic effect in anemia and leukemia by oral, intramuscular or even intravenous routes, the first detonation of a nuclear device on July 16, 1945 provided the major stimulus for research in bone marrow transplantation. This research was initially heavily supported by various government agencies that had an interest in understanding the pathophysiology of ionizing radiation injury in man.

II) EARLY HISTORY

Protection from the toxicity of total body irradiation (TBI) by shielding hematopoietic tissues in animals was initially observed by the Danish investigator Fabricious-Moeller in 1922 and later confirmed with more extensive investigation by Jacobsen in 1949. This provided the rationale for further experiments of Jacobsen and colleagues in 1951 and Lorenz and colleagues in the same year, that showed that the systemic injection of spleen cells or marrow cells would provide the same protection from radiation lethality afforded by the shielding of hematopoietic tissue.

III) ACUTE GRAFT-VERSUS-HOST DISEASE (aGVHD)

In 1955 English workers reported that mice given syngeneic hematopoietic cells following otherwise lethal TBI were afforded long lasting protection but that mice given allogeneic cells engrafted, but died with a characteristic illness before 100 days. Initially this was called secondary disease. These observations were confirmed and extended by a number of workers and by the late 1950s and early 1960s the typical involvement of the skin, gut and liver was characterized as acute GVHD. The small lymphocyte later identified as a T lymphocyte was eventually implicated as the initiator of this iatrogenic disease.

It was recognized early that blood product transfusion could, because of its allogeneic T lymphocyte content, cause clinical and fatal aGVHD in individuals who were at risk by virtue of their stage of development (i.e., fetus), disease (i.e., severe combined immunodeficiency) or treatment (i.e., in some cases immunosuppressed because of chemotherapy of a malignancy). The simple measure of exposing the blood products to sufficient x-irradiation was able to prevent this problem.

Early workers noted that following marrow transplantation a marked immunodeficiency occurred but that a redevelopment of the immune system by donor cell precursors proceeded in a pattern similar to a recapitulation of ontogeny. This was quite clear in syngeneic transplants but the immunologic recovery was severely impaired in allogeneic animal transplants. A more complete understanding of this state of immunodeficiency came later in clinical studies.

Investigations in the late 1950s and early 1960s in mice and extending to the early 1970s in canines indicated that methotrexate (MTX) administered following transplantation could markedly decrease severe aGVHD. For many years it was standard practice to use MTX following transplantation to decrease the severity of GVHD. Subsequently prophylaxis for acute GVHD has been replaced with the use

Bone Marrow Transplantation, edited by Richard K. Burt, H. Joachim Deeg, Scott Thomas Lothian, George W. Santos. © 1996 R.G. Landes Company.

1.1

of cyclosporine alone or with MTX or by various forms of T cell depletion of the infused marrow.

IV) **TOLERANCE**

In early studies it became evident that a number of animals would eventually recover from aGVHD. Further investigations indicated that donor lymphoid tissues were specifically tolerant to host tissues. Strong evidence was presented in experimental models that this donor to host tolerance was mediated by specific T lymphocyte suppressor cells of donor origin.

V) **CHRONIC GVHD (cGVHD)**

In the mid 1970s, the first cases of human cGVHD were reported in recipients of HLA-identical sibling marrow allografts. Clinical manifestations were systemic involving dermal, oral, ocular, hepatic, gut and pulmonary tissues. Many patients presented with severe skin and subcutaneous fibrosis with contractures, severe wasting and frequent infections. The disease has been associated with a number of autoimmune phenomena. The incidence of this disease has been reported to be between 25-50%. Initially, without treatment only about 20% of patients with extensive cGVHD became long term survivors with a reasonable quality of life. Earlier diagnoses and advances in treatment have markedly decreased the morbidity and mortality of this complication.

VI) **TRANSPLANT PREPARATIVE REGIMENS**

Before the early 1970s, TBI alone was employed prior to marrow transplantation in end stage leukemia. The Johns Hopkins group developed the rationale for the use of high dose cyclophosphamide (CY) as a preparative regimen. Because of the high relapse rate with both regimens they were abandoned for patients with malignancies. The CY regimen with slight modification was adopted for transplantation in severe aplastic anemia and has remained the most frequently used regimen for marrow transplantation in that disease. In 1976 the Seattle group reported the use of CY (60 mg/kg) given on 2 successive days followed by 1000 cGy of TBI on the third day. Subsequently the TBI was modified as fractionated single daily doses. With minor modifications this has been the most widely used regimen. A few other groups, however, have further modified this regimen by the use of chemotherapeutic agents other than CY preceding the fractionated TBI or by employing hyperfractionated TBI.

A few transplant regimens have been employed involving only chemotherapeutic agents. The most widely employed has been a combination of busulfan and CY. In randomized studies this regimen has been shown to be as effective as the combination of CY and TBI in chronic myelogenous leukemia and in acute myelogenous leukemia. The regimen has also been adopted in the treatment of certain genetic diseases including thalassemia and sickle cell disease.

VII) **DONORS**

Initially the majority of marrow donors were genotypical HLA-identical siblings. This provided compatible donors to only about a quarter of patients. When phenotypic matched or one HLA antigen mismatched family members were employed as donors 30-35% of the patients were able to have a donor. Although there is some increase in the incidence or severity of aGVHD with these donors, the event-free survivals in the hematopoietic malignancies are the same. With greater HLA disparity, however, the results are significantly worse. In children, closely matched but unrelated donors (primarily phenotypically identical) have given reasonable results (Fig. 1.1.1).

1.1

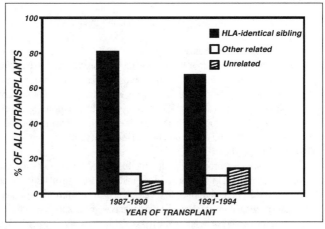

Fig. 1.1.1. Donor type for allotransplants. Reprinted with permission from IBMTR/ ABMTR 1995; 2:4.

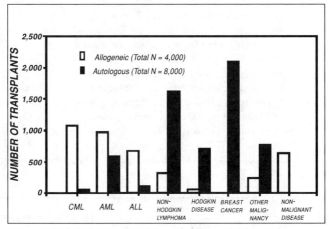

Fig. 1.1.2. Indications for blood and marrow transplants in North America, 1994. Reprinted with permission from IBMTR/ABMTR 1995; 2:3.

Although syngeneic donors are uncommon, a number of series have reported acceptable results in the hematopoietic neoplasms and in severe aplastic anemia. Encouraged by these results, the use of autologous bone marrow transplantation has been pursued with renewed and increasing vigor, at first in the lymphomas, later with the acute leukemias and more recently with various solid tumors. Autologous bone marrow transplantation in carcinoma of the breast is currently the most prevalent type of transplant for solid tumors (Fig. 1.1.2).

Although the antitumor effect demonstrated in allogeneic transplants for hematologic diseases is absent with syngeneic and autologous transplants, the hazards of morbidity and mortality associated with acute and chronic GVHD and the attempts in prevention and treatment are absent. At present the increase in relapse rate seen with autologous or syngeneic transplants is balanced against the decrease

1.1

in toxicity associated with allogeneic transplants. In general, the overall survival for autologous and allogeneic transplants in lymphoma and acute myelogenous leukemia are the same.

Peripheral blood hematopoietic stem cells have more recently been shown to be a satisfactory means of reconstituting the hematopoietic system after marrow ablative therapy (Fig. 1.1.3). This approach offers at least two potential advantages: 1) it can be collected in patients where marrow cannot be harvested because of prior damage from radiotherapy or infiltration by malignant disease and 2) it appears to produce a faster marrow reconstitution than marrow. This results in a shorter period of neutropenia and thrombocytopenia.

VIII) CLINICAL STUDIES

Despite a few promising results, the 10 years of the late 1950s to the late 1960s were a period of frustration and disappointment for marrow transplantation. Most transplants were performed in terminally ill patients who died too soon for evaluation of engraftment. Many of these patients were transplanted with marrow from individuals who were not tissue typed or who were tissue typed with unreliable methods. In cases where engraftment was successful it was followed by lethal acute GVHD and fatal viral and fungal infections.

A dramatic improvement in the therapeutic results of allogeneic transplantation began in the early 1970s when the definition of the HLA antigen system and more reliable methods of tissue typing were beginning to be established. In addition, patients with acute leukemia were being transplanted while in a complete remission rather than in florid relapse or in an infected state.

Therapeutic outcomes continued to improve during the 1970s and 80s as experience with the clinical management of transplants was acquired and advances in supportive care, including the use of newer antibacterial, antifungal and antiviral antibiotics, employment of hematopoietic growth factors and methods to control serious acute and chronic GVHD were developed. Currently one might expect cure rates of 50-90% for patients with severe aplastic anemia and various genetic diseases, 20-70% for patients with acute leukemia, 20-80% for patients with chronic

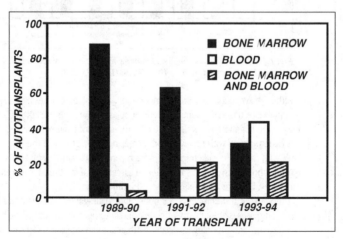

Fig. 1.1.3. *Stem cell source for autotransplants. Reprinted with permission from IBMTR/ABMTR 1995; 2:4.*

myelogenous leukemia or lymphoma depending on the age, clinical status of the patient at the time of transplantation and the type and degree of tissue matching of the donor.

IX) CONCLUSION

In the relatively short time of 3.5 decades, bone marrow transplantation has evolved from a highly experimental laboratory activity to a well established and curative therapeutic modality for the treatment of malignancy, diseases of marrow failure and selected genetic diseases.

SUGGESTED READING

1. Santos GW. History of Bone Marrow Transplantation. Clin Haem 1983; 12:611-39.
2. Santos GW. Historical background. In: Atkinson K, ed. Clinical Bone Marrow Transplantation: A Reference Textbook. Cambridge University Press, 1994; 1-9.

Bone Marrow Transplantation Unit

Philip L. McCarthy Jr., Lori A. Williams

I) REQUIREMENTS

Although autologous peripheral blood stem cell transplants may be safely performed on a general hematology/oncology unit, even more recently entirely on an outpatient basis, allogeneic bone marrow transplantation (BMT) currently remains an investigational therapy with high mortality and should not be performed on a regular medical or oncology floor. A BMT unit is a specialized area of the hospital specifically designed to prevent infectious complications and provide care by a group of nurses and physicians specifically trained in complications of intensive chemoradiotherapy, fever, neutropenia and graft-versus-host disease. The unit may be designed and staffed so that critically ill patients, who may require invasive hemodynamic monitoring, mechanical ventilation and renal dialysis, can be cared for. A BMT transplant unit is designed primarily for allogeneic patients but can be used for autologous transplants. The use of peripheral blood as a source of hematopoietic stem cells has changed the character of autologous and allogeneic transplantation. The term BMT unit will be used interchangeably for all types of stem cell transplants. Criteria for the establishment of a transplant unit are listed in Table 1.2.1. In the future, BMT units may require certification to assure minimum standards of care. Later in this chapter, we will describe recent advances that have allowed for the outpatient treatment of autologous patients.

A) NURSING STAFF

The first requirement for a BMT unit is a dedicated nursing staff who are trained in the management of pancytopenic patients and are able to recognize the complications of transplantation. The nurse to patient ratio will vary from as low as 1:1 for a patient with multiple problems to as high as 1:3 to 1:4 for patients who are relatively uncomplicated. Nurses must be familiar with the administration and complications of intensive dose chemotherapy and radiation therapy. Complications include veno-occlusive disease, renal failure, interstitial pneumonitis, cardiac arrhythmias, hemorrhage, congestive heart failure, severe stomatitis and dermatitis as well as bacterial, fungal and viral infections. Allogeneic transplantation also includes complications related to immunosuppression and graft-versus-host disease, in particular organ damage to skin, liver and gastrointestinal tract. Experienced nursing staff is critical to the consistent management of BMT patients. This is especially true in units where houseofficers and hematology/oncology fellows rotate for short periods of time. Standard orders for the management of routine problems such as fever, administration of blood products, chemotherapy and intravenous hydration facilitate the care of these complex patients and help to eliminate disorientation for rotating physicians. The lack of consistent application of protocols and implementation of procedures leads to unnecessary confusion and is demoralizing for staff and others who participate in the transplant patient care.

Bone Marrow Transplantation, edited by Richard K. Burt, H. Joachim Deeg, Scott Thomas Lothian, George W. Santos. © 1996 R.G. Landes Company.

B) PHARMACY

A dedicated pharmacist and pharmacy department that is aware of the unique needs of the transplant patient is important for the delivery of appropriate medications and for drug adjustments as clinical situations dictate. Many immunosuppressive drugs such as cyclosporine and recently FK 506 are used primarily in bone marrow or solid organ transplants. Other drugs such as busulfan, thiotepa, melphalan and cyclophosphamide are used in doses that are unfamiliar to most pharmacists. The dispensing pharmacy must be located near the transplant unit or have a transport system that provides rapid delivery of drugs to the BMT unit. A stock of commonly used medications for emergent use should be available on the BMT unit at all times, in particular antibiotics, vasopressor agents, antihypertensives, diuretics, steroids and electrolyte solutions.

C) CONSULTANTS

Allied health support is essential for a successful BMT unit including social services, dietary/nutritional support, chaplaincy, and physical and occupational therapy. Major transplant centers have physicians specialized in one or more particular areas of transplant-related complications: gastroenterology, pulmonary disease, cardiology, infectious disease, nephrology, dermatology, etc. Consultants who are familiar with transplant complications are essential for management of clinical complications that arise from intensive therapy and GVHD.

D) EXPERTISE

A certain number of transplants must be performed if a level of expertise is to be attained and maintained in the management of transplant patients. The National Marrow Donor Program, Southwest Oncology Group, Eastern Cooperative Oncology Group, Cancer and Leukemia Group B and the World Marrow Donor Association have minimum criteria for participation in allogeneic

---

Table 1.2.1. Criteria for the performance of bone marrow transplantation from the American Society of Hematology and American Society of Clinical Oncology

1) Patient volume 10-20 transplants per year (10 allogeneic if applicable, 10 autologous)
2) New unit up to this level in 2 years
3) Unit is never empty
4) Two or more designated transplant beds
5) Cryopreservation facilities
6) Histocompatibility laboratory (allogeneic)
7) Isolation procedures
8) Airhandling
9) 24-hour laboratory and radiology support including full blood bank services
10) Radiotherapy unit
11) Experienced transplant physicians
12) Consultant physicians
13) Dedicated nursing team
14) Institutional support
15) Bone marrow transplant coordinators
16) Social work support
17) Publication of results
18) Data reporting (Cooperative oncology or transplant group)
19) Pathologist experienced in histology of graft-versus-host disease

1.2

or autologous transplant protocols. In centers that do less than ten transplants a year, maintaining expertise will be difficult unless there are also significant numbers of clinically complex patients such as those with acute leukemia. Attaining and maintaining a consistent level of expertise is critical for physicians, nurses, pharmacists and all health care personnel who participate in the care of the transplant patient.

E) LOCATION

The BMT unit should be located in an area of low personnel and visitor traffic. Since BMT patients, especially patients undergoing allogeneic transplants are highly immunosuppressed, exposure to infected non-BMT patients must be minimized. Many major allogeneic BMT units are equipped for full ICU support including cardiac monitoring and ventilator management. The incidence of transplant-related cardiopulmonary compromise will vary depending on the conditioning regimen, type of transplant (allogeneic, autologous) and the underlying status of the patient before transplant. If there is a low incidence of complications (10% or less) requiring ventilator care, then complete ICU training for all BMT nursing personnel may not be cost-effective or feasible.

F) BLOOD BANK

A successful transplant unit must have a dedicated blood bank that will provide blood products including red cells, platelets and plasma emergently and in a timely manner. The cryopreservation facilities must be near the BMT unit or else a mechanism for the prompt delivery of thawed stem cell products must be guaranteed.

G) AIR QUALITY

The air quality of the unit is an important issue especially in older buildings, buildings under frequent renovation or in areas with warm, humid climates. These are conditions that favor the release or propagation of fungi. The quality of air ranges from no filtration, to standard filtration, to high efficiency particulate air (HEPA) and to laminar air flow (LAF). Standard filtration is found on most heating or air conditioning units and will filter out particles of 10-15 microns or larger. HEPA filtration will remove particles of 0.3 microns or larger with a 99.97% efficiency. LAF rooms use continuously exchanged HEPA filtered air and position the patient to receive optimal air flow. The patient is isolated by a plastic curtain or other barrier through which medical and nursing staff examine and care for the patient. Usually individuals entering the room must wear some form of gown, gloves (± sterile), mask and cap. In strict isolation LAF rooms, all individuals entering the room must be in sterile dress. This level of isolation is unusual and most BMT units have relaxed restrictions on patient isolation. LAF rooms were used initially and are still used for patients at high risk for the development of fungal infections, especially allogeneic transplants. The use of LAF rooms varies from center to center. Several large BMT centers do not use LAF rooms but will use so-called clean rooms with or without HEPA filtered air. HEPA filtered rooms provide clean air without the restrictions of the LAF room. The room air is exchanged approximately 10 to 15 times per hour. Positive air pressure establishes an air flow pattern from the patient's room to the hallway or anteroom. The corridor air in a HEPA filtered unit will be exchanged approximately six times per hour with positive pressure between the unit and adjacent areas of the hospital.

1.2

H) ROOM CLEANING

Patient rooms are designed to be easily cleaned daily with minimal dust accumulation. A full room cleaning is done after each patient stay. Air filters are often changed more infrequently depending on the type of filter and the degree to which they collect dust particles. Usually some monitoring system for air quality will determine the timing for filter changes. Seamless floors are easier to maintain than tile floors in patient rooms and bathroom facilities. Some centers use bedside commodes only and require the patients to take sponge baths instead of showers. While this is easier for maintenance, psychologically it is difficult for the patient who is already enduring the rigors of transplant conditioning and the loss of privacy associated with hospitalization.

I) WATER AND DIET

The water supply can be a source of bacterial contamination especially if used for oral intake. Tap water can be routinely screened for pathogens or bottled water may be used. Some units use filtered tap water. Diet is another variable. Many allogeneic patients will receive parenteral feeding and will have poor oral intake. However, in strict isolation units sterile food is the only food provided, whereas with less stringent protocols sterile food is not required. Most units will not allow the patient to eat raw fruit and vegetables while neutropenic. Gut decontamination with antibiotics is standard in many transplant centers and may be more important than the use of sterile food. Nonabsorbable antibiotics such as vancomycin and neomycin are used for selective gut decontamination along with an antifungal such as oral nystatin or amphotericin. Alternatively, absorbable antibiotics such as quinolines and antifungals such as fluconazole or intravenous amphotericin may be used for antimicrobial prophylaxis.

J) VISITORS

Visits by family and friends are important for the patient's psychological well being. Visiting hours and the degree of isolation will vary from center to center. Visitors who are ill should not visit the patient. Most centers will allow the family to enter the patient's room but will require some form of procedure that is designed to emphasize the patient's immunosuppressed state and to protect the patient. This can range from handwashing only to full sterile dress with gowns, gloves, mask and cap. Visitation by young children is at the discretion of the transplant center and can be a major issue for patients who wish to see their children while they are in isolation. The issues of isolation are even more acute in a pediatric setting. In pediatric centers, one parent often will stay with the child, sometimes in the room or nearby on the nursing unit. Most centers are relaxing the requirements for family member visitation, which makes the transplant stay easier psychologically for the patient and family.

K) PHYSICAL ACTIVITY

Exercise is encouraged although a patient who is confined to an isolation room will have limited choices for exercise. Thrombocytopenia will limit the ability of the patient to perform certain exercises that might put the patient at risk for bleeding. Stationary bicycle riding or exercises with small weights are reasonable alternatives for an isolated patient. In transplant units where there is less physical restriction, ambulation in the hall is permitted and encouraged, especially if the hallway is HEPA filtered.

1.2

L) INFECTION CONTROL

Infection prophylaxis guidelines vary considerably between institutions, but a consistent implementation of infection control procedures is crucial in the management of these patients. For HEPA filtered and LAF units, routine maintenance, cleaning and replacement of the filtration systems are critical for air quality. Air quality is measured by particulate quantification and by bacterial and fungal culture. Culture analysis and interpretation are controversial and are useful primarily as adjuncts to evaluating the accuracy of particulate measurements. Routine surveillance cultures of skin, mucous membranes, stool and urine are performed less often than in the past due to the lack of consistent efficacy in management of the patient. As mentioned above, many patients receive gut decontamination and are aggressively treated with intravenous antibacterial, antifungal and antiviral agents. The use of these agents has lowered the incidence of certain endogenous pathogens. Pathogens such as human herpes virus 6 (HHV-6), adenovirus, BK virus and invasive *Aspergillus*, which have little response to current antimicrobials, are major pathogens in allogeneic transplant patients who remain immunosuppressed for prolonged periods of time.

II) OUTPATIENT AUTOLOGOUS BMT

Recent advances in supportive care as well as economic pressures have radically changed where and how autologous bone marrow transplants are done. The use of outpatient infusion therapy, growth factors, peripheral blood stem cells and portable air filtration units have allowed for the early discharge of autologous transplant patients after chemotherapy or for minimal hospitalization of these patients during the transplant course. Autologous BMT patients may now stay in a hotel or motel room or be cared for at home. This can significantly decrease the cost of transplantation without compromising patient care. Many transplant centers, including our own, routinely perform outpatient autologous hematopoietic stem cell transplants. Requirements for this type of transplant include a compliant patient, a caregiver who can accompany the patient at all times, rapid engraftment, a conditioning regimen that causes minimal mucositis or other major toxicities, living quarters near the transplant center and specialized home infusion and outpatient services that guarantee prompt intervention for medical emergencies. This type of outpatient management requires more dedication and effort on the part of the physicians and outpatient nursing staff. Laboratory test results have to be rigorously checked by medical personnel and critical values must be addressed immediately. While not yet considered standard care, outpatient nursing units, laboratory support and home intravenous infusion therapy has allowed for the early discharge and outpatient monitoring of autologous neutropenic patients after transplantation. Due to the higher morbidity and mortality of allogeneic BMT, outpatient protocols have not routinely been attempted in the allogeneic setting.

SUGGESTED READING

1. The American Society of Clinical Oncology and American Society of Hematology recommended criteria for the performance of bone marrow transplantation. Journal of Clinical Oncology 1990; 8:563-564.
2. ASCO/ASH recommended criteria for the performance of bone marrow transplantation. 1990; 75:1209.
3. Rowe JM, Ciobanu N, Ascensao J, Stadmauer EA, Weiner RS, Schenkein DP, McGlave P, Lazarus HM and the Eastern Cooperative Oncology Group. Recommended guidelines for the

management of autologous and allogeneic bone marrow transplantation. Annals of Internal Medicine 1994; 120:143-158.

4. Goldman JM for the WMDA Executive Committee. A special report: bone marrow transplants using volunteer donors—recommendations and requirements for a standardized practice throughout the world -1994 update. Blood 1994; 84 Vol 19:2833-2839.

5. Treleaven J, Barrett J, eds. Bone Marrow Transplantation in Practice. Edinburgh: Churchill Livingstone, 1992.

6. Deeg HJ, Klingemann HG, Phillips GL, eds. A Guide to Bone Marrow Transplantation, 2nd ed. Berlin: Springer-Verlag, 1992.

1.2

Patient and Donor Evaluation

Lori A. Williams, Philip L. McCarthy Jr.

1.3

I) INTRODUCTION

A patient being considered for high-dose cytoreductive therapy (HDCT), with or without a hematopoietic stem cell transplant (HSCT), must undergo a thorough evaluation and explanation of the proposed treatment. The rigors of the therapy will make harsh demands on all organ systems, as well as on the psychological and social resources of the patient and family. The patient and significant others should be well informed of the potential risks and benefits of this form of treatment and about alternate therapies.

II) PATIENT EVALUATION

A) GENERAL HISTORY

1) Presenting signs and symptoms
2) Method of initial diagnosis and staging
3) Previous therapies especially chemotherapy and/or radiotherapy and responses
4) Complications
5) Relapses or progressions
6) Current disease status
7) Childhood illnesses and infectious disease exposure
8) Major illnesses, chronic illnesses, and recurring illnesses
9) Transfusion history
10) Surgical history with complications
11) Routine health screenings with dates, i.e. dental, Pap smear, and mammography
12) Allergies
13) Current medications
14) Usual weight

B) FOR FEMALE PATIENTS

1) Menarche
2) Pregnancies and outcomes
3) Date of last menstrual period
4) Onset of menopause

C) SOCIAL HISTORY

1) Educational and work history
2) Living arrangements
3) Availability of support systems
4) Financial resources and medical insurance
5) Sexual and contraceptive history
6) Smoking history
7) Alcohol and recreational drug use
8) Reactions to previous treatments for the current illness and other major health problems

D) FAMILY HISTORY

1) Composition of family
2) Health status of parents, siblings, grandparents, and children

Bone Marrow Transplantation, edited by Richard K. Burt, H. Joachim Deeg, Scott Thomas Lothian, George W. Santos. © 1996 R.G. Landes Company.

1.3

3) If testing for allogeneic hematopoietic stem cell donors is contemplated, patients who have siblings should be questioned to determine if the siblings are full, identical, or half siblings.

E) **REVIEW OF SYSTEMS**

Areas often overlooked are:

1) Upper respiratory tract symptoms
2) Visual and auditory problems
3) Dental problems
4) Fever blisters or frequent mouth sores (herpes simplex infections)
5) Indigestion
6) Constipation or diarrhea
7) Urinary incontinence
8) Nocturia
9) Peripheral edema or neuropathy
10) Muscular and skeletal pain
11) Current or previous central venous access devices should be noted.

F) **PHYSICAL EXAMINATION**

1) Vital signs, temperature, height, and weight
2) Performance status (PS) (Table 1.3.1) by one of several parameters, such as Karnofsky, Eastern Cooperative Oncology Group (ECOG), Southwest Oncology Group, or American Joint Committee on Cancer
3) Mental status
4) Eyes, ears, nose, mouth, throat, and neck
5) Breasts, chest wall, heart, and lungs
6) Abdomen, liver, and spleen
7) Rectum and external genitalia
8) Skin, extremities, and lymph nodes
9) Neurologic and musculoskeletal systems.
10) If the patient has adequate white blood and platelet counts, a digital rectal exam should be performed, and in females, a pelvic exam.

III) **PRIOR OR OUTSIDE LABORATORY TESTS**

All tissue samples should be obtained for review to confirm the diagnosis. It is also helpful to review all laboratory and diagnostic studies from the initial diagnosis, staging, and any restaging. Other available laboratory studies and diagnostic tests should be reviewed to complete the picture of the patient's current status.

IV) **REQUIREMENTS FOR HDCT**

A) **AGE**

1) **Autologous**

Autologous patients are eligible for HDCT from several months of age to 65 years or more. Some centers will consider older patients in good physical condition.

2) **Allogeneic**

Due to increasing risk of severe GVHD with older age, allogeneic HSCT have generally been restricted to patients under the age of 55, especially in the case of an unrelated or mismatched related donor. However, allogeneic sibling HSCTs have been performed on selected patients up to 70 years of age.

B) **PERFORMANCE STATUS**

PS is also an important consideration with a usual minimum requirement of a PS of less than or equal to 2 on the ECOG scale (Table 1.3.1).

1.3

Table 1.3.1. Karnofsky and ECOG performance status scale

Karnofsky Scale %	Definition of Karnofsky Scale	ECOG Scale	Definition of ECOG Scale
100	asymptomatic	0	asymptomatic
90	minor signs/symptoms but able to work normal	1	symptomatic fully ambulatory
80	some signs/symptoms, normal activity with effort	1	
70	unable to perform normal work or activity but cares for self	2	symptomatic, in bed < 50% of the day
60	requires occasional care for self needs	2	
50	requires considerable assistance and frequent medical care	3	symptomatic, in bed more than 50% of the day
40	disabled, requires special care	3	
30	severely disabled hospitalization indicated	4	bedridden
20	hospitalization and active treatment necessary	4	
10	fatal process, progressing rapidly	4	
0	dead		

C) **RESPONSE TO CHEMOTHERAPY**

A second consideration is the responsiveness of malignant disease to standard dose chemotherapy. In general, malignant disease should show at least a partial response to standard dose chemotherapy, although an exception is hematologic malignancies being treated with an allogeneic HSCT where a graft versus leukemia effect may occur.

D) **CREATININE**

Serum creatinine less than 1.5 to 2.0, or creatinine clearance > 50 ml/min.

E) **TRANSAMINASES**

Serum transaminases and bilirubin less than 2 to 4 times the upper limit of normal. The risk of venocclussive disease of the liver increases with transaminases > 2 x normal.

F) **EJECTION FRACTION**

Cardiac ejection fraction ≥ 40-45%.

G) **EKG**

No evidence of a potential life-threatening cardiac arrhythmia by EKG.

H) **PULMONARY FUNCTION TESTS**

Forced vital capacity, forced expiratory volume, and corrected diffusing capacity ≥ 50% of predicted.

I) **BACTERIAL / FUNGAL INFECTIONS**

No evidence of active sinusitis, pneumonia or other infections.

J) **VIRAL INFECTIONS**

Patients that are HIV positive are considered for HSCT only in very rare circumstances in the context of investigational protocols. Patients with active viral hepatitis or CMV infection generally are not considered for HDCT because of the extremely high risk of mortality, unless the infection has been treated or controlled. Patients serologically positive for hepatitis B or C, but without evidence of active disease are eligible for transplant, but have a slightly higher risk of hepatic toxicity.

K) PREGNANCY

Females must not be pregnant or breast feeding.

L) OTHER ILLNESSES

No other uncontrolled serious or life-threatening illnesses.

M) SOURCE OF STEM CELLS

1) Autologous

If autologous hematopoietic stem cells are to be used, the marrow, in general, must be without histologic evidence of malignant disease. Exceptions to this may be made for multiple myeloma, chronic myelogenous leukemia or if a purging procedure is being used. In general, transplants involving tumor-containing autologous stem cells are performed only in the context of investigational protocols. For an autologous bone marrow harvest, a bone marrow cellularity of < 20% will usually not produce an adequate cell harvest for transplantation. Mobilized peripheral blood stem cells are used more often than bone marrow as a source of hematopoietic stem cells.

2) Allogeneic

A patient being considered for an allogeneic HSCT must have a satisfactorily matched available donor. Donors may be of any age, ranging from approximately 6 months to 60 years old. Older related donors in good health are considered, especially if there is no alternative related donor. The donor must have no conditions which could make the process of procurement of cells permanently disabling or life-threatening to the donor. Some centers have refused to harvest extremely obese donors. If adequate care is taken with anesthesia (epidural is nearly mandatory for an obese patient) and positioning, donors as heavy as 190 kilograms have been successfully harvested using very long harvest needles. When considering a harvest on an allogeneic donor that presents unusual problems, such as compliance or question of ability to procure an adequate number of cells, it may be prudent to harvest the donor prior to starting the HDCT on the patient and cryopreserve the allogeneic marrow to ensure that it is available when it is needed. People who are HIV or HTLV-1 positive or who have a history of certain cancers are not suitable donors. Individuals who have had localized neoplasms, such as cervical or basal cell carcinoma, or who are disease-free several years after cancer treatment may be potential candidates for donation, especially if there is no alternative donor. Pregnant or breast feeding females are not eligible to be donors. People who are positive for hepatitis B or C may be donors, but the patient should be warned of the danger of contracting hepatitis from the transplant and the risk-benefit ratio should be carefully evaluated. If several donors are available, all equally good HLA matches, donors of the same sex, blood type and infectious serology (either positivity or negativity) as the patient are preferred. Donors who have never had blood product transfusions or been pregnant are also preferred. Related donors are preferred to unrelated donors due to closer HLA matching (genotypically), and because they are

1.3

1.3

more likely to be accessible for further donations of blood products and additional hematopoietic stem cells, if necessary. Allogeneic peripheral blood stem cell transplantation is becoming more common. There have been reports of cardiovascular problems in donors receiving G-CSF for stem cell mobilization. A careful cardiac evaluation should be performed on these donors.

N) SOCIAL SUPPORT

Patients must have a minimum amount of social support, which should include at least one primary care giver who could be available to provide continuous care if necessary. Transplantation can be a rigorous experience for the patient and his/her family.

V) FINANCIAL SCREENING AND EVALUATION

Once it is determined from the initial evaluation that a patient is medically a potential candidate for HDCT with or without HSCT, a financial screening evaluation must be undertaken before any HDCT-related testing is done. A written statement of specific payer benefits and the willingness of the payer to apply those benefits to the patient's proposed treatment must be obtained. This will most likely require submission of a history and current evaluation of the patient and a letter documenting the medical necessity of the proposed treatment. Other financial resources that the patient has available to cover treatment costs and additional living expenses during and after HDCT must be determined. If an allogeneic transplant is being considered, determination of financial responsibility for donor search and harvest costs must also be made.

VI) PRETREATMENT EVALUATION PARAMETERS

Tests and evaluations that should be performed prior to an autologous bone marrow harvest are found in Table 1.3.2.

A) END ORGAN FUNCTION

Tests which evaluate cardiac, renal, pulmonary and hepatic reserve are listed in Table 1.3.2. (See above requirements for HDCT.)

B) SEROLOGY (TABLE 1.3.2)

Testing allogeneic HSCT recipients for prior exposure to certain infectious organisms such as CMV is useful in selection of blood products, risk for subsequent disease and prophylactic therapy.

C) BONE MARROW BIOPSY/ASPIRATE

1) Autologous

Bilateral bone marrow aspirates and biopsies are performed to determine if disease is present in the bone marrow and if there is adequate cellularity (> 20%) for a successful autologous harvest.

2) Allogeneic

Bone marrow biopsy and aspirate of the recipient is necessary to evaluate disease. The healthy donor does not require a preharvest bone marrow biopsy or aspirate provided peripheral blood counts are normal.

D) CYTOGENETICS

Chromosomal markers for disease or clonality are present in leukemias, lymphomas and some solid tumors. Even if cytogenetic abnormalities are not expected, recent reports have documented a 5-10% incidence of myelodysplastic syndromes developing after transplant, and some investigators routinely obtain cytogenetic analysis on the patient's bone marrow before an autologous transplantation.

Table 1.3.2. Pretreatment testing and evaluation schedules

Tests and Evaluations	Autologous Bone Marrow Harvest	Autologous Peripheral Blood Stem Cell Collection	High-Dose Cytoreductive Therapy[1]	Allogeneic Hematopoietic Stem Cell Donor Evaluation
History, ROS, PE	X	X	X	X
Chest x-ray	X	X	X	X
12-lead EKG	X		X	X
CBC, diff	X	X	X	X
Serum chemistry	X	X	X	X
Urinalysis	X		X	X
PT, PTT	X		X	X
HIV by ELISA	X	X	X	X
Hepatitis B and C serology	X	X	X	X
Pregnancy test for females	X	X	X	X
Bilateral bone marrow biopsies and aspirate	X	X	X	
Cytogenetic analysis	Physician discretion	Physician discretion	Physician discretion	
2-D echo or MUGA scan			X	
PFTS with DLCO and ABGs			X	
Sinus x-ray			X	
ABO Rh blood typing			X	X
Cholesterol and triglycerides			X	
24 hour urine creatinine clearance			X	
CMV serology			X	X
Restaging evaluation			X	
Sperm / oocyte banking			X	
Nutrition evaluation			X	
Radiation oncology evaluation			X TBI conditioning	
Dental evaluation			X	
HLA typing			X allogeneic transplant	X
HTLV-1, HSV, EBV, toxoplasma titers			X allogeneic transplant	X
Family conference	X	X	X	X

[1] High-dose cytoreductive (conditioning or preparative) regimens done prior to infusion of autologous or allogeneic stem cells (marrow, peripheral blood or cord blood)

1.3

E) RESTAGING

The restaging evaluation, conducted after the completion of all other radiation therapy and systemic chemotherapy, may include:

1) Other x-rays
2) Computerized axial tomography or magnetic resonance imaging scans
3) Nuclear medicine scans
4) Tumor-associated antigens
5) Bone marrow aspirations and biopsies
6) Cerebrospinal fluid analysis

F) DENTAL

Depending on the underlying disease for which the patient is receiving treatment, the dental evaluation should include full mouth x-rays and cleaning, if this has not been done within the last 6 months or if the patient is symptomatic. Dental repairs should also be made. If major dental restorations are needed, serious consideration should be given to extracting the diseased teeth. Extractions should be completed at least 1 week prior to the start of the preparative regimen to allow for healing.

G) HLA TYPING

Patients receiving an allogeneic HSCT require human leukocyte antigen (HLA) typing for donor selection. At a minimum, this must include the identification of A, B and DR antigens by serologic testing. Many centers also include C,DQ and DP molecular antigen typing, and may routinely identify D-region antigens by molecular typing. Molecular screening of all HLA antigens may become standard practice in the future. Extended HLA testing is especially important for patients who are to receive transplants from unrelated or mismatched related donors. HLA typing of parents may need to be done to confirm the HLA typing of the patient and potential related donors. Mixed lymphocyte cultures (MLCs) to confirm tissue compatibility have not been an absolutely reliable adjunct to DR typing, but it is performed by some transplant centers.

VII) PRETREATMENT PROCEDURES

A) VENOUS ACCESS DEVICE PLACEMENT

1) Peripheral Blood Stem Cell Collection

A patient who is to undergo a PBSC collection should have some type of double lumen pheresis catheter placed for stem cell collection. After this catheter is placed, diligent care must be exercised to insure that both lumens of the catheter remain patent for easy infusion of fluids and withdrawal of blood. Urokinase should be inserted, left to dwell, and then aspirated to restore patency if either lumen of the catheter fails to function properly.

2) HDCT and Stem Cell Transplant

Arrangements must be made for a patient to have adequate venous access during HDCT. In general, the use of subcutaneous implanted venous access ports is not recommended during periods of prolonged neutropenia and thrombocytopenia because of the difficulty of removing the catheter if it becomes infected. In the past, removal of these ports prior to starting HDCT had been recommended. However, recent data have shown that there is no significant difference in complications from implanted catheters between patients who had their implanted ports removed prior to

1.3

starting HDCT and patients who had their implanted ports flushed, and then not accessed until their blood counts had returned to normal after therapy. Multilumen external central venous catheters are recommended during HDCT. These catheters may or may not be tunneled.

B) **REPRODUCTIVE CELL STORAGE**

Many patients undergoing HDCT are young. Prior to the start of therapy, the patient should be advised that the therapy may render them permanently sterile. If maintaining reproductive potential is important to the patient, arrangements for sperm or oocyte banking should be discussed with the patient. Because this storage may require several weeks to months, this subject should be discussed with the patient at the initial transplant consultation or soon thereafter.

C) **RADIOTHERAPY EVALUATION**

For a patient who will receive total body or total lymph node irradiation as part of the HDCT, a radiotherapy consultation should be obtained in a timely manner prior to the planned start of the HDCT. During this consultation, the patient's suitability for the planned therapy will be determined, risks and benefits will be discussed and special problems or needs will be addressed and planned for.

D) **AUTOLOGOUS BLOOD AND BACKUP HEMATOPOIETIC STEM CELL STORAGE**

1) **Autologous Blood Storage**

Prior to an autologous bone marrow harvest and depending on the patient's hemoglobin, hematocrit and treatment plans, 1 or 2 units of autologous blood or packed red blood cells (PRBCs) may be stored for use on the day of the harvest, or the patient may be typed and crossmatched for homologous PRBCs on the day of the harvest. If the patient's Hgb is above 10.4, autologous blood or PRBCs can be stored weekly up to 1 week prior to the harvest.

2) **Autologous Marrow Backup**

Autologous hematopoietic stem cells may be stored as a backup prior to the start of HDCT in allogeneic HSCT patients with an increased risk of graft failure. These patients include those who receive T cell depleted HSCTs or who receive transplants from unrelated or mismatched related donors.

E) **SMOKING**

It is generally recommended that the patient who smokes, stops smoking at least 2 weeks before the start of HDCT. This may be facilitated by using a nicotine transdermal patch (e.g. Nicoderm or Habitrol patch applied topically once a day. Patches come as 21 mg, 14 mg and 7 mg; generally start at 21 mg/day and taper over a period of weeks).

VIII) **DONOR EVALUATION**

A) **DONOR SCREENING**

Potential allogeneic hematopoietic stem cell donors undergo HLA typing. This typing may be done in stages to reduce costs. Initially, only serologic typing for HLA-A and -B antigens may be performed. If these antigens match those of the patient, then DR antigen typing will be done. Depending on the confidence of the typing and the relation of the donor to the patient, further HLA antigen typing or molecular typing as well as MLCs may be performed (see above discussion). Once an allogeneic hematopoietic stem cell donor is

1.3

identified, a complete medical history, review of systems and physical examination must be performed on the donor. Screening tests that are performed on the donor within 3 weeks of the harvest but prior to the start of the patient's HDCT are found in Table 1.3.2. In evaluating a potential donor, it is especially important to note:

1) History of serious or chronic illnesses
2) History of hematologic problems including bleeding tendencies
3) Cancer history
4) Transfusion history
5) Adverse anesthesia reactions
6) Current medications
7) Allergies
8) Risk factors for HIV or viral hepatitis infection
9) Pregnancy history for females

B) DONOR SELECTION

Donors will be deferred for obvious conditions, such as HIV positivity or malignancy. However, if more than one HLA matched sibling is available, donor selection can also be dependent on viral serologic status (e.g. CMV, hepatitis B or C). Preference being given to donors who are the same sex as the recipient and CMV negative donors.

C) DONOR STEM CELL HARVEST

Allogeneic hematopoietic stem cell donors should have 1-3 units of autologous blood or PRBCs stored prior to a bone marrow harvest (depending on the quantity of bone marrow to be harvested). The donor should be typed and screened for possible homologous PRBC transfusion during the harvest, but homologous blood should be used in these donors only in a life-threatening emergency. The bone marrow may be harvested under general or epidural anesthesia depending on donor and transplant center preference. The donor may stay overnight in the hospital or be discharged on the day of the procedure. Recovery from the harvest can take several days due to local trauma from multiple aspirations from the posterior (and occasionally) the anterior iliac crests.

Recently, allogeneic peripheral blood stem cell transplants have been undertaken in selected cases with success, especially with hematopoietic stem cells that have been mobilized with cytokines, such as granulocyte-colony stimulating factor (G-CSF). This is a new source of hematopoietic stem cells and may be used alone or in combination with bone marrow. The same screening criteria are used for allogeneic peripheral blood stem cell donors as for allogeneic bone marrow donors. In addition, venous access is a critical issue in allogeneic peripheral blood stem cell donors who may be required to donate on multiple occasions. Donors with poor peripheral venous access may require central venous pheresis catheters for stem cell collection, adding another potential donor complication.

Bone Marrow Transplantation from Alternate Donors

1.4

James Gajewski, Richard Champlin

I) INTRODUCTION

For patients who do not have an HLA-matched sibling, alternative donors must be considered. The opportunity of HLA-matched unrelated donor marrow transplantation provides a potentially curative treatment option for selected patients with otherwise fatal hematologic diseases. At present, this approach is limited by logistical difficulties in identifying a compatible donor and a relatively high rate of morbidity and treatment-related mortality. Decisions regarding the use of unrelated donor marrow transplantation must consider the likely length of the search process and the interests of the donor as well as the recipient. Patients with far advanced leukemia and those in unstable general medical condition are unlikely to benefit and may well suffer more as a result of the high-dose therapy or graft-versus-host disease. Volunteer unrelated individuals should only be asked to undergo general anesthesia and marrow harvest if there is a reasonable chance of success.

A) HLA-MISMATCHED RELATED DONOR

Approximately 5% of patients have a relative that is matched for 5 of the 6 HLA, -A, -B and DR serologically defined antigens. Transplants from these individuals are associated with a greater risk of both graft rejection and GVHD, but survival is comparable to transplants from an HLA-identical sibling. Transplants from relatives matched for only 3 or 4 loci have been successful in children with severe combined immune deficiency, but have been more controversial in leukemia patients because of graft failure and severe acute GVHD.

B) UNRELATED DONOR

The majority of unrelated transplants have been performed in patients with leukemia. Transplant in these patients is associated with increased GVHD, graft rejection and lower survival when compared to HLA matched sibling donor transplants. It is difficult to compare the results of HLA-identical sibling transplants with those from matched unrelated donors since unrelated transplants have generally been performed in patients with more advanced disease who tend to become more debilitated during the protracted search process. The degree of histocompatibility matching necessary for a successful outcome remains controversial. Unrelated donor and recipient will not be completely genotypically matched; phenotypic matching is a matter of degree. DNA sequencing will reveal differences between donor and recipient not fully detectable by either serologic, cellular or even the newer molecular typing techniques. In several, but not all studies, use of a serologic mismatched donor has been demonstrated to be detrimental to outcome. Patients for whom a serologic mismatched donor was selected have uncommon HLA types for which a better match could not be identified.

II) UNRELATED DONOR REGISTRIES

The development of the Anthony Nolan Foundation in the United Kingdom and subsequently the National Marrow Donor Program in the United States, as well as

Bone Marrow Transplantation, edited by Richard K. Burt, H. Joachim Deeg, Scott Thomas Lothian, George W. Santos. © 1996 R.G. Landes Company.

1.4

other large registries of potential donors in other countries, has made it possible to evaluate the use of HLA-closely matched unrelated donor transplants (Section III in Appendix I).

III) HOW TO BECOME A VOLUNTEER DONOR

Every registry has specific requirements and restrictions for persons they will allow to be considered as an unrelated bone marrow donor. All registries require that the donors be volunteers, and not be coerced in any fashion to be bone marrow donors. The NMDP requires all of its donors to meet the following basic standards:

1) Be an adult between the age of 18 and 55
2) Be in good general health
3) Not be excessively overweight
4) Have read "What you should know about becoming a marrow donor" and have a general understanding of the process of being a marrow donor and the transmission of infections, in particular HIV, through blood, marrow and plasma donation.

Contraindications to donating marrow include: HIV positive, hepatitis B surface antigen positive, hepatitis C antibody positive, IV substance abuse, cancer, sickle cell anemia, active asthma, diabetes and history of heart problems. Temporary deferrals are given for pregnancy, treatment of syphiliis and gonorrhea, blood transfusions, tattoos and hepatitis immune globulin.

Bone marrow donors are recruited through the following groups: apheresis and whole blood donors, family members tissue-typed for patients, corporations and businesses, military employees and their dependents, community groups and racial/ethnic minority populations through targeted recruitment campaigns. Most prospective donors are initially HLA-A and -B locus typed. Only targeted minority populations have typing cost funded by government grants. The cost of most typing of donors is funded by charity. Donors identified as potential HLA-A and -B matches are then asked to donate a specimen for DR locus typing. When a prospective donor is identified as a potential HLA-A, -B and DR match for a given recipient, the donor is checked for potential transmissible infectious diseases, including CMV. The donor donates marrow close to their home and the marrow is shipped by courier to the recipient. The cost of both the evaluation for donating marrow and the actual donation are borne by the recipient. The chances of being asked to donate are small. For the 1.5 million donors in the NMDP, only 3000 will be asked to donate this year.

If you have a question about donating in the United States or Canada, the NMDP maintains a toll free public education hotline—1-800-MARROW-2.

IV) UNRELATED DONOR SEARCH PROCESS

The search process involves an initial computer analysis of all donor HLA types. Potentially matching donors who are typed only for HLA-A and -B have another blood sample drawn for DR typing. HLA-A, -B and DR identical donors have a second sample drawn and tested to confirm the serologic typing and perform oligonucleotide typing for DR-beta alleles (refer to chapter on HLA). A compatible donor then undergoes a rigorous informed consent process, medical evaluation and collection of autologous blood before marrow donation.

A) DONOR POOL SIZE

A major logistical problem is identification of an HLA-compatible donor. Due to the tremendous polymorphism of the HLA gene complex, the chance of two unrelated individuals being identical for the 6 HLA-A, -B and DR loci is approximately 1/100,000. The total pool of potential donors worldwide includes approximately 2,500,000 persons. It is now possible to identify an HLA-A, -B and DR serologically matched unrelated marrow donor for up to 50% of patients depending on their racial and ethnic group.

B) RACIAL AND ETHNIC CONSIDERATIONS

The distribution of HLA phenotypes is not random; because of linkage disequilibrium, common haplotypes exist which vary among racial and ethnic groups (refer to chapter on HLA). Linkage disequilibrium results in certain A, B and DR haplotypes occurring in high frequency while others are extremely rare. HLA haplotype frequencies vary considerably among different races. Common haplotypes have been identified for major ethnic groups such as Native Americans, African Blacks, Japanese, Ashkenazi Jews and Asians which vary from Caucasians. Patients with two common haplotypes have approximately a 1 in 2000 chance of matching another individual, but a large fraction of patients have apparently unique haplotypes and will likely not find a matched unrelated donor regardless of the number of available donors. Since most donors in the major registries are Caucasians, the chance of identifying a matched donor is greater for Caucasians than for other racial or ethnic groups which have a different distribution of HLA haplotypes. A major goal of the NMDP is to recruit additional donors from racial and ethnic minorities and form reciprocal agreements with foreign registries in order to provide equal opportunity for all groups to identify a matched donor. Patients of mixed racial origin with uncommon haplotypes are the most difficult to match.

C) TIME CONSTRAINTS

This process involves multiple steps and considerable time and coordination between the transplant and donor centers. The median time from initiation of the search to performing a transplant is approximately 6 months. This is sufficient for patients with chronic myelogenous leukemia in chronic phase. This prolonged interval is, however, a major barrier to the effective use of unrelated BMT for patients with acute leukemia or bone marrow failure who are likely to deteriorate rapidly. Because of the period necessary to conduct the search, it is important for physicians to begin the search process early in the course of the patient's disease.

D) COST

Another important factor is the cost of the search process which is often not covered by third party medical insurance carriers, and is often a barrier to patient access. The cost of a donor search varies depending on how common a patient's HLA type is. The average cost of a donor search at our institution has ranged between $6,000-$10,000.

V) HISTOCOMPATIBILITY CONSIDERATIONS

The HLA-antigen system is the human major histocompatibility complex and the major immunologic barrier to transplantation (refer to chapter on HLA). One haplotype is derived from each parent and encodes a set of the class I (HLA-A,-B, -C) and class II (DR,-DQ,-DP) antigens. Both class I and II HLA loci are effective stimulators and targets for graft rejection and graft-versus-host reactions.

A) CLINICALLY SIGNIFICANT HLA TRANSPLANT LOCI

The clinically significant loci at this point in time are HLA-A, -B and DR. Mismatches may predict for both GVHD and graft rejection. Recent reports are

1.4

suggestive that C and DQ may be significant. It is doubtful that DP will be clinically significant. The clinical significance of specific loci may depend on the type of GVHD prophylaxis used.

B) **PHENOTYPE VERSUS GENOTYPE**

HLA-identical siblings inherit the same parental chromosomes and are "genotypically" identical. Unrelated donors are selected to be "phenotypically" identical as defined by serologic testing (i.e. antibody recognition of a small peptide fragment of the HLA molecule). Therefore, the majority of unrelated marrow donors do not have the same HLA genes as the recipient, only similar antibody recognition of a small peptide sequence on the HLA protein (refer to chapter on HLA).

C) **BROAD SPECIFICITY VERSUS SPLITS**

Some HLA specificities are closely related and are cross-reactive with many typing sera (refer to chapter on HLA). It is uncertain whether results from typing for broad specificities are sufficient or more specific typing for these more specific "splits" is required to prevent graft rejection and graft-versus-host disease. If precise matching for splits and structural variants is required, much larger registries will be required to identify a compatible donor for the majority of patients.

D) **MINOR HISTOCOMPATIBILITY ANTIGENS**

Minor histocompatibility antigens are the immunologically active but as yet unidentified antigens that mediate GVHD and graft rejection in the setting of an HLA genotypic identical sibling bone marrow transplant. Results of unrelated donor transplants are likely influenced by minor histocompatibility differences as well, which are likely to be greater between unrelated than related individuals.

VI) **RESULTS OF UNRELATED DONOR BMT**

A) **GRAFT FAILURE**

There is a greater risk of graft rejection following transplants from matched unrelated donors compared to transplants from HLA-identical siblings (Table 1.4.1).

Approximately 5-20% of patients receiving bone marrow transplants from matched unrelated donors experience graft failure with the highest incidence in cases using T cell depletion to prevent GVHD. This risk of graft failure is higher than seen in matched siblings and is similar to transplants from HLA-partially matched related donors.

B) **GVHD**

There is a greater risk of GVHD following transplants from matched unrelated donors compared to transplants from HLA-identical siblings (Table 1.4.1). Despite maximal conventional posttransplant immunosuppressive therapy with cyclosporine, methotrexate +/- corticosteroids, acute graft-versus-host disease > grade 2 has occurred in 60-80% of matched unrelated bone marrow transplant recipients with more than half having severe, grade 3 or 4 involvement. This compares to a total incidence of 30-40% in HLA-identical siblings receiving similar therapy with < 15% developing grade 3 or 4 manifestations. Chronic GVHD has occurred in approximately 40% of surviving patients. Given the greater risk of GVHD, many centers have intensified immunosuppressive therapy with T cell depletion to prevent GVHD.

1.4

C) GRAFT-VERSUS-LEUKEMIA

Data from many centers and the NMDP confirms a lower rate of acute and chronic GVHD in unrelated T cell depleted transplants. The risk of graft failure was increased, however, but this may possibly be overcome by more intensive immunosuppressive preparative therapy. There has not been a marked increase in leukemia relapse, possibly due to the greater degree of genetic disparity with unrelated transplants which may require fewer effector cells to mediate the GVL effect. In the initial analysis of the NMDP, use of T cell depletion was associated with a trend toward improved short term survival, but this issue remains controversial and the role of T cell depletion has not been established.

D) INFECTIONS

Patients receiving unrelated donor BMT appear to have a more severe posttransplant immunodeficiency and have a relatively high incidence of infectious complications. CMV infections are common and use of prophylactic ganciclovir is reported to reduce the incidence and mortality from this agent. Fungal infections are also common, particularly in patients receiving corticosteroids for graft-versus-host disease.

E) RELAPSE AND SURVIVAL

1) AML in Remission, ALL in Remission, CML in Chronic Phase

In these patients survival beyond 1 year is 10-15% lower than recently reported for matched siblings (Table 1.4.1). Early mortality related to rejection, GVHD and infections appear to increase with unrelated donor BMT. Preliminary data indicate transplant related mortality of 30-50% within the first 180 days. The major prognostic factors are age, early versus advanced disease status and CMV status of the donor and recipient.

2) Relapsed Acute Leukemia, CML in Transformation

Patients with more advanced leukemia have approximately a 30% disease free survival. Because of the perceived higher risk of unrelated donor bone

---

Table 1.4.1. Graft-versus-host disease, graft failure and disease-free survival for transplants with sibling matched or alternate donors

Degree of HLA Match	Acute GVHD Grade III or IV	Chronic GVHD	Graft Failure	DFS AML or ALL in Remission	DFS CML in Chronic Phase	DFS AML or ALL Relapse	DFS CML in Transformation	DFS AA
SIBLING 6/6	7-15%	30-35%	< 2%	50-60%	60-80%	20%	10-35%	78-90%
RELATED 5/6	25-30%	50%	7-9%	40-60%	60-80%	20%	10-35%	25-40%
RELATED 4/6	45-50%	50%	21%	10-40%		10%	10-30%	
HAPLO-IDENTICAL 3/6	50-100%	> 50%	> 20%	10-40%		10%	10-30%	
UNRELATED 6/6	45-50%	55%	6%	45%	40%	20%	20%	30-40%

AA=aplastic anemia, AML=acute myelogenous leukemia, ALL=acute lymphocytic leukemia, CML=chronic myelogenous leukemia, DFS=disease free survival, GVHD=graft-versus-host disease

1.4

marrow transplants, most centers reserve their use for patients without other effective treatment options, such as those with relapsed acute leukemia, chronic myelogenous leukemia in late chronic phase or with signs of acceleration.

3) Aplastic Anemia

Unrelated donor transplantation is also effective for nonmalignant diseases, although fewer patients have been treated and the optimal use of this approach is less well defined. Patients with aplastic anemia have a higher rate of graft rejection and require a more intensive preparative regimen than necessary for matched sibling grafts; survival was only 30% in the initial studies, probably due to the debilitated state of patients at the time a donor can be identified and the relatively high rate of graft failure, GVHD and infections.

4) Congenital Immunodeficiency

Excellent results have been reported with use of unrelated donor transplants for children with inborn errors of metabolism and congenital immunodeficiency syndromes. Preliminary reports of the NMDP indicate 3-year survival rates in excess of 60%. Neurologic deficits are arrested, but not reversed at the time of transplant. Therefore if bone marrow transplant is the only therapeutic option for a child where neurologic impairment is the adverse outcome, then the donor search and the subsequent transplant must be pursued and scheduled with all deliberate speed.

SUGGESTED READING

1. Kernan N, Bartsch G, Ash R et al. Analysis of 462 transplants from unrelated donors facilitated by the NMDP program. N Engl J Med 1993;328:593.
2. McGlave P, Bartsch G, Anasetti C et al. Unrelated donor transplantation therapy for CML; initial experience of the NMDP. Blood 1993; 81:543.
3. Gajewski J, Ho W, Feig S et al. Bone marrow transplantation using unrelated donors for patients with advanced leukemia or bone marrow failure. Transplant 1990; 50:244.
4. Beatty P, Hansen J, Longton G et al. Marrow transplantation from HLA-matched unrelated donor for treatment of hematologic malignancies. Transplant 1991; 51:443.
5. Stroncek D, Bartsch G, Perkins H et al. National marrow donor program. Transfusion 1993; 33:567.
6. Stroncek D, Holland P, Bartch G et al. Experiences of the first 493 unrelated marrows in the national marrow donor program. Blood 1993; 81:1940.

Costs of Care for Bone Marrow Transplantations for Malignant Hematopoietic Diseases

Ilana Westerman, Charles Bennett

I) INTRODUCTION

A limited number of studies have been conducted on the costs and cost-effectiveness of bone marrow transplantation for hematologic malignancies. A review of three signature papers will follow, each of which includes a distinct analytic framework. The type of resources included in the cost analyses, the state of technology at the time of analysis, the duration of follow-up and the clinical and quality of life benefits contribute to the analytic framework of these studies. For example, in 1975 allogeneic bone marrow transplantation for acute leukemia may not have been cost-effective when compared to intensive chemotherapy. However, due to medical advances in transplantation and supportive care, allogeneic bone marrow transplants are now cost-effective for many patients with acute leukemia.

II) WELCH STUDY

The first study of bone marrow transplantation for a hematologic malignancy, acute nonlymphocytic leukemia, was conducted by Welch et al in 1989. This study evaluated 41 patients who participated in a prospective trial of allogeneic bone marrow transplantation versus intensive chemotherapy. All of the patients had induction chemotherapy at the Fred Hutchinson Cancer Center in Seattle with all transplants supervised by the same team of doctors. Seventeen patients had an HLA matched donor and therefore received an allogeneic bone marrow transplant, while 19 patients did not have a suitable donor and received two courses of consolidation chemotherapy followed by monthly maintenance chemotherapy.

Resource-based measures were used to estimate the cost of each procedure. Data were collected for 5 resources: number of non-intensive care unit days spent in hospital, number of days spent in the intensive care unit, number of lab tests performed, number of x-rays and the number of operating room procedures. Because of the large discrepancy between charges and costs, hospital charges were not directly included. A charge estimate equation was derived to estimate the average expected cost of each procedure.

Clinical results of the study showed that the overall survival rate for bone marrow transplant patients was much higher than that of chemotherapy patients. Of the 17 patients in the bone marrow transplant group, 10 were alive 5 years later and all but 1 had a high Karnofsky score. For the 19 chemotherapy patients, 5 were alive 5 years later and they also had high Karnofsky scores. Patients treated with chemotherapy alone averaged 7 hospitalizations, while bone marrow transplant patients had 4.6 hospitalizations. However, although chemotherapy patients spent 10% more time in the hospital, bone marrow transplant (BMT) patients spent most of their hospital time in ICU centers (57% for BMT patients versus 5% for chemotherapy patients). Over a 5-year period, patients who survived had lower costs than patients who did not. For example, the average cost of a chemotherapy

1.5

survivor was $79,000 versus $157,000 for a nonsurvivor, and the average cost of a bone marrow transplant survivor was $166,000 versus $232,000 for a nonsurvivor.

Although chemotherapy costs were lower than bone marrow transplantation costs, bone marrow transplantation for acute nonlymphocytic leukemia had a favorable incremental cost-effectiveness (ICE) ratio relative to standard chemotherapy. The authors defined incremental cost- effectiveness as the difference in cost (BMT vs chemotherapy) divided by the difference in effectiveness (the number of years patients lived after BMT vs chemotherapy):

$$ICE = \frac{\text{Cost (BMT-Chemotherapy)}}{\text{\# of years of survival after therapy (BMT-Chemotherapy)}}$$

The average age of patients receiving allogeneic bone marrow transplants was 30, and all patients in remission 5 years later were expected to live to the end of their life span (47 more years). Patients receiving bone marrow transplants were in remission after 5 years at a greater rate that patients receiving chemotherapy treatment alone. Therefore, the bone marrow transplant procedure, while more expensive than chemotherapy, cost only $10,000 more than chemotherapy per life year gained which is even lower than the incremental costs of treatment of moderate hypertension in middle-aged men ($13,500 per life year gained).

Allogeneic bone marrow transplants became less cost effective as the practice of mismatched donors proliferated, which created a lower survival rate, and as the age of recipients increased. For example, if the upper age limit were increased to 55 years, a 125% rise in bone marrow transplant dollars results. Finally, the amount of time bone marrow transplant patients spend in intensive care units may be shortened as changes in practice patterns from 1989 occur, and the use of step-down units will greatly increase the cost-effectiveness estimates of bone marrow transplants. The authors concluded that even though an allogeneic bone marrow transplant for acute nonlymphocytic leukemia (ANLL) is more costly than chemotherapy alone, the procedure is cost-effective.

III) **VIENS-BITKER STUDY**

A 1989 paper entitled, "Cost of allogeneic bone marrow transplants in four diseases" analyzed the costs of allogeneic bone marrow transplants for: acute myelogenous leukemia (AML), severe combined immunodeficiency, severe aplastic anemia and chronic granulocytic leukemia (this review will not explore the results of the severe combined immunodeficiency study).

Data were collected for 12 months after the initial diagnosis because most acute clinical complications occur within 6 months to 1 year of transplantation. The study was conducted in 3 Parisian hospitals where the same treatment protocol for allogeneic HLA-identical BMT was administered for each disease. Cost components included: pharmaceuticals and blood products, disposable medical supplies, laboratory tests, radiological imaging, medical and nursing care, HLA typing and donor costs and outpatient care. Also, the costs of laboratory tests and radiology imaging include indirect costs (such as overhead) and salary costs.

Patients with acute nonlymphocytic leukemia were admitted to the hospital at least twice. The first admission prior to transplantation was for chemotherapy (vincristine, daunorubicine, cytarabine, lomestine) and supportive care (with antibiotics for febrile episodes, transfusions of irradiated blood and platelets and oral de-

1.5

contamination). Patients then were readmitted for the bone marrow transplantation. After preparation with total body irradiation and cyclophosphamide, methotrexate was given to prevent GVHD and other medications were used as necessary. Costs associated with the most common posttransplant complications were calculated: severe thrombocytopenia lasting 3 months, interstitial pneumonia and localized zoster infection. The standard costs of bone marrow transplantation and the costs with complications for bone marrow transplantation for patients with acute myelogenous leukemia are listed in Table 1.5.1. Uncomplicated cases cost an estimated $40,923, and complicated cases cost an estimated $55,839.

A second series of estimates were derived for transplantations for patients with severe aplastic anemia. Patients with severe aplastic anemia were admitted to the hospital on average 40 days prior to transplantation. They were treated with chemotherapy (cyclophosphamide), radiation therapy (6 Gy thoraco-abdominal irradiation), ketaconzole and antibiotics. Cyclosporin A and methotrexate were administered to prevent GVHD. Patients remained in the hospital for 35 days on average after transplantation. Complications secondary to GVHD almost tripled the cost of the procedure. All costs are listed in Table 1.5.2 including costs for complications (primarily representing the costs for GVHD treatment). Uncomplicated cases cost an estimated $84,537 and complicated cases cost an estimated $232,007.

Patients with chronic granulocytic leukemia had very similar treatment as patients with severe aplastic anemia, with the major differences being that patients were admitted to the hospital only 10 days prior to transplantation when immunotoxins could be used for T cell depletion of the bone marrow. Costs for this procedure are reported in Table 1.5.3. Uncomplicated cases cost an estimated $64,937.

The cost for a bone marrow transplantation for hematologic malignances appears to vary depending on the underlying pathology and the extent of complications. The cost of an uncomplicated bone marrow transplant for acute myelogenous leukemia was $40,923, for severe aplastic anemia $84,538 and for chronic granulocytic leukemia $64,938. This study did not include the costs of hospitaliza-

Table 1.5.1. Allogeneic bone marrow transplantation costs in adult acute myelogenous leukemia (in US dollars)

Cost Components	Direct Standard Cost		Cost with Complications	
Medical supplies	2,188	5.4%	2,569	4.6%
Pharmaceutical products	6,841	16.7%	9,027	16.2%
Blood products	6,467	15.8%	11,514	20.6%
Laboratory tests and x-ray	12,002	29.3%	13,803	24.7%
Nursing care	11,266	27.5%	15,327	27.5%
Physicians' time	2,159	5.3%	3,599	6.4%
Total	**40,923**	**100%**	**55,839**	**100%**

*Note table lists costs not patient charges
**From Viens-Bitker C, Fery-Lemonnier E, Blum-Boisgard C et al. Cost of allogeneic bone marrow transplantation in four diseases. *Health Policy* 12:309-317, 1989.
***French francs converted to US dollars at the 1989 rate of .16339 $/FF (an average of January, June and December rates).

1.5

tion, therefore, blood products were the largest factor in the difference in costs. However, other studies have concluded that either pharmacy costs or days in the hospital account for the largest percent of total costs and largest differences in cost estimates.

IV) BENNETT STUDY

Bennett et al reported in 1995 on the costs of care and outcomes for high-dose chemotherapy for autologous transplantation. A major goal of the study was to determine if costs improved over time. Data for the study were collected from two databases at the University of Nebraska: The University of Nebraska Medical Center Lymphoma Autologous Bone Marrow Transplant Team (for clinical data) and The Medical Center Financial Systems Database (for financial data). The authors

Table 1.5.2. Allogeneic bone marrow transplantation costs in severe aplastic anemia (in US dollars)

Cost Components	Direct Standard Cost		Cost with Complications	
Medical supplies	6,189	7.3%	11,786	5.1%
Pharmaceutical products	18,982	22.5%	69,587	30.0%
Blood products	17,184	20.3%	81,835	35.3%
Laboratory tests and x-ray	26,267	31.1%	42,150	18.1%
procedures	14,149	16.7%	22,702	9.8%
Nursing care	1,766	2.1%	3,947	1.7%
Total	**84,537**	**100%**	**232,007**	**100%**

*Note table lists costs not patient charges
**From Viens-Bitker C, Fery-Lemonnier E, Blum-Boisgard C et al. Cost of allogeneic bone marrow transplantation in four diseases. *Health Policy* 12:309-317, 1989.
***French francs converted to US dollars at the 1989 rate of .16339 $/FF (an average of January, June and December rates).

Table 1.5.3. Allogeneic bone marrow transplantation costs in chronic granulocytic leukemia (in US dollars)

Cost Components	Direct Standard Cost	
Medical supplies	3,351	5.1%
Pharmaceutical products	16,751	25.8%
Blood products	9,984	15.4%
Laboratory tests and x-ray procedures	23,842	36.7%
Nursing care	9,787	15.1%
Physicians' time	1,222	1.9%
Total	**64,937**	**100%**

*Note table lists costs not patient charges
**From Viens-Bitker C, Fery-Lemonnier E, Blum-Boisgard C et al. Cost of allogeneic bone marrow transplantation in four diseases. *Health Policy* 12:309-317, 1989.
***French francs converted to US dollars at the 1989 rate of .16339 $/FF (an average of January, June and December rates).

1.5

reported results as costs not charges, because charges were not representative of true costs. Medicare's cost to charge ratios were used to convert hospital charges to costs. All patients were treated at the University of Nebraska Medical Center between 1987 to 1991 and received high-dose chemotherapy in conjunction with either autologous bone marrow transplantation or peripheral stem cell transplantation. Patients who died in the hospital were not included in the study because their costs differed markedly from those of survivors.

One hundred seventy-eight autotransplantations were performed for patients with Hodgkin's disease. The mortality rate decreased steadily (Fig. 1.5.1). The sharp decrease in mortality during the 1987-1988 year is indicative of a learning curve effect. The costs of autotransplantations, for both autologous bone marrow transplantations and peripheral stem cell transplantations, also decreased between 1987-1991. In 1987 the average cost was $96,000, and in 1991 average costs had decreased to $55,000 (Fig. 1.5.2), mostly due to differences in the number of days patients spent in the hospital. In 1987 a patient stayed in the hospital for 51 days on the average, and in 1991 the average had decreased to 32 days while changes in costs per day changed very little over time.

One hundred forty-nine autotransplantations were performed for patients with non-Hodgkin's lymphoma. In hospital mortality rate decreased steadily over time (Fig. 1.5.3). Other clinical factors which affected survival were elevated LDH level at the time of transplantation and having a history of 4 or more previous chemotherapy treatments prior to autologous stem cell transplantation.

Costs for non-Hodgkin's lymphoma also decreased over time. In 1987 the average cost for an autotransplantation was $91,000 and in 1991 costs had decreased to $74,000, again mostly due to a decrease in hospital days. In 1987, a patient averaged 45 days in a hospital. By 1991, the average days spent in the hospital had decreased to 38 days. The average cost of an autologous bone marrow transplant for a patient who had received intensive chemotherapy without total body irradiation was $73,360, and the average cost of a peripheral stem cell transplantation for a patient who had received intensive chemotherapy without total body irradiation

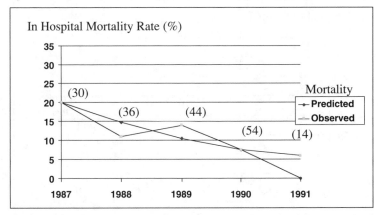

Fig. 1.5.1. Hodgkin's disease. Reprinted with permission from Bennett CL, Armitage JL, Armitage GO. Costs of care and outcomes for high-dose therapy and autologous transplantation for lymphoid malignancies: results from the University of Nebraska 1987 through 1991. JCO 13(4):969-973, 1995.

1.5

was $69,870. However, the average cost of treatment for a patient who had received total body irradiation and an autologous bone marrow transplant was $82,520, and for the patients receiving peripheral stem cell transplantation and total body irradiation, the average cost was $102,560.

Factors associated with decreasing costs were hospital staff improvements due to a learning curve effect, improvements in technology and the process of selective referral of better patients to the transplant center over time. Examples of the learning curve effect which reduced costs were addition of ICU trained nurses and addition of a specialist who was trained in both infectious diseases and oncology to the transplant team. Also, hematopoietic growth factors are an example of new

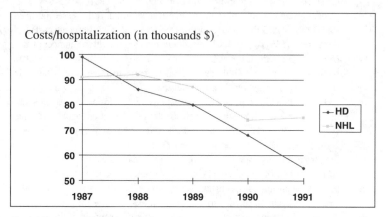

Fig. 1.5.2. Average hospitalization costs. Reprinted with permission from Bennett CL, Armitage JL, Armitage GO. Costs of care and outcomes for high-dose therapy and autologous transplantation for lymphoid malignancies: results from the University of Nebraska 1987 through 1991. JCO 13(4):969-973, 1995.

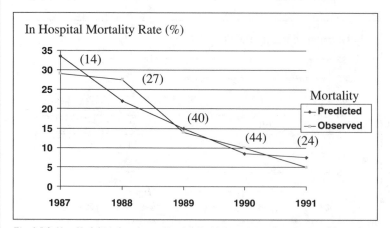

Fig. 1.5.3. Non-Hodgkin's lymphoma. Reprinted with persmisison from Bennett CL, Armitage JL, Armitage GO. Costs of care and outcomes for high-dose therapy and autologous transplantation for lymphoid malignancies: results from the University of Nebraska 1987 through 1991. JCO 13(4):969-973, 1995.

technologies that affected costs. Organizations also become more efficient over time due to organizational changes. The transplant service determined which personnel were best suited for the transplant team, which patients were most appropriate for alternative treatment regimens, where to care for patients and what equipment to use. These issues can be very different at different hospitals.

V) ITEMIZED CHARGES
FOR BONE MARROW TRANSPLANTATION

It is difficult to itemize the institutional costs and hospital charges for bone marrow transplantation due to regional variations, differences in technology, complications and duration of hospital stay. However, for your reference, approximate charges associated with bone marrow transplantation procedures are listed in Table 1.5.4.

Table 1.5.4. Approximate adult BMT charges per item (in dollars)

Item	Charge per Unit (dose or procedure or day)	Number of Units	Total Charge
Pharmacy			
acetominophen	$ 3.80	14 doses	$ 53.20
acyclovir 500 mg q8	$ 155.26	84 doses	$13,041.00
amikacin 750 mg q8	$ 277.85	42 doses	$11,669.70
amphotercin B 50 mg/day	$ 63.93	14 days	$ 895.02
ceftazidime 2 g q8	$ 101.92	42 doses	$ 4,279.80
ciprofloxacin 400 mg q12	$ 109.28	28 days	$ 3,059.84
compazine 25 mg/BID	$ 21.80	56 doses	$ 1,220.80
cyclosporine 5 mg/kg/d IV	$ 151.76	30 days	$ 4,552.80
decadron 20 mg/d	$ 42.60	5 doses	$ 213.00
diphenhydramine 25 mg	$ 21.02	14 doses	$ 294.28
droperidol QID	$ 21.20	56 doses	$ 1,187.20
famotidine 20 mg q12	$ 40.79	14 doses	$ 571.06
fluconazole 400 mg/d	$ 322.45	30 days	$ 9,673.50
foscarnet 40 mg/kg q8	$ 65.45	42 doses	$ 2,748.90
furosemide 20 mg	$ 20.90	30 doses	$ 627.00
gangciclovir 5 mg/kg/q12	$ 646.91	28 doses	$18,085.48
gentamicin 80 mg q8	$ 38.63	42 doses	$ 1,622.46
G-CSF 5 µg/kg/d	$ 559.69	14 days	$ 7,820.26
GM-CSF 250 µg/m²/d	$ 521.69	14 days	$ 7,303.66
hydrocortisone 100 mg IV	$ 22.22	14 doses	$ 311.08
itraconazole 200 mg/d	$ 27.90	24 days	$ 699.60
IVIG 400 mg/kg/week-30 g	$2100.00	12 weeks	$25,000.00
mesna 1500 mg/d			$ 3,886.94

1.5

Table 1.5.4. Approximate adult BMT charges per item (in dollars) (continued)

Item	Charge per Unit (dose or procedure or day)	Number of Units	Total Charge
morphine drip 100 mg/d	$ 70.65	7 days	$ 494.55
normal saline + 20 mEq KCL q8	$ 35.65 per liter	28 doses	$ 998.20
odansetron 32 mg	$ 445.85	3 doses	$ 1,337.65
ortho-novum (21 pack)	$ 70.00	1	$ 70.00
pentamidine 300 mg/month	$ 176.25	1	$ 176.25
peridex (chlorhexidine)	$ 40.75	10 days	$ 407.50
salt/soda	$ 9.00	10 days	$ 90.00
silver sulfadiazine cream 50 g	$ 20.70	3 tubes	$ 62.10
TPN 1 L/d	$ 200.00	14 days	$ 2,800.00
vancomycin 1 g q12	$ 66.15	28 doses	$ 1,852.20
Conditioning agents[1] ATG 30 mg/kg/day	$5685.00	3 days	$17,052.75
busulfan 16 mg/kg			$ 1,856.02
carboplatinum 800 mg/m^2			$ 6,989.68
cyclophosphamide 120 mg/kg			$ 1,219.44
etoposide 60 mg/kg			$ 1,854.70
ifosphamide 16 g/m^2			$ 3,527.50
melphalan 180 mg/m^2			$ 1,974.13
TBI 1200 cGy			$ 9,000.00
Hospital room/nursing	$1200.00	30 days	$36,000.00
Central line placement surgeon fee			$ 1,050.00
operating room-nursing fee			$ 1,000.00
HLA typing serologic class I	$ 900.00	2 people	$ 1,800.00
serologic class II	$ 900.00	2 people	$ 1,800.00
Stem cell marrow harvest-operating room -recovery-nursing			$10,000.00
peripheral blood stem cell collection	$4000.00	3 collections	$12,000.00
surgeon			$ 2,750.00
cyropreservation			$ 750.00
reinfusion			$ 692.00
Blood products PRBC 1 unit	$ 390.00	10 units	$ 3,900.00
platelets random–6 units	$ 209.00	10 units	$ 2,090.00
platelet single donor pheresis –6 units	$ 875.00	10 pheresis	$ 8,750.00

1.5

Item	Charge per Unit (dose or procedure or day)	Number of Units	Total Charge
Laboratory/radiology			
CBC c diff	$ 79.00	30	$ 2,370.00
PT	$ 35.00	3	$ 105.00
PTT	$ 55.00	3	$ 165.00
chem-7	$ 132.00	30	$ 3,960.00
chem-20	$ 214.50	30	$ 6,435.00
magnesium	$ 56.50	20	$ 1,130.00
urinalysis	$ 35.50	5	$ 177.50
chest x-ray PA/lateral	$ 122.50	3	$ 367.50
portable chest x-ray	$ 161.50	5	$ 807.50
EKG	$ 98.50	2	$ 197.00
echocardiogram	$ 561.00	1	$ 561.00
stool culture	$ 82.80	5	$ 414.00
urine culture	$ 82.50	5	$ 412.50
blood culture	$ 82.50	10	$ 825.00
C difficile toxin	$ 77.90	5	$ 389.50
gentamicin level	$ 140.00	5	$ 700.00
tobramycin level	$ 188.00	5	$ 940.00

[1] = conditioning agents cost based on 70 kg patient

SUGGESTED READING

1. Welch GH, Arson BE. Cost effectiveness of bone marrow transplantation in acute nonlymphocytic leukemia. N Eng J Med 1989; 32(12):807-812.
2. Bennett CL, Armitage JL, Armitage GO et al. Costs of care and outcomes for high-dose therapy and autologous transplantation for lymphoid malignancies: results from the University of Nebraska 1987 through 1991. JCO 1995; 13(4):969-973.
3. Finkler SA. The distinction between cost and charges. Ann Intern Med 1982; 96:102-109.
4. Veins-Bitker C, Fery-Lemonnier E, Blum-Boisgard C et al. Cost of allogeneic bone marrow transplantation in four diseases. Health Policy 1989; 12:309-317.

2

Transplantation Procedures

Suzanne G. Demko

I) **BONE MARROW ASPIRATE/BIOPSY**

A) INDICATIONS

   1) Aspirate–to assess morphology for diagnosis of hematologic malignancies, aplastic anemia, etc.; obtain samples for cytogenetic, DNA, and flow cytometry evaluation

   2) Biopsy–to assess cellularity, infiltration by malignancy, storage diseases, granuloma, vascular lesions; special stains can also be performed on sample for specific diagnoses

   3) Monitor engraftment after bone marrow transplantation

B) CONTRAINDICATIONS

   1) Infection near the intended site of access

   2) Third trimester pregnancy

   3) Prior radiation to intended site (causes fibrosis)

C) MATERIALS

   1) Minor procedure basic tray (sterile towels/drapes, 4 x 4" gauze sponges, 2 x 2" gauze sponges, povidone-iodine prep solution (Betadine), 70% isopropyl rubbing alcohol, 5 ml syringes, 19 and 25 gauge needles)

   2) Sterile gloves

   3) Lidocaine HCl 1%

   4) Sodium Bicarbonate 8.4%

   5) #11 sterile scalpel blade

   6) Bone marrow aspirate needle (We use "I" type disposable, Manan Medical Products, 16 gauge RW x 3/8"-1 7/8" with adjustable depth stop, but any other similar product is acceptable.)

   7) Bone marrow biopsy needle (We use "J" type disposable, Manan Medical Products, 11 gauge RW x 4", but any other similar product is acceptable.)

   8) 2-3 60 ml syringes

   9) Preservative-free Heparin

   10) Pressure dressing

(There are a number of companies now selling bone marrow aspirate and biopsy kits that may be substituted for the above list of materials.)

D) ANATOMIC CONSIDERATIONS

   The best approach to access the marrow cavity is just inferior to the external lip of the posterior, superior iliac crest, which is easily palpable in most patients. Care must be taken to avoid the sacro-iliac joint. The anterior iliac crest may also be utilized in patients who have been harvested recently, or whose posterior crests have been radiated as part of their treatment. The sternum may also be utilized, but is generally the last site chosen owing to the minimal depth of the marrow cavity in this site as well as the importance of, and relative ease of injuring underlying structures.

E) PREMEDICATION

   Lorazepam 2 mg IV given about 15 minutes prior to performing the procedure will offer enough temporary amnesiac effect to perform the procedure without difficulty. A combination of medications may also be used, such as

Bone Marrow Transplantation, edited by Richard K. Burt, H. Joachim Deeg, Scott Thomas Lothian, George W. Santos. © 1996 R.G. Landes Company.

2

lorazepam 1 mg IV and Demerol 50 mg IV or fentanyl citrate 10 μg IV and midazolam hydrochloride 2.5 mg IV.

F) PROCEDURE

1) Place the patient in the prone position and expose the entire posterior pelvis. Some prefer to position the patient on one side in the lateral decubitus position.

2) Palpate the posterior, superior iliac crest and determine the most accessible area for the procedure.

3) Open the basic tray, put on sterile gloves and prep the area with povidone-iodine solution in a circular motion, making increasingly larger concentric circles as you prep. Allow some time for the prep solution to dry, then wipe clean with rubbing alcohol a central area overlying the crest. This is done to allow for collection of viable marrow cells, as povidone-iodine solution is cytotoxic to bone marrow cells.

4) If collecting samples for cytogenetic, flow cytometry or DNA analysis, rinse the syringes to be used for collection with approximately 1-2 ml of preservative-free heparin. No heparin is necessary for the syringe in which the initial sample is to be collected.

5) Using a 25 gauge needle and 1% lidocaine HCl with an 8.4% sodium bicarbonate buffer (in a 3:1 dilution), raise a skin weal over the crest. Anesthetize the deeper tissue with a 19 gauge needle being certain to reach the periosteum. Then anesthetize a circular area of the periosteum in a clockface at 12, 3, 6 and 9, as well as center.

6) If performing a bone marrow aspirate only, examine the aspirate needle for defects, then insert it into the skin weal applying downward pressure as you go through the layers of subcutaneous tissue and muscle until you reach the anesthetized periosteum. If you have miscalculated the angle of entry and have reached an area of periosteum that is not anesthetized, the patient will make you aware of this very quickly. When you have determined that you are firmly anchored in anesthetized periosteum, continue to apply downward pressure using a back and forth semi-circular motion with your dominant hand as you go through cortical bone and enter the medullary cavity. As you gain proficiency with the procedure, you will develop a "feel" for this, but until that time, you need only advance the needle approximately 1-2 cm. With the needle firmly anchored, withdraw the stylet, attach a 60 ml syringe, and pull up firmly on the plunger until you have aspirated 1-3 ml of bone marrow. Transfer this immediately to a petri dish from which the laboratory technician will prepare the slides. It is important in performing this procedure that you obtain bone spicules as you aspirate, especially in the first sample. The vacuum generated by the syringe, if adequate, will insure that you obtain an appropriate sample.

7) Repeat the procedure, turning the bevel of the needle approximately 90° between each pull, and obtain one or more samples for the following as the patient's diagnosis dictates:
   1) Cytogenetic analysis
   2) DNA analysis
   3) Immunophenotyping
   4) Viral or other cultures

8) If performing a bone marrow biopsy, make a small incision in the weal

2

with a #11 scalpel blade. Take the biopsy needle, enter the incision and apply downward pressure as you go through the subcutaneous tissue and muscle, and reach the periosteum. When you have identified an area of anesthetized periosteum, advance the needle beyond the cortex, remove the stylet and advance approximately 1-2 cm. With the needle well anchored, begin to gently rock the needle until you have adequately loosened the bone specimen. Remove the needle, and use the stylet supplied to transfer the specimen to a petri dish, if touch preps are to be made; or to a specimen cup, if the specimen is to be sent directly to a pathologist.

9) Apply pressure to the area for hemostasis, and apply a pressure bandage.

10) Instruct the patient to remove the bandage the next day and leave the site to air to facilitate healing.

G) COMPLICATIONS

Needles may bend or break. There have been incidents reported where pliers have had to be autoclaved to be used to remove a broken bone marrow needle, and cut downs to retrieve needles have also been reported. Pulmonary embolism is an uncommon complication of this procedure, but can occur; therefore patients should be instructed to contact you immediately should they develop any symptoms whatsoever. In a particularly obese patient, a spinal needle may have to be utilized to properly anesthetize the underlying structures and to reach the periosteum. Pain is a common complication. Should post anesthesia pain occur, instruct the patient to take acetaminophen 650 mg PO for analgesia. If the procedure was particularly difficult, narcotic analgesia may be necessary. Bleeding of any significance occurs only rarely, is more likely with a biopsy, and is generally the result of continuing cutaneous or deeper structure hemorrhage.

If bleeding should occur, instruct the patient to apply pressure or ice to the site until the bleeding stops. Infection, including osteomyelitis can occur, but does only rarely. Remind the patient to keep the site clean and dry, especially for the first 24 hours.

II) SKIN BIOPSY

A) INDICATIONS

Diagnosis of acute or chronic graft-versus-host disease (GVHD), drug eruptions, etc. Evaluation of therapy for GVHD.

B) CONTRAINDICATIONS

Coagulopathy (relative)

C) MATERIALS

1) Minor procedure basic tray (see description above)

2) Sterile gloves

3) Lidocaine HCl 1% with epinephrine 1:100,000 (not to be used if biopsy to be taken from fingers, toes, face or genitalia)

4) Sodium Bicarbonate 8.4%

5) 3, 4 or 6 mm punch biopsy

6) Tissue specimen containers (10% formalin and transport media (Michelle's solution))

7) #11 sterile scalpel blade

8) Needle driver (Halstead forceps)

9) Suture scissors

10) Tissue forceps

2

11) 4-0 silk or nylon suture material with cutting needle
12) Antibiotic ointment
13) Bandaid (or pressure dressing)

D) CLINICAL CONSIDERATIONS

There are almost no circumstances that would preclude a skin biopsy from being performed if it is needed for diagnostic purposes. If the area for biopsy does not lend itself to the use of a punch biopsy (i.e. shin, palm of hand, sole of foot, face, etc.), it is generally preferable that the patient be referred to a dermatologist or plastic surgeon for an excisional biopsy. In most cases where a skin biopsy is needed after bone marrow transplantation, however, the thorax is the intended site of biopsy, and lends itself to the use of a punch biopsy. In the case of severe thrombocytopenia (platelets < 20,000), a platelet transfusion may be given prior to or during the procedure. In the majority of cases, the use of lidocaine HCl with epinephrine will provide sufficient hemostasis for this procedure, even in a patient who is severely thrombocytopenic.

E) PROCEDURE

1) Position the patient to provide maximum exposure to the area chosen for biopsy.
2) Open the basic tray, put on sterile gloves, and prep the area with povidone-iodine solution in a circular motion, making increasingly larger concentric circles as you prep. Allow some time for the prep solution to dry, then wipe clean the intended biopsy site with rubbing alcohol.
3) Using a 25 gauge needle, 1% lidocaine HCl with epinephrine and an 8.4% sodium bicarbonate buffer (in a 3:1 dilution), raise a skin weal. Using the same needle, anesthetize the deeper tissues in an area larger than the size of the punch biopsy to be utilized.
4) Take the punch biopsy between the thumb and index finger of your dominant hand and apply downward pressure to the anesthetized area while rotating the punch back and forth between your fingers. Continue in this fashion until the depth of the cut being made is at the hub of the punch. Lift the skin with the tissue forceps, then cut at skin level with the suture scissors. Apply a piece of gauze to the site, and divide the tissue vertically into two equal parts, placing one half into transport media, and the other into formalin.
5) Using the needle driver, tissue forceps and suture material, suture the wound.
6) Apply antibiotic ointment to the suture site and place a bandaid over it.
7) Instruct the patient, or nursing staff, to clean the sutures daily with a solution of 50% H_2O and 50% hydrogen peroxide. After 24 hours, the bandaid may be removed, and the site may be left open to air. Instructions should also be given regarding the signs and symptoms of infection and who should be informed in the event that these occur.
8) The sutures may be removed in 5-7 days, but should remain in longer, approximately 10 days, if the patient is on steroids.

F) COMPLICATIONS

Pain may occur locally after the anesthetic has worn off. Acetaminophen 650 mg PO may be taken for analgesia should this occur. In patients who are severely thrombocytopenic (< 20,000 platelets) continued bleeding/oozing may occur in spite of sutures having been placed. If this is the case, a pressure bandage may be necessary. Should the site become infected, and this occurs only rarely, a culture of the area should be performed and appropriate antibiotic therapy initiated.

III) BONE MARROW HARVEST

A) INDICATIONS

Collection of hematopoietic stem cells in sufficient numbers to reconstitute the entire hematopoietic system.

B) CONTRAINDICATIONS

Any medical condition that might put a normal, healthy, allogeneic or autologous donor at increased surgical risk. Insufficient peripheral cell counts.

C) ANATOMIC CONSIDERATIONS

The majority of marrow harvests are performed by bilateral access of the posterior iliac crests; however, should cell yields be inadequate, the anterior crests may also be harvested. The sternum can also be used, but this is not done routinely.

D) PRE-OP/ANESTHESIA

There is no "anesthesia of choice" for this procedure. General anesthesia is used commonly, as are epidural, spinal and caudal anesthesia. The patient or donor can be admitted to the hospital the night before or the day of harvest for evaluation and anesthesia consult. Pre-op orders should include NPO, CBC, PT, PTT and typing and cross matching for 2 or 3 units of packed red blood cells. In the case of an allogeneic BMT donor, arrangements should have been made in advance of the harvest for a unit of autologous blood to have been donated, and this should be on call to the operating room. An autologous unit of blood may be donated a minimum of 3 days prior to the date of harvest and can be stored without freezing for no longer than 21 days.

E) PROCEDURE

1) The patient is placed in the prone position on the operating table, and the entire area of the posterior pelvis is exposed.
2) The patient is prepped using povidone-iodine solution (Betadine), and the area is then cleaned using sterile saline.
3) The operative field is draped, being certain to allow for adequate lateral exposure.
4) Prior to beginning collection, a solution of 100 ml of tissue culture media, Plasma-Lyte, or other appropriate collection media to which 10,000-16,000 units of preservative-free heparin is added, is mixed. All harvest needles and collection syringes are rinsed with this solution. A container with 50 ml of the solution is maintained on the instrument table for the purpose of rinsing the collection syringes between each aspiration of marrow. This rinse solution is added to the marrow collection system at the end of the harvest prior to filtration. The remainder of the solution (50 ml) is placed in a collection beaker, if an open system is being used; or bag, if a commercially-available closed system is being used. When necessary for processing purposes, additional anticoagulant, such as ACD, may be added.
5) Marrow is aspirated using large bore, ball top, Rosenthal-type needles, by making several skin puncture sites on each side of the crest and multiple punctures of the bone through each skin puncture. Two to three hundred puncture sites are made in the crest by the time the procedure has been

completed.

6) No more than 10 ml should be aspirated from any given access site in order to minimize peripheral blood contamination.

7) Using the traditional method of bone marrow harvest, the collected marrow is expelled into an open stainless steel beaker, containing heparinized culture media in a collection stand (Fig. 2.1). After harvesting has been completed, fat and bone particles are removed by passing the marrow through syringes containing graded polycarbonate mesh screens (Fig. 2.2). The marrow is then transferred into a collection bag for transport. A disposable, closed collection system manufactured by Baxter, Fenwall Division is also available. With this system, collected marrow is expelled directly into a 1200 ml plastic reservoir bag in which an 850 micron pre-filter is contained. After harvesting has been completed, the marrow is passed through two additional in-line graded filters, 500 and 200 microns, respectively, and into a collection bag for transport (Figs. 2.3, 2.4, 2.5).

8) The volume of marrow harvested varies, but an average volume is approximately one liter. The number of nucleated cells generally deemed to be adequate for the purposes of engraftment varies from institution to institution, but is thought to be approximately $2.0\text{-}4.0 \times 10^8$ cells/kg recipient ideal body weight for an allogeneic recipient and $1.0\text{-}3.0 \times 10^8$ cells/kg recipient ideal body weight for an autologous recipient. The National Marrow Donor Program requests a marrow nucleated cell count of $> 2 \times 10^8$/kg of recipient actual body weight when requesting marrow of an unrelated donor.

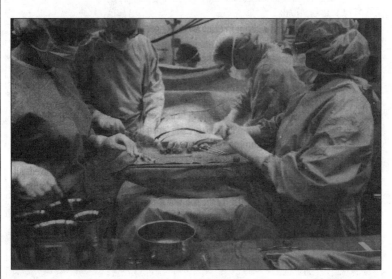

Fig. 2.1. Operative room bone marrow harvest. Two physicians, one on each side, aspirate marrow from the patient's posterior, superior iliac crest. Two scrub nurses in the forefront assist. Anesthesiologist is behind sterile drape at patient's head. Marrow is collected in stainless steel beakers (lower left).

2

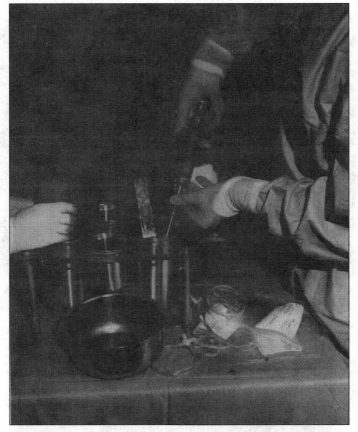

Fig. 2.2. Marrow which was collected in a stainless steel beaker is being expelled through a syringe containing mesh screens (which removes bone particles) into another sterile open stainless steel beaker.

F) POSTOPERATIVE NOTE

This should include pre- and postoperative diagnosis, a description of the harvest procedure, the harvest kit lot number and expiration date when a harvest collection kit is used, names of the surgeons performing the procedure, type of anesthesia with beginning and ending times, a description of any complications and a description of the condition of the patient after the procedure is competed. In addition, total volumes of the bone marrow specimen, tissue culture media, heparin, other anticoagulants and intraoperative fluids administered should be noted (Table 2.1).

G) POSTOPERATIVE ORDERS

Special consideration should be taken with regard to pain control, volume loss and anesthesia reactions when writing these orders. In addition, donors should be discharged with a prescription for $FeSO_4$ 325 mg PO TID x 25 days (Table 2.2).

2

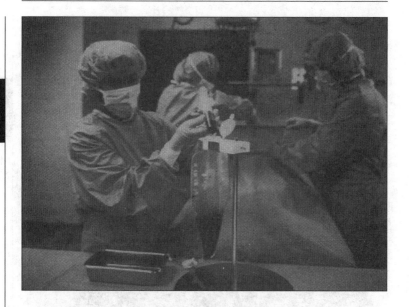

Fig. 2.3. Marrow is collected directly into a Fenwall collection bag.

Fig. 2.4. Marrow is passed by gravity from the Fenwall collection bag through in-line filters to a transfer pack.

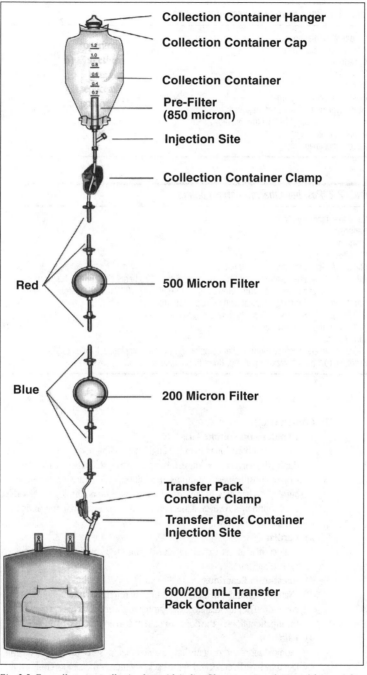

2

Collection Container Hanger

Collection Container Cap

Collection Container

Pre-Filter
(850 micron)

Injection Site

Collection Container Clamp

Red

500 Micron Filter

Blue

200 Micron Filter

Transfer Pack
Container Clamp

Transfer Pack Container
Injection Site

600/200 mL Transfer
Pack Container

Fig. 2.5. Fenwall marrow collection bag with in-line filters to remove bone and fat particles. After filtering marrow into the transfer pack, the pack is labeled, clamped and transported at room temperature to the marrow processing lab.

Table 2.1. Bone marrow harvest operative note

Pre-operative diagnosis:
Postoperative diagnosis:
Procedure:
Surgeons:
Anesthesia:
Anesthesia beginning time: ending time:
Fluids (type and volume) administered:
Blood products (e.g. PRBC) administered:
Complications:
Marrow specimen:

Table 2.2. Post bone marrow harvest orders

Admit to—recovery room
Diagnosis:
Condition:
Allergies:
Vital signs—per recovery room routine
Notify M.D. for: temperature > 38.2°C; pulse < 56 or > 120 beats per minute; respirations
    < 7 or > 32 per minute; systolic blood pressure < 80 or > 180 mmHg; diastolic blood
    pressure > 110 mmHg; unusual pain or respiratory distress
IV fluids: D5 $^1/_2$NS at 150 ml/hour until returned to floor
Diet—clear liquids when alert
Transfusions:
CBC when enters recovery room or after completing PRBC transfusion, call result to M.D.
Morphine (0.1 mg/kg, maximum 5 mg) IM or IV prn pain

H) COMPLICATIONS

1) Hypotension/Volume Loss

Severe hypotension is an exceedingly rare complication which can occur during the harvest procedure. Intraoperative crystalloids are used liberally to avoid post-harvest symptoms. In addition, colloids are used liberally in autologous harvests. In the case of allogeneic donors, autologous blood is used routinely; however, allogeneic blood has been used in the setting of significant symptoms.

2) Cardiac

Full cardiac arrest can also occur during the harvest procedure, but is exceedingly rare.

3) Anesthesia Reactions

Nausea, vomiting, headache and urinary retention are among the most common. They are managed symptomatically in the recovery room or on the inpatient floor, if the donor or patient is to be admitted to the hospital.

4) Pain

Some donors and patients have likened their post-harvest pain to feeling like they have been kicked in the back by a mule, others report only a mild backache. Suffice it to say, the level of discomfort should dictate type of analgesia, but the majority require some form of narcotic analgesia.

5) Hemorrhage

Occurs in some form in the majority of cases, ranging from mild subcutaneous purpura (common), to muscular hemorrhage (uncommon) or retroperitoneal bleed (rare).

6) Infection

Occurs only infrequently. Both cellulitis and osteomyelitis can occur and septicemia has also been reported. The most common pathogens are gram positive organisms and treatment strategies in the event of infection should take this into account.

7) Embolism

Pulmonary embolus has been reported after bone marrow harvest, but is a rare complication.

IV) BONE MARROW RE-INFUSION

A) INDICATIONS

Patients who have received radiation and/or chemotherapy in doses lethal to the bone marrow in an attempt to treat their hematologic malignancies or solid tumors.

B) MATERIALS

Bone marrow product

Emergency medications and equipment available at bedside to include:

Benadryl 50 mg IV

Epinephrine 1 mg IV (1:1000)

Solu-Cortef 200-500 mg IV

Oxygen, suction and other airway equipment

Cardiac monitor

C) SPECIAL CONSIDERATIONS

Bone marrow is a blood product and can precipitate reactions like any other transfusion product, including anaphylaxis. Because this product is intended to reconstitute the entire hematopoietic system and any manipulation after processing can injure or remove vital cells, bone marrow is **never irradiated or infused via IV infusion pump**. Special care must be taken with bone marrow products that have been processed and contain cryoprotectants. Infusions of such products must take place slowly, in limited volume aliquots, followed by normal saline to flush the IV line. The marrow should always be infused via a dedicated IV line that is not cycled through an infusion device or pump.

D) PROCEDURE FOR INFUSION OF AUTOLOGOUS MARROW

The patient should be placed on a cardiac monitor during the infusion and for 1 hour thereafter. Vital signs should be monitored at routine intervals as should urine output. In the case of autologous or peripheral blood stem cell product infusions which are generally cryopreserved with DMSO, hydration with NaHCO$_3$ should be used to alkalinize the urine and minimize the effects of DMSO-mediated histamine release. Premedications including Solu-Cortef 50-250 mg IV and Benadryl 50 mg IV should be given 15-30 minutes prior to the start of the infusion. Because autologous bone marrow and peripheral blood stem cell products have been processed and cryopreserved, they are thawed in a water bath at 37° C over a period of 1-2 minutes and infused in small aliquots (30-60 ml) over a period of a few minutes each. After each aliquot is infused, the tubing is flushed with normal saline prior to the infusion of the next aliquot.

2

E) **PROCEDURE FOR INFUSION OF ALLOGENEIC MARROW**

Allogeneic bone marrow products are infused, generally, without prior manipulation or cryopreservation with DMSO. For this reason, pre-infusion hydration and premedication are not necessary. However, in the event the product is a major ABO mismatch, red cells are removed prior to re-infusion, and the volume of marrow to be infused tends to be smaller. The infusion in either case begins slowly, at a rate of 5-10 ml/minute for the first few minutes in order to monitor for reactions. If the patient tolerates the infusion without difficulty, the rate may be increased up to a limit of 10 ml/kg/hour.

F) **COMPLICATIONS**

1) **Anaphylaxis**

Hypersensitivity reactions can occur with bone marrow products, as with any blood product, but is rare. Premedication with Solu-Cortef 50-250 mg IV and Benadryl 50 mg IV tends to ameliorate the chances of this complication occurring, but emergency airway equipment should be close at hand. If a reaction occurs, therapy includes epinephrine 1 mg (1:1000) IV, Solu-Cortef 200-500 mg IV and Benadryl 50 mg IV.

2) **Hypertension/Hypotension**

Blood pressure swings can and do occur with variable frequency and should be treated based on the presumed underlying cause. Marrow infusion may have to be slowed or temporarily interrupted pending stabilization of blood pressure.

3) **Pulmonary Edema**

Pulmonary edema is generally associated with a high volume product or red cell lysis by-products.

4) **Pulmonary Emboli**

Since marrow is filtered to remove fat and bone spicules, clinical symptoms from pulmonary emboli are rare.

5) **Dysrhythmias**

Cardiac dysrhythmias occur infrequently, but are of enough concern that intra-infusional cardiac monitoring should be employed and aggressive treatment initiated when complications of this nature do occur.

6) **Infection**

If a product is known to be contaminated, antibiotic prophylaxis is the better part of valor.

7) **Nausea, Vomiting, Abdominal Cramping**

Gastrointestinal symptoms occur relatively infrequently.

8) **Other**

It is interesting to note that often at autopsy, bone emboli are reported in bone marrow transplant recipients. These are incidental findings in the lung and other organs. It is postulated that these pieces of bone escaped filtration after harvesting.

V) **PERIPHERAL BLOOD STEM CELL COLLECTION**

A) **INDICATIONS**

Patients who have received radiation and/or chemotherapy in doses lethal to the bone marrow in an attempt to treat their malignancies or who have a hypocellular or fibrotic marrow. Patients whose marrow is infiltrated heavily with disease. Patients who have received prior pelvic radiation.

B) **CONTRAINDICATIONS**

History of severe reaction to any product used in the apheresis process.

C) **MATERIALS**

1) Central intravenous double lumen catheter
2) Anticoagulants (typically heparin and ACD)
3) Apheresis device with modifications for peripheral blood stem cell (PBSC) collection. (There are a number of different devices being used at present, for example: Aminco centrifuge, Hemonetics Model V50, Cobe Spectra, Fenwall CS 3000.)

D) **PROCEDURE/MOBILIZATION OF PBSC**

PBSCs are collected by stem cell apheresis using either a discontinuous flow device or a continuous flow blood cell separator. A predetermined cell volume is collected in one or more apheresis procedures. A variety of so-called "priming" regimens consisting of chemotherapy (e.g. Cytoxan $4 g/m^2$) or growth factors (e.g. G-CSF, GM-CSF) may be employed to optimize the cell yield prior to the patient undergoing apheresis. Using such priming tactics, it is important to note the timing of pheresis. With a 5-10 liter pheresis, daily pheresis for a period of 2-4 days should begin 4 days after growth factors begin, and 16 days after the start of chemotherapy, or when the patient's absolute neutrophil count reaches $1000/\mu l$.

E) **ADVANTAGE OF PBSCs**

In certain, if not all diseases, it is thought that peripheral blood collection offers the advantage of decreased tumor cell burden when compared to bone marrow. Additionally, one does not need an unirradiated pelvis for this procedure, nor is there the same anesthesia risk or postoperative pain. Perhaps most importantly, the period of aplasia following this procedure is decreased thereby decreasing infectious complications.

F) **DISADVANTAGE OF PBSCs**

The collection of PBSCs can require multiple apheresis procedures that are time and labor intensive for both medical personnel and the patient. There is some risk of anesthesia owing to the need for insertion of an indwelling central catheter. However, temporary pheresis catheters may be placed under local anesthesia. For allogeneic transplants, it appears that the severity of chronic GVHD is worse with PBSCs compared to bone marrow.

G) **COMPLICATIONS**

1) **Infection**

Infection occurs infrequently.

2) **Anaphylaxis**

Anaphylaxis occurs infrequently. Prophylactic premedications may be given to attempt to prevent this complication.

3) **Hypocalcemia**

Hypocalcemia can occur in virtually all apheresis procedures ranging from mild symptoms (tingling of the lips and fingers) to moderate (cramping of fingers, feet and legs) or severe (nausea, seizures or cardiac dysrhythmias), and can be avoided if replacement solutions containing calcium are added during the apheresis process.

3

**3.0 Transfusion Medicine in Marrow
and Blood Cell Transplantation**

3

Transfusions

Michele Cottler-Fox

I) BLOOD DONATION

Blood for transfusion can be donated either by an unpaid volunteer donor (allogeneic) or by the same individual who will later receive it (autologous). Directed allogeneic donations may be made for a specific recipient. Marrow transplant recipients generally receive allogeneic products due to the large number required during the course of their hospitalization. Normal marrow donors, related or unrelated, should donate autologous blood prior to harvest, to be reinfused as needed during or after the harvest. Donation of a unit of packed red blood cells (PRBC) involves removal of approximately 450 ml of whole blood via the antecubital vein, and 1-2 units of PRBC are usually adequate for an average adult marrow harvest. In a healthy person, this is generally without complications except for an occasional vasovagal episode which is treated by terminating the donation, and placing the donor in the supine or Trendelenberg position. Although rare, other potential complications include seizures related to a severe vasovagal episode and, in older donors or those with unrecognized cardiac conditions, cardiopulmonary decompensation. After donation, the donor is encouraged to drink fluids to help compensate for the blood volume lost. Contraindications to donation other than anemia (females: hemoglobin < 12.5 g/dl, males: hemoglobin < 13.5 g/dl) include among other things, factors which would generally make the person unsuitable as a marrow donor, i.e. pregnancy, a medical history of malignancy, cardiac disease and HIV infection.

II) BLOOD COMPONENTS

After donation, allogeneic whole blood is processed into separate components, including PRBC, platelets and various plasma products (fresh frozen plasma (FFP), cryoprecipitate, albumin and immunoglobulins) (Table 3.1.1). Alternatively, plasma and platelets, as well as leukocytes, may be collected by apheresis.

A) RED BLOOD CELLS

Donor red blood cells must be compatible with the recipient's ABO and Rh type (special considerations regarding allogeneic transplant between ABO or Rh-mismatched donor and recipient pairs will be discussed separately later; see also Tables 3.1.3 and 3.1.4). As a rule of thumb, infusion of 1 unit of PRBC should increase the hematocrit by 3% and the hemoglobin by 1 g/dl.

1) Whole Blood

Except for trauma patients with massive hemorrhage, whole blood is rarely used today. A unit of whole blood consists of ~450 ml of blood collected into an anticoagulant solution and has a hematocrit of 36-44%.

2) Packed Red Blood Cells (PRBC)

This product is prepared by centrifuging whole blood or allowing the red blood cells to sediment in the cold and removing the 200-250 ml of plasma. Since most medical patients are euvolemic, PRBC are preferred to whole blood, especially since this also makes it possible to transfuse group O red blood cells into A, B and AB recipients by minimizing the isohemagglutinins (immunoglobulins of IgG and IgM types, i.e. alloantibodies, directed at red blood cells) capable of causing hemolysis.

Bone Marrow Transplantation, edited by Richard K. Burt, H. Joachim Deeg, Scott Thomas Lothian, George W. Santos. © 1996 R.G. Landes Company.

3.1

Table 3.1.1. Blood components

1. Whole Blood
Use: acute hemorrhage (> 30% of blood volume lost)
ABO: compatibility needed
Disease potential: may transmit virus, bacteria, parasites
Comments: rarely needed ouside of trauma surgery

2. PRBC
Use: restore or maintain oxygen-carrying capacity
ABO: compatibility needed
Disease potential: may transmit virus, bacteria, parasites
Comments: only exceptional circumstances need transfusion to hemoglobin > 10 g/dl

3. Platelets
Use: maintain hemostasis in setting of thrombocytopenia or abnormal native platelet function
ABO: compatibility may be helpful, but not required
Disease potential: may transmit virus, bacteria, parasites
Comments: apheresis platelets contain significant plasma clotting factors, isohemagglutinins and proteins to cause allergic reactions

4. FFP
Use: replace multiple clotting factor deficiency or as fluid for plasma exchange
ABO: compatibility required
Disease potential: may transmit virus
Comments: should not be used to expand blood volume or to replace immunoglobulins

5. Cryoprecipitate
Use: to replace Factor VIII, von Willibrand factor (except in von Willibrand type IIb), Factor XIII, fibrinogen
ABO: compatibility required if large volumes are to be used
Disease potential: may transmit virus
Comments: may be used with topical thrombin as a weak glue, or to make a stronger glue by concentration (fibrin glue) for use in surgical repair

6. Factor VIII
Use: hemophilia A therapy
ABO: compatibility not a problem
Disease potential: may transmit virus
Comments: available as moderate, high, ultra-high purity concentrates

7. Factor IX
Use: hemophilia B (Christmas disease) therapy
ABO: compatibility not a problem
Disease potential: may transmit virus
Comments: available in a complex concentrate associated with thrombotic complications or as a high purity concentrate

8. Albumin
Use: acute volume replacement or expansion
ABO: compatibility not a problem
Disease potential: may transmit virus, may acquire bacterial contamination
Comments: may lead to volume overload

9. IVIG
Use: prophylaxis for acute GVHD, or infections in deficiency states; as part of CMV therapy
ABO: compatibility not a problem
Disease potential: may transmit virus
Comments: may cause hemolysis if isohemagglutinin titers high, interfere with IgG serology tests

3.1

3.1

Table 3.1.2. Transfusion reactions

1. Acute Hemolytic Transfusion Reaction
Cause: ABO incompatibility
Signs/symptoms: restlessness, fever, chest or low back pain, flushing, hypotension, oliguria, anuria
Treatment: transfer to intensive care unit immediately for aggressive management of hypotension, fluid
balance, renal function

2. Delayed Hemolytic Transfusion Reaction
Cause: development of an alloantibody to transfused red blood cell antigens
Signs/symptoms: mild jaundice
Treatment: identify new alloantibody and provide red blood cells for transfusion of appropriate phenotype

3. Febrile, Nonhemolytic Transfusion Reaction
Cause: cytokines in blood component, antibodies to leukocytes, HLA antibodies, plasma proteins
Signs/symptoms: fever, rigors, nausea, vomiting, dyspnea, urticaria
Treatment: treat symptoms as they appear, consider washing blood product if premedication does not work

4. Anaphylaxis
Cause: allergy to plasma protein, especially IgA
Signs/symptoms: urticaria, wheezing, dyspnea, change in voice, laryngeal obstruction
Treatment: subcutaneous epinephrine, steroids, diphenhydramine, airway protection as needed, test
recipient for IgA deficiency

5. Bacterial Contamination
Cause: bacteria acquired from donor
Signs/symptoms: fever, rigors, hypotension
Treatment: as for bacteremia or septic shock

6. Transfusion-associated GVHD
Cause: passenger lymphocytes in a transfused blood component which has not been irradiated
Signs/symptoms: fever, rash, liver dysfunction, diarrhea, marrow aplasia
Treatment: once aplasia occurs the only chance for survival is a marrow transplant

Table 3.1.3. Selection of recipient PRBC, platelets and plasma after allogeneic BMT

Recipient	Donor	PRBC	Platelets	Plasma
O	O	O	O,A,B,AB	O,A,B,AB
O	A	O*	A,AB#	A,AB
O	B	O*	B,AB#	B,AB
O	AB	O*	AB#	AB
A	A	A or O*	A,AB#	A,AB
A	O	O*	A,AB#	A,AB
A	B	O*	AB#	AB
A	AB	O*	AB#	AB
B	B	B or O*	B,AB#	B,AB
B	O	O*	B,AB#	B,AB
B	A	O*	AB#	AB
B	AB	O*	AB#	AB
AB	AB	AB or O*	AB#	AB
AB	O	O*	AB#	AB
AB	A	O*	AB#	AB
AB	B	O*	AB#	AB

*consider using Adsol units to minimize out of group supernatant plasma volume
#consider volume reducing to minimize out of group plasma volume

Table 3.1.4. Donor/recipient ABO incompatibility

Major ABO Incompatibility	Minor ABO Incompatibility	Major and Minor ABO Incompatibility
Recipient has antibody to donor	Donor has antibody to recipient	Recipient has antibody to donor and donor has antibody to recipient
Immediate hemolysis –prevent by RBC depletion of marrow –some centers use prophylactic recipient plasma exchange prior to marrow infusion if titer 1:526 or greater	Immediate hemolysis –prevent by plasma depletion of marrow –some centers use prophylactic recipient RBC exchange with group O RBC.	Immediate hemolysis –prevent by RBC and plasma depletion of marrow –some centers use prophylactic recipient group O RBC exchange and prophylactic plasma exchange when recipient isohemagglutinin titer 1:526 or greater
Delayed hemolysis –occurs 2-4 weeks after BMT + direct antiglobulin test –risk increased with high recipient isohemagglutinin titer	Delayed hemolysis –occurs day 9-16 after BMT + direct antiglobulin test –risk increased with T cell depleted marrow	Delayed hemolysis + direct antiglobulin test
Delayed erythropoiesis (duration months to years) plasma exchange, erythropoietin, steroids	Delayed erythropoiesis (duration months to years) plasma exchange, erythropoietin, steroids	Delayed erythropoiesis (duration months to years) plasma exchange, erythropoietin, steroids
Transfusion recommendations **RBC**-Give recipient type RBC or group O RBC that is plasma depleted, e.g. adsol **Plasma, platelets**-Give donor plasma and platelets	*Transfusion recommendations* **RBC**-Give group O RBC **Plasma, platelets**-Give recipient plasma and platelets	*Transfusion recommendations* **RBC**-Give group O RBC **Plasma, platelets**-Give AB plasma and platelets

3.1

3.1

3) Leukocyte-Depleted Red Blood Cells

Leukocytes express HLA antigens which may cause alloimmunization in the recipient, thereby increasing the risk of graft rejection or making it difficult to transfuse platelets, which also bear class I HLA antigens. Many centers, therefore, remove the leukocytes as a buffy coat at the time of initial centrifugation. Other centers go a step further and routinely use special leukocyte depletion filters for marrow recipients and patients who will need prolonged transfusion support, such as those with leukemia and aplastic anemia. Leukocyte-reduced products are also thought to be the equivalent of CMV-negative products, since the virus is transmitted by leukocytes. Since there is now also good data suggesting that febrile non-hemolytic transfusion reactions are the result of cytokines released into the stored blood component by contaminating leukocytes, it is likely that blood banks will begin to leukocyte deplete all blood components shortly after collection in the near future.

B) PLATELETS

A unit of platelets is defined as 5×10^{10} platelets. Platelets may be obtained from a unit of whole blood (random donor platelets) or from a plateletapheresis (single donor platelets). The expected absolute increment at 1 hour in a stable patient receiving fresh platelets is 6,000-8,000/ml per unit of platelets transfused, with a corrected count increment of 10,000-20,000/ml. The number of units in a single apheresis collection varies depending on the donor and apheresis machine, but averages 5-6 units and may contain 10 units or more. Platelets derived from whole blood are generally transfused in pools derived from 5-10 donors. Since this results in 5-10 donor exposures for the transfusion recipient, with an increased risk of alloimmunization and transmission of viral disease, many centers prefer to use single donor apheresis platelets for marrow transplant recipients. Platelets, like red blood cell collections, are contaminated by leukocytes and may also be subjected to leukocyte depletion.

Although platelets carry ABO antigens, their survival is little affected by transfusion into a mismatched recipient. However, platelets are collected with donor plasma and isohemagglutinins in the plasma may lead to hemolysis in a recipient whose red blood cells carry the corresponding antigen. For this reason ABO-mismatched platelets may, in some centers, be volume reduced by centrifugation to remove the plasma. Volume reduction, however, also reduces the number of platelets to a variable but significant degree. As the plasma also contains some contaminating red blood cells, many centers also make an effort to provide Rh negative platelets to Rh negative recipients, or suggest the use of Rho immune globulin when necessary to avoid the possible formation of anti-D. The need for Rho immune globulin in the setting of BMT has not been demonstrated.

C) LEUKOCYTES

White blood cell products may include granulocytes, lymphocytes, monocytes and hematopoietic progenitor stem cells. Only granulocytes are recognized (and regulated) by the American Food and Drug Administration as a licensed product to date. Granulocytes, lymphocytes, monocytes and peripheral blood derived stem cells (PBSC) are collected by apheresis. Other sources of hematopoietic stem cells include cord blood and bone marrow at present.

Granulocyte transfusions have been considered prophylactically for the prevention of infection in a neutropenic patient, or for therapy of a severe infection unresponsive to antibiotic therapy. Early prospective, randomized studies showed a survival benefit of therapeutic granulocyte infusions in patients with bacteremia. Similar prospective, randomized studies of prophylactic granulocytes failed to show a survival benefit, although a decrease in incidence of infection was noted. As complications of granulocyte transfusions were not negligible, as it was difficult to collect adequate numbers of granulocytes, and as empiric broad spectrum antibiotic therapy for fever in neutropenia became accepted therapy, prophylactic granulocyte transfusions were not felt to be worth pursuing. Recently, however, as a result of the ability of G-CSF to increase the number of granulocytes available for collection in a normal volunteer donor, and to increase their survival both in vitro and in vivo, granulocyte transfusions have again become an area of research interest in oncology centers.

3.1

Monocytes have recently been tried at the National Institutes of Health Clinical Center as therapy for fungal infections unresponsive to antifungal therapy. Monocytes and lymphocytes have also been used as components of adoptive immunotherapy trials, and as carriers of marker or therapeutic genes for trials of gene therapy. Lymphocytes expanded in cell culture and directed against EBV, CMV and HIV are currently undergoing clinical trials at several centers. The role of monocytes and lymphocytes in future transfusion therapy remains to be determined.

D) PLASMA-DERIVED COMPONENTS

Plasma contains many different proteins, which can be categorized as transport proteins (albumin), immunoglobulins (including isohemagglutinins), lipoproteins, complement factors, clotting factors and proteinase inhibitors. Centrifugation is used to separate plasma from fresh or outdated whole blood. Alternatively, plasma may be collected by plasmapheresis. Plasma is then processed into further components: fresh frozen plasma (FFP), cryoprecipitate, coagulation factor concentrates, albumin and immunoglobulin preparations.

1) Fresh Frozen Plasma (FFP)

FFP is defined as plasma separated from whole blood within 6 hours of collection and rapidly frozen at -18°C or greater. This rapid cryopreservation preserves the activity of the most labile clotting factors, Factor V and VIII. Indications for infusion of FFP are pathophysiologic processes leading to increased consumption or decreased production of clotting factors, including in BMT recipients such processes as DIC from sepsis, hemolytic uremic syndrome related to cyclosporine, or hepatic insufficiency related to veno-occlusive disease (VOD) and GVHD. Since FFP also contains isohemagglutinins, it must be ABO compatible with the recipient. For selection of FFP in the setting of ABO-mismatched allogeneic transplant, see Table 3.1.3.

2) Cryoprecipitate

When fresh frozen plasma is thawed in the cold (1-6°C) a layer of white precipitate forms (cryoprecipitate). This cryoprecipitate is removed by centrifugation. Cryoprecipitate contains high concentrations of von Willebrand factor/Factor VIII, Factor XIII, fibrinogen and fibronectin. It is used to treat patients with specific congenital factor defects (hemophilia A, von Willebrand disease, dysfibrinogenemia or afibrinogenemia) and defects related to disease processes (DIC) or chemotherapy (L-asparaginase). Since no isohemagglutinins are present, it is not necessary for cryoprecipitate to be ABO compatible with the recipient.

3) Clotting Factor Concentrates

Hemophilia A and B are treated with Factor VIII and Factor IX concentrates, respectively. These factor concentrates are made from plasma pooled from thousands of donors or by recombinant DNA technology, and consequently come in intermediate, high and ultra-high grade purity products which are life-saving. However, these products are not without adverse consequences such as infectious complications (hepatitis B,C, HIV), hemolysis (due to isohemagglutinins), allergic reactions, development of factor inhibitors by the recipient and venous thrombosis (especially with Factor IX complex). Recent improvements in manufacturing procedures have improved the safety of these products.

4) Albumin

Albumin is available for clinical use as a 5% solution in saline, a 25% solution in distilled water or as a plasma protein fraction (PPF; a 5% protein solution consisting of 88% albumin and 12% α and β globulins). These preparations are heat-treated for 10 hours at 60°C to inactivate blood-borne viruses. PPF or a 5% albumin solution is used in treatment of hemorrhagic and hypovolemic shock, as replacement for plasma exchanged during plasmapheresis (except in TTP, where FFP or cryoprecipitate-poor plasma is used) and to prime cardiopulmonary by-pass pumps. The infusion of 25% albumin is used to relieve the symptoms of hypoalbuminemia related to severe liver disease (sometimes including VOD), protein malnutrition, nephrotic syndrome and protein-losing enteropathies. Adverse reactions are rare, but include circulatory overload with rapid infusion and, in the past, hypotension attributed to presence of prekallikrein activator (not a problem with current preparations).

5) Immunoglobulins

Immunoglobulins are prepared from plasma by the Cohn-Oncley cold ethanol fractionation method. Solutions currently in use are composed almost entirely of IgG with only traces of IgM and IgA. Both intramuscular and intravenous preparations are available. These preparations have been used to prevent viral infections in healthy people (e.g. hyperimmune globulin for hepatitis B) as well as in patients with agamma- and hypogammaglobulinemia. Rho immune globulin is used to prevent sensitization in rhesis-negative persons exposed to Rh (D) antigen-bearing red blood cells. In allogeneic marrow transplant recipients the intravenous product (IVIG) has been used to decrease the rate of acute GVHD, as part of prophylaxis and treatment of CMV, and to potentially decrease the severity of bacterial infections and the number of platelet products needed for transfusion.

Systemic reactions to IVIG include urticaria, pruritus, chills, fever, tremor, flushing, malaise, nausea, vomiting, back pain, joint pain, dyspnea, reversible renal dysfunction, bronchospasm and hypotension. Fortunately, all of the potential adverse reactions except headache are rare; they occur most commonly with the first infusion, and are less likely on repeat infusion. The hypotension is related to the rate of infusion, and resolves when the rate is decreased. Hemolytic anemia has occasionally been seen with the administration of IVIG. Hemolysis has been related to the presence of high titers of anti-A, anti-B, anti-D or anti-Kell blood group antibodies in a particular lot of IVIG. Of note in the setting of allogeneic marrow transplant is the fact that high-dose IVIG causes the recipient's direct antibody screening test (direct Coomb's test) to be weakly reactive, and gives rise to a positive IgG titer for most viruses. For this reason, viral antigenemia tests or cultures are needed to document an infection when the transplant recipient is receiving IVIG.

III) STORAGE OF BLOOD COMPONENTS

Whole blood is collected into a sterile bag containing an anticoagulant/preservative solution (citrate, which acts as an anticoagulant by binding calcium, plus buffers and nutrients) to maintain red blood cell function. Common solutions include CPD (citrate-phosphate-dextrose), CPDA (the A is for adenine), and Adsol or Sagmannitol (adenine-dextrose-saline-mannitol, which is added to CPD). PRBC units prepared in Adsol have less plasma than those collected in other solutions since red blood cells survive better in this solution and less plasma is therefore needed.

PRBC are stored at 1-6°C, and the shelf life depends on the preservative solution used. Shelf life for CPD is 21 days, for CPDA is 35 days and for Adsol is 42 days. For this reason, it is important that a blood collection center know the date on which an autologous unit of PRBC is needed for a marrow harvest. Although red blood cells may be frozen in glycerol for later use, this is not recommended for autologous units meant for a marrow harvest, since a unit may be lost during thawing and processing to remove the glycerol.

Platelets are stored at 20-24°C on a platelet rocker for a maximum of 5 days. FFP and cryoprecipitate are stored at -18°C for up to 1 year. Leukocytes for transfusion, be they granulocytes, monocytes or lymphocytes, are transfused as soon after collection as possible and always within 24 hours. Allogeneic hematopoietic progenitor stem cells are commonly infused directly after collection (after being processed if necessary to remove plasma or red blood cells), while the autologous product is more likely to be cryopreserved and stored in a mechanical freezer (-70°C) short term or in liquid nitrogen long term. However, both marrow and blood (stem) cell collections have been used after liquid storage at 4°C for 24 hours up to 7 days, with the blood (stem) cells declining from essentially 100% at 24 hours to around 70% at 7 days.

IV) ADMINISTRATION OF BLOOD PRODUCTS

Units of PRBC are processed as required by the Food and Drug Administration (FDA) or American Association of Blood Banks (AABB) regulations. Before being issued by the blood bank, the unit must undergo a clerical check by at least two people to insure that it is properly labeled, and is appropriate to a given recipient. Before infusion each unit must again be checked to see that it is the appropriate group and type for a recipient, and that it is correctly labeled as being meant for that specific recipient. It cannot be over emphasized that the most common reason for a severe hemolytic transfusion reaction is a failure to identify the unit and recipient properly.

Leukocytes are cross matched before being issued only if there are contaminating red blood cells present. Otherwise, confirmation of ABO/Rh by typing, and an antibody screen (direct and indirect Coomb's tests) are adequate.

PRBC and platelets are transfused through standard blood infusion tubing, which contains an in-line microaggregate filter to remove clumps. Leukocyte reduction filters, which remove leukocytes primarily by adsorption may be used in addition. Only marrow or blood-derived stem cells, and leukocytes (e.g. lymphocytes used

3.1

for graft-vs-tumor effect post-marrow transplant) are infused without a filter. Many centers consider leukocyte-depleted PRBC and platelets as equivalent to CMV-negative products, an important point when trying to provide CMV-negative products for a large marrow transplant center's population.

V) TRANSFUSION REACTIONS

Adverse reactions to blood products may be minor or life threatening. Signs and symptoms of such reactions must be recognized, and appropriate interventions made in a timely fashion (Table 3.1.2).

A) HEMOLYTIC TRANSFUSION REACTIONS

Acute hemolytic transfusion reactions occur when ABO or Rh incompatible red blood cells are given to a recipient with preformed antibodies (isohemagglutinins). In the laboratory there may be evidence of intravascular or extravascular hemolysis. The symptoms may vary in clinical intensity but include fever, chest pain, hypotension, flushing, low back pain, oliguria and anuria. These symptoms have been thought to be related to release of activated proteins of the complement system. Recently, in vitro data have also shown that cytokines are released by mononuclear phagocytes in response to red blood cells coated with Rh alloantibodies. If clinical data support the in vitro findings, cytokine antagonists and receptor blocking agents currently in phase I trials for other purposes may eventually become available for treatment of the clinical effects related to acute hemolytic transfusion reactions. Current treatment, however, requires aggressive management of hypotension, fluid balance and renal function and is best done in the intensive care unit.

Delayed hemolytic transfusion reactions occur when a transfusion recipient makes a red blood cell antibody after transfusion and then hemolyzes the transfused cells. Symptoms are usually minor, and it is not uncommon for the blood bank to notify the care team of the presence of a new red blood cell antibody before the care team has become aware of a problem. Once the antibody has been identified, future red blood cell transfusions are from donors lacking the problematic antigen.

B) FEBRILE NONHEMOLYTIC TRANSFUSION REACTIONS

Febrile nonhemolytic transfusion reactions are more common than hemolytic reactions, occurring in 1-2% of recipients. Such reactions may occur in response to leukocytes contaminating the red blood cell or platelet product, to plasma proteins, as a result of the presence of HLA or anti-granulocyte antibodies in the donor or recipient, or as a result of cytokines (principally IL-6) secreted into the component during storage by leukocytes. Clinical symptoms range from low-grade fever to rigors, nausea, vomiting, shortness of breath or full blown anaphylaxis. For patients with frequent but minimal reactions, premedication for transfusion with acetaminophen and diphenhydramine may suffice.

For more serious or more frequent reactions, premedication with hydrocortisone may be tried, or maneuvers which remove leukocytes (washing, filters) or plasma (washing, resuspension) may be tried. For patients with urticarial reactions, diphenhydramine prior to transfusion may be required, although it should be recognized that since such reactions often do not reappear many patients receive unnecessary medication. However, for IgA deficient patients with allergic reactions due to anti-IgA antibodies, products from rare IgA deficient donors may be needed if reactions occur despite washing.

C) **BACTERIAL CONTAMINATION**

Bacterial contamination of red blood cells, primarily by cold-loving bacteria such as *Yersinia*, is a relatively rare but potentially fatal occurrence. The contamination of platelets, which are stored at 20-24°C for up to 5 days, has recently been recognized as an increasing problem. Evaluation of a newly febrile patient should always include the question of whether the patient has recently received a transfusion.

D) **TRANSFUSION-ASSOCIATED GRAFT-VS-HOST DISEASE (TA-GVHD)**

TA-GVHD is a rare but almost uniformly fatal complication of transfusion, resulting from infusion of immunocompetent lymphocytes which the recipient is unable to destroy. Cases have been reported in children with severe combined immunodeficiency disease, in patients with neuroblastoma, glioblastoma multiforme, rhabdomyosarcoma and small cell lung cancer, as well as in marrow transplant recipients and in normal recipients who receive blood from a haploidentical donor (who shares one of the recipient's HLA haplotypes either because the donor is a relative or because the gene pool in a population is relatively limited as, e.g. in Japan). Because the risk of an HLA haploidentical donor is greatest within a family, AABB standards require irradiation of directed donations by family members.

3.1

E) **ALLOIMMUNIZATION**

Alloimmunization to HLA or granulocyte antigens may occur as a result of red blood cell or platelet transfusions, since these components are contaminated by passenger leukocytes. Alloimmunization may increase the risk of a febrile or leukoagglutination type reaction to further transfusions, or make it necessary to support a patient with HLA-matched platelet products. Third generation leukocyte reduction filters (used at the time of transfusion) which reduce leukocyte contamination by 3 $\log_{10}$ or fourth generation filters (used at the time of blood component collection) which reduce contamination by 4 $\log_{10}$ are useful in reducing but not eliminating alloimmunization.

Alloimmunization to red blood cell antigens, leading to delayed hemolytic transfusion reactions, may occur if a patient is transfused with red blood cells bearing a surface antigen which is lacking on the surface of the recipient's cells. Since red blood cell antigens are present in high frequency in one racial group and in low frequency in others, some authors have argued that patients who will need prolonged transfusion support should receive transfusion from phenotyped donors or that they should simply receive blood from racially matched donors. However, this remains a point of considerable controversy.

F) **IMMUNOSUPPRESSION**

Immunosuppression as a result of transfusion is a controversial subject of increasing research interest. Data exist suggesting that oncology patients who are most heavily transfused may have a decreased survival rate, possibly due to increased rates of tumor recurrence or infection resulting from immunosuppression. There are also data suggesting increased decline of the immune system of HIV positive patients who are transfused. Recently it has been suggested that transfusion-related immunosuppression may be an effect of contaminating passenger leukocytes. If this is borne out by further studies, it may become reasonable to routinely remove leukocytes from components by filtration.

VI) TRANSFUSION-TRANSMITTED DISEASES

Transfusion-transmitted viral diseases for which testing is currently performed routinely in the United States include hepatitis B and C, as well as the retroviruses HTLV-I and HIV. Donor questioning and surrogate markers are used to eliminate those with hepatitis A as an adult, and anticipated new but as yet unknown hepatitis viruses. Although the risk of acquiring a transfusion-transmitted virus due to failure of the testing methods used is now exceptionally small, but not zero, methods of inactivating viruses (e.g. pasteurization, detergent treatment, photoinactivation) are used for components derived from plasma and are under development for use with other cellular components.

Other infectious agents which may be transmitted by transfusion include the viruses CMV and EBV, bacteria and protozoa (*Yersinia, Staphylococcus epidermidis, Enterobacteriaceae,* syphilis, malaria, babeziosis, trypanosomiasis, brucellosis, toxoplasmosis, Lyme disease, among others). Donors are routinely screened for syphilis. Although tests are available for the other diseases, testing is only done for CMV with any degree of frequency since it is important to transfuse CMV seronegative products to seronegative transplant recipients. Other agents are commonly eliminated from the blood supply through donor questioning and observation of the component for changes in color or consistency, but concern remains about the need for direct testing of the other potential problems.

A) MICROBIAL CONTAMINATION OF HARVESTED BONE MARROW OR PERIPHERAL BLOOD STEM CELLS

In general, microbial cultures are obtained at two or three stages before infusing marrow or mobilized stem cells into a patient: 1) immediately after harvest from the transfer bag, 2) after processing and prior to cryopreservation and 3) from the infusion bag after thawing. In approximately 1-3% of specimens, the culture is positive. The organisms are usually skin flora (*Staphylococcus epidermidis, Proprionobacterium, Corynebacterium,* etc.), but on rare occasions enteric pathogens are identified. Given the expense and risk or inability to repeat a harvest, the contaminated stem cell products have usually been infused into the patient. In general, no adverse clinical sequela have occurred. However, each case should be considered individually, and the patient should be preemptively covered with appropriate antibiotics.

VII) SPECIAL CONSIDERATIONS FOR TRANSFUSION IN THE BMT PATIENT

A) DIRECTED DONORS

Family members should not be used for transfusion support pretransplant if an allogeneic transplant is contemplated, since this may lead to sensitization to minor transplant or HLA antigens, potentially leading to graft rejection. Posttransplant, however, family members may be preferred donors for a recipient who does not respond well to platelet transfusion. Granulocytes may also be collected preferentially from the marrow donor either prophylactically or for a neutropenic recipient with an infection.

B) IRRADIATION

All cellular blood products, except stem cell grafts and lymphocytes given for graft-vs-tumor effect, must be irradiated to 2500 cGy to prevent TA-GVHD by abrogating the ability of lymphocytes to respond to antigenic stimulation. There are now data showing that at least viable stem cells may be recovered from FFP. It may become necessary to consider FFP as a cellular product, although this has not been the case to date. Ultraviolet light, which has been shown to abrogate both the ability of lymphocytes to stimulate as well as to respond to an antigenic stimulus at a dose which spares hematopoietic stem cells, may eventually also be used to prevent TA-GVHD but is not currently available for clinical use outside of an experimental study.

C) **AUTOLOGUS PRBC DONATION PRIOR TO BMT HARVEST**

At the time of harvest, 10-20 ml of marrow/kg of recipient body weight is commonly removed (the amount will vary depending on how the marrow is to be processed). Because an autologous marrow donor will require many transfusions during transplant (commonly performed directly after harvest), and is often anemic from recent chemotherapy, it is rare to collect autologous PRBC. However, for the patient in remission who is simply having marrow stored and for normal allogeneic marrow donors, 1-2 units of autologous PRBC should be collected within the month prior to harvest. If the harvest is delayed the units may, if necessary, be frozen for later use. Donors are usually given iron supplementation (ferrous sulfate 324 mg PO TID) for 1-3 months post-harvest. Even when available, autologous PRBC units should not be ordered for transfusion unless there is a clinical indication, since clerical error by the transfusion service may lead to allogeneic blood being transfused, with the attendant risk of infectious disease transmission, alloimmunization and possibly GVHD if the component is not irradiated.

D) **PROPHYLACTIC TRANSFUSIONS DURING THE TRANSPLANT PERIOD**

The general approach for transfusing PRBC during the transplant period is to maintain a hemoglobin > 9 g/dl. The prophylactic platelet transfusion trigger is more controversial, with some centers maintaining a level > 10,000/ml for stable patients and others maintaining a level > 20,000/ml.

E) **ABO/RH-MISMATCHED TRANSPLANTS**

Transfusion product support for major, minor or mixed major/minor mismatched transplants is detailed in Table 3.1.3 and Table 3.1.4. A major mismatch is defined as the expression of ABO antigens on donor red blood cells which are not present on recipient cells; i.e. the recipient has isohemagglutinins against the donor red blood cells which will cause hemolysis (e.g. A,B or AB donor and O recipient). If allogeneic marrow containing donor red blood cells is infused in the setting of a major mismatch, acute hemolysis occurs. Therefore, it is now common practice to remove donor red blood cells from the graft. The alternative approach of plasma exchange to reduce recipient isohemagglutinins has also been used. A major mismatch commonly results in delayed hemolysis or delayed red blood cell engraftment. Less commonly, delayed neutrophil, lymphocyte and platelet engraftment may occur. Rarely, pure red cell aplasia may occur.

A minor mismatch is defined as the presence of recipient red blood cell ABO antigens which are not present on donor cells; i.e. the donor has isohemagglutinins against recipient red blood cells which will cause hemolysis if infused (e.g. O donor and A,B or AB recipient, or A1 donor and A2 recipient). If allogeneic marrow containing donor plasma is infused in the setting of a minor mismatch, acute hemolysis will occur, for which reason it is now common practice to remove donor plasma from the graft. A minor mismatch may also result in delayed or prolonged hemolysis, especially if mixed chimerism occurs post-transplant (more frequent after T cell depletion).

Mixed major/minor mismatches may occur, as well as mixed ABO/Rh mismatches, in which case it may be necessary to remove both donor plasma and red blood cells from the graft. Minor mismatches in the Rh system have been reported to carry a 10-15% chance of hemolysis. Mismatches within other red blood cell antigen systems resulting in hemolysis also occur but rarely (Kidd and MNSs have been reported).

The approach to documenting the recipient's original blood group/type (historical type) and the change to donor's group/type (actual type) is not uniform among blood banks, nor is the policy regarding a switch from recipient to donor type blood products. Some centers switch only after the blood group/type changes completely, others switch after 100 days, and others always use group O red blood cell products. Differentiating between historical and actual blood group/type is of particular importance in those recipients who develop HUS/TTP posttransplant and require plasma exchange therapy.

3.1

F) CMV NEGATIVE BLOOD PRODUCTS

CMV infection may be transmitted to an allogeneic or autologous CMV negative transplant recipient by transfusion of a blood component from a CMV seropositive donor. Although not all cases of infection result in CMV disease (primarily pneumonitis), the mortality in the allogeneic setting prior to the advent of ganciclovir combined with IVIG was over 90%, and even with this therapy mortality may be as high as 50% if disease has progressed to the point of causing changes on chest x-ray. It is, therefore, routine to request CMV negative blood products for CMV negative allogeneic recipients. Since no more than 50% of blood donors (at best) are likely to be seronegative, provision of seronegative products becomes a problem in large transplant centers. Fortunately, it has recently been demonstrated that products which have undergone leukodepletion with a third generation filter are equivalent to seronegative products, easing the pressure on blood banks to find seronegative donors.

Provision of CMV seronegative blood products to autologous BMT patients has been more controversial, since only 1-9% have been reported to develop CMV disease after infection and allogeneic recipients had priority on the limited supply. Now that third generation leukodepletion filters are in common use however, many centers also supply seronegative products to seronegative autologous recipients.

G) FAILURE TO RESPOND TO PLATELET TRANSFUSIONS

Transplant recipients may become refractory to platelet transfusions for multiple reasons, including but not limited to: splenic sequestration, fever, infection, diffuse intravascular coagulation (DIC), antibiotics, HLA alloimmunization, anti-platelet antibodies and VOD. The first step in determining whether a patient responds appropriately to platelet transfusion is to acquire posttransfusion platelet counts. Although this is commonly ordered for 1 hour posttransfusion, the blood sample may be drawn shortly after the transfusion is completed, since those who are alloimmunized will already have destroyed the platelets at that point, and for those who are not alloimmunized there is little difference between the count at 15 minutes and 1 hour posttransfusion. If the expected increment is not seen, but the count does increase, HLA alloimmunization is less likely but possible. If a decrement in the platelet count is seen, alloimmunization is more likely.

Many algorithms have been suggested for dealing with a patient who responds poorly to platelet transfusions. In a stable patient it is possible to try single donor apheresis platelets if these have not been tried (rather than pooled platelets from whole blood collections), followed by ABO matched or fresh platelets, then cross matched platelets if they are available or HLA-matched or compatible apheresis platelets. However, the transplant recipient who is responding poorly to platelets is rarely stable long enough to proceed through such algorithms. For a transplant patient, the fastest approach to the problem is to look for anti-HLA antibodies in the patient's serum, and if these are present to use either HLA-compatible platelets (if only a limited number of antibodies are present) or HLA-matched platelets (if many such antibodies are present). If the transplant center does not have a platelet apheresis collection program, matched platelets must be requested from the nearest blood center, which may take 1-3 days. Alternatively, the original marrow donor or family members may be used as platelet donors since they share many transplant antigens and the recipient is more likely to do well with their platelets. High dose IVIG has been tried in platelet-refractory patients with some degree of success, both in those with and without HLA antibodies. It remains unclear who is most likely to respond to IVIG, how much to give and for how long to give it. However, a total of 2 g/kg recipient body weight is the usual dose tried, either as a single dose or divided over up to 5 days. If the underlying problem is infection, fever, DIC or VOD, resolution of the problem leads to improved response to platelets. It should be pointed out that many believe that undergoing marrow transplantation is in itself the greatest risk factor for poor response to platelets.

H) VOD

VOD is the result of small venule blockage in the liver, of unclear etiology but thought to be related to toxicity from the conditioning regimen. It leads to fluid retention and third-spacing of fluids due to decreased albumin levels as the liver damage increases, potentially resulting in hepatorenal syndrome and renal failure. Early efforts to deal with the problem led to the practice of hypertransfusing red blood cells to a hemoglobin of 12-15 g/dl, since it was thought the red blood cells would be more likely to remain in the intravascular space, and the renal failure might be avoided. More recent work has focused on the decrease in coagulation factors, particularly proteins C and S, early in the disease process. Attempts to diagnose VOD early by measuring plasminogen activator inhibitor (PAI-1) and to treat VOD with plasma components, fibrinolytic agents such as tissue plasminogen activator and/or anticoagulants such as heparin are currently in progress.

I) LYMPHOCYTE INFUSIONS

Recently, marrow donor-derived lymphocyte transfusions have been given post-transplant in an effort to achieve a graft-vs-tumor effect and decrease the risk of relapse. The optimal dose and timing of the transfusions post-transplant remain to be determined. While it is clear that graft-vs-tumor effect can be achieved, including induction of a remission after relapse of CML, lymphocyte transfusions are not without risk. Lymphocyte transfusions are not irradiated, as this would abrogate the desired effect, and thus they are capable of causing severe GVHD, as well as marrow aplasia (presumably due to the fact that no donor marrow remains after the leukemia has recurred). Since the recipient will die from aplasia and its complications (infection, bleeding) without a new hematopoietic stem cell infusion, the donor must be informed of this possibility before the lymphocytes are collected and agreement reached on the possible need for blood (stem) cell collection with growth factor mobilization or repeat marrow harvest.

3.1

SUGGESTED READING

1. Rossi EC, Simon TL, Moss GS, eds. Principles of Transfusion Medicine. Baltimore: Williams and Wilkins, , 1991.
2. Branch DR, Dzik WH, Judd WJ, Kakaiya R, McMican A, Ness P, Polesky HF, Rolih SD, Vengelen-Tyler V, eds. American Association of Blood Banks Technical Manual, 11th Edition. Bethesda: 1993.
3. Atkinson K, ed. Clinical Bone Marrow Transplantation: a reference textbook. Cambridge University Press: 1993.

Hematopoietic Progenitor Stem Cells

Michele Cottler-Fox

I) INTRODUCTION

A variety of assay methods are available for identifying hematopoietic progenitor cells (HPC), a term defined by the International Society for Hematotherapy and Graft Engineering as including both primitive pluripotent stem cells capable of self-renewal as well as maturation into any of the hematopoietic lineages, and the committed and lineage-restricted progenitor cells. However, correlation of these assay results with the rate of reconstitution seen in vivo after transplantation is difficult, since many factors other than HPC number also play a role in hematopoietic reconstitution. Nevertheless, in evaluating the quality of a graft the first important characteristic which must be defined is the number of HPC present. Eventually it may also become necessary to define the kind of HPC present.

II) DEFINITION AND CHARACTERIZATION OF HPC

The gold standard used to define HPC is their ability to reconstitute marrow function after lethal irradiation. This sustained hematopoietic reconstitution after transplant is referred to as engraftment. The relative numbers of the various cell types used for graft evaluation is shown in Table 3.2.1.

A) NUCLEATED CELL COUNT

Initially, human marrow grafts were evaluated on the basis of the number of nucleated cells given per kg recipient body weight. When using total nucleated cells (NC) as a surrogate marker for HPC, an allogeneic marrow graft has been considered to be acceptable if it contains $2\text{-}4 \times 10^8$ NC/kg recipient body weight. An autologous marrow graft has been considered adequate to provide sustained hematopoiesis if it contains $1\text{-}3 \times 10^8$ NC/kg.

B) COLONY-FORMING UNITS

In the 1970s it became possible to evaluate grafts on the basis of the number of colonies of various types grown in semisolid culture medium, either agar or methylcellulose. These colonies were named on the basis of the cells they contained, which were progeny of the founding HPC, i.e. CFU-GM (colony-forming units-granulocytes/macrophages), CFU-GEMM (colony-forming units granulocytes/erythrocytes/macrophages/monocytes), BFU-E (blast-forming units-erythrocytes). However, graft evaluation on the basis of clonogenic (colony-forming unit) assays had two major drawbacks for clinical use in transplantation: 1) the assays took 2 weeks to complete, so that results were available only after transplantation; and 2) murine studies suggested that the majority of clonogenic cells were not responsible for long-term hematopoietic reconstitution.

C) CD34 CELL COUNT

In the 1980s flow cytometry became useful for graft evaluation when the cell surface marker, CD34, was identified as being present on all known varieties of HPC. While the optimal number of CD34+ cells in a graft remains unclear, there is a growing consensus that $> 2 \times 10^6$ CD34+ cells/kg recipient weight is usually associated with recovery of granulocyte and platelet production to

Bone Marrow Transplantation, edited by Richard K. Burt, H. Joachim Deeg, Scott Thomas Lothian, George W. Santos. © 1996 R.G. Landes Company.

Table 3.2.1. Average relative frequencies of cell types in an autologous graft

Nucleated cells	8×10^8/kg recipient body weight
Mononuclear cells	6×10^8/kg recipient body weight
CD34+ cells	$1\text{-}8 \times 10^6$/kg recipient body weight
CFU	$3\text{-}6 \times 10^5$/kg recipient body weight
LTC-IC	$0.1\text{-}1 \times 10^4$/kg recipient body weight

CFU=colony-forming units, LTC-IC = long-term culture initiating cell

3.2

clinically important levels (ANC > 500, platelets > 20,000) within 14 days of transplant. Morphological examination of cells separated by cell sorting on the basis of CD34 demonstrated that the majority of such cells have the appearance of small lymphocytes or monocytes. For this reason, evaluation of grafts on the basis of mononuclear cell content as a surrogate for flow cytometry has been suggested. Inability to agree on whether the term "mononuclear cell" should be applied only to lymphocytes and monocytes in a graft, or whether very early myeloid precursor cells should also be included, has prevented enumeration of mononuclear cell content of grafts from becoming widespread.

D) CD34 SUBSETS

Shortly after CD34 entered use as a marker for cells capable of engrafting and reconstituting hematopoiesis, it became clear that there were subsets of CD34+ cells identifiable on the basis of other cell surface markers, e.g. CD33, CD38, CD71, HLA-DR, Thy-1, and the ability of the cell to stain with the dye rhodamine. The pleuropotential subset of CD34+ cells are rhodamine dim (do not stain with rhodamine), lineage negative (CD33-, CD38-, CD71-, HLA-DR-), and Thy-1+. CD34+ cells comprise only 1% of a marrow harvest. CD34+, CD38-, HLA-DR- cells comprise less than 0.01% of a marrow harvest.

E) LONG-TERM CULTURE INITIATING CELL

Cell separation studies and animal transplant studies also suggested that some CD34+ cells might be responsible for early reconstitution, while others might be responsible for long-term reconstitution. The long-term reconstituting cells are known as long-term culture initiating cells (LTC-IC). While study of LTC-IC has resulted in an improved understanding of hematopoiesis, their enumeration through culture takes at least 5 weeks, making them useless as a means of evaluating an allogeneic graft prior to transplant. Monoclonal antibodies capable of defining LTC-IC or true "stem" cells are currently the goal of several research teams, and may eventually replace CD34 for graft evaluation.

III) SOURCES OF HEMATOPOIETIC PROGENITOR CELLS (HPC)

Current sources of HPC for transplantation include marrow, peripheral blood and cord blood. Use of alternative sources such as fetal liver and xenogeneic marrow is being examined experimentally. The choice of HPC source is based on a number of considerations discussed below.

A) MARROW

The logistics of marrow collection are simple, allowing the harvest center to schedule an operating room for harvest in advance at a convenient time point. The major drawbacks to allogeneic harvest are the risk of anesthesia for the normal donor, and the discomfort associated with large numbers of marrow aspirates. If no donor is available, autologous marrow may be harvested.

B) PERIPHERAL BLOOD

Blood-derived HPC ("peripheral stem cells") were originally considered as an alternative to autologous marrow harvest in the patient unable to undergo anesthesia, with marrow infiltrated by tumor or where marrow was difficult to aspirate as a result of hard bone or fibrosis related to the underlying disease process or radiochemotherapy. More recently it has been demonstrated that allogeneic blood cell transplant may also be used, either with or without T cell depletion.

Autologous blood-derived HPC may be collected in steady state, after mobilization with chemotherapy, or after mobilization with chemotherapy followed by colony-stimulating factors such as G-CSF, GM-CSF or erythropoietin. Simultaneous or sequential (e.g. IL-3+G-CSF or IL-3 followed by G-CSF) combination of growth factors with or without chemotherapy has also been used to mobilize autologous HPC. The relative fold increase of HPC collected by apheresis with the different strategies is shown in Table 3.2.2.

1) **Steady State Collection**

The advantage of collecting in steady state is that apheresis may be scheduled in advance at a time of convenience to donor and apheresis unit. Disadvantages to steady state collections include the fact that far fewer HPC are collected in this manner, requiring as many as 10 collections to achieve an adequate graft; and no advantage is seen to using these cells relative to marrow.

2) **Mobilized Collection**

After mobilization with chemotherapy more HPC are found in the blood and their use is associated with a shorter period of neutropenia and thrombocytopenia after transplant. However, mobilization with chemotherapy alone is usually associated with some degree of uncertainty regarding scheduling of apheresis, since not all patients will recover white blood cell (WBC) counts at the same rate, and apheresis is usually not started until the WBC count has reached some pre-defined number (usually 1000 WBC/μl). Chemotherapy is also associated with a significant risk of fever and infection during the period of induced neutropenia. Addition of growth factors, either singly or in combination, to chemotherapy shortens the period of risk for neutropenia and fever/infection during the mobilization period, increases further the number of HPSC mobilized into the blood and is associated with a shorter period of neutropenia and thrombocytopenia after transplant. However, predicting when to schedule

Table 3.2.2. Relative fold increase in autologous CFU-C collected by apheresis utilizing different mobilization strategies

Steady state	1x
Chemotherapy	10-20x
Growth factor alone (G-CSF most common)	10x
Chemotherapy plus growth factor (G-CSF most common)	1000x

apheresis becomes more difficult as patients will vary in their rate of recovering WBC counts as a function of previous chemoradiotherapy, growth factor exposure and disease. Mobilization with growth factors alone has the advantage of no risk of neutropenia or thrombocytopenia and apheresis may be scheduled after a fixed number of days (usually starting day 3, 4 or 5 after initiating growth factor with apheresis daily until collection of greater than 2-3 x 10^6 CD34+ cells/kg recipient weight). Use of a single growth factor such as G-CSF mobilizes fewer HPC than chemotherapy plus G-CSF, but combinations of growth factors may be as effective as chemotherapy plus a single growth factor.

3.2

C) CORD BLOOD

To date, more than 100 cord blood transplants have been performed, and emerging data suggest that such transplants may carry a decreased risk of GVHD. Cord blood may be collected from both the umbilical cord and placenta after birth, preferably using 20 ml of citrate-phosphate-dextrose as the anticoagulant, since this solution can anticoagulate blood over a wider volume range and the range of cord blood collected has varied from 50-200 ml. With the placenta still in utero, if the cord is clamped within 30 seconds of vaginal delivery it is reported that an experienced harvester can acquire up to 100 ml of cord blood, with no ill effects to the newborn or its mother noted. After delivery of the placenta, an additional 5-15 ml cord blood can be obtained by sterile puncture and aspiration of the engorged placental vessels on the fetal side. Cord blood has generally not been manipulated after collection for fear of losing significant numbers of HPC. However, with the advent of cord blood banks it has become important to develop methods which reduce the component volume in order to reduce the number of freezers required for storage after cryopreservation. Both Ficoll and gelatin separation procedures have recently been reported to permit recovery of 86-92% of the HPC present in umbilical cord blood.

Suggested exclusion criteria for collection of cord blood have included congenital malformations, known hereditary diseases of the newborn and absence of informed consent from the mother. With cord blood collected for use in unrelated transplant, it has also been suggested that medical histories of both the biologic mother and father should be obtained, and that the infant from whom the cells were obtained should be evaluated between 6 and 12 months of age for evidence of genetic or infectious disease before the cells may be used for transplant. In all cases the usual transfusion transmissible diseases are tested for. Microbiological examination of the cord blood collection should also be obtained, since contamination rates have been reported from 3-15%, depending on whether a closed or open collection system has been used.

IV) SEPARATION OF HPC (ENRICHMENT)

Methods for HPC enrichment of a graft are divided into techniques based on physical, immunological or pharmacological methods (Table 3.2.3). Pharmacologic methods are generally aimed at removing tumor cells from an autologous graft, rather than enriching for HPC since most pharmacologic purging methods result in loss of, or damage to, HPC. Physical methods use cell size, shape and density to separate general cell populations such as lymphocytes and CD34+ cells, but separation is imperfect and often fewer than 50% of the starting HPC population remains in the final product. CD34+ cells originally lost from the product may be

Table 3.2.3. Methods used either alone or in combination to enrich for hematopoietic progenitor cells via positive or negative cell selection

Physical methods
    density gradients
    counterflow centrifugal elutriation
Immunological methods
    mAb + complement
    mAb + immunomagnetic beads
    mAb + immunoaffinity column
    mAb + toxin
    mAb + radioactive
    mAb bound to a flask (panning)
Pharmacological methods
    4-hydroperoxycyclophosphamide
    mafosfamide (ASTA-Z)
    etoposide (VP-16)
    vincristine
    alkyl-lysophospholipids (ALP)
    methylprednisolone

abbreviations: mAb = monoclonal antibody

3.2

reclaimed by adding an immunological step to a procedure which starts with a physical method. Immunologic methods may be used in a positive or negative selection step, i.e. either to acquire specifically the population of choice (CD34+ cells) or to remove unwanted cell populations (T cells or tumor cells). Immunologic methods based on mAb bound to a solid phase require that the cells bound to the mAb either be released from the mAb or from the solid phase support system. The combination of a physical and immunologic method (e.g. elutriation followed by positive selection of HPC based on a CD34 mAb) has the advantage of reducing the expense of the mAb used by reducing the number of cells subjected to selection. Nonetheless, combination of techniques always results in substantial HPC losses, and adds to the time spent in processing.

V) **EX VIVO EXPANSION OF HPC**

Unlimited ex vivo expansion of HPC is the Holy Grail of transplantation—greatly desired but tantalizingly out of reach, at least for the present. It is currently possible to expand HPC ex vivo to at least three times their starting number, but after this the pluripotent stem cells seem to be exhausted. For this reason, present work is aimed at expanding only a portion of the original harvest, to provide earlier production primarily of neutrophils to reduce the risk of infection immediately post-transplant.

VI) **CLINICAL RESULTS OF AUTOLOGOUS BLOOD-DERIVED HPC TRANSPLANTATION**

Potential anticipated advantages of blood-derived HPC transplant over autologous marrow include: 1) more rapid neutrophil and platelet recovery; 2) avoidance of an operative procedure under general anesthesia; 3) ability to transplant a patient with unharvestable marrow due to obesity, impenetrably hard bone, marrow fibrosis or pelvic bone disease; 4) reduced risk of tumor contamination and 5) acceleration in the rate of immune reconstitution.

A) ACUTE MYELOID LEUKEMIA

At present, data suggest that for standard risk AML, the overall probability of disease-free survival at 2 years after autologous blood-derived HPC transplant is not statistically different from autologous marrow transplant, although there may be a trend for increased relapse in those receiving autologous blood-derived HPC transplant compared to purged autologus marrow.

B) CHRONIC MYELOGENEOUS LEUKEMIA

Investigators at the Hammersmith Hospital in London first used autologous blood-derived HPC collected in chronic phase to support patients after myeloablative therapy for blast crisis. In their series and a later one from France, a chronic phase was restored at least transiently, a small proportion of patients lost the Philadelphia chromosome in marrow following transplant and alpha interferon after transplant appeared to prolong survival. Cell culture studies later revealed that normal HPC clones coexisted with leukemic HPC clones. There are data suggesting that cells collected by apheresis on recovery from intensive chemotherapy may be predominantly Philadelphia chromosome negative, and this has been utilized as a strategy for collecting blood-derived HPC for autologous transplant. Clinical results using this approach are promising but preliminary at present.

C) HODGKIN'S AND NON-HODGKIN'S LYMPHOMA

In relapsed Hodgkin's disease, it has been shown that blood-derived HPC are a suitable alternative to marrow, with long-term event-free survivors even among those with evidence of marrow involvement by disease at the time of transplant. However, relapse has been seen as late as 70 months after treatment, suggesting that the curative potential of this approach remains to be determined. Data from prospective randomized trials comparing marrow and blood-derived HPC for transplant are not available at present.

In aggressive non-Hodgkin's lymphoma, one retrospective analysis has shown a statistically significant superior event-free survival for those who received autologous blood-derived HPC relative to autologous marrow. At least three reasons have been proposed for this possible advantage: presence of greater numbers of cytotoxic effector cells, improved immune reconstitution and presence of fewer clonogenic tumor cells.

D) MYELOMA

A randomized French trial comparing standard treatment with autologous marrow transplant has shown a significant increase in response rate and overall disease-free survival for the transplant arm. While a direct comparison of marrow and blood-derived HPC has not been done, the addition of blood-derived cells to marrow has resulted in a shorter duration of neutropenia than with autologous marrow alone, decreasing treatment-related morbidity and mortality.

E) BREAST CANCER

High dose chemotherapy with autologous marrow support appears to be an effective therapy for chemotherapy-responsive metastatic or locally advanced breast cancer, although the number of patients treated in a consistent manner for whom long-term follow-up has been reported are few. A number of studies, none prospective and randomized, have compared hematopoietic recovery and toxicity in patients receiving marrow alone vs growth factor mobilized blood-derived HPC alone or in combination with marrow. In all cases there was a more rapid recovery of hematopoiesis, fewer antibiotic days, transfusion requirements and shorter length of hospital stay in those receiving the mobilized blood-derived HPC relative to historical marrow recipients. Early results suggest that there may be no significant difference in response rates and progression-free survival for the marrow vs blood-derived HPC groups.

F) OTHER SOLID TUMORS

Although breast cancer and lymphoma are the most common indications for autologus marrow transplant at present, over 200 autologous transplants are performed in the United States each year for other solid tumors (ovarian carcinoma, germ cell tumors, neuroblastoma, small cell lung cancer, sarcoma, melanoma and brain tumors). While autologous blood-derived HPC are readily collected in these malignancies no randomized prospective studies looking at the results of using these cells as opposed to autologous marrow have been reported.

VII) CLINICAL RESULTS OF ALLOGENEIC BLOOD-DERIVED HPC TRANSPLANTATION

The ability to safely mobilize HPC using recombinant human growth factors alone enables syngeneic and allogeneic donors to have HPC collected in only 1-2 apheresis procedures. Peripheral blood HPC collections contain a log more T cells than bone marrow, raising concerns about an increased risk of graft-versus-host disease (GVHD). However, preliminary data indicate no increase in the incidence or severity of acute GVHD with cryopreserved peripheral blood HPC. The risk of extensive chronic GVHD with allogeneic peripheral blood HPC is unknown.

SUGGESTED READING

1. Juttner CA, To LB, Haylock DN, Branford A, Kimber RJ. Circulating autologous stem cells collected in very early remission from acute nonlymphoblastic leukaemia produce prompt but incomplete haemopoietic reconstitution after high dose melphalan or supralethal chemoradiotherapy. J Haematol 1985; 61:739-745.

2. Kessinger A, Armitage JO, Landmark JD, Smith DM, Weisenburger DD. Autologous peripheral hematopoietic stem cell transplantation restores hematopoietic function following marrow ablative therapy. Blood 1988; 71:723-727.

3. To LB, Shepperd KM, Haylock DN et al. Single high doses of cyclophosphamide enable the collection of high numbers of hemopoietic stem cells from the peripheral blood. Exp Hematol 1990; 18:442-447.

4. Siena S, Bregni M, Brando B, Ravagni F, Bonnadonna G, Gianni AM. Circulation of CD34+ hematopoietic stem cells in the peripheral blood of high-dose cyclophosphamide-treated patients: enhancement by intravenous recombinant human granulocyte-macrophage colony-stimulating factor. Blood 1989; 743:1905-1914.

5. Weaver CH, Buckner CD, Longin K et al. Syngeneic transplantation with peripheral blood mononuclear cells collected after the administration of recombinant human granulocyte colony-stimulating factor. Blood 1993; 82:1981-1984.

3.2

Purging of Bone Marrow and Peripheral Blood Stem Cells

Ellen Areman, Chitra Rajagopal

I) INTRODUCTION

When bone marrow (BM) or peripheral blood stem cells (PBSC) are transplanted, the patient usually does not receive an infusion of a pure population of hematopoietic cells. Depending on the processing method, varying numbers of red blood cells, granulocytes, T and B lymphocytes, monocytes, NK cells and committed myeloid and erythroid progenitor cells also accompany the hematopoietic stem cells into the recipient's circulation.

A) ALLOGENEIC TRANSPLANTS

The assorted populations of normal cells present in collections from allogeneic donors can cause problems in certain transplant settings. Red blood cells can cause an immediate hemolytic transfusion reaction in an ABO incompatible transplant, especially when the recipient possesses antibodies to red blood cell antigens of the donor. Immunocompetent donor T cells transplanted into an immunosuppressed recipient can initiate a profound and often fatal graft-versus-host disease (GVHD). Other types of cells such as granulocytes and red blood cells can interfere with efficient purging of the graft when treated with some of the pharmacologic and immunologic methods described below.

B) AUTOLOGOUS TRANSPLANTS

In autologous transplants, undetectable numbers of clonogenic tumor cells may contaminate the marrow or peripheral stem cells collected from autologous donors. To minimize this risk, marrow is usually collected from patients in remission. However, recent studies utilizing molecular diagnostic assays, tissue culture, gene marking and immunohistochemical techniques have demonstrated that occult tumor cells can be detected in histologically normal marrow and peripheral blood. Therefore, various techniques have been developed for purging the undesirable cells before infusion.

II) PURGING OF AUTOLOGOUS MARROW/PERIPHERAL BLOOD STEM CELLS (TABLE 3.3.1)

A) SIGNIFICANCE OF PURGING TUMOR CELLS FROM GRAFT

No one has performed randomized clinical trials comparing disease-free survival in patients receiving purged versus unpurged bone marrow transplants due to the number of patients required for such a study. However, many investigators believe that purging can prevent the reintroduction of potentially clonogenic tumor cells during autologous transplantation. A recent study suggests that the ability to successfully purge autografts of lymphoma cells as measured by negative PCR correlates with an improved disease-free survival after ABMT. Brenner et al have demonstrated with elegant gene marking studies that cells from both unpurged and purged bone marrow can be found in tumors of patients with recurrent disease after autologous transplantation. These and similar studies indicate that some type of tumor cell purging is probably necessary, although the optimum methods are yet to be determined. To be successful, a purging technique must eliminate clonogenic tumor cells

Bone Marrow Transplantation, edited by Richard K. Burt, H. Joachim Deeg, Scott Thomas Lothian, George W. Santos. © 1996 R.G. Landes Company.

from the graft while sparing the stem cells necessary for hematopoietic reconstitution. Whether any of the purging methods currently in use or under study are capable of removing all clonogenic cells from autologous grafts is not yet known. Until recently there were no methods available for detection and measurement of minimal residual disease in vivo and occult tumor cell contamination in vitro. Sophisticated cell culture and molecular techniques can now detect tumor cells in "remission" bone marrow of patients with leukemia and lymphoma as well as breast cancer cells in the bone marrow of patients with even early stage disease. When these assays are standardized and widely available, investigators will be able to measure the efficacy of various purging techniques and determine which are most effective at removing tumor cells.

3.3

B) TUMOR CONTAMINATION IN PBSC VERSUS MARROW

The presumed minimal risk of tumor contamination in PBSC has been one important factor in favor of using this type of hematopoietic progenitor cell for autologous transplantation. However, Ross et al have demonstrated in

---

Table 3.3.1. Purging methods for autologous bone marrow and peripheral stem cells

Pharmacologic
    4-hydroperoxycyclophosphamide (4-HC)
    mafosfamide
    etoposide (VP-16)
    methylprednisolone
    vincristine
    alkyl-lysophospholipids (ALP)
    photo-activated dyes
    guanine arabinoside (Ara-G)
    phenylalanine methylester

Immunologic-monoclonal antibodies
    complement
    immunotoxins
        ricin A chain
        pseudomonas exotoxin
    magnetic polymer microspheres
    photosensitive antibody-directed liposomes
    fluorescence-activated cell sorting (FACS)
    lectins
    radioisotopes

Mechanical
    hyperthermia
    freezing

Culture/immunotherapy
    IL-2 activation
    long-term marrow culture
    colony-stimulating factors

Combination techniques
    multiple drug treatment
    chemical + immunological
    cytokines + chemical agents
    ether lipids + cryopreservation
    cytokines + toxins
    chemical + hyperthermia

patients with localized breast cancer, no tumor contamination of PBSC compared with tumor cells detected in 50% of bone marrow specimens. In patients with metastatic disease, 22% of PBSC and 70% of BM samples contained tumor cells. Techniques for mobilization of PBSC may actually increase the number of circulating tumor cells. Brugger et al demonstrated that after chemotherapy and growth factor priming for stem cell collection in patients with stage IV breast cancer, there was an increase of tumor cells (100%) in peripheral blood, compared with 20% tumor contamination in the steady state. Although PBSC tumor cell contamination may be less than that of BM, the numbers are still significant. Therefore both PBSC and BM should be considered for purging techniques.

C) METHODS TO PURGE TUMOR CELLS

Most autologous purging involves negative selection techniques in which contaminating cells are removed from the graft, while the remaining cell populations are infused. In the more recently developed positive selection procedures, the hematopoietic progenitor cells which carry the CD34 antigen are removed from the cell suspension and infused, while the remaining cell populations are usually discarded. Combinations of positive and negative cell selection can be used as an effective methodology for purging unwanted cells.

1) **Pharmacologic Purging**

The first attempts at tumor cell removal took advantage of the fact that certain cytotoxic chemicals preferentially damage clonogenic leukemic cells while sparing at least some uncommitted hematopoietic progenitor cells. The most commonly used agents are 4-hydroperoxycyclophosphamide (4-HC) (an active metabolite of cyclophosphamide), mafosfamide and etoposide, although other chemotherapeutic drugs have been used, both singly and in combination. The efficiency of 4-HC depends on the concentration of red blood cells in the suspension, as the aldehyde from these cells can inactivate the drug. Preparation of a mononuclear fraction permits the use of a lower drug concentration. The laboratory incubates the cell suspension with the drug or drugs for the appropriate time period, followed by washing and cryopreservation. This technique has been shown to eliminate up to 4 logs of tumor cells at a drug concentration tolerated by hematopoietic cells.

This type of purging is relatively nonspecific and can lethally damage committed myeloid progenitors, often resulting in extended periods of neutropenia and thrombocytopenia with an accompanying increase in morbidity. Questions have been raised regarding the possible mutagenic potential of stem cells exposed to high concentrations of alkylating agents. Pharmacologic purging (e.g. 4-HC) is most frequently used in AML.

2) **Antibody Purging**

Certain cell surface markers have been identified which appear exclusively or predominately on certain tumor cells. Antigens expressed by clonogenic leukemic progenitor cells would appear to be the most appropriate targets for immunologic purging of marrow autografts, although these antigens have not been well described. Techniques have been developed which use one or more monoclonal antibodies against antigens such as CD14, CD15 and CD32 to remove or destroy contaminating tumor cells.

Although the concept of immunologic selection is attractive, there are some problems with this procedure. Specific antigens for all tumor cells have not been identified. Antigenic modulation by tumors may alter their phenotypes, making specific antibody purging unreliable. Furthermore, some tumor antigens cross-react with antigens on early hematopoietic progenitor cells. Immunologic purging with monoclonal antibodies is the most common method of purging for ALL.

a) *Complement Lysis*

The original immunologic purging method was a manual procedure in which antibody-sensitized target cells were lysed following multiple incubations with complement. This technique can be effective in cell removal, but is also extremely time consuming and labor-intensive. Ball and coworkers developed an automated method of antibody purging with complement by using a cell separator to constantly introduce fresh complement into the cell suspension, so that a single complement treatment is sufficient.

b) *Toxin-Bound Antibodies*

Monoclonal antibodies can also be bound to immunotoxins such as the ricin A chain, toxic lectin and pseudomonas exotoxin. The antibody binds to the target cell which is then poisoned by the attached toxin. These methods can eliminate from 3-5 logs of malignant cells from human marrow. Because of the specificity of the antibody, there is very little risk of damaging the hematopoietic stem cells.

c) *Magnetic Separation*

An efficient technique for removal of targeted cells which does not adversely affect hematopoietic cells, involves the use of magnetic polymer microspheres bound to one or more monoclonal antibodies. The antibody-coated beads attach to the cells of interest during an incubation period, after which the cell suspension is exposed to a strong magnet. The magnet holds the cells which have bound to the beads, allowing the unbound cells to be removed from the suspension.

3) Positive Selection

Immature progenitor cells possess a surface marker (CD34) which gradually diminishes as the cells differentiate and mature. If the patient's tumor cells are negative for this CD34 antigen, a more efficient method of tumor cell removal may be to isolate and infuse only those cells which bear this marker. This method should remove heterogeneous populations of tumor cells more effectively than does the "shotgun" approach of chemical purging or the negative selection techniques using "cocktails" of antibodies which may or may not bind to antigens on every clonogenic tumor cell. A fairly simple immune adsorption technique consists of incubating the cells with an anti-CD34 monoclonal antibody bound to biotin. The cell suspension is then passed over an avidin-coated column (Fig. 3.3.1) to which the biotinylated antibody attaches. The CD34-negative cells which run through the column are collected and discarded. Mechanical shaking of the column detaches the antigen-positive cells which are collected, cryopreserved and ultimately reinfused. Up to 7 logs of contaminating tumor and other cell types can be depleted with this technique. Because

3.3

Fig. 3.3.1. The Cell Pro CEPRATE system involves the treatment of bone marrow cells with a biotinylated monoclonal antibody to the CD34 antigen. The cells are then passed over a column of avidin-coated beads to concentrate the CD34+ cells. The concentrated cells are removed from the column by agitating the beads with a magnetic stirrer.

ancillary cells such as T helper cells are also removed, a delay in engraftment can occur. Furthermore, since the purity of the selected cells is less than 100%, it is not possible to guarantee that all tumor cells will be removed.

4) Other Purging Techniques

a) Cytotoxic Lymphocytes

Charak et al have shown that bone marrow and peripheral stem cells, when incubated with interleukin-2 (IL-2) for 24 hours, can generate cytotoxic effector cells which have a potent anti-tumor effect in vitro and in vivo. In vitro data suggest that these activated NK cells may preferentially suppress or kill malignant committed leukemic progenitors while sparing normal hematopoietic progenitors. Clinical trials are currently underway to determine whether infusion of these cells followed by a course of intravenous IL-2 can increase disease-free survival of patients with solid tumors and hematologic malignancies. These investigators have also shown that long-term culture with IL-2 can completely eliminate leukemia cells in vitro. This technique is currently in clinical trials as a means of autologous purging in various chronic and acute leukemias, multiple myeloma and lymphoma.

b) Temperature

Some tumor cells are more sensitive to temperature extremes than are normal hematopoietic cells. Freezing alone has been shown to preferentially preserve normal cells, while hyperthermic treatment of marrow grafts damages some tumor cells to a greater degree than normal hematopoietic cells.

c) Long-Term Cultures

A number of centers are investigating the use of long-term incubation in various culture systems as a method of autologous marrow purging. Placing bone marrow from patients with chronic myelogenous leukemia in culture for 7 days or more generally results in a decrease in the number of cells possessing the Philadelphia chromosome. However, approximately 30% of these cells may survive and contaminate the graft.

d) Antisense Oligonucleotides

Antisense oligonucleotides directed against several proliferative proteins such as P53 are currently being investigated in preclinical trials as purging tools.

3.3

III) PURGING OF ALLOGENEIC BONE MARROW AND PERIPHERAL BLOOD STEM CELLS (TABLE 3.3.2)

Purging of allogeneic marrow is not designed to remove tumor cells, since the donor must be healthy and without a malignancy. Instead, this type of purging is directed at removing red cells in ABO/Rh incompatible allografts or T lymphocytes for decreased risk of graft-versus-host disease (GVHD).

A) RED BLOOD CELL REMOVAL

The presence of incompatible red blood cells in either the infused marrow or in the circulation of the transplant recipient can result in a hemolytic transfusion reaction. When the recipient has circulating antibodies to antigens on the donor's red blood cells, the red blood cells in the graft can be removed by sedimentation, buffy coat preparation followed by density gradient centrifugation or separation by an apheresis instrument. When the donor has antibodies to the recipient's red blood cell antigens, plasma in the graft can be removed by centrifugation.

B) T CELL DEPLETION

In allogeneic transplantation it is thought that GVHD arises from infusion of donor lymphocytes. This reaction is especially severe in the mismatched related and matched unrelated settings, where administration of immunosuppressive drugs alone may not be sufficient to prevent this often fatal complication.

1) Nonselective Removal of T Lymphocytes

This depletion can be done chemically, mechanically or immunologically (Table 3.3.2). However, because some of these cells appear to be necessary for marrow engraftment and anti-tumor activity, nonselective T cell depletion can also increase the incidence of graft failure and disease recurrence compared to unmanipulated allogeneic grafts. Several clinical trials with nonselective T cell depletion, although decreasing the incidence of GVHD, have resulted in increased disease recurrence and decreased leukemia-free survival. Nonselective depletion eliminates all T cell subsets and natural killer (NK) cells, which may be responsible for a graft-vs-leukemia (GVL) effect. Increase in the incidence of infections and Epstein-Barr Virus (EBV)-associated lymphomas which may be secondary to removal of immunocompetent T cells have also been noted. To solve these problems donor lymphocytes have been added back to the recipient in the immediate posttransplant period. This has, however, been complicated by chronic GVHD, especially of the liver.

Table 3.3.2. Purging methods for allogeneic bone marrow and peripheral stem cells

Red blood cell antigen incompatibility
    Recipient has antibody to donor RBC
        red cell depletion
    Donor has antibody to recipient RBC
        plasma removal

T Cell Depletion
    Pharmacologic
        1-leucyl-1-leucine methyl ester
        methylprednisolone/vincristine

    Immunologic—Monoclonal antibodies
        complement
        immunotoxins
            ricin A chain
            pseudomonas exotoxin
        magnetic polymer microspheres
        fluorescence-activated cell sorting (FACS)
        immunoadsorption
        lectins + E-rosette separation
        positive selection

    Mechanical
        counterflow centrifugal elutriation

3.3

2) **Selective Removal of T Lymphocytes**

Protocols are being undertaken to selectively deplete marrow of T cell subsets before reinfusion. For example, at M.D. Anderson, CD8 T cells are being selectively depleted from allografts before reinfusion. Other centers are developing methods of functional depletion to selectively remove recipient reactive donor lymphocytes from the allograft. It is hoped that selective depletion of lymphocyte subsets will decrease GVHD but not be plagued by the increase in disease relapse or infection which accompanies nonselective T cell depletion.

3) **Methods of T Cell Depletion**

    a) Lectin T Cell Depletion

O'Reilly and coworkers developed the first method of T cell depletion by using the phenomenon that T cells incubated with soybean lectin will form rosettes around sheep red blood cells. These rosettes can then be removed by density gradient centrifugation. Although this technique is tedious and takes many hours to perform, it is effective and is still in use in a number of centers.

    b) Mechanical T Cell Depletion

Lymphocytes can be separated from bone marrow by counterflow centrifugal elutriation (CCE), taking advantage of the significant differences in size between red blood cells, lymphocytes and lymphocyte-poor mononuclear cells. The media flow rate increases as each progressively larger cell type is harvested. The final fraction, containing the majority of the progenitor cells, is collected with the centrifuge stopped and the media still flowing ("rotor off" fraction) (Fig. 3.3.2).

3.3

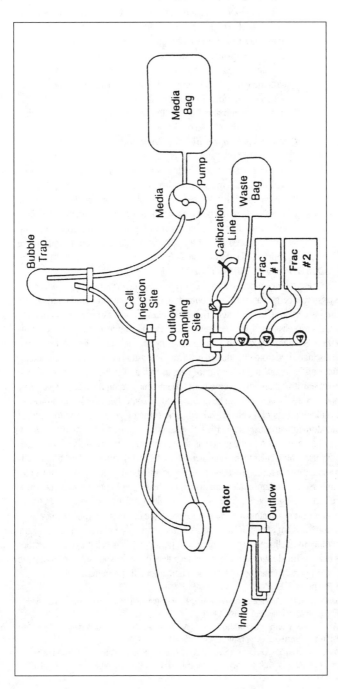

Fig. 3.3.2. Beckman JE5.0 rotor system and attachments for bone marrow elutriation. With medium flowing through rotor, marrow cell suspension is injected. As media flow rate is increased, cells of larger sizes are eluted and different flow rates containing different sized cells are directed into different bags such as "Frac 1" (lymphocyte fraction) and "Frac 2" monocyte/stem cell fraction). Reprinted with permission from Areman E, Deeg HJ, Sacher RA, eds. Bone Marrow and Stem Cell Processing: a manual of current techniques. Philadelphia: F.A. Davis. 1992:197.

c) Immunologic T Cell Depletion

Just as monoclonal antibodies to specific antigens on tumor cells can be used to remove those cells from the autologous marrow graft, antibodies to T cell antigens can perform the same function in allogeneic transplantation. Methods have been developed which employ antibodies to such antigens as CD3, CD4, CD5, CD6 and CD8 in conjunction with complement, immunotoxins and magnetic microspheres to remove T cells from allogeneic grafts. Plastic tissue culture flasks with anti-T cell antibodies bound to their inner surface are also used for T cell depletion.

d) Other Techniques of T Cell Depletion

A number of pharmacologic agents such as 1-leucyl-1-leucine methyl ester and methylprednisolone or vincristine have been used which are selectively toxic to lymphocytes. There are some data that freezing and thawing reduces the alloreactivity of allogeneic bone marrow. This may be a reason why allogeneic peripheral blood stem cell transplants, in which the donor cells are collected and cryopreserved before transplantation, do not seem to induce the anticipated level of acute GVHD.

IV) EVALUATION OF PURGING TECHNIQUE

In order to determine the adequacy of any purging technique, appropriate bioassays must be available. Flow cytometry is a rapid method of enumerating cells carrying certain surface markers. However, this method is not sensitive enough to detect rare cells which may be significant, especially tumor cells in the autologous setting. Furthermore, cells that are damaged by chemical agents and toxins cannot be distinguished from undamaged cells. Various systems have been developed for culturing leukemia and other tumor cells as well as T cells. These methods are move sensitive than flow cytometry but require incubation periods of 2 weeks or more before the colonies can be enumerated. In some T cell depletion protocols, specific numbers of T cells are added back to the depleted marrow to facilitate engraftment. Obviously, immediate results are necessary to determine the number of cells to return to the graft, and flow cytometry is the method which is generally used. Molecular techniques may be useful in detecting residual tumor cells in some cases, but the procedures are expensive and the methodology still quite cumbersome. Immunocytochemical methods have recently been introduced which are sensitive and specific for some tumor cells. Until acceptable quality control techniques are available, we will not be sure of the makeup of any bone marrow or peripheral stem cell graft. Regardless of the method employed, the goal of any autologous or allogeneic bone marrow or peripheral stem cell purge is to remove as many of the designated cells from the graft as possible while still conserving the greatest possible number of functional hematopoietic progenitor cells.

SUGGESTED READING

1. Areman E, Deeg HJ, Sacher RA, eds. Bone Marrow and Stem Cell Processing: a manual of current techniques. Philadelphia: F.A. Davis, 1992.
2. Gee AP, Gross S, Worthington-White DA, eds. Advances in Bone Marrow Purging and Processing, Fourth International Symposium. New York:Wiley-Liss, 1994.
3. Charak BS, Areman EM, Dickerson SA et al. A novel approach to immunomodulation of frozen human bone marrow with interleukin-2 for clinical application. Bone Marrow Transpl 1993; 11:147-154.

4. Brenner MK, Rill DR, Moen RC et al. Gene-marking to trace origin of relapse after autologous bone marrow transplantation. Lancet 1993; 341:85-86.
5. Yeager AM, Kaiser H, Santos GW et al. Autologous bone marrow transplantation in patients with acute nonlymphocytic leukemia, using ex vivo marrow treatment with 4-hydroperoxycyclophosphamide. N Eng J Med 1986; 315:141-147.
6. Reisner Y, Kapoor N, Kirkpatrick D et al. Transplantation for acute leukemia with HLA-A and -B nonidentical parental marrow cells fractionated with soybean agglutinin and sheep red blood cells. Lancet 1981; 2:327-336.
7. Ball ED, Mills LE, Cornwell III GG et al. Autologous bone marrow transplantation for acute myeloid leukemia using monoclonal antibody-purged bone marrow. Blood 1990; 75:1199-1206.
8. Kvalheim G, Sorensen O, Fodstad O et al. Immunomagnetic removal of B-lymphoma cells from human bone marrow: a procedure for clinical use. Bone Marrow Transpl 1988; 3:31-41.
9. DeWitte T, Raymakers R, Plas A. Bone marrow repopulation capacity after transplantation of lymphocyte-depleted allogeneic bone marrow using counterflow centrifugation. Transplantation 1984; 37:151-155.
10. Barnett MJ, Eaves CJ, Phillips GL et al. Successful autografting in chronic myeloid leukaemia after maintenance of marrow in culture. Bone Marrow Transpl 1989; 4:345-351.

3.3

Marrow Processing and Cryopreservation

Michele Cottler-Fox

I) INTRODUCTION

Hematopoietic progenitor stem cell grafts are currently obtained either from human marrow, peripheral blood or cord blood. In the future it may also be possible to derive a graft from fetal liver, xenogeneic hematopoietic cells or through tissue culture expansion of a minimal cell dose obtained from one of the above named sources. It is hypothesized that by using cord blood, fetal or xenogeneic cells, it will be possible to minimize or avoid the occurrence of GVHD since cells from these sources may not respond to or stimulate alloresponses as do cells from a normal mature host.

Historically, transplant physicians have aimed to acquire $2\text{-}4 \times 10^8$ nucleated cells/kg recipient body weight for an unmanipulated allogeneic marrow graft, and $1\text{-}3 \times 10^8$ nucleated cells/kg for an unmanipulated autologous graft. In general, this number of cells may be acquired with a collection of 10-20 ml of marrow/kg recipient body weight. However, it is imperative to know prior to collection whether a graft is to undergo any manipulation, and if so what the cell loss associated with the processing is likely to be, in order to harvest a sufficient number of cells such that an adequate graft remains afterwards.

II) MARROW TRANSPORT

Marrow arrives from the operating room to the processing laboratory in a standard blood transfer bag of 600-2000 ml volume, bearing a label identifying the donor and recipient (if it is a related donor), as well as the fluid and anticoagulant solution into which it has been collected. Unrelated donor marrow is labeled with an identifying number, rather than the donor name, and is hand carried by a courier from the operating room at the harvest center to the processing laboratory at the transplant center. Unrelated donor marrow may be transported on wet ice or without coolant in an insulated container, and must arrive at the transplant center within 24 hours of harvest. Since marrows are collected and transported worldwide, it is not unusual for the marrow courier to arrive outside of regular working hours. This means that if processing is required, the receiving laboratory may need to have personnel on call 24 hours a day.

III) VOLUME REDUCTION OF MARROW

In many centers an ABO-matched allogeneic marrow is transported directly from the operating room after harvest to the bedside, where it is infused through a central line as if it were a red blood cell transfusion. In such cases the only potential complications to the recipient are related to the volume infused, and the possibility that, despite filtering the marrow after collection, cell clumps may cause respiratory difficulty. In cases where the volume collected would be enough to cause problems with volume overload, the volume is easily reduced by centrifuging the product in a blood transfer bag and expressing off the plasma to give a reasonable final volume. This will increase the hematocrit of the final product, but should not present a clinical problem as long as the marrow is infused over an appropriate time period (3-4 hours). In cases of ABO-minor mismatch allogeneic

Bone Marrow Transplantation, edited by Richard K. Burt, H. Joachim Deeg, Scott Thomas Lothian, George W. Santos. © 1996 R.G. Landes Company.

transplant (where donor plasma contains a clinically significant ABO isoagglutinins or other red blood cell antibodies against the recipient), the marrow or peripheral blood stem cells are volume reduced by centrifugation to remove donor plasma (Table 3.4.1).

Although the amount of heparin used as an anticoagulant in allogeneic marrow harvests varies from center to center, it is generally not enough to cause bleeding problems. Nonetheless, the amount and type of anticoagulant contained in a marrow harvest should be on the transfusion component label of the infused bag, and should be checked prior to beginning the infusion.

3.4

Table 3.4.1. Standard marrow component processing

Procedure	Indication	Method
Plasma removal	Minor ABO incompatibility–donor plasma contains antibody to recipient RBCs	**1) Manual method**–First centrifugation then express off plasma into transfer bag– decreases plasma by 75%, loss of nucleated cells < 5% **2) Automated method**–Cobe 2991 blood cell processor uses centrifugation to remove plasma and RBC, 15% of nucleated cells lost, 20-30 ml of donor RBCs remain– sufficient to cause hemolysis if incompatible
RBC removal	Major ABO incompatibility–recipient plasma contains antibody to donor RBC	**1) Starch sedimentation**–25% loss of nucleated cells, 5-25 ml RBC remain **2) Double buffy coat method**–buffy coat obtained with Cobe 2991 is resuspended in a unit of recipient compatible irradiated leukocyte-depleted RBCs and a second buffy coat obtained with Cobe 2991–25% loss of nucleated cells, < 10 ml incompatible cells which should not cause hemolytic reaction **3) Density gradients**–e.g. Ficoll, albumin, percoll (also removes platelets and neutrophils) 20-30% loss of mononuclear cells, virtually no RBCs remain **4) Mononuclear preparation** see below
Buffy coat [#1]	Concentrate marrow for cryopreservation	**1) Manual method**–as above, remove plasma by manual method then aliquot into 50 ml tubes, centrifuge and pipette off buffy coat–25% loss of nucleated cells **2) Centrifuge**–Cobe 2991 blood cell processor
Mononuclear cell preparation[#2]	Depletion of RBCs and plasma– first step to prepare for T cell depletion, tumor cell purging or CD34 positive selection	**1) Density gradient**–Ficoll, albumin, etc. 20-30% loss of progenitor cells **2) Automated**–Fenwall CS 3000 or Cobe Spectra cell separators, > 50% of starting mononuclear cells remain

1–Buffy coat contains white blood cells, i.e. neutrophils, platelets, mononuclear cells.
2–Mononuclear cells are T and B lymphocytes, monocytes, NK cells and hematopoietic progenitor cells.

IV) RED BLOOD CELL DEPLETION

In the case of an ABO-major mismatched allogeneic transplant (where the recipient has isoagglutinins against a donor red blood cell antigen), the simple removal of red blood cells will prevent acute hemolysis. In the past, it was feared that this marrow manipulation might interfere with engraftment, for which reason recipient isoagglutinins were removed by vigorous plasmapheresis, by plasma exchange over an immunoadsorbent column, or by infusing a small amount of incompatible red blood cells to adsorb out the recipient isoagglutinins in vivo just prior to marrow infusion. However, as it has become clear that progenitor stem cells do not carry A, B or Rh antigens and are therefore not prone to hemolysis by isoagglutinins, and as red blood cell removal has been shown to involve minimal white blood cell (i.e. stem cell) loss, it is now more common to remove red blood cells using one of the following methods (in cases where further processing is contemplated, methods aimed at isolating white blood cells rather than removing red blood cells are preferred). Red blood cells removed from an allogeneic or autologous marrow may also be reinfused into the donor after the harvest.

A) STARCH SEDIMENTATION

Bone marrow containing heparin is mixed with 6% hydroxyethyl starch at a ratio of 8:1, after which the red blood cells are allowed to sediment as a function of gravity with the blood transfer bag in the inverted position (entry port-side down). After 30-180 minutes, the sedimented red blood cells are removed into a secondary bag. The original bag contains approximately 75% of the original nucleated marrow cells for transplantation, as well as 5-25 ml of the original red blood cells. This volume of incompatible red blood cells is still capable of causing a hemolytic transfusion reaction, for which reason other techniques are commonly preferred to sedimentation.

B) REPEATED DILUTION WITH COMPATIBLE RED BLOOD CELLS

After a white blood cell-rich cell concentrate has been prepared by sedimentation (or centrifugation with, e.g. a Cobe 2991), the remaining incompatible red blood cells may be decreased by diluting the cells with a unit of recipient-compatible, irradiated, leukocyte-depleted red blood cells and 150 ml of recipient-compatible plasma (from the marrow donor if possible). The product is then allowed to resediment (or recentrifuge), reducing the total of incompatible red blood cells to < 10 ml while retaining 85% of nucleated white blood cells. This procedure is sufficient to prevent hemolytic reactions but is time consuming, and most processing laboratories today would choose instead to go directly to a method aimed at mononuclear cell separation (see below).

V) BUFFY COAT PREPARATIONS

Buffy coat preparation is the white blood cell layer obtained when centrifuging the marrow collection. It contains neutrophils, lymphocytes, monocytes and progenitor cells. A standard blood bank centrifuge (e.g. Cobe 2991) may be used for this procedure. Approximately 80% of nucleated cells from the marrow are retained in this way while simultaneously reducing the volume and red blood cell concentration by 80%. This is an adequate method of volume reduction for cryopreservation, and a good preliminary step towards a mononuclear cell separation. It is not recommended where red blood cell contamination may interfere with chemical or immunological purging methods. It is also not sufficient to prevent a hemolytic reaction from ABO-incompatible RBCs.

3.4

VI) MONONUCLEAR CELL SEPARATION

This is usually the first step towards additional processing of allogeneic collections for T cell depletion, and of autologous products for tumor cell purging or cryopreservation. Mature myeloid elements may be removed from a buffy coat preparation using manual or automated methods, with or without density gradient materials. Each of these agents removes essentially all red blood cell and myeloid elements, leaving only mononuclear (morphological lymphocyte+monocyte populations, including progenitor cells) portions of the original graft.

A) MANUAL METHOD

Density gradient materials such as albumin, Ficoll and Percoll have been used successfully by many different centers. While this is of great importance to the economic feasibility and success of purging protocols for autologous transplantation, and of T cell depletion for allogeneic transplantation, it is also true that with density gradients upwards of 30% of the original progenitor stem cell elements of a graft may be lost. In order to offset this loss, and to maintain adequate numbers of such cells after separation procedures, it is necessary to collect considerably more than are needed for the final graft.

B) AUTOMATED METHOD

Alternatively, a mononuclear cell preparation can be done in a more automated fashion using any of the apheresis devices currently available (e.g.Cobe Spectra, Fenwall CS 3000) (Fig. 3.4.1). In these sterile closed systems, marrow is pumped into the instrument and components separated into 3 layers by size and density using centrifugation: an outer RBC layer, a mononuclear buffy coat layer in the center, and an inner layer of platelet rich plasma. The separated components exit the centrifuge by different outlet channels.

3.4

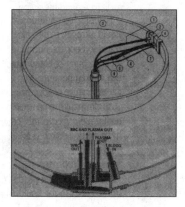

Fig. 3.4.1. Cobe Spectra system for automated collection of bone marrow or peripheral blood mononuclear cells. Centrifugation separates marrow inside the rotor into 3 products: RBCs on the outside, white blood cells in the center and platelet rich plasma on the outside. The interface is held constant by adjusting plasma flow.

VII) CRYOPRESERVATION

There is little published data on the use of cryopreserved allogeneic marrow, although many centers have used this option when it was impossible for a related donor to be available at the planned time of transplant. Collection of marrow from cadaveric donors is feasible, and cadaveric donor marrow has been documented to engraft in the related recipient.

Autologous marrow and blood-derived stem cells are usually cryopreserved for later use, but some centers maintain it at 4°C and infuse it after a brief conditioning regimen. Viability of hematopoietic cells is greater than 95% after 24 hours at 4°C and may be as great as 70% for up to 7 days. Cells to be cryopreserved may be aliquoted into small vials, or they may be placed into freezing bags with a capacity of 100-200 ml. If the bag to be frozen requires heat sealing, it is imperative to remove all air prior to sealing, as air in the bag may cause it to explode on thawing, since the air rapidly expands before the bag becomes flexible.

A) ADDITION OF CRYOPROTECTANT

Dimethylsulfoxide (DMSO) is the most widely used cryoprotective agent used to cryopreserve hematopoietic stem cells at present. DMSO is a universal solvent capable of stabilizing cell membranes under rapidly changing conditions, preventing intracellular ice crystal formation during freezing and heat release during the period of phase transition. DMSO has been described as toxic to stem cells at room temperature, for which reason most protocols have emphasized the need to add it at 4°C to cells prior to controlled rate freezing. However, more recent data suggests that DMSO may not be as toxic as previously thought. DMSO is reported to be capable of degranulating mast cells and has also been described as hepatotoxic, for which reason efforts have been made to reduce the amount used for freezing. A final concentration of 10% DMSO with albumin or human serum is commonly used, with some centers using hydroxyethyl starch to help stabilize cell membranes and reduce the amount of DMSO used to 5%. Data on hepatotoxicity of DMSO is, however, sparse and even its half-life in humans remains to be determined.

B) FREEZING TECHNIQUE

Freezing has traditionally been performed in a controlled-rate freezer, using liquid nitrogen to decrease the temperature at a rate of 2°C/min until a phase transition occurs, during which heat is given off by the solution, followed by an extra burst of liquid nitrogen to prevent an increase in temperature. After this, the temperature drop is adjusted to 5°C/min until the mixture has reached a temperature of approximately -120°C.

An alternative to controlled rate freezing is so-called "dump" freezing. In this technique a freeze mix containing DMSO, albumin or human serum and hydroxyethyl starch is added to cells and the product is placed into a -80°C freezer.

C) STORAGE CONDITIONS

After freezing, the products are removed from the freezing apparatus and stored either in the liquid (-196°C) or vapor (-156°C) phase of a liquid nitrogen freezer, or in a mechanical (-80°C)freezer. Marrow frozen with a controlled-rate freezer and infused after more than 9 years in liquid nitrogen storage has resulted in good hematopoietic reconstitution, but products which have been subjected to "dump" freezing and stored at -80°C have been shown to deteriorate within 6-12 months of being collected.

VIII) THAWING AND INFUSION

A number of studies have shown that rapid thawing is desirable for the survival of hematopoietic progenitor stem cells. Thawing may be performed in a temperature controlled water bath, or in a sterile basin using sterile saline heated to 40°C prior to use. Once thawed, the cells are commonly infused, directly and rapidly, via central venous access. Each bag, usually less than 100 ml in volume, is infused over no more than 15 minutes to avoid the supposed DMSO toxicity to stem cells. Immediate side effects associated with rapid infusion of stem cell products may include: volume overload; bradycardia (the result of cold cardioplegia); nausea and vomiting (the result of the unpleasant taste of DMSO via direct nerve stimulation from the product in blood); fever, tachycardia, hypo- or hypertension (the result of lysed granulocytes if the product is not a clean mononuclear product); and allergic reactions ranging from urticaria to anaphylaxis (related to plasma proteins and/or DMSO). Most if not all of these problems may be avoided by volume reducing the products after collection, and giving antiemetics, diphenhydramine and/or hydrocortisone prior to infusing stem cells. Renal dysfunction related to red cell hemolysis is a delayed problem which may be seen with infusion of stem cell products, for which reason some centers prophylactically give mannitol and lasix along with stem cell infusions.

3.4

Alternatively, some centers process the thawed product to remove DMSO prior to infusion, and to reduce the volume. This is not a widely accepted maneuver, however, due to concern for stem cell loss. Still other centers rapidly thaw and then dilute the product with a sterile buffered saline solution prior to infusion, then infuse the product slowly at room temperature. While this allows the processing laboratory to deal with potential bag breakage in the most efficient manner, it increases the volume infused and generally requires the use of DNAse to avoid cell clumping.

IX) PRODUCT ASSESSMENT/QUALITY ASSURANCE

A) STEM CELL DOSE

Inappropriate freezing can lead to cell death, either immediately or at the time of thawing. Since there is thought to be a threshold dose for stem cells, above which rapid hematopoietic reconstitution can be expected, it is important to evaluate graft quality. While all processing laboratories evaluate grafts for cell dose prior to cryopreservation, some have argued that grafts should again be evaluated after thawing. While this is a logical suggestion, it is not commonly performed due to the expense and the fact that data are not evaluable until after the graft has been infused.

1) Nucleated Cell Dose

On the basis of allograft data from patients with aplastic anemia, a marrow dose of 3×10^8 nucleated cells/kg was originally proposed as a threshold dose. Fewer nucleated cells are believed to be acceptable in autologous transplantation. However, nucleated cells are a surrogate for the cells which reconstitute hematopoiesis, and much effort is currently being expended to refine a threshold dose in more accurate ways.

2) Colony Forming Unit

Until recently, the most commonly used assay for reconstituting ability of a graft was the growth of the day 14 granulocyte-macrophage colony forming unit (CFU-GM) in semisolid culture media. Threshold doses for CFU-GM differ widely, but may be in the range of $0.1\text{-}1 \times 10^4$/kg for marrow, and

1-5 x 10^5/kg for blood-derived stem cells. Unfortunately, culture conditions vary among laboratories, making comparisons difficult. Further, since the cultures cannot be counted until 14 days after a harvest, real time evaluation of a graft is impossible.

3) CD34 Cell Count

Progressively more centers are now using flow cytometry based on the CD34 cell surface marker, which is found on all progenitor stem cells capable of engraftment, to determine the acceptability of a graft for transplantation. Here, technical difficulties make comparisons between centers difficult, and the threshold doses in the literature differ. However, it seems likely at present that more than 2.0 x 10^6 CD34+ cells/kg will lead granulocyte and platelet recovery to clinically acceptable levels within approximately 14 days of transplantation.

B) MICROBIOLOGY

Microbiological evaluation of a graft for contamination resulting from the collection or processing procedures may be performed. While some feel that all grafts should be subjected to bacterial and fungal cultures both initially and after processing, others have pointed out that: 1) these results are often not available until after the infusion; 2) these cultures are expensive and 3) due to the small volumes most processing laboratories are willing to sacrifice for cultures, the results may be invalid.

SUGGESTED READING

1. Areman E, Deeg HJ, Sacher RA, eds. Bone Marrow and Stem Cell Processing: a manual of current techniques. Philadelphia: F.A. Davis, 1992.
2. Sacher RA, McCarthy LJ, Smit Sibinga C, eds. Processing of Bone Marrow for Transplantation. Bethesda: American Association of Blood Banks, 1990.

4.0 Cytoreduction

4

Conditioning (Preparative) Regimens

Richard K. Burt, Wyndham H. Wilson

I) GENERAL PRINCIPLES

Over the past 25 years, a large number of intensive conditioning regimens requiring hematopoietic stem cell transplantation (HSCT) have been developed. The earliest regimens were developed for the allogeneic transplantation of acute leukemia, and primarily employed total body irradiation (TBI)/chemotherapy combinations. However, over the past decade, a proliferation of mostly chemotherapy-based regimens has occurred for use with autologous HSCT of solid tumors, lymphomas and AML. The design of conditioning regimens has generally been empirical and few randomized trials have established the superiority of one regimen over another. Thus, when choosing a conditioning regimen, a physician should consider a number of factors: is the regimen effective in the disease; does the patient have prior treatment or preexistent organ damage which may be exacerbated by specific cytotoxic drugs/TBI; and/or is immunosuppression needed to facilitate engraftment.

A) ALLOGENEIC

HSCT conditioning regimens must provide effective tumoricidal activity and suppress host immunity to prevent rejection of the allograft. Following engraftment, an allogeneic graft-versus-leukemia (GVL) effect provides additional and important antitumor activity. Commonly used cytotoxic agents include TBI, cyclophosphamide (CY), busulfan (BU), cytarabine (ara-c) and etoposide (VP). Immunosuppressive agents include TBI, cyclophosphamide, anti-lymphocyte antibodies (anti-thymocyte globulin (ATG)), steroids, cyclosporine A (CsA) and methotrexate. An exception to the need for conditioning regimens is sibling matched BMT for severe combined immune deficiency (SCID), where no host immunity or tumor exists.

Specific problems occur during transplantation in patients who receive T cell depleted (TCD) stem cells. TCD BMT is done to decrease early morbidity and mortality from GVHD. These patients have an increased risk of failure to engraft compared to conventional (unmanipulated) sibling HLA-identical BMT. In addition, TCD BMT has an increased risk of leukemia relapse. Therefore, more aggressive regimens which include higher doses of TBI (1320-1440 cGy), ATG and/or a second myeloablative drug (e.g. thiotepa or cytosine arabinoside) are commonly combined with cyclophosphamide.

B) AUTOLOGOUS

Since HSCT conditioning regimens are administered to eradicate tumor cells, immunosuppression is unnecessary and detrimental. In autologous HSCT of solid tumors, TBI is generally avoided since effective tumoricidal doses exceed extramedullary dose-limiting toxicity (DLT). Selection of cytotoxic agents is based on efficacy, steep dose response, lack of cross resistance with other agents in the regimen and low extramedullary DLT. Combination regimens (> 2 agents) are generally more effective than single agent regimens.

Bone Marrow Transplantation, edited by Richard K. Burt, H. Joachim Deeg, Scott Thomas Lothian, George W. Santos. © 1996 R.G. Landes Company.

4.1

II) CONDITIONING REGIMENS

For discussion in this chapter, conditioning regimens are placed into one of five major categories: CY/TBI, BU/CY, non-CY single drug/TBI, derivatives of the BACT regimen, and solid tumor regimens.

A) CY/TBI (CYCLOPHOSPHAMIDE AND TOTAL BODY IRRADIATION)

This is one of the most commonly used conditioning regimens for allogeneic HSCT of hematologic and lymphoid malignancies. Multiple variations in the dose and fractionation of TBI (Table 4.1.1), dose of CY, and addition of other agents have appeared in the literature. For high risk leukemias, increasing TBI to 1575 cGy decreases tumor relapse but overall survival is unchanged due to increased toxicity (especially interstitial pneumonitis). Attempts to improve CY/TBI have also been made by adding a second drug such as cytarabine, etoposide or thiotepa. The addition of cytarabine, thiotepa or ATG is associated with decreased graft rejection in T cell depleted allogeneic HSCT. Unfortunately, there are few randomized studies to determine if any combination is superior to CY/TBI alone. Thus, CY/TBI is still commonly used at a dose of 120 mg/kg CY (60 mg/kg days -5,-4), and 1200 cGy TBI (200 cGY BID days -3,-2,-1).

4.1

Table 4.1.1. Examples of some cyclophosphamide plus total body irradiation (CY/TBI) regimens

Regimen	Total Dose	Daily Dose	Schedule
CY/TBI			
cyclophosphamide	120 mg/kg	60 mg/kg/d	day -5,-4
total body irradiation	1200 cGy	200 cGy BID	day -3,-2,-1
CY/TBI			
cyclophosphamide	120 mg/kg	60 mg/kg/d	day -8,-7
total body irradiation	1200 cGy	200 cGy/d	day -6,-5,-4,-3,-2,-1
CY/TBI			
cyclophosphamide	200 mg/kg	50 mg/kg/d	day -8,-7,-6,-5
total body irradiation	1200 cGy	300 cGy /d	day -4,-3,-2,-1
HDAC/CY/TBI			
cytarabine	18 g/m²	3 g/m² BID	day -8,-7,-6
cyclophosphamide	90 mg/kg	45 mg/kg/d	day -5,-4
total body irradiation	1200 cGy	200 cGy BID	day -3,-2,-1
TBI/VP/CY			
total body irradiation	1320 cGy	120 cGy TID	day -8,-7,-6,-5*
etoposide	60 mg/kg	60 mg/kg/d	day -4
cyclophosphamide	120 mg/kg	60 mg/kg/d	day -3,-2
TBI/TT/CY/ATG			
total body irradiation	1375 cGy	125 cGy TID	day -9,-8,-7,-6*
thiotepa	10 mg/kg	5 mg/kg/d	day -5,-4
cyclophosphamide	120 mg/kg	60 mg/kg/d	day -3,-2
antithymocyte globulin	120 mg/kg	30 mg/kg/d	day -5,-4,-3,-2

* TBI given twice only on this day

B) BU/CY (BUSULFAN/CYCLOPHOSPHAMIDE)

This regimen is used in patient groups similar to CY/TBI, including AML, CML, ALL, multiple myeloma and lymphoma. In some situations, TBI may be contraindicated such as in patients who have received prior radiation, or unavailable such as in hospitals which lack adequate radiation facilities, equipment or training to administer TBI. In BU/CY, busulfan is administered orally but, because oral administration is associated with erratic absorption, bioavailability varies. Veno-occlusive disease (VOD) of the liver which may be lethal, correlates with increased oral bioavailability of busulfan. Several BU/CY combinations exist (Table 4.1.2). BU/CY$_4$, also called "big" BU/CY, administers 16 mg/kg BU and 200 mg/kg CY. In an attempt to decrease toxicity, BU/CY$_2$ was developed; also called "little" BU/CY, which employs a lower CY dose (120 mg/kg). There is no convincing evidence that either regimen is superior, so institutional preference generally determines the regimen of choice. A number of other cytotoxic agents have been combined with BU/CY, a discussion of which is beyond the scope of this book. However, nonrandomized studies suggest that the addition of etoposide to BU/CY may be particularly promising for the treatment of lymphoid and hematologic malignancies.

There are some clinical data comparing BU/CY and CY/TBI. For AML and CML, nonrandomized studies and almost all randomized comparisons suggest that BU/CY and CY/TBI produce equivalent engraftment, relapse and survival. However, a single randomized French trial in patients with AML found a significantly lower leukemia-free survival and increased relapse with BU/CY$_2$ compared to CY/TBI. In patients at risk for radiation pneumonitis (e.g. prior mediastinal radiation) and in children, BU/CY is generally preferred over CY/TBI due to less toxicity and problems with growth and development. However, newer information indicates that growth and development may be equally affected by BU/CY and CY/TBI.

C) NON-CY SINGLE AGENT/TBI REGIMENS

Following the success of CY/TBI, other combinations of single chemotherapy drugs with TBI were developed for hematologic malignancies (Table 4.1.3). The combination of etoposide and TBI (VP/TBI), is a reasonable alternative to CY/TBI or BU/CY. A randomized comparison of VP/TBI and BU/CY$_2$ in patients with hematologic malignancies undergoing allogeneic HSCT showed the regimens to have similar toxicity, relapse rates and overall survival.

Table 4.1.2. Examples of busulfan-based regimens

Regimen	Total Dose	Daily Dose	Schedule
"Big" BU/CY (BU/CY$_4$)			
busulfan	16 mg/kg	1 mg/kg every 6 hrs	day -9,-8,-7,-6
cyclophosphamide	200 mg/kg	50 mg/kg/d	day -5,-4,-3,-2
"Little" BU/CY (BU/CY$_2$)			
busulfan	16 mg/kg	1 mg/kg every 6 hrs	day -7,-6,-5,-4
cyclophosphamide	120 mg/kg	60 mg/kg/d	day -3,-2

Table 4.1.3. Examples of some noncyclophosphamide single agent chemotherapy plus total body irradiation regimens

Regimen	Total Dose	Daily Dose	Schedule
TBI/VP			
total body irradiation	1320 cGy	120 cGy TID	day -7,-6,-5,-4*
etoposide	60 mg/kg	60 mg/kg	day -3
HDAC/TBI			
cytarabine	36 g/m²	3 g/m² every 12 hrs	day -9,-8,-7,-6,-5,-4
total body irradiation	1200 cGy	200 cGy BID	day -3,-2,-1
MEL/TBI			
melphalan	140 mg/m²	140 mg/m²	day -3
total body irradiation	1000 cGy	200 cGy BID	day -2,-1, 0⁺

*TBI given twice on this day   ⁺ TBI given only once on this day

4.1

High-dose cytarabine (HDAC), a very effective drug in myelogenous leukemia, has also been combined with TBI, and has the added benefit of good central nervous system penetration. Nonrandomized trials of HDAC/TBI in allogeneic HSCT suggest that it is an acceptable alternative to CY/TBI or BU/CY. Melphalan and TBI have been combined for the treatment of multiple myeloma, using a variety of doses; melphalan (120-200 mg/m²) and TBI (800-1300 cGy). The optimal combination is unknown since no randomized comparisons have been performed.

D) **BACT-DERIVED REGIMENS**

Some 25 years ago, investigators at the National Cancer Institute (NCI)(U.S.) began a HSCT program which used a conditioning regimen with carmustine (BCNU), cytarabine, cyclophosphamide and 6-thioguanine (BACT) for the treatment of lymphomas and hematologic malignancies. Although BACT was never fully evaluated as a conditioning regimen, a number of derivatives including BAVC, CBV, LACE and BEAM were developed at other institutions and are currently in use (Table 4.1.4). However, due to an absence of randomized prospective studies, it remains unclear which regimen(s) produces the best overall survival.

BACT derivative regimens are generally used with autologous HSCT for the treatment of Hodgkin's and non-Hodgkin's lymphomas. There has been extensive experience with CBV, and like other regimens, a number of drug doses and schedules have been explored. In regard to lymphomas, non-TBI containing regimens such as these are used with autologous HSCT, while CY/TBI and BU/CY are generally used with allogeneic HSCT. Nonrandomized studies of CY/TBI and BACT-derived regimens in patients undergoing autologous HSCT for lymphoid malignancies show similar outcomes, so toxicity considerations should guide the choice of regimens.

E) **REGIMENS FOR SOLID TUMORS**

In general, drugs are selected for conditioning regimens based on their single agent activity against the tumor being treated, steep dose response curves,

Table 4.1.4. Examples of BACT-derived regimens

Regimen	Total Dose	Daily Dose	Schedule
BACT			
carmustine	200 mg/m²	200 mg/m²	day -6
cytarabine	800 mg/m²	200 mg/m²	day -5,-4,-3,-2
cyclophosphamide	200 mg/kg	50 mg/kg	day -5,-4,-3,-2
6-thioguanine	800 mg/m²	200 mg/m²	day -5,-4,-3,-2
BAVC			
carmustine	800 mg/m²	800 mg/m²	day -6
amsacrine	450 mg/m²	150 mg/m²	day -5,-4,-3
etoposide	450 mg/m²	150 mg/m²	day -5,-4,-3
cytarabine	900 mg/m²	300 mg/m²	day -5,-4,-3
CBV			
carmustine	300 mg/m²	100 mg/m²	day -8,-7,-6
etoposide	2400 mg/m²	800 mg/m²	day -8,-7,-6
cyclophosphamide	7.2 g/m²	1.8 g/m²	day -5,-4,-3,-2
LACE			
lomustine	200 mg/m²	200 mg/m²	day -7
etoposide	1.0 g/m²	1.0 g/m²	day -7
cytarabine	4.0 g/m²	2.0 g/m²	day -6,-5
cyclophosphamide	5.4 g/m²	1.8 g/m²	day -4,-3,-2
BEAM			
carmustine	300 mg/m²	300 mg/m²	day -6
etoposide	800 mg/m²	200 mg/m²	day -5,-4,-3,-2
cytarabine	800 mg/m²	200 mg/m²	day -5,-4,-3,-2
melphalan	140 mg/m²	140 mg/m²	day -1

extramedullary DLT at high drug doses and nonoverlapping toxicity with other agents in the regimen. Synergistic activity with other agents in the regimen is desirable but of unproved clinical benefit. As a drug class, alkylating agents possess many of these characteristics and form the backbone of most solid tumor conditioning regimens (Table 4.1.5). Agents derived from natural products such as anthracyclines and taxanes (paclitaxel) are very active in solid tumors but have significant extramedullary toxicity at relatively low drug doses. Nevertheless, less toxic analogs like mitoxantrone or less toxic administration schedules such as 3 hour infusions of paclitaxel are being incorporated and tested in conditioning regimens. To maximize dose intensity by reducing extramedullary toxicity and to potentially increase efficacy, continuous infusion administration schedules are being investigated. A number of conditioning regimens such as ICE and CTCb administer one or more drugs by continuous infusion (Table 4.1.5). Another way to increase dose intensity is double autologous transplantation, an approach tested in patients with breast and testicular cancers, multiple myeloma, neuroblastoma, AML and CML. The second transplant is usually given within 3 months of the first BMT. This allows time for recovery of hematopoiesis (ANC > 1500 and platelets > 100,000). It remains to be shown that this approach increases survival compared to a single transplant.

III) SPECIFIC DISEASES

Conditioning regimens for which there is experience in specific diseases are shown in Table 4.1.6. Genetic or acquired diseases, such as severe combined immunodeficiency (SCID), aplastic anemia, Fanconi's anemia and thalassemia, in which hematopoiesis is abnormal are necessarily treated with allogeneic HSCT.

4.1

Table 4.1.5. Examples of solid tumor conditioning regimens

Regimen	Total Dose	Daily Dose	Schedule
BCC[1]			
cisplatin	165 mg/m^2	55 mg/m^2	day -6,-5,-4
cyclophosphamide	5625 mg/m^2	1875 mg/m^2	day -6,-5,-4
carmustine (BCNU)	600 mg/m^2	600 mg/m^2	day -3
CTCb (STAMP V)			
thiotepa	500 mg/m^2	125 mg/m^2	day -7,-6,-5,-4
cyclophosphamide	6.0 g/m^2	1.5 g/m^2	day -7,-6,-5,-4
carboplatin	800 mg/m^2	200 mg/m^2	day -7,-6,-5,-4
CVP[2]			
cyclophosphamide	5.25 g/m^2	1.75 g/m^2	day 1,2,3
cisplatin	180 mg/m^2	60 mg/m^2	day 1,2,3
etoposide	1.2 g/m^2	400 mg/m^2	day 1,2,3
MVT			
mitoxantrone	30 mg/m^2	30 mg/m^2	day -5
etoposide	1.2 g/m^2	400 mg/m^2 divide q12 hrs	day -5,-4,-3
thiotepa	750 mg/m^2	250 mg/m^2	day -5,-4,-3
CTB			
cyclophosphamide	7.5 g/m^2	2.5 g/m^2	day -6,-4,-2
thiotepa	675 mg/m^2	225 mg/m^2	day -6,-4,-2
carmustine	450 mg/m^2	150 mg/m^2	day -6,-4,-2
ICE			
ifosfamide	16 g/m^2	4.0 g/m^2	day -6,-5,-4,-3
carboplatin	1.8 g/m^2	600 mg/m^2	day -6,-5,-4
etoposide	1.5 g/m^2	500 mg/m^2 divided q12 hrs	day -6,-5,-4
CBDA/VP			
carboplatin	1.5 g/m^2	500 mg/m^2	day -7,-5,-3
etoposide	1.2 g/m^2	400 mg/m^2	day -7,-5,-3
MCC			
mitoxantrone	75 mg/m^2	25 mg/m^2	day -8,-6,-4
carboplatin	1.5 g/m^2	300 mg/m^2	day -8,-7,-6,-5,-4
cyclophosphamide	120 mg/m^2	40 mg/m^2	day -8,-6,-4

[1]: day of marrow infusion defined as day 1
[2]: day of marrow infusion defined as day 7

In general, when an HLA-matched sibling is available, the preferred transplant approach for AML, CML and ALL is also allogeneic HSCT. Multiple myeloma, Hodgkin's disease and non-Hodgkin's lymphoma are traditionally treated with autologous HSCT. In those patients with an HLA-matched sibling, a small but increasing number of patients are being considered for allogeneic HSCT because of the increased relapse-free survival associated with the GVL effect. Solid tumors such as testicular, ovarian and breast cancer are treated with autologous HSCT.

Table 4.1.6. Examples of some conditioning regimens for specific diseases

Disease	Allogeneic Regimen	Autologous Regimen
SCID	None	
Aplastic anemia	CY/ATG, CY, CY/CsA, CY/TBI	
Fanconi's anemia	cyclophosphamide (20 mg/kg)/ thoracoabdominal irradiation (400-500 cGy)	
Thalassemia	BU/CY	
Leukemia (AML, ALL, CML)	CY/TBI, BU/CY, VP/TBI, Mel/TBI, HDAC/TBI,	CY/TBI, BU/CY, BAVC, CBV,
Myelodysplastic syndrome	BU/CY/TBI	BU/VP
Multiple myeloma	Mel/TBI, Mel, CY/TBI, BU/CY	Mel/TBI, Mel, CY/TBI, BU/ CY
Chronic lymphocytic leukemia	CY/TBI	CY/TBI
Myelofibrosis	CY/TBI, BU/CY	
Lymphoma BEAC,	CY/TBI, BU/CY	CY/TBI, BU/CY, CBV, BEAM, LACE, ICE, BAVC
Testicular		CBDCA/VP, CBDCA/ VP/IFOS, PEC, ICE
Breast		CVP, BCC, MVT, CTCb, CTB, ICE
Ovarian		CCT, MCC
Neuroblastoma	Mel/TBI	Mel, Mel/TBI

BAVC = carmustine/amasacrine/etoposide/cytarabine; BCC =carmustine/cisplatin/cyclophosphamide; BEAM = carmustine/etoposide/cytarabine/melphalan; BEAC = carmustine/etoposide/cytarabine/ cyclophosphamide; BU/CY = busulfan/ cyclophosphamide; BU/CY/TBI = busulfan/cyclophosphamide/ total body irradiation; BU/VP = busulfan/etoposide; CBDCA/VP = carboplatin/etoposide; CBDCA/VP/ IFOS = carboplatin/etoposide/ifosfamide; CBV = cyclophosphamide/carmustine/etoposide; CCT = cyclophosphamide/cisplatin/thiotepa; CVP = cyclophosphamide/etoposide/cisplatin; CTCb = cyclophosphamide/thiotepa/carboplatin; CTB = cyclophosphamide/thiotepa/carmustine; CY = cyclophosphamide; CY/ATG = cyclophosphamide/antithymocyte globulin; CY/CsA = cyclophosphamide/cyclosporine; CY/TBI = cyclophosphamide/total body irradiation; HDAC/TBI = cytarabine/ total body irradiation; ICE = ifosfamide/cyclophosphamide/etoposide; LACE = lomustine/cytarabine/cyclophosphamide/etoposide; MCC = mitoxantrone/carboplatin/ cyclophosphamide; Mel = melphalan; Mel/TBI = melphalan/total body irradiation; MVT = mitoxantrone/ etoposide/thiotepa; PEC = cisplatin/etoposide/carboplatin; VP/TBI = etoposide/total body irradiation

A) **SCID**

Severe combined immunodeficiency (SCID) is a term used for a set of diseases caused by profound T and/or B lymphocyte deficiency. For the purposes of transplantation, SCID is any disorder (e.g. reticular dysgenesis, adenosine deaminase deficiency, classic SCID, SCID with B cells) in which unmanipulated bone marrow from an HLA-matched sibling will engraft in the absence of conditioning or immunosuppression. However, if the donor is an HLA-mismatched sibling or a haploidentical parent, T lymphocyte depletion of the bone marrow graft to reduce GVHD and treatment with conditioning regimens and anti-lymphocyte antibodies (ATG, anti-LFA-1) to increase engraftment is recommended.

B) **WISKOTT-ALDRICH OR COMMON IMMUNODEFICIENCY (CID)**

Immune deficiency syndromes other than SCID such as Wiskott-Aldrich syndrome or common immunodeficiency (CID) have sufficient functional lymphocytes to prevent marrow engraftment from HLA-matched siblings and require conditioning with regimens such as BU/CY.

C) **APLASTIC ANEMIA**

Aplastic anemia is usually caused by autoimmune destruction of hematopoietic stem cells. Consequently, conditioning is recommended before transplantation to eliminate the autoimmune process and ensure engraftment.

1) **Syngeneic Donor**

If a monozygotic twin is available, single agent CY (50 mg/kg x 4 days) may be required for conditioning. Approximately half of the patients with an identical twin engraft promptly without any preparatory measures.

2) **HLA-Matched Sibling**

For HLA-matched sibling donors, the Seattle regimen containing CY (50 mg/kg/d IV on days -5, -4, -3, -2) and ATG (30 mg/kg/day hr IV on days -5, -4, -3) results in a high rate (> 90%) of sustained engraftment with little conditioning regimen toxicity.

3) **Unrelated Donor**

If an unrelated or HLA-mismatched relative is the donor, further immunosuppression is required with combination cyclophosphamide (120 mg/kg) and radiation (TBI, TAI (thoracoabdominal) or TNI (total nodal irradiation)) administered at a total dose of 600-1000 cGy. The ideal conditioning regimen for this donor-recipient combination still remains to be developed.

4) **Aplastic Anemia with Clonal Abnormalities**

Patients with aplastic anemia in whom chromosomal abnormalities are found should be considered to have a hypocellular myelodysplastic syndrome and receive leukemia-type conditioning regimens such as CY/TBI (1200cGy) or BU/CY with the intent of eradicating the abnormal clone.

D) **FANCONI'S ANEMIA (FA)**

FA is caused by an inherited disorder of DNA repair which results in increased DNA damage from castogenic stress (i.e. DNA cross-linking agents). When transplanting such patients, one must be aware that the hypersensitivity of Fanconi's cells to the DNA-damaging agents used in conditioning regimens requires a lower dose. Most experience in this setting is with cyclophosphamide (5 mg/kg/d x 4 days) combined with 400-500 cGy of thoracoabdominal irradiation.

4.1

E) THALASSEMIA

Thalassemia is a congenital abnormality which affects the synthesis of one or more hemoglobin chains and results in anemia and a hypercellular bone marrow due to ineffective erythropoiesis. Most allogeneic transplant experience has been with a conditioning regimen of busulfan, 14 mg/kg administered 1.0 mg/kg q6 hours and cyclophosphamide, 50 mg/kg/day x 4 days.

F) LEUKEMIAS (ALL, AML, CML)

Leukemias may be treated with a variety of preparative regimens (Table 4.1.6). Unfortunately, few randomized trials have compared conditioning regimens, and even fewer have enrolled large patient numbers or stratified for relapse risk factors such as duration of first remission, number of relapses, patient age, or severity of GVHD. A comparison of nonrandomized studies in patients receiving transplants with different conditioning regimens from HLA-matched siblings show surprisingly similar relapse-free and overall survival. This suggests that the anti-leukemia activity of the major conditioning regimens has been maximized and that further therapeutic gains will be made by reducing transplant-related toxicity and by increasing graft-versus-leukemia activity of the allogeneic grafts. Thus, choice of a conditioning regimen should be dependent on the experience of the physician with a particular regimen and on considerations of toxicity. Conditioning regimens are generally the same regardless of leukemia type. A possible exception to this is ALL, where the radiosensitivity of lymphocytes and capability to simultaneously treat sanctuary sites such as the CNS argues in favor of a TBI-containing regimen. Nevertheless, BU/CY has been used for ALL with similar results.

In most instances, sibling matched allogeneic HSCT is the procedure of choice for hematopoietic reconstitution following conditioning therapy in CML, AML, and ALL. However, autologous HSCT is increasingly being used in AML and ALL for patients who lack a suitable allogeneic donor. Nonrandomized comparison of autologous and allogeneic HSCT in AML suggests that both procedures have equivalent overall survival, but differ in toxicity and relapse rates. The role of autologous BMT in CML is being explored in clinical studies.

G) MULTIPLE MYELOMA (MM) AND CHRONIC LYMPHOCYTIC LEUKEMIA (CLL)

These diseases affect older patients. Despite being essentially incurable with standard chemotherapy they are chronic disorders. For this reason, transplantation of these diseases has lagged behind that of leukemias and lymphomas. Melphalan is a component of most conditioning regimens used for multiple myeloma and it may be administered alone (180-200 mg/m^2), or in combination (120-200 mg/m^2) with TBI (800-1300 cGy). Although no randomized studies have been performed, the results with melphalan alone are considered inferior to those achieved with melphalan/TBI combinations. Generally, autologous HSCT is used for multiple myeloma, although there is limited but increasing experience with allogeneic HSCT. Despite a median duration of remission of roughly 2 years, there is little evidence to support the curability of multiple myeloma with autologous transplantation. Allogeneic BMT for multiple myeloma results in cure of approximately 40% of patients.

A limited number of carefully selected patients with CLL have undergone transplantation. In general CY/TBI regimens have been used in combination with either autologous or allogeneic HSCT. Complete remissions are achievable but there is little evidence that CLL is curable with autologous transplantation. The results of allogeneic BMT are more promising and may result in cure for some patients.

H) MYELOFIBROSIS

Primary myelofibrosis is not curative by conventional therapies. Although only anecdotal case reports are available, allogeneic HSCT is the only curative therapy. Despite marrow fibrosis, a hallmark of the disease, hematopoietic engraftment has been documented to occur after allogeneic transplantation, followed by resolution of marrow fibrosis. Conditioning has been with either BU/CY or TBI conditioning regimens.

I) LYMPHOMAS

Lymphomas are usually treated with autologous HSCT, although allogeneic HSCT is used as well. TBI-based conditioning regimens are used interchangeably with chemotherapy-based regimens for non-Hodgkin's lymphomas, while Hodgkin's disease is mostly treated with chemotherapy-based conditioning regimens. A number of conditioning regimens are commonly used, including CBV, BEAM and CY/TBI (Table 4.1.6). Comparison of nonrandomized trials suggest that these conditioning regimens produce similar long term disease-free survival. Due to the antitumor benefit of graft-versus-tumor, allogeneic HSCT produces higher disease-free survival than autologous HSCT, but the gain is offset by greater morbidity and mortality related to GVHD. Allogeneic HSCT should be seriously considered for younger patients who have poor prognostic tumor features and an HLA-matched sibling donor.

4.1

J) SOLID TUMORS

Solid tumors are generally treated with autologous HSCT and are maximally debulked with standard chemotherapy prior to HSCT. Autologous HSCT regimens that originated at the Dana Farber Cancer Institute in Boston have been given the acronym STAMP (solid tumor autologous marrow program).

1) Breast Cancer

The most common solid tumor treated with high-dose chemotherapy and autologous HSCT is breast cancer. Clinical experience suggests that 10-20% of patients with stage IV (metastatic) disease and 70% of patients with local disease and more than 10 nodes positive (high risk of relapse with conventional therapies) will have long term disease-free survival following autologous HSCT. The long term DFS after BMT for stage IV breast cancer increases to 30-40% if the patient is in a complete remission at the time of transplant. No single conditioning regimen appears to be superior (Table 4.1.6). The most common regimens include CPB, CVP and CTCb with regimens such as ICE and PCC still under early investigation.

2) Ovarian Cancer

Autologous HSCT is increasingly being used in ovarian cancer but without proof of benefit. Conditioning agents are similar to those used in breast cancer and include regimens such as MCC and CCT (Table 4.1.5).

3) Germ Cell Tumor

Autologous HSCT is used in relapsed and refractory germ cell/testicular tumors with 15-20% long term disease-free survival. Active conditioning regimens usually contain carboplatin, etoposide and/or ifosfamide.

4) Neuroblastoma

This childhood tumor is initially chemosensitve, usually disseminated at presentation, and less than 20% are curable by standard chemotherapy programs. The European neuroblastoma study group in a randomized study showed a significantly higher disease-free survival for patients randomized to autologous BMT. Conditioning for neuroblastoma has usually been with melphalan alone or combined with TBI. Allogeneic BMT which is unusual treatment for a solid tumor has also been tried in some patients with neuroblastoma. It was hoped that a GVL effect combined with absence of tumor in the infused allograft marrow would decrease relapse. Currently, there is no evidence in neuroblastoma of benefit from allogeneic BMT.

IV) TOXICITY OF CONDITIONING AGENTS

Extramedullary toxicities of the most commonly used conditioning agents are listed in Table 4.1.7. In most cases, drug doses are limited by mucositis and/or major organ toxicity (e.g. heart, lung, kidney or CNS). When combining drugs in a conditioning regimen, particular attention must be given to overlapping toxicities such as mucositis. For example, abnormal mucosa is an important portal of entry for bacteria and fungi which may lead to life-threatening infections. Drug clearance may be seriously reduced by renal or hepatic damage which in turn can result in higher drug levels and further end organ toxicity. Thus, when evaluating a patient's appropriateness for high-dose chemotherapy and selecting a conditioning regimen, the patient's end organ function must be normal and not overly susceptible to toxicity; e.g. prior cisplatin may increase ifosfamide renal toxicity, and prior bleomycin may increase radiation pulmonary toxicity.

A) CYCLOPHOSPHAMIDE

This alkylating agent must undergo hepatic metabolism to form cytotoxic species. It is highly immunosuppressive, but unlike other alkylating agents like busulfan and nitrosoureas, it is not a potent stem cell poison and hence is not myeloablative at transplant doses up to 200 mg/kg. Hemorrhagic myocarditis is the principal dose-limiting extramedullary toxicity. This toxicity occurs most often in patients who receive more than 200 mg/kg, are older (> 50 years) and in those with a previous history of congestive heart failure. Clinically, cardiac toxicity may be manifest by the rapid onset of intractable heart failure, pericardial effusion and death in the most extreme cases. The fulminant presentation is uncommon and most patients with cardiac toxicity will develop electrocardiographic voltage loss and transient increased heart size. Minor voltage loss, ST-T changes or supraventricular arrhythmias are not predictive of subsequent heart failure, and should not prevent completion of the preparative regimen. Pretransplant ejection fraction correlates in some, but not all studies with CY-induced cardiac injury, so it should be normal (> 45%) in patients undergoing CY-based conditioning. Prevention of cyclophosphamide-related hemorrhagic cystitis is discussed under ifosfamide.

B) IFOSFAMIDE

Ifosfamide, a closely related analog of CY, is also a prodrug which must undergo hepatic metabolism. Unlike CY, however, the DLT of ifosfamide is renal and toxic encephalopathy. The risk of renal insufficiency, caused by renal tubular damage, appears to be increased in patients with preexistent renal damage, particularly those with prior exposure to cisplatin and/or ifosfamide. Unfortunately, there are no well characterized algorithms for adjusting the dose of ifosfamide based on creatinine clearance. However, in patients receiving ICE chemotherapy, the incidence of drug-induced renal toxicity is low in patients with creatinine clearances > 45 ml/min/m^2. A common side effect of ifosfamide administration is metabolic acidosis due to accumulation of the metabolic by-products, chloroacetaldehyde and chloracetic acid, and to an ifosfamide-induced proximal renal tubular acidosis. This complication is best monitored by the serum HCO_3^-. When it falls below 18 mEq/L, patients should receive sodium bicarbonate at 15-30 mEq/hour which is tapered off when the serum HCO_3^- is > 22 mEq/L.

Hemorrhagic cystitis caused by urinary excretion of toxic metabolites (probably acrolein) of cyclophosphamide and ifosfamide, can be a life-threatening

Table 4.1.7. Toxicity of some conditioning regimen drugs

Drug / Dose	Extramedullary Dose-Limiting Toxicity	Other Toxicities
BCNU (carmustine) (250-600 mg/m²)	Interstitial pneumonitis	Renal insufficiency, encephalopathy, nausea, vomiting, VOD
Busulphan (12-16 mg/kg)	Mucositis, VOD	Seizures, rash, hyperpigmentation, nausea, vomiting, pneumonitis
CCNU (lomustine) (200-500 mg/m²)	Interstitial pneumonitis	Renal insufficiency, encephalopathy, nausea, vomiting, VOD
Cyclosphamide (120-200 mg/kg)	Heart failure	Hemorrhagic cystitis, SIADH, nausea, vomiting, pulmonary edema, interstitial pneumonitis
Cytarabine (4-36 g/m²)	Mucositis, CNS ataxia	Pulmonary edema, conjunctivitis, rash, fever, hepatitis, toxic epidermal necrolysis
Cisplatin acidosis, (150-180 mg/m²)	Renal insufficiency, peripheral neuropathy	Nausea, vomiting, renal tubular hypomagnesemia
Carboplatin hypomagnesemia, (450-1500 mg/m²)	Ototoxicity, hepatitis	Renal insufficiency, peripheral neuropathy
Etoposide (450-2400 mg/m²) (60 mg/kg)	Mucositis	Nausea, vomiting, hemorrhagic cystitis, pneumonia, hepatitis
Ifosfamide (12-16 g/m²)	CNS—encephalopathy, renal insufficiency	Hemorrhagic cystitis, renal tubular acidosis
Melphalan (120-200 mg/m²) insufficiency	Mucositis	Nausea, vomiting, hepatitis, SIADH, pneumonitis, renal
Mitoxantrone (30-75 mg/m²)	Cardiac	Mucositis
Taxol (625 mg/m²)	Mucositis	Peripheral neuropathy, bradycardia, anaphylaxis
Thiotepa (10 mg/kg) (500-800 mg/m²)	Mucositis	Intertriginous rash, hyperpigmentation, nausea, vomiting

4.1

complication. Cystitis usually occurs within 2 weeks of receiving these drugs, but other causes of cystitis should be excluded such as bacterial, fungal or viral cystitis (adenovirus, BK virus). There are several methods to prevent cyclophosphamide induced cystitis:

1) Alkaline Hydration

Toxic metabolites can be rapidly washed out of the bladder by frequent urination and hydration with $\frac{1}{2}$ NS (normal saline) + 100 mEq sodium bicarbonate/L with 10 mEq KCL/L at 3 ml/kg/hr, beginning 12 hours before cyclophosphamide is infused and continuing for 24 hours beyond the last dose. Cyclophosphamide may cause hyponatremia due to water retention, so if urine output is < 1.5 ml/kg/hour or the body weight increases > 2 kg, furosemide should be administered. Electrolytes and volume status should be closely monitored.

2) MESNA

2-mercaptoethane sulfonate (MESNA) is a sulfhydryl compound which conjugates with alkylating species in the urine and prevents cystitis. MESNA may be administered at 20 mg/kg IV beginning 1 hour before CY and repeated every 4 hours at 10-20 mg/kg for 24 hours beyond the last dose. There are several variations for the scheduling of MESNA but all should use continuous or frequent administrations due to its 20 minute half-life.

3) Irrigation

Bladder irrigation may be considered in the event that cystitis has occurred despite the above interventions to keep the urethra from occluding with clotted blood. A Foley catheter should be inserted and the bladder irrigated with NS at 250-1000 ml/hour. Since bladder irrigation does not protect the upper urinary tract, it is generally not recommended as a prophylactic measure in an uncomplicated patient.

C) THIOTEPA AND MELPHALAN

Both of these agents are alkylating agents. Mucositis is dose-limiting for both drugs, and when severe, it can be life-threatening secondary to airway obstruction, sloughing of bowel and/or uncontrolled sepsis. Medical management is largely supportive, consisting of bowel rest, total parental nutrition, good oral hygiene and analgesia.

D) BUSULFAN

This alkylating agent, unlike nitrogen mustards (cyclophosphamide, ifosfamide and melphalan), has a more marked effect on myeloid cells than on lymphoid cells. Clinically, this is manifest by prolonged aplasia following its administration. The extramedullary DLT of busulfan is mucositis, with pneumonitis and VOD occurring intermittently at high drug doses. Busulfan rapidly enters the CNS and may cause seizures. Consequently, patients should receive prophylactic phenytoin (with therapeutic levels) begun before busulfan and continued for 24 hours after the last dose.

Because busulfan is only available in an oral formulation, vomiting can seriously interfere with its administration. To minimize vomiting, busulfan is given before other cytotoxic drugs along with an aggressive anti-emetic regimen. If vomiting occurs within half an hour of a dose or if pill fragments are present in the emesis, some institutions repeat the dose. There is a wide intra- and interindividual variation of absorption and metabolism of busulfan. In addition, pharmacokinetic differences exist between age groups; children show an increased rate of drug clearance.

E) BCNU AND CCNU

Nitrosourea alkylating agents have clinical activity against a number of tumors. At high dose, pulmonary and hepatic toxicity is dose-limiting. The incidence of pneumonitis increases after cumulative doses of BCNU exceed 1200-1400 mg/m^2, while single doses of 450 mg/m^2 have been reported to cause pulmonary infiltrates, hyaline membrane formation and fibrosis within 3 weeks of treatment. When given in combination with other agents, BCNU doses of 300-600 mg/m^2 are associated with acute or late pneumonitis in at least 20% of patients. Patients who receive such doses of BCNU (e.g. CPB) should be told at discharge to monitor their exercise tolerance. If exercise tolerance diminishes, further evaluation with resting and exercise PO$_2$, pulmonary diffusion capacity and chest x-ray should be performed. If early evidence of drug-induced pneumonitis is documented, prednisone should be started at 2 mg/kg/day and then tapered slowly over 2-3 months; early intervention reverses most cases of BCNU pneumonitis. Hepatic toxicity occurs with nitrosoureas, and their administration is associated with an increased incidence of VOD.

F) PLATINUM COMPOUNDS

Cisplatin and carboplatin covalently bind to DNA bases and disrupt DNA function. These two analogs currently used in conditioning regimens, have different toxicity profiles. At high doses, cisplatin is limited by renal insufficiency and peripheral neuropathy, while carboplatin is limited by hepatitis and ototoxicity. Renal insufficiency secondary to cisplatin may be reduced by normal saline loading and by maintaining good urine output with mannitol 25-50 mg (mixed with the cisplatin) and furosemide. Theoretically, a high chloride concentration within the renal tubules prevents aqueation or activation of cisplatin.

G) CYTARABINE (ARA-C)

Cytarabine is an analog of deoxycytidine and has multiple effects on DNA synthesis. At high dose, cytarabine causes neurologic toxicity, mucositis and chemical conjunctivitis. The neurologic toxicity, manifest by cerebral and cerebellar dysfunction, is more common in older patients (> 50 years) and in patients receiving high doses (3 gm/m^2 x 12). This toxicity may present as slurred speech, unsteady gait, dementia and coma, and in most cases it is reversible but occasionally may be fatal. If neurologic symptoms arise, cytarabine should be immediately terminated. The conjunctivitis is responsive to topical steroids which should be used prophylactically. Patients may occasionally develop idiosyncratic pulmonary edema after receiving high-dose cytarabine. Pulmonary symptoms may begin from 4-30 days after starting cytarabine and when it occurs, it is often fatal.

H) ETOPOSIDE AND MITOXANTRONE

These natural products inhibit topoisomerase II. Etoposide may be administered at very high doses (25- to 50-fold over standard dose) with mucositis and hepatic toxicity the major limiting toxicities. In contrast, the dose of mitoxantrone is largely limited by cardiac toxicity.

V) DOSE AND METHOD OF ADMINISTRATION

By convention, most conditioning regimens for leukemias are dosed by body weight (kg), while most solid tumor preparative regimens are dosed by surface area (m^2). When dosing conditioning agents by weight, some centers use actual

4.1

weight while others utilize ideal weight. There is no consistency in dosing of cyclo-phosphamide by ideal or actual body weight. We recommend dosing by ideal body weight since this drug and its metabolites are not fat soluble and the risk of hem-orrhagic myocarditis increases over 200 mg/kg (ideal body weight). However, some centers administer cyclophosphamide based on ideal body weight plus 20-25% of the actual body weight.

High-dose chemotherapy used in BMT is based on normal end organ function. The pharmacokinetics in situations with impaired organ function are not under-stood, and there are no algorithms for dose adjustment due to renal, liver, pulmo-nary or cardiac dysfunction. Patients with comorbid diseases involving organ func-tions are generally excluded as candidates for BMT.

4.1

REFERENCES FOR CONDITIONING REGIMENS:

CY/TBI
    Blood 1991; 77(8):1660-1665.
    Blood 1992; 79(10):2578-2582.
BU/CY
    BU/CY$_4$ – NEJM 1983; 309:1347-1353.
    BU/CY$_2$ – JCO 1992; 10(2):237-242.
    BU/CY IVP – VMT 1989; 4:559-565.
Non-cyclophosphamide/TBI
    Blood 1987; 69(4):1015-1020.
    Blood 1993; 81(8):2187-2193.
    Blood 1988; 71(4):888-893.
BACT-derived
    BAVC – Blood 1990; 75(12):2282-2285.
    BEAM – Blood 1993; 81(5):1137-1145.
    CBV – JCO 1993; 11(7):1329-1335.
    LACE – BMT 1990; 5 suppl 2:124.
Solid tumor
    CBDA/VP – Ann Int Med 1992; 117(2):124-128.
    ICE – JCO 1992; 10(11):1712-1722.
    MVT – JCO 1990; 8(11):1782-1788.
    CTCb – JCO 1992; 10(1):102-110.
    MCC – JCO 1994; 12(1):176-183.
    BCC – JCO 1988; 6(9):1368-1376.
    CTB – JCO 1992; 10(11):1743-1747.
    CVP – JCO 1990; 8(7):1207-1216.

Total Body Irradiation

Shalina Gupta-Burt, Paul G. Okunieff

I) INTRODUCTION

An integral component of several conditioning regimens for bone marrow transplantation (BMT) is total body irradiation (TBI). This single modality is effective in both immunosuppression and tumor eradication. However the therapeutic ratio of TBI is small. Significant potentially lethal and nonlethal complications may be noted acutely or chronically, including enhanced immunologic toxicity such as graft-versus-host disease. The unique advantages of this modality include:

1) Absence of cross-reactivity with other agents
2) Dose delivery independent of vascular supply
3) Effectiveness in sanctuary sites (testes, CNS)
4) Independence of metabolic pathways (i.e. liver and kidney)
5) Ability to tailor dose distribution by using selective shielding or boosting areas of high risk or large volume disease

4.2

II) DEFINITIONS

A) X-RAYS AND GAMMA RAYS

These are both identical photon radiations but differ in their origin. X-rays are produced in a linear accelerator by high energy electrons bombarding a target, whereas gamma rays originate from the nuclear decay of radioactive materials such as cobalt 60.

B) ROENTGEN (R)

Exposure is measured in units of roentgen, which is useful in calibration but has little use in quantitating the biologically significant absorbed dose. The SI unit for roentgen is the sievert.

C) GRAY

The basic SI unit for absorbed radiation dose is the gray (Gy) which equals 1J/kg.

D) RAD

Another unit in usage for absorbed dose which is slowly being abandoned is the rad (radiation absorbed dose) which is equal to 0.01 Gy (100 rad = 1 Gy = 100 cGy).

E) DOSE RATE (CGY/MINUTE)

A term used to describe the absorbed dose per unit time. Low dose rate is arbitrarily defined as $\leq$ 4-10 cGy/min and high dose rate as $\geq$ 10 cGy/min.

F) STANDARD FRACTIONATION

Treatment plan which uses a single 180-200 cGy fraction per day, 5 days a week.

G) HYPERFRACTIONATION

This treatment regimen uses a smaller fraction size (e.g. 120 cGy) given 2-3 times in a 24 hour period. The time interval between fractions should be a minimum of 6 hours to allow sufficient repair of sublethal damage. The total treatment duration will be similar to standard fractionation. Hyperfractionation regimens in principle allow higher total doses with the same or less normal tissue toxicity (due to a smaller fraction size).

Bone Marrow Transplantation, edited by Richard K. Burt, H. Joachim Deeg, Scott Thomas Lothian, George W. Santos. © 1996 R.G. Landes Company.

H) ACCELERATED FRACTIONATION

   The fraction size is similar to standard fractionation (180-200 cGy) but multiple fractions are given in a 24 hour period. The total treatment time will be decreased, which may be advantageous in rapidly proliferating tumors by decreasing tumor repopulation. However, acute toxicity may be greater; late toxicity should be the same as standard fractionation.

I) LARGE FIELD RADIATION

   Total and half body irradiation may be used in a variety of clinical situations besides BMT. Table 4.2.1 summarizes the different types of large field irradiation. A number of different TBI regimens are currently in use today. The most commonly used regimens are outlined in Table 4.2.2.

III) NORMAL TISSUE RADIOSENSITIVITY

   Single dose (one large fraction) regimens which are of historical importance, have a high rate of toxicity and do not have a role in clinical practice today. Repair of sublethal damage between fractions favors normal tissue. Consequently, fractionated regimens allow delivery of a higher total dose with equivalent or less normal tissue toxicity, thus enhancing the therapeutic ratio. Therefore, the majority of centers currently use fractionated regimens. Table 4.2.3 reviews the normal tissue tolerance for single and fractionated whole organ irradiation.

IV) HEMATOPOIETIC CELL RADIOSENSITIVITY

   Hematopoietic stem cells (HSC) are extremely radiosensitive and are markedly depleted with TBI. The radiation survival curves for HSC in four species including

---

Table 4.2.1. Large field irradiation-definitions

Total Nodal Irradiation: Radiation portal which encompasses all nodal areas; fields include mantle, para-aortic and inverted-Y. Indicated primarily for malignant lymphomas.

Total Lymphoid Irradiation: Radiation portals which encompass all nodal areas (TNI), spleen, thymus and often Waldeyer's ring and/or liver. It may be used as primary treatment in lymphomas or as an adjunct to immunosuppression in solid organ transplantation, autoimmune diseases, aplastic anemia and multiple sclerosis.

Total Body Irradiation: By definition radiation field encompasses the entire body with the intention of irradiating the whole marrow cavity. Indications include autologous or allogeneic bone marrow transplantation for leukemias, lymphomas and solid tumors.

Total Abdominal Irradiation: Radiation portals which encompass the contents within the entire peritoneal reflection and the inguinal-femoral nodes. Indications include neoplastic diseases with bulky abdominal masses or peritoneal dissemination.

Hemibody Irradiation: Sequential upper and lower hemibody irradiation usually given 6-8 weeks apart; used in palliation of diffuse skeletal metastasis.

Table 4.2.2. Most commonly employed TBI regimens

Fraction Size (cGy)	#of Fraction(s)/ Day	Total Dose (cGy)	Total Treatment Time (days)
200	1	1200	6
175-200	2 (6 hours apart)	1200-1400	3.5
120	3 (6 hours apart)	1320	3.5

mouse, rat, rhesus monkey and man have been determined. Similar radiation sensitivities of HSC between species were noted, however there were different HSC concentrations per body unit. Larger species have a lower HSC concentration per unit body mass, and therefore are more susceptible to bone marrow damage by radiation. The required amount of marrow cells for successful rescue after myeloablation is approximately 10 times higher for man than mouse.

V) LYMPHOCYTE RADIOSENSITIVITY-IMMUNOSUPPRESSION

The degree of immunosuppression as determined by rate of engraftment is a function of total dose, dose rate and fractionation (Table 4.2.4). In animals, consistent engraftment occurs with a single fraction dose of 800 cGy, whereas with a fractionated regimen, a total dose of 1200-1500 cGy is needed. In addition, these studies suggest that TBI is more immunosuppressive than cyclophosphamide.

A) HLA-MATCHED SIBLING ALLOGRAFTS
Graft failure with sibling HLA-matched grafts is generally 1-2% with 12 Gy fractionated 200 cGy BID.

B) HLA-MISMATCHED GRAFTS
In HLA-mismatched grafts, the rate of graft failure increases with the degree of genetic disparity. For patients with leukemia who have been conditioned with 1200 cGy in fractions of 200 cGy BID, graft failure for 0, 1 and 2 HLA-mismatched loci is 7%, 9% and 21%, respectively. However, a 10 Gy single fraction regimen (rarely used secondary to high rate of normal tissue toxicity) or fractionated regimens with higher total dose improve engraftment.

C) T CELL DEPLETED ALLOGRAFTS
The risk of graft failure is further increased if T cells are depleted from the donor marrow. T cell depleted marrow engraftment has been improved with the use of higher doses of TBI, addition of total lymphoid irradiation (TLI) or adding agents (antithymocyte globulin, cytarabine) to the conditioning regimen

4.2

Table 4.2.3. Whole organ irradiation normal tissue tolerance

Estimates of the minimal tolerance dose ($TD_{5/5}$) and maximal tolerance dose ($TD_{50/5}$) indicate a 5% and 50% probability of severe or life-threatening complication(s) occurring within 5 years after irradiation. Data is based on uniform irradiation of the entire organ, without exposure to adjuvant drugs or surgical manipulations, in the adult (nonpediatric, nongeriatric) patient. Fractionated dose refers to conventional fractionation of 180-200 cGy/fraction, 5 fractions/week.
(Modified from Rubin P: The law and order of radiation sensitivity, absolute vs relative. In: Vaeth JM, Meyer JL, eds. Radiation Tolerance of Normal Tissues, vol 23. Basel: Karger, 1989.)

Organ	Single Dose (cGy)		Fractionated Dose (cGy)	
	$TD_{5/5}$	$TD_{50/5}$	$TD_{5/5}$	$TD_{50/5}$
Lung	700	1000	2000	3000
Kidney	1000	2000	2000	3000
Gastrointestinal	500	1000	5000	6000
Liver	1500	2000	3500	4000
Brain	1500	2000	6000	7000
Eye (lens)	200	1000	600	1200
Ovary	200	600	600	1000
Testes	200	1000	100	200
Heart	1800	2000	4000	5000

Table 4.2.4. Effect of dose, dose fractionation and dose rate on clinical parameters important in bone marrow transplantation

| Factor | Effects | | Toxicities | |
	Marrow Ablation	Immuno-suppression	Early	Late
Increased total dose	↑	↑	↑	↑
Dose fractionation	No change	↓	No change	↓
Increased dose rate	↑	↑	↑	↑

4.2

Reprinted with permission from Appelbaum F. WB Saunders Periodicals. Seminars in Oncology. 1993; 20(4):9.

to increase suppression of host T cell activity. Engraftment with T cell depleted marrow using a single fraction TBI dose of 7.5-10 Gy is superior to fractionated regimens with a total dose of 10-12 Gy. In a review by the International Bone Marrow Transplant Registry (IBMTR) of T cell depleted HLA-matched sibling transplants, two factors associated with significantly better engraftment were higher total doses (≥ 11 Gy) and higher dose rates (≥ 14 cGy/min).

VI) **LEUKEMIA CELL RADIOSENSITIVITY**

Leukemic cells maintain the radiosensitivity of the normal hematopoietic cells. Studies of radiation sensitivities in a variety of normal and malignant human cell lines and experimental animal models show a 4 log tumor cell kill with maximum tolerated dose of TBI. An additional 6-8 logs of tumor cell kill may be achieved with the addition of cytoxan. TBI factors which determine the leukemic cell kill include dose, dose rate and fractionation schedule (Table 4.2.4).

A) **FRACTIONATION AND LEUKEMIA CELL KILL**

A single dose of 10 Gy has a similar relapse rate to a regimen with a total fractionated dose of 12-13 Gy, but the toxicity is markedly less for a fractionated regimen. Among the fractionated regimens the relapse rate is clearly lower with higher total doses.

B) **DOSE RATE AND LEUKEMIA CELL KILL**

In a single fraction regimen, a low dose rate (≤ 4 cGy/min), results in relapse rates of 30-40% as compared to 10-20% with a high dose rate regimen. However, due to normal organ toxicity single fraction regimens are rarely used, and it is important to note that dose rate differences are much less significant with fractionated TBI as compared to single dose TBI. The majority of institutions employ a multifractionated regimen with dose rates ≤ 20 cGy/min.

C) TOTAL DOSE AND LEUKEMIA CELL KILL

As total dose increases, the risk of leukemic relapse becomes less. The Seattle group has evaluated the overall survival, relapse rates and toxicity of two different TBI regimens: 200 cGy fractions daily for a total dose of 1200 cGy, versus 225 cGy fractions daily for a total dose of 1575 cGy. The survival data were the same for both groups, but the leukemia relapse and leukemia death rate was higher for the group with lower total dose, whereas death due to regimen toxicity was greater for the higher total dose group. The optimum dose has not yet been defined. In general higher total doses are needed with smaller fraction sizes.

VII) TBI TECHNIQUES

There is no standardized system for delivering or reporting dose or treatment specifications such as dose rate, beam energy/quality, dose profiles, dose homogeneity and patient positioning. This has made comparisons of different regimens difficult. Nonetheless, in a given institution the technique of TBI used needs to be understood by all physicians involved.

A) BEAM ENERGY/QUALITY

The type of treatment unit defines the beam energy/quality. The most commonly used machines are cobalt 60 teletherapy units and low to medium energy (4-10 MV) linear accelerators (linacs). In addition, some institutions use higher energy linacs (up to 25 MV). The decision regarding which machine to use is dependent on multiple factors including equipment availability, ability to make required modifications, acceptable dosimetry, size of maximum field and reproducibility of treatment set-up. Few institutions have dedicated facilities specifically designed for large field treatments. The majority of institutions modify conventional treatment facilities to provide the required treatment geometry or develop unconventional treatment geometry to obtain the desired field specifications.

1) Cobalt 60 Units

These units are more economical, low energy machines, usually with lower dose rates in the range of 2-6 cGy/min. The 1.25 MeV gamma rays of cobalt 60 have limited penetration and greater risk for larger dose variation within the patient.

2) Linear Accelerators (LINAC)

High energy machines have greater uniformity of depth and provide higher dose rates of 10-50 cGy/min, but require methods to minimize the substantial build-up inhomogeneity region at the patient's surface causing significant under dosage of the skin. This can be compensated for by placing tissue equivalent material on the surface or using a beam spoiler to provide adequate dose to the skin.

B) BEAM PROFILES

The dose variation across the field at a specified depth is represented by a beam profile. For a given technique of TBI, beam profiles are determined using water or anthropomorphic phantoms. This data is used in the modification of the unit (e.g. design of flattening filter) to improve beam homogeneity. In addition, in vivo measurements using thermoluminescent dosimeters (TLDs) or diodes should be done to record and verify the dose distribution. Midplane, entrance and exit doses should be measured at several different points (e.g. head, neck, shoulders, abdomen, hips, ankles) on the patient.

C) DOSE HOMOGENEITY

The target volume in TBI is the entire patient, which has significant consequences to the technical aspects of TBI. The two primary issues are how to fit the entire patient in the field and how to obtain a homogenous dose distribution in an irregular shaped body. The majority of centers use a single large field technique which has better dose homogeneity. Although rare, some institutions employ a multiple field technique which has the added disadvantage of dosimetric problems related to field junctions. Alternative techniques include a moving field technique by rotation of the source head or moving the couch with a fixed beam. Patients are usually treated in the supine or prone position. Figure 4.2.1 shows the different techniques currently in use.

Dose inhomogeneity due to the patient's thickness and nonuniformity is to a certain degree unavoidable. Methods to improve homogeneity include patient positioning (see below), use of high energy machines and use of tissue compensators and/or bolus materials. The longer the treatment distance the lower the dose variation. A larger dose variation is also noted in patients with larger diameters. Tissue compensators may be used in thin areas of the body (neck, head, feet) to even out the dose distribution and are especially important in lateral opposed techniques. They are produced individually for each patient, and usually require 3 days preparation time. Shielding may be used judiciously to protect normal structures and keep the dose below tolerance levels. Lung shields are used in techniques which use the AP/PA position. In

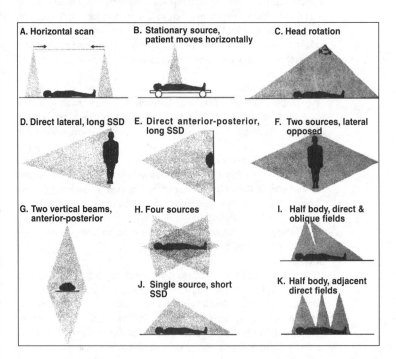

Fig. 4.2.1. Alternative techniques for total and hemibody irradiation. SSD = Source—skin—distance

lateral opposed techniques, the patients arm may be used to shield the lung or alternatively additional compensators may be used.

The recommended dose specification is to the mid pelvis. Rarely some institutions will specify dose to the mid abdomen. A dose variability of 5% is excellent; a 10% dose variability is acceptable.

D) PATIENT POSITIONING

In centers where large field sizes can be attained, patients are usually treated anteriorly/posteriorly while standing or lying on a stretcher, or laterally in a seated or reclining position. A technique using an anterior/posterior position is preferable to a lateral or sideways position, because there is less inhomogeneity with the AP/PA position. Nonuniformity could also be decreased by placing the patient in a more compact position, (semi-erect, fetal, or squatting position). Compact positions are also required at centers where large treatment distances cannot be achieved. These compact positions need to be maintained by the patient for the duration of the treatment period, approximately 15-30 minutes, and in general are not well tolerated. Finally, it is important to develop a set-up which is reproducible throughout the treatment course.

4.2

VIII) TOXICITIES AFTER ACCIDENTAL EXPOSURE TO RADIATION

Historically, with accidental whole body exposure to radiation, three syndromes have been described.

A) CEREBROVASCULAR SYNDROME

This syndrome occurs at very high doses, > 5000 cGy, and manifests as severe nausea and vomiting within minutes of exposure, followed by disorientation, muscular incoordination, respiratory distress, diarrhea, seizures, coma and finally death within hours.

B) GASTROINTESTINAL SYNDROME

At levels of 500-1200 cGy, a gastrointestinal syndrome is noted with nausea, vomiting and prolonged diarrhea. Lethargy and anorexia is apparent. Without supportive care death occurs 3-10 days later.

C) HEMATOPOIETIC SYNDROME

Leukopenia, thrombocytopenia and anemia develop at doses of 300-500 cGy after a latent period of 2-3 weeks. If bone marrow regeneration does not occur death follows in 3-8 weeks. For humans the lethal dose fatal to 50% of the exposed population in 60 days ($LD_{50/60}$) is 300-450 cGy. In animals and most likely in humans, the LD_{50} may be raised by a factor of two with judicious use of antibiotics and fluids.

IX) TOXICITY AFTER TBI AND HSC RESCUE

Normal tissue toxicity may be seen early, within days to months after radiation, or late, years after treatment. The primary extramedullary dose limiting toxicities of TBI are pneumonitis, veno-occlusive disease of the liver, renal impairment and mucositis.

A) ACUTE TOXICITIES

Early (acute) toxicity is noted primarily in tissues with rapid growth and intact repair mechanisms and is dependent on total dose, fraction size and total treatment time. It is usually self-limited with significant or complete resolution within days to months after treatment completion. Notable exceptions include interstitial pneumonitis which may progress to pulmonary fibrosis, and veno-occlusive disease of the liver which is associated with a high mortality rate. Table 4.2.5 summarizes the potential acute toxicities associated with TBI.

1) Nausea and Vomiting

Virtually all patients develop nausea and vomiting toward the end of the irradiation period. Emesis may be quite severe in the initial period, and subsides 3-4 days following irradiation. Tolerance has been markedly improved with the liberal use of antiemetics such as ondansetron, compazine or torcan. These medications should be given prior to and on schedule throughout the TBI course.

4.2

Table 4.2.5. TBI-acute toxicities

System	Symptoms/ Signs	Time of Onset	Comments
Gastro-intestinal	Nausea, vomiting, diarrhea	24-48 hours	Most common after single fraction of 800-1000 cGy; decreases with increased fractionation or decreased dose rate
Salivary gland	Parotitis	24-48 hours	Responds to corticosteroids
Mucosal tissues	Dry mouth, sore throat, oral mucositis, decreased lacrimation	10 days	Mucositis may be increased with methotrexate
Pancreas	Acute pancreatitis		Rare
Skin	Erythema; Reversible alopecia	5-7 days; 7-14 days	Usually transient; may be severe if combined with cytarabine, etoposide, busulfan, methotrexate, bleomycin or adriamycin
CNS	Syncope	24 hours	Usually with orthostatic blood pressure changes; increased with low hemoglobin; increased with phenothiazines
Lung	Pneumonitis	1-3 months	50% with single large fraction TBI; 30% mortality overall; 10-20% incidence with fractionation or low dose rate; predisposing conditions: preexisting pulmonary disease, previous chest XRT, GVHD, increased age, infections, prior bleomycin
Liver	Veno-occlusive disease, elevated liver enzymes	6 days	Less incidence with increased fractionation

2) Diarrhea

TBI alone or when combined with antibiotics and/or chemotherapy especially methotrexate may cause severe diarrhea and should be treated as the clinical situation warrants. In general gastrointestinal complications are self-limited, symptoms persisting without evidence of improvement, 5-7 days after TBI should be evaluated for other underlying etiologies.

3) Fever

Most patients will develop low grade fever which may be associated with chills at the end of irradiation. This usually subsides within 24 hours. The mechanism may be related to release of pyrogens from leukemic or normal cells.

4) Parotitis

Parotitis presents as an acute swelling of the parotid glands associated with jaw pain and generally subsides over 2-3 days. If needed, a short course of steroids is effective for pain relief.

5) Skin

Generalized erythema with or without dry desquamation, which is generally mild, occurs a few days postirradiation. The intensity of this reaction may be increased in patients who receive chemotherapy. Alopecia is expected at 10-14 days after irradiation and is generally reversible, with regrowth noted within 3-6 months.

4.2

6) Mucositis

Oral mucositis is more a result of the conditioning chemotherapy, however it can be aggravated by irradiation. In addition, some patients have a mild, transient decrease in saliva and tear production.

7) Interstitial Pneumonitis

Pneumonitis is a common complication after TBI. The clinical signs and symptoms include dyspnea, pyrexia, cough and cyanosis. It is vital to rule out potential reversible causes of pneumonitis in this immunocompromised population, including infections, cardiovascular compromise or drug reaction/toxicity. Radiographically there is diffuse bilateral shadowing and increased interstitial markings. Pulmonary function studies reveal a restrictive pattern with a low carbon monoxide diffusing capacity (DLCO).

The pneumonitis phase usually occurs 1-3 months postirradiation. The incidence of pneumonitis is 25-55% in patients receiving single dose TBI, 10-12% with fractionated regimens and 6-9% in patients receiving chemotherapy alone. About 25% of interstitial pneumonitis may be directly attributed to TBI or chemotherapy, and 40-45% are associated with cytomegalovirus.

Dose factors which affect the incidence of pneumonitis include a 15-20% increased transmission of dose through lung tissue with the use of supervoltage equipment, especially with cobalt 60 units. For this reason lung shielding is crucial to reduce pneumonitis. The use of a single high dose of 850 cGy, 930 cGy and 1100 cGy causes a 5%, 50%, 80% incidence of lethal radiation pneumonitis, respectively. This incidence is much less when fractionation or reduced dose rates are used.

Risk factors for interstitial pneumonitis include preexisting pulmonary abnormalities, previous chest irradiation, advanced age, prior use of certain chemotherapeutic agents and concomitant viral infections.

Chemotherapeutic drugs which may enhance radiation damage include bleomycin, busulfan, carmustine, actinomycin D and methotrexate. The combination of any one of these drugs with radiation may increase the incidence, severity and intensity of reaction. In addition a recall phenomenon may be noted when these drugs are used after radiation.

Long term evaluation reveals gradual stabilization, and in some instances improvement in PFTs. Factors which may influence progression of pulmonary dysfunction include continued tobacco use, age over 50 years, male gender, persistent or recurrent viral infections and progression of graft-versus-host disease. In symptomatic patients a course of high-dose corticosteroids is indicated. In addition asymptomatic patients who are receiving corticosteroids for other indications, require a cautious, slow taper of corticosteroids; acute respiratory deterioration in previously asymptomatic patients has been noted with rapid tapering of corticosteroids.

8) Liver Toxicity

The risk of liver complications is multifactorial, and is related to a history of prior liver disease, use of hepatotoxic drugs (e.g., cytarabine, busulfan), concurrent infections, leukemic infiltrates, graft-versus-host disease and TBI. Transient asymptomatic elevations in LDH, SGOT and alkaline phosphatase are common after TBI. A rise is noted approximately 10-15 days post TBI with a peak level around 25-30 days and gradual normalization at 40-50 days post-TBI. However, some patients develop veno-occlusive disease of the liver presenting clinically with hepatomegaly, ascites and elevation in liver function studies, primarily bilirubin and alkaline phosphatase. The Seattle group and the Institut Gustave-Roussy have shown that fractionated TBI regimens markedly reduce the incidence of veno-occlusive disease of the liver.

B) CHRONIC TOXICITY

Total dose and most importantly the fraction size determines the extent of late damage. Maneuvers which will decrease the fraction size will in principle decrease late toxicity. Table 4.2.6 summarizes the potential chronic toxicities associated with TBI.

Table 4.2.6. TBI chronic toxicities

Cataracts	80% with single large fraction
	20% with fractionation usually develops within 5 yrs
	18% with chemotherapy alone
Bone growth	Decreased bone growth velocity noted in pediatric patients
Gonadal failure	Adults have rare recovery of reproductive function. Children have delayed puberty; can be induced by hormone replacement
Thyroid Failure	45% of patients develop subclinical hypothyroidism
Renal dysfunction	Poorly defined; may have persistent proteinuria and hematuria
Second malignancies	May be in part related to chemotherapy

1) Ocular Manifestations

a) Cataracts

Low dose rate regimens are associated with less cataract formation as compared to high dose rate regimens. The severity of cataract formation is greater with single dose versus fractionated treatment. The minimal estimated dose to induce cataracts in man is 200 cGy in single fraction, 400-550 cGy in multiple fractions. An incidence of 80% cataract induction has been reported with single dose TBI. This rate can be decreased to less than 20% with fractionated regimen. Approximately 50-60% of patients with cataracts after single dose TBI require surgical repair as compared to 20% of patients with cataracts after fractionated TBI. It is noteworthy that the incidence of cataracts is 18% in patients with aplastic anemia conditioned with a non-TBI containing regimen. The latency period for cataracts ranges from 6 months to 3-4 years.

Additional factors which may contribute to cataract formation include acute and chronic graft-versus-host disease and steroid use. The increased risk of cataract formation in patients with acute lymphoblastic leukemia may be related to the use of steroids in the initial treatment of these patients.

In general cataract prophylaxis with eye shielding is not employed because the eye is a potential site for relapse. Treatment of cataracts is a lens implant. Contact lenses may be preferred for the pediatric population.

b) Keratoconjunctivitis Sicca

This is a rare complication of TBI and is more often noted in patients receiving a radiation boost to the lacrimal apparatus (e.g. patients who present with orbital leukemia or lymphoma). The major factor related to the development of keratoconjunctivitis sicca, however, is chronic graft-versus-host disease.

2) Gonadal Function and Fertility

Prior to the initiation of the conditioning regimen it is imperative to discuss the option of sperm or oocyte banking with patients who desire to have children. After TBI, gonadal failure is common but the time course for the development of infertility is erratic. Therefore, patients should be informed that contraception is recommended for a minimum of 1 year after irradiation to decrease the potential of teratogenic and mutagenic complications.

a) Males

In males a dose of 250 cGy may cause transient sterility for a period of 1-2 years. A single fraction of 600 cGy or a dose of 1500 cGy in 10 fractions will cause permanent sterility. Despite the loss of spermatogenesis, hormonal levels are not altered enough to cause impotency or loss of libido. Thus, in general, potency is maintained, but fertility is lost. The Seattle group reported on 16 prepubescent boys treated with TBI. Five developed secondary sexual characteristics appropriate for their age, whereas 11 had delayed onset of maturation. Thirty-six of 41 adult men treated with TBI had normal

4.2

luteinizing hormone levels and 10 had normal follicle stimulating hormone levels. However only 2 of 41 had spermatogenesis.

b) Females

The dose required for permanent sterility in adult females is related to age, and may be due to the number of oocytes remaining. A single dose of 400 cGy or a fractionated regimen of 1500 cGy is sufficient to produce permanent sterility. Hormonal effects are significant, resulting in premature menopause. Seattle has reported that after TBI, 15 out of 15 prepubescent girls developed primary ovarian failure and failed to develop secondary sex characteristics or normal menstrual cycles. In postpubertal women, only 3 of 38 retained normal gonadotropin levels and normal menstrual cycles. Provided there are no medical contraindications, exogenous hormones are recommended for women who undergo premature menopause.

3) Thyroid Dysfunction

The majority (40-55%) of patients develop thyroid dysfunction after TBI. The most common clinical presentation is compensated hypothyroidism with elevated TSH and normal T_4 levels. Replacement hormone may be indicated in patients who are symptomatic or cannot adequately compensate (i.e. elevated TSH levels and low T_4 or a persistent rise of TSH).

4) Growth and Development

The majority of children who receive TBI will have decreased growth rates on longitudinal growth velocity curves. A reduction in sitting height due to a disproportionate reduction in spinal growth has been noted. In general, the growth spurt noted during puberty has a larger gain in sitting height than in leg length. It is this spinal height as well as long bone growth which is impaired after TBI. A University of Minnesota study noted that on average a height decrement of 4.0 cm in girls and 4.5 cm in boys could be predicted at 2 years post-TBI. The mechanism(s) of growth suppression with TBI is(are) unknown. Damage of the epiphyseal growth centers in the spine and long bones, directly or indirectly, may alter their response to growth mediators. TBI may also damage the liver, causing decreased somatomedin production. The hypothalamic-pituitary axis may be altered causing abnormal production or release of growth hormone. Other endocrinologic abnormalities which may enhance the growth suppression may be abnormal thyroid hormone production and primary gonadal failure, preventing the bone growth response to pubertal hormone level changes. The Seattle group reported that approximately one-third of their pediatric population who received TBI developed subnormal adrenal cortical function and growth hormone levels.

Other factors associated with growth suppression include severe chronic graft-versus-host disease, use of steroids or methotrexate alone and to a lesser degree use of the combination of anti-thymocyte globulin, methotrexate and steroids. The association with severe chronic graft-versus-host disease may be related to decreased liver synthesis of somatomedins, since hepatic involvement is common with chronic graft-versus-host disease. Prior cranial irradiation was not independently associated with excess growth suppression.

5) **Renal Toxicity**

Renal toxicity after TBI is uncommon. Acute radiation renal injury is a clinical syndrome manifesting hypertension, edema and uremia with an onset 6-12 months postirradiation. Urinalysis reveals proteinuria, microscopic hematuria and low specific gravity; mild to moderate elevations in serum creatinine and blood urea nitrogen are noted. An anemia, which may be due to intravascular hemolysis may be seen. Hemolytic uremic syndrome associated with TBI has been described.

Factors which may increase renal damage include accelerated fractionation, insufficient time interval between fractions (minimum recommended interval is 6 hours), high fraction size, use of concurrent nephrotoxic agents (e.g. cis-platinum, cyclophosphamide, cyclosporin) and young age.

Adequate data regarding long term follow up are lacking. In general, the majority of patients will have stabilization of renal function with only minimal elevations in serum creatinine and BUN. Many will continue to have intermittent proteinuria and microscopic hematuria.

6) **Second Malignancies**

The carcinogenic potential of radiation is considered to be a stochastic effect, it is an all or none effect which does not have a dose threshold. However, the probability of developing cancer from exposure to radiation is dependent on dose and fraction size. The latent period for the development of secondary leukemia is several years with a peak at 5-12 years which subsequently returns to baseline risk. Solid tumors have a latency of 10-40 years with a persistent elevation in risk.

A direct association of TBI with malignancy remains controversial. Reports of one tumor registry including 9,880 patients revealed 127 patients who developed second malignancies. The distribution of the malignancies was similar to other organ transplant patients who did not receive irradiation as a part of their conditioning regimen. In contrast, Seattle has reported a 3.9-fold increased risk of second malignancy after TBI. The Seattle experience with 1-12 years of follow up reveals 3 out of 300 hundred patients developing a second malignancy.

Early cancer screening (mammography, CXR, routine physical exam including yearly pelvic and PAP smear) should be initiated appropriately. In addition counseling against tobacco and caution regarding alcohol use must be emphasized.

X) **PRE-TBI EVALUATION**

In addition to the routine history and physical exam, data regarding prior exposure to radiation, chemotherapy or biologic response modifiers must be ascertained.

A) **PRIOR OR CURRENT DRUG EXPOSURE**

Specific concerns include prior or current use of bleomycin, cytarabine, etoposide, adriamycin, actinomycin D, semustine, cis-platinum and interferons—all agents which may enhance radiation toxicity and/or decrease the radiation tolerance of certain organs.

B) **PRIOR RADIATION**

Parameters which are important include: area treated, field size, type/energy of treatment machine, size and number of fractions, total dose, treatment

4.2

duration, interval since prior radiation, side effects/reactions. A patient may not be a candidate for TBI if the normal tissue tolerance dose of a vital organ has already been reached.

C) **LABORATORY STUDIES**

In addition to routine laboratory data, CXR, EKG, creatinine clearance and pulmonary function studies including DLCO should be obtained.

D) **PEDIATRIC PATIENTS**

In the pediatric population baseline and prospective measurements of hormone levels should be done. In addition growth curves should be carefully maintained. Baseline cognitive and psychomotor testing should be considered in children who may receive cranial irradiation in addition to TBI.

E) **RELATIVE CONTRAINDICATIONS TO TBI**

History of symptomatic pulmonary disease, history of heavy tobacco use, two or more cycles of bleomycin, DLCO of < 75% for allogeneic or < 50% for autologous, age > 50 years, prior large field or mediastinal irradiation and prior spinal irradiation to a dose > 45 Gy are relative contraindications to TBI.

XI) **FOLLOW-UP GUIDELINES**

Routine evaluation in the asymptomatic posttransplant patient includes physical examination, hematologic and chemistry profiles. These should be done at least monthly for 6 months then every 3 months for the next year. CXR every 3-6 months is recommended. Additional studies include PFTs and thyroid function studies every 6 months for the initial 2 years, annually thereafter if the patient remains stable and asymptomatic. The pediatric population may require further endocrinologic evaluations.

SUGGESTED READING

1. Cosset JM, Socie G, Dubray B et al. Single dose versus fractionated total body irradiation before bone marrow transplantation: radiobiological and clinical considerations. IJROBP 1994; 30:477-492.

2. Deeg HJ, Strob R, Thomas ED. Bone marrow transplantation: a review of delayed complications. Br J Haematol 1984; 57:185-208.

3. Lin H, Drzymala RE. Total-body and hemibody irradiation. In:Perez C, Brady L, eds. Principles and Practice of Radiation Oncology. 2nd ed. Philadelphia: JB Lippincott Company, 1992:256-264.

4. Van Dyk J, Galvin JM, Glasgow, GP et al. The physical aspects of total and half body photon irradiation. A report of the Task Group 29 Radiation Therapy Committee: AAPM Report No 17. New York, American Association of Physicists in Medicine, 1986:1-55.

High-Dose Alkylating Agent Pharmacology/Toxicity

William P. Petros, Colleen J. Gilbert

I) RATIONALE

Alkylating drugs are the primary class of cytotoxics used in BMT ablative regimens because they possess the principles necessary for successful high-dose chemotherapy (Table 4.3.1). The dose response against many types of malignant cells is steep, with tumor cell viability decreasing in a logarithmic manner with a linear increase in drug dose. There is also evidence that a substantial increase in the dose of alkylating agents may circumvent some mechanisms of acquired drug resistance. The dosages of many active agents are limited by myelosuppression, even with the use of colony-stimulating factors. Thus, use of hematopoietic stem cell support allows for increased drug dosage and combination therapy with agents that would typically produce an unacceptable degree of myelosuppression. Combinations of alkylating agents are selected for nonoverlapping extramedullary toxicities. Alkylating drugs commonly used in BMT conditioning regimens are: busulfan, carmustine, carboplatin, cisplatin, cyclophosphamide, ifosfamide, melphalan and thiotepa.

II) DOSE SELECTION

The ratio of the maximally-tolerated dose with/without ABMT is typically 4- to 10-fold; with some agents such as thiotepa it is 30-fold (Table 4.3.2). The doses are often 20-40% lower when alkylating agents are combined, compared to use as a single agent in BMT. This is due to the propensity of combinations to cause multisystem organ failure.

III) MECHANISMS OF ACTION OF ALKYLATING AGENTS

Alkylating agents are thought to primarily act via covalent binding of alkyl groups to cellular molecules such as DNA and RNA, although other mechanisms of cytotoxicity such as stimulation of programmed cell death (apoptosis) are probably also important. Drugs in this class vary in their specific binding sites and in their ability to crosslink DNA. Mechanisms of cellular resistance and repair following exposure to alkylating agents depend on the drug's specific chemical subclassification (e.g. nitrosoureas, platinum analogs). Consequently, preclinical data have demonstrated lack of cross-resistance among these subclassification types and in addition, therapeutic synergy. Inhibitors of topoisomerase II (e.g. etoposide) may act synergistically with alkylating agents, perhaps by modulating DNA repair.

IV) TOXICOLOGIC CONSIDERATIONS

Table 4.3.3 lists adverse effects which may be attributable to individual agents when used in high doses. Specific toxicities include:

A) ACUTE RENAL FAILURE

Cisplatin is dose-limited by acute renal toxicity characterized by reduced glomerular filtration rate, electrolyte wasting and elevations of BUN and serum creatinine. Adequate hydration with careful monitoring of hourly urine output (100-200 cc/hr minimum) is essential. Hypomagnesemia, hypokalemia, hypophosphatemia and, less often, hypocalcemia are manifestations of renal tubular damage and require daily supplementation. A chronic phase of renal toxic-

4.3

Bone Marrow Transplantation, edited by Richard K. Burt, H. Joachim Deeg, Scott Thomas Lothian, George W. Santos. © 1996 R.G. Landes Company.

ity with reduced creatinine clearance may persist for weeks after therapy. Carboplatin is less renal toxic but causes similar clinical effects when nephrotoxicity occurs. Ifosfamide is associated with increased BUN and serum creatinine, and reduced serum bicarbonate which may lead to metabolic acidosis.

B) GASTROINTESTINAL

Direct effects of the high-dose regimen, perturbations in endogenous cytokines, use of ancillary drugs, GVHD, infections and mucositis are contributors to acute and delayed (3-6 weeks) gastrointestinal irritation. Specific toxicities include nausea, vomiting, anorexia and diarrhea. Antidiarrheals may be used after stool culturing to eliminate infectious etiologies. Uncontrolled diarrhea and/or vomiting may contribute to electrolyte wasting and increase requirements for KCl or KPhos supplementation.

4.3

Table 4.3.1. Pharmacologic principles of high-dose chemotherapy

1. In vitro the specified type of cancer cells demonstrate a dose-response effect toward alkylating agents.
2. Drugs with documented single agent activity in the malignancy should form the basis of therapeutic regimens.
3. Utilization of noncross-resistant chemotherapeutic compounds in the induction and high-dose phases will increase the likelihood of killing resistant cells.
4. Drugs with different nonhematologic dose-limiting toxicities should be combined in maximal doses.
5. Treatment of early stage disease will lessen the likelihood of developing either biochemical or physical resistance to therapy.

Table 4.3.2. Dose-limiting toxicity of selected chemotherapeutic agents

Drug	Standard Dose		High Dose–BMT	
	Toxicity	Dose	Toxicity[1]	Dose
Busulfan	Myelosuppression	.06 mg/kg/day	Anorexia, hepatic	0.8-1 mg/kg x16
Carmustine	Myelosuppression	200-225 mg/m^2	Pulmonary, hepatic	300-600 mg/m^2
Carboplatin	Myelosuppression	300-360 mg/m^2	Hepatic	600-1200 mg/m^2
Cisplatin	Renal myelosuppression	50-100 mg/m^2	Renal, neurologic	150-180 mg/m^2
Cyclophosphamide	Myelosuppression	40-50 mg/kg	Cardiac	120-200 mg/kg
Ifosfamide	Myelosuppression	3-6 g/m^2	Neurologic, renal	12-16 g/m^2
Melphalan	Myelosuppression	8-40 mg/m^2	Gastrointestinal, hepatic	120-200 mg/m^2
Thiotepa	Myelosuppression	15-20 mg/m^2	Mucositis, neurologic, hepatic	500-800 mg/m^2
Etoposide	Myelosuppression	140-300 mg/m^2	Mucositis, gastrointestinal	600-2700 mg/m^2

Significantly higher dosages are given when used as a single agent.
[1] Extramedullary toxicity

C) CARDIAC TOXICITY

Cyclophosphamide is the most frequent cause of severe cardiac toxicity which can be evident in approximately 5-10% of patients. Dose-related hemorrhagic myocarditis may present acutely and is thought to possibly be prevented by the prophylactic use of platelet transfusions. Almost all patients treated with high-dose cyclophosphamide have minor ECG changes. ECG voltage loss or pericarditis can occur within several weeks of drug administration. One to several weeks following high-dose cyclophosphamide, a reversible and typically mild cardiomyopathy may occur which has been correlated to individual alterations in drug metabolism. Treatment is based on symptoms and managed with pharmacologic agents as well as fluid optimization.

D) BLADDER TOXICITY

Use of either high-dose cyclophosphamide or ifosfamide will result in virtually all patients experiencing hemorrhagic cystitis if precautionary measures are not undertaken; however, with adequate protective measures this side effect can be minimized. Hematuria may be evident almost immediately or could first present several weeks following drug administration. The clinical picture ranges from self-limited micro- or macroscopic hematuria to gross hematuria with clot retention and obstruction. Preventive measures undertaken during and following drug infusion include: forced diuresis, continuous

4.3

Table 4.3.3. Relatively common adverse effects of individual high-dose drugs

Effect	BU	BCNU	Cbpt	CDDP	CY	Ifos	L-PAM	TT	Etop
Arrhythmias		X[A]			X				
CHF					X				
Hypotension		X							X
Proteinuria							X		
Reduced CrCl			X	X		X			
Pulmonary	X	X			X				
Hepatic	X	X	X				X	X	
Mental status changes	X[B]					X		X	
Mucositis	X						X	X	X
N/V	X	X	X	X	X	X	X	X	X
Peripheral neuropathy					X				

A: Chest pain with transient ECG changes occurs acutely and resolves quickly
B: generalized seizures (seizure prophylaxis, dilantin, commonly used)
Abbreviations: BCNU carmustine; BU busulfan; Cbpt carboplatin; CDDP cisplatin; CHF congestive heart failure; CY cyclophosphamide; Etop etoposide; Ifos ifosfamide; L-PAM melphalan; N/V nausea, vomiting; TT thiotepa

bladder irrigation with urologic saline (e.g. 1 L/hr), and/or systemic administration of MESNA (mercaptoethane sulfonate) at doses approximately equivalent to those of ifosfamide, administered frequently over the expected duration of the cytotoxic exposure, e.g. q4hr or by continuous infusion for 24 hours. Severe hemorrhagic cystitis is treated by correction of thrombocytopenia, hydration and bladder irrigation. The latter is particularly useful for prevention of urethral obstruction by clots which could lead to acute renal failure. Invasive measures (electrocoagulation, intravesicular therapies, surgery) have also been utilized. Oxybutynin chloride (Ditropan®) 5 mg PO TID may be used to control bladder spasms.

E) PULMONARY TOXICITY

A noninfectious pneumonia typically characterized as diffuse alveolar damage or interstitial pneumonia may occur posttransplant and seems to be related to the type of induction regimen (particularly those containing carmustine or busulfan) and extent of chest radiation therapy. The typical presentation includes dyspnea, diffuse pulmonary infiltrates, nonproductive cough and hypoxemia. Bronchoalveolar lavage is often used to differentiate this syndrome from CMV pneumonia. The onset of these effects are also thought to be regimen-specific and can range from acute (< 4 weeks) to delayed (> 6 weeks). Treatment includes use of glucocorticoids (e.g. 1-2 mg/kg/d prednisone) but may require supplemental oxygen and mechanical ventilation. Oxygen should be used sparingly since it is thought to produce synergistic toxic effects in patients who have received drugs such as carmustine.

F) STOMATITIS

Mucosal toxicity is regimen-dependent, occurring in a majority of patients treated with agents such as thiotepa, melphalan, etoposide or busulfan; whereas, it is rare in those treated with regimens which do not include these agents. Unfortunately, preventative measures are generally not useful. Contributing factors include oral HSV and fungal infections, both of which are targets of prevention strategies in high-risk patients. Treatment of stomatitis involves the use of analgesics, local anesthetics and if severe, intubation. Parenteral administration of ancillary drugs is preferred over oral due to alterations in drug absorption in patients with moderate to severe stomatitis. Parenteral nutrition may be indicated in patients with prolonged, severe stomatitis.

G) CUTANEOUS EFFECTS

Use of cytarabine, thiotepa, carmustine, busulfan, etoposide, melphalan or some recombinant cytokines may result in significant skin toxicity. Toxic metabolites of thiotepa are thought to be secreted in perspiration, thus dressing changes and cleansing may be useful following drug administration. Rashes vary considerably in appearance and irritability. Alkylating agents, particularly thiotepa, busulfan and melphalan, may produce a "tanning effect" or hyperpigmentation which eventually fades. Mild cutaneous erythema is often seen when leukocyte counts are recovering following autologous BMT. More severe late onset rashes may be attributable to antibiotics or other supportive drugs.

H) HEPATIC VENO-OCCLUSIVE DISEASE (VOD)

The triad of hepatomegaly, weight gain and jaundice in the first several weeks following BMT constitutes VOD. Liver tenderness and increased size typically

4.3

occur first (8-10 days from initiation of therapy) with hyperbilirubinemia (serum bilirubin > 2 mg/dl) occurring later. Multiple ablative regimens have been associated with VOD and the drug sequence may be important. Busulfan seems to be one of the worst individual offenders, which has led to strategies for individualizing busulfan dosing by use of pharmacokinetic monitoring. Several other therapeutic measures have been tested for prevention and/or treatment of VOD using TPA +\– heparin (100-150 units/kg/d) or PGE1; however, the dose and schedule of TPA is quite variable (see chapter 13.1) and no generally accepted approach is established. Standard treatment is primarily supportive and focused on management of sodium and water excess as well as thrombocytopenia.

I) **SIADH**

A syndrome of inappropriate anti-diuretic hormone secretion or "water intoxication" is associated with high-dose cyclophosphamide or ifosfamide. This syndrome is characterized by weight gain, decreased urinary output with high urine osmolality and decreased serum osmolality with low serum sodium. It is usually self-limiting and can be conservatively managed by electrolyte replacement and fluid restriction.

4.3

V) **PHARMACOKINETICS**

The pharmacokinetic disposition of some chemotherapeutic drugs is altered when doses are escalated to the levels used in BMT regimens. These effects are mediated by dose-dependent (i.e. saturable) processes and/or sites of inappropriate drug accumulation as outlined in Table 4.3.4.

Pharmacokinetic data are important in determining the duration of time necessary between the last infusion of drug and the infusion of hematopoietic stem cells in order to avoid potential damage to the infused cells. Most centers use arbitrary definitions of adequate liver and renal function to predetermine those who are eligible for therapy and thus the disposition and effects may vary if these are not followed. These still leave wide margins for error and most of the tests utilized are not necessarily very accurate estimates of true drug disposition potential. Table 4.3.5 lists average half-life and major excretion routes for some drugs commonly used in ablative regimens.

Interpatient variability in pharmacokinetics of cytotoxic drugs explains at least part of the heterogeneity in toxic effects experienced following high-dose chemotherapy. One drug in which the toxicity profile seems to be more predictable (at

---

Table 4.3.4. Pharmacokinetic issues of particular importance in high-dose chemotherapy

Pharmacokinetic Construct	Potential Implication
Saturable (nonlinear)	Exaggerated adverse effects for drug clearance drugs which are detoxified by metabolism
Saturated protein binding	Exaggerated toxic effects or altered pharmacokinetics of drugs highly protein bound (i.e. > 90%)
Residual drug in depot sites (e.g. ascites, effusions)	Extended duration of cytotoxic drug concentrations in plasma secondary to drug leaching out of third space fluid leading to enhanced toxicity and possibly damaging the transplanted cells

Table 4.3.5. Plasma half-life and primary excretion route for typical drugs used in high-dose regimens

Drug	Average Plasma Half-life (hrs)	Major Excretion Routes
Busulfan	2-4	Hp
Carmustine[δ ε]	0.5	Hy/?Hp
Carboplatin	2-4	R >> B
Cisplatin	0.5	B > R
Cyclophosphamide	3-6*	Hp >>> R
Ifosfamide	6-8*	Hp > R*
Melphalan	0.75-2	Hp$^{\Phi}$
Thiotepa	1.5*	Hp (?R)
Etoposide[ε]	7.0	Hp; R

δ Spontaneous hydrolysis occurs in solution
ε Vehicle-induced hypotension
* Dose-dependent
Φ Possible alteration in patients with renal dysfunction
Abbreviations: Hp hepatic; Hy hydrolysis; R renal; B irreversible protein binding

4.3

least in the pediatric setting) with pharmacokineticly-guided dose individualization is busulfan. Many centers use the patient's individual blood concentrations in order to titrate busulfan dosing to a desired exposure.

Use of test dose strategies (i.e. small doses given prior to the high-dose regimen in order to predict individual pharmacokinetic disposition) have been pursued by some with drugs such as melphalan, but this is not commonly practiced.

VI) ADMINISTRATION ISSUES

A) SCHEDULE

Dependency may be very important in the high-dose setting, especially in regard to toxicity. The reasons behind these effects could be due to alterations in pharmacokinetics or a more complex pharmacodynamic interaction. High-dose chemotherapy preparation, infusion rate and schedule are defined by individual protocols which will vary between institutions. It is recommended that the infusion rate and sequencing not be modified outside a formal therapeutic trial situation.

B) BODY SIZE

The dosages of drugs used in ablative regimens are typically based on the patient's body surface area (BSA). Despite this measure to decrease variability, it is common to see a 2- to 10-fold range in the systemic exposure (area under the concentration-time curve) between patients. Centers vary with their approach to dosing in the grossly obese patient (i.e. twice ideal body weight) since formal studies generally are not available. Practices range from dosing on ideal weight, to use of an "adjusted body weight", and finally some will give doses based on total weight.

VII) POTENTIAL DRUG INTERACTIONS/CONTRAINDICATIONS

One has to be very careful of potential drug interactions (DI) since the doses of drugs used in ablative regimens are typically close to those considered maximally tolerated. Importantly, DI may not always be predicted based on studies in the standard-dose situation. Thus one should assume that DI are highly likely unless proven otherwise. Interactions which have been reported or are possible with the agents used in high-dose regimens are listed in Table 4.3.6.

Drugs such as cyclophosphamide and ifosfamide require hepatic enzyme activation for their therapeutic efficacy, therefore both metabolic induction and inhibition are concerns. In addition, high doses of drugs such as cyclophosphamide may accelerate the metabolism of other multiple drugs in a dose-dependent manner. This is particularly important in patients stabilized on anticonvulsants (e.g. phenytoin).

Administration of ancillary drugs which may be nephrotoxic should be avoided during the infusion of agents such as cisplatin; however, these can be included in the posttransplant care regimen. Hydration regimens commonly used with cisplatin

4.3

Table 4.3.6. Drugs which may interact with ablative regimens

Interacting Drug(s)	Effect on Chemotherapy Efficacy and/or Toxicity
P-450 Enzyme Inhibitors [e.g. cimetidine, chloramphenicol, valproate, ciprofloxacin, cyclosporine]	–Decrease drugs requiring activation [e.g. cyclophosphamide, ifosfamide] –Increase drugs detoxified by metabolism [e.g. busulfan, carmustine, etoposide] ? Parent drug and metabolite are active [e.g. thiotepa]
P-450 Enzyme Inducers [e.g. steroids, phenytoin, carbamazepine, rifampin, aminoglutethamide, barbiturates]	–Increase drugs requiring activation [e.g. cyclophosphamide, ifosfamide] –Decrease drugs detoxified by metabolism [e.g. busulfan, carmustine, etoposide] ? Parent drug and metabolite are active [e.g. thiotepa]
Inhibitors of aldehydedehydrogenase (ALDH) [e.g. chloral hydrate, disulfiram, oral contraceptives]	Increase drugs detoxified by ALDH [e.g. cyclophosphamide, ifosfamide]
NSAIDs [e.g. indomethacin]	May increase potential for SIADH by inhibition of prostaglandins
Succinylcholine	Cyclophosphamide may potentiate neuromuscular blockade by inhibition of pseudocholinesterase
Warfarin	Either etoposide or ifosfamide may increase INR
Disulfiram	Some chemotherapy diluents contain sufficient ethanol content to cause reaction [e.g. carmustine]
Anticonvulsants [e.g. phenytoin]	Some ablative regimens may increase clearance of anticonvulsants

may alter the pharmacokinetics of drugs which are primarily eliminated by the kidney and distributed in body water such as the aminoglycosides. Serum aminoglycoside concentrations will be lower than expected in this situation but patient volume status often changes over time and drug concentrations may increase later. Monitoring of trough and peak aminoglycoside concentrations at least weekly and with significant serum creatinine changes is indicated. Patients who have been heavily pretreated with cisplatin may also be more susceptible to renal toxic effects of ifosfamide in high dose.

Although multiple drug regimens are available to treat/prevent nausea and vomiting effects, one should exercise caution in selection due to the possibility of altering the pharmacology of the cytotoxic drugs. Examples of antiemetics which have been associated with altered cytotoxic drug disposition in this setting include corticosteroids and ondansetron.

Drugs with potentially significant negative effects on hematopoiesis are generally avoided following BMT (if possible), until hematopoietic recovery occurs. Several common drugs on this list include: co-trimoxazole, H2 receptor antagonists, heparin and flucytosine.

Ablative regimens which cause mucosal toxicity may alter the absorption of drugs with marginal bioavailability or narrow therapeutic ranges (e.g. digoxin), thus patients should be carefully assessed as to changes in the desired outcome or blood concentration.

SUGGESTED READING

1. Chabner BA, Longo DL, eds. Cancer Chemotherapy and Biotherapy. Philadelphia: JB Lippincott Co., 1996.
2. Dorr RT, Von Hoff DD, eds. Cancer Chemotherapy Handbook, 2nd Edition. Norwalk: Appleton & Lange, 1994.

5

MHC and Bone Marrow Transplantation

Nancy Hensel

I) DEFINITION AND CLASSIFICATION OF MHC

The major histocompatibility complex (MHC) is a term given to a family of genes that play a pivotal role in immune recognition of tissue compatibility. This is because MHC molecules are responsible for binding and presentation of peptides to T lymphocytes. In humans, the MHC genes have been designated HLA (Human Leukocyte Antigens). HLA molecules occur on most nucleated human cells and are not limited to leukocytes. HLA genes are among one of the most polymorphic genetic systems known and on chromosome 6 are divided into three distinct gene clusters, class I, class II and class III (see Fig. 5.1).

A) HLA CLASS I

Class I molecules are composed of 2 noncovalently bound polypeptides, a β chain (β_2 microglobulin) which does not vary and an α chain which is highly polymorphic and may be from several different loci. The classical class I (α chain) loci are HLA-A, -B, -C. Nonclassical class I (α chain) loci include HLA-E, -F, -G, -H. The distribution and functions of the nonclassical HLA class I genes are not well understood. HLA-A, -B and -C molecules are present on the surface of all nucleated cells and are recognized by cytotoxic T lymphocytes (CTL). Class I molecules present intracellular (endogenous) proteins that have been processed into peptides of 7-9 amino acids.

B) HLA CLASS II

Class II molecules are also composed of two polypeptide chains, however, unlike class I, both chains are highly polymorphic. Class II HLA loci are HLA-D/DR, DP, and DQ. Three class II region genes, DPβ2, DPα2 and DRβ2 (see Fig. 5.1), are pseudogenes which are not expressed due to mutations that prevent gene transcription or translation. Therefore, DP molecules are the product of DPα1 and DPβ1. DQ molecules are the product of DQα1 and DQβ1 genes. DR molecules derive alpha chains from DRα but the beta chains are coded either by DRβ1 (called DR allele), DRβ3 (DR52 alleles) or DRβ4 (DR53 allele). DRβ5 occurs only in some haplotypes (DR2).

Class II molecules are expressed on the surface of antigen presenting cells such as activated T cells, B cells, macrophages, Langerhans cells, Kupffer cells, dendritic cells, monocytes and some endothelial cells. Class II molecules present foreign (exogenous) proteins which have undergone endocytosis and have been processed into 10-15 amino acid peptides. These peptides are recognized by T lymphocytes when presented by class II HLA molecules.

C) HLA CLASS III

HLA class III genes code for components of the complement cascade (BF, C2, C4A, C4B) and will not be discussed further.

Bone Marrow Transplantation, edited by Richard K. Burt, H. Joachim Deeg, Scott Thomas Lothian, George W. Santos. © 1996 R.G. Landes Company.

II) STRUCTURE OF HLA MOLECULES

HLA molecules are cell surface glycoproteins structured in such a way that they form folds and helices which result in a peptide-binding groove. Conceptually, peptide presentation by the HLA molecule may be viewed as a "hot dog in a bun". Class I and class II molecules are quite similar in the way that they present peptides (see Fig. 5.2).

A) CLASS I STRUCTURE

HLA Class I α chain is a 44 kD heavy chain with three domains, α1, α2 and α3. This heavy chain is associated with β_2-microglobulin. The gene for β_2-microglobulin is located on chromosome 15.

B) CLASS II STRUCTURE

The class II molecule is a heterodimer of two glycoproteins, alpha and beta. The alpha chain is 28 kD and the beta chain is 33 kD. Each chain is divided into two extracellular domains, α1, α2, β1 and β2.

III) LABORATORY METHODS

Several assays such as complement-dependent lymphocytotoxicity, mixed lymphocyte culture, primed lymphocyte typing, sequence specific oligonucleotide probes, sequence specific primers and isoelectric focusing are available to identify an individual's HLA type. Refer to Figure 5.3.

A) LYMPHOCYTOTOXICITY

The complement-dependent lymphocytotoxicity assay is the most common serologic method of identifying class I and class II gene products. Peripheral

5

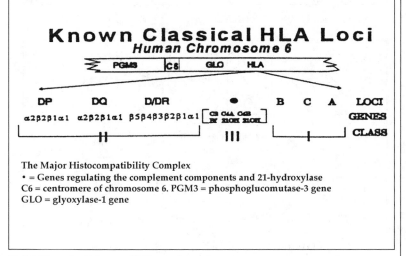

Fig.5.1. The major histocompatibility complex.

blood T and B lymphocytes express HLA class I. To identify HLA class I antigens, a peripheral blood lymphocyte suspension is added to a Terasaki tray (small 60 or 72 well microtray) which contains 1 microliter of antisera per well. Each well of antisera has a known HLA specificity; some sera are monospecific and others are multispecific. After incubation, complement is added to the wells. Where specific antibody-antigen binding has occurred, complement is activated and lymphocyte lysis and death results. A vital dye such as trypan blue or ethidium bromide is added to visualize the dead cells. The pattern of

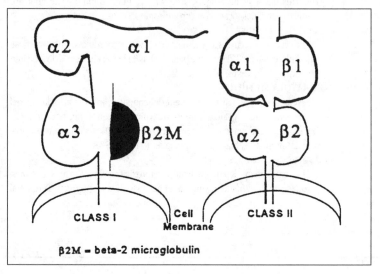

Fig. 5.2. *Schematic of class I and class II antigens.*

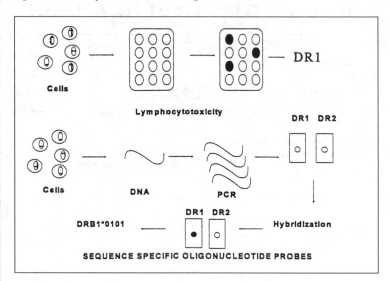

Fig. 5.3. *HLA typing techniques, lymphocytotoxicity and sequence specific oligonucleotide probes.*

cell death is analyzed to assign HLA specificity. For example, if only cells in wells containing anti-HLA-A2 were lysed, then the specificity assigned is HLA-A2.

This technique is more tedious for HLA class II because class II molecules are normally expressed on B but not on T lymphocytes. B lymphocytes are only 10-20% of the peripheral blood lymphocyte population. Therefore, the B cells must be isolated from the peripheral blood suspension using techniques such as nylon wool or panning ("sticky" B lymphocytes adhere to plastic). More recently, B lymphocytes have been isolated using magnetic beads coated with an anti-B lymphocyte antibody (anti-HLA-DR or anti-CD19).

B) MIXED LYMPHOCYTE CULTURE (MLC)

This assay was originally used to identify class II gene products of the HLA-Dw locus. Actually, HLA-Dw is not a specific group of genes but a composite response of the contributions from the HLA-DR, -DQ, -DP and perhaps other class II loci. A mononuclear cell suspension is tested for proliferation against a panel of known irradiated stimulator cells. The panel of stimulator cells are irradiated so they remain viable but cannot proliferate. The stimulator cells (called homozygous typing cells or HTCs) are homozygous for a particular Dw specificity (e.g., Dw1, Dw1). Proliferation of the responding cells is dependent on the degree of Dw allele disparity between stimulating and responding cells. Responding cells which show reduced or no proliferation when stimulated must share specificity with the stimulator cells (HTCs). Although MLC testing is sensitive, it is not specific. MLC testing is still widely used as a cellular cross-match in related donor bone marrow transplantation, but it is rarely used to identify the HLA-Dw specificity. Unfortunately, MLC does not accurately predict for graft-versus-host disease since the assay does not give information as to what cell subset is responsible for the proliferation detected. Also there are many variables that are difficult to control.

C) PRIMED LYMPHOCYTE TEST (PLT)

The PLT was originally used to define HLA-DP specificities. It is a modification of the MLC that amplifies weak responses. After the initial stimulation, responder cells are allowed to maximally proliferate. Thus, the cell population is enriched for specific T cells that have responded to disparate class II molecules on the stimulators. These T cell responders are now "primed" and after restimulation with the same stimulator cells, peak proliferation occurs quickly (2-3 days). The MLC and PLT assays are tedious, time consuming and depend on normal immunologic function of responder cells. Samples from patients in leukemic relapse or failed induction may not respond normally. HLA laboratories in large transplantation centers are minimizing the use of the classic MLC and PLT techniques. They have developed molecular genetic techniques such as sequence specific oligonucleotide probes (SSOP) and sequence specific primers (SSP) for class II typing.

D) MOLECULAR METHODS

Molecular genetic techniques are now the preferred method of determining HLA class II molecules. Sequence specific oligonucleotide probes (SSOP) and sequence specific primers (SSP) are sensitive and specific methods capable of detecting a single nucleotide difference between two class II alleles. Molecular genetic techniques are currently being developed for HLA class I identification.

1) Sequence Specific Oligonucleotide Probes (SSOP)

HLA typing by SSOP is done by extracting genomic DNA from peripheral blood cells. Then by using specific oligonucleotide primers, the DNA can be amplified using the polymerase chain reaction (PCR) technique. The amplified DNA is transferred to a membrane in a drop wise fashion called dot blotting. A labeled probe specific for a HLA class II gene will bind to the sample DNA on the membrane, if the gene is present. The probe, depending on how it was labeled, may be detected by several methods including radiographic, colorimetric or chemiluminescent techniques.

2) Sequence Specific Primers (SSP)

SSP is almost as cost effective as serologic testing but has far greater sensitivity and specificity. Instead of generic primers used in SSOP which amplify the entire DR region, specific primers for the unique HLA class II polymorphisms are used in the PCR amplification step. Therefore, SSP eliminates the need for the probes required in SSOP. The major problem with SSOP and SSP is that new and rare alleles will not be detected since only known DNA sequences can be used for the probes and primers.

E) Isoelectric Focusing (IEF)

Many allelic variants of class I and II genes which are not distinguishable serologically may be identified by this technique which depends on separation of proteins based on their isoelectric point (the pH at which a protein has no charge).

IV) HLA NOMENCLATURE

HLA nomenclature is defined by the traditional serologic and more recent molecular techniques. Molecular techniques determine an actual HLA gene nucleotide sequence, while serology defines antigenic determinants (epitopes) on the HLA gene product (i.e. the HLA glycoprotein molecule).

A) Nomenclature Rules

HLA terminology combines both serologic and molecular methods used to define the HLA system (Table 5.1). The nomenclature rules are:

1) A capital letter designates HLA class I locus (A, B, C, D, E, F, etc.). Two capital letters designates HLA class II locus (DR, DQ, DP, DN, DO, etc.). Different serologically defined class I or II HLA molecules are identified by a number after the locus designation (e.g. HLA-A2). The number assigned is an indication of the order in which the allele was officially recognized. For class I, HLA-A1 was the first allele named followed by HLA-A2, HLA-A3, HLA-Bw4, HLA-B5, etc.

2) On chromosome 6, each class I loci (A, B, C, etc.) is a single gene. However, class II loci (DR, DQ, DP, etc.) each have several genes ($\alpha 1$, $\alpha 2$, $\beta 1$, $\beta 2$, etc.) (See Fig. 5.1). The specific HLA class II gene is designated after identification of the class II locus: e.g. DR$\beta 1$, DQ$\alpha 1$, etc.

3) Since HLA is highly polymorphic, each gene has many alleles in a population. Specific alleles are designated by an asterisk (*) followed by a two digit number indicating the most closely associated serologic specificity, followed by another two digit number that defines a unique allele (identified by molecular techniques). For example, the serologically defined HLA-A2 molecule is actually 10 distinct alleles. These alleles are defined as HLA-A*0201 through HLA-A*0210. Therefore, HLA-A2 is a broad serologic

specificity but HLA-A*0201 is a unique allele whose protein product reacts with the antibody that is used to identify HLA-A2.

4) If alleles differ by a single nucleotide substitution but the amino acid sequence is the same, a fifth digit is added as in HLA-Cw0202<u>1</u> and HLA-Cw0202<u>2</u>.

5) Serologically defined products of the DRβ1 locus are known simply by the allele name omitting the β1*. For example, DRβ1*0103 is known as DR103.

B) SEROLOGY

HLA molecules are composed of over 200 amino acids. Therefore, the HLA molecule has many amino acid sequences theoretically capable of inducing a humoral response. In reality antibodies tend to be produced against the exposed external surface of the HLA molecule, especially around the peptide binding region. The region that an antibody recognizes is termed an epitope. There are two general types of HLA epitopes defined serologically, private epitopes and public epitopes.

1) Private Epitopes

Private epitopes occur only on a single gene product. For example, HLA-A1.

2) Public Epitopes

Public epitopes are common to more than one gene product. An example of high frequency public epitopes are Bw4 and Bw6 which are epitopes present on all HLA-B and some HLA-A and HLA-C molecules. Examples of these can be found in Table 5.2.

3) Cross-Reactive Groups (CREGs)

Antibodies that bind or cross react with more than one HLA molecule have been used to place HLA gene products into major cross-reactive groups (CREGs). For patients awaiting organ transplantation or requiring

5

Table 5.1. Examples of HLA nomenclature

Serological Specificity	Gene Loci	Polymorphic or Variant Alleles	Number of Gene Products
HLA-A1	HLA-A β₂M	A*0101	1
HLA-B7	HLA-B β₂M	B*0701 B*0702	2
HLA-Cw7	HLA-Cw β₂M	Cw*0701 Cw*0702	2
HLA-DR1	HLA-DRβ1 HLA-DRα	DRβ1*0101 DRβ1*0102	2
HLA-DR52	HLA-DRβ3 HLA-DRα	DRβ3*0101 DRβ3*0201 DRβ3*0202 DRβ3*0301	4
HLA-DQ2	HLA-DQβ1 HLA-DQα1	DQβ1*0201 DQα1*0201	1

repetitive platelet transfusions, antibodies that recognize a CREG are important since the presence of such an antibody can be devastating in terms of finding a compatible donor. Some major HLA class I CREGs are listed in Table 5.3.

4) Splits

Refer to Table 5.4. The goal of HLA typing is to define discrete HLA alleles. What is actually detected serologically, however, are antigenic determinants (epitopes) on HLA molecules. This is straight forward if an epitope is specific for a single HLA gene product (private epitope). However, when an epitope occurs on more than one HLA gene product (public epi-

Table 5.2. Examples of high frequency public epitopes

Bw4	Bw6	DR51	DR52	DR53
B5,B13,B17,B27	B7,B703,B8,B14	DR2	DR3	DR4
B37,B38,B44	B18,B22,B35	DR15	DR5	DR7
B47,B51,B5102	B39,B3901	DR16	DR6	DR9
B5103,B52,B53	B3902,B40		DR11	
B57,B58,B59	B4005,B41,B42		DR12	
B63,B77	B46,B48,B50		DR13	
A9,A23,A24	B54,B55,B56		DR14	
A2403,A25	B60,B61,B62		DR1403	
A32	B64,B65,B67		DR1404	
	B70,B71,B72		DR17	
	B73,B75,B76		DR18	
	B7801			

Table 5.3. Cross-reactive groups

Some Major HLA Class I CREGs		
Major CREG	Occurrence on HLA Molecules	Approximate Frequency (%)
1C	A1,A36,A3,A9,A10,A11,A29,A19	79
2C	A2,A28,A9	66
5C	B5,B17,B18,B35,B53,B70	50
7C	B7,B22,B27,B42,B40,B41,B13,B47,B48	54
8C	B8,B14,B16,B18	38
12C	B12,B21,B13,B40,B41	44
4C	A9,A25,A32,Bw4	79
6C	A11,Cw1,Cw3,Cw7,Bw6	87

tope), groups of alleles rather than specific alleles are defined. This phenomenon has led to the concept of "splitting". For example, when HLA-B12 was recognized as a new specificity, it was considered to be a distinct HLA gene product. But soon antisera were discovered that reacted with only a specific subset of those cells defined as HLA-B12. These subsets (B44 and B45) were referred to as splits or subtypic specificities included in the broad B12 specificity. Splits are discrete HLA gene products that share common serologically defined epitopes. The process of splitting is continually occurring as more discriminating antibodies are discovered for HLA gene products.

Table 5.4. Broad splits and associated antigens

Original Broad Specificities	Splits and Associated Antigens
A2	A203#,A210#
A9	A23,A24,A2403#
A10	A25,A26,A34,A66
A19	A29,A30,A31,A32,A33,A74
A28	A68,A69
B5	B51,B52
B7	B703#
B12	B44,B45
B14	B64,B65
B15	B62,63,75,76,77
B16	B38,B39,B3901#,B3902#
B17	B57,B58
B21	B49,B50,B4005#
B22	B54,B55,B56
B40	B60,B61
B70	B71,B72
Cw3	Cw9,Cw10
DR1	DR103#
DR2	DR15,DR16
DR3	DR17,DR18
DR5	DR11,DR12
DR6	DR13,DR14,DR1403#,DR1404#
DQ1	DQ5,DQ6
DQ3	DQ7,DQ8,DQ9
Dw6	Dw18,Dw19
Dw7	Dw11,Dw17

Associated antigen which is a variant of the original broad specificity and not a split as previously defined.

5

C) Molecular Techniques

Compared to serology, molecular assays such as SSOP and SSP (see section III above) are a more direct and specific method of determining unique HLA alleles. Molecular techniques are the "gold standard" for identifying a unique HLA gene and are currently complementing but will eventually replace serologic methods.

V) HLA GENETICS

Population and family studies have provided strong evidence that inheritance of HLA genes follows Mendelian principles. The HLA system shows linkage disequilibrium; i.e. alleles from separate HLA loci are inherited together more commonly than would be predicted by chance.

A) Linkage Disequilibrium

Certain combinations of genes occur on HLA haplotypes within the population far more frequently than expected based upon the gene frequencies. This phenomenon is referred to as linkage disequilibrium. For example, in Caucasian populations with northern European origins, the actual occurrence of the A1-B8 haplotype may be as high as 8% which far exceeds the expected frequency of 1.6% (frequency of HLA-A1 is 16% and frequency of HLA-B8 is 1% so 0.16[A1] x 0.1[B8] = 0.016 or 1.6%). Linkage disequilibrium usually affects the entire haplotype involving class III as well as class I and II MHC encoded genes. For class II genes in particular, association of certain DQ, DR52 or DR53 with specific DR alleles can be close to 100% in some ethnic populations (see Table 5.5 and Table 5.2). A patient with a haplotype displaying a high degree of linkage disequilibrium (A1-B8 or A3-B7) will have a much easier time finding a compatible donor than a patient with haplotypes which do not show linkage disequilibrium (such as A24-B13).

B) Population Variance

There are three levels of population variance in the distribution of HLA genes and specific HLA haplotypes. Refer to Table 5.6. First, the same genes may occur in all populations but at different frequencies. Second, certain HLA genes may be restricted to a specific population (see Table 5.7). Finally, certain haplotypes, because of linkage disequilibrium, appear to be population restricted as shown in Table 5.8.

C) Polymorphism

After the Eleventh International Histocompatibility Workshop in 1991 in Japan (workshop meets every four years to update data), 14 new specificities were recognized. See Table 5.9 for the list of the locus and allele number and number of serological specificities. The list continues to grow as epitopes are defined more precisely. The apparent discrepancy between the number of alleles and the number of serologically defined specificities means that many alleles are serologically "silent" and have been defined by T cell clones, isoelectric focusing or sequencing of DNA. A complete listing of the recognized HLA specificities is shown in Table 5.10.

VI) REPORTING OF HLA TYPING DATA

A person's HLA type may be reported as their phenotype or genotype. The phenotype is determined by HLA typing an individual. The genotype indicates which MHC chromosomes are inherited from each parent, and is usually determined by analyzing the phenotypes of the parents and offspring. Therefore, when selecting a potential donor for BMT, the patient, all siblings, **and** the parents should have their HLA phenotype determined.

Table 5.5. HLA-DR, DQ, Dw and Dβ1 associations

HLA-DR	HLA-DQ	DRβ1	HLA-Dw
DR1	DQ5(1)	DRβ1*0101	Dw1
DR1	DQ5(1)	DRβ1*0102	Dw2
DR103	DQ5(1)	DRβ1*0103	D'BON'
DR15	DQ6(1)	DRβ1*1501	Dw2
DR15	DQ6(1)	DRβ1*1502	Dw12
DR15		DRβ1*1503	NONE
DR16	DQ5(1)	DRβ1*1601	Dw21
DR16	DQ7(3)	DRβ1*1602	Dw22
DR17	DQ2	DRβ1*0301	Dw3
DR18	DQ4	DRβ1*0302	D'RSH'
DR18		DRβ1*0303	NONE
DR4	DQ7(3),DQ8(3)	DRβ1*0401	Dw4
DR4	DQ7(3),DQ8(3)	DRβ1*0402	Dw10
DR4	DQ7(3),DQ8(3)	DRβ1*0403,DRβ1*0407	Dw13
DR4	DQ8(3)	DRβ1*0404,DRβ1*0408	Dw14
DR4	DQ4	DRβ1*0405	Dw15
DR4	DQ3	DRβ1*0406	D'KT2'
DR11	DQ7(3)	DRβ1*1101	Dw11
DR11	DQ7(3)	DRβ1*1102	D'JVM'
DR12	DQ7(3)	DRβ1*1201	DB6
DR13	DQ6(1)	DRβ1*1301	Dw18
DR13	DQ6(1)	DRβ1*1302	Dw19
DR13	DQ7(3)	DRβ1*1303	D'HAG'
DR14	DQ5(1)	DRβ1*1401	Dw9
DR14	DQ7(3)	DRβ1*1402	Dw16
DR7	DQ2,DW9(3)	DRβ1*0701	Dw17
DR8	DQ4,DQ7(3),DQ6(1)	DRβ1*0801,DR$_B$1*0802, DRβ1*0803	Dw8
DR9	DQ9(3)	DRβ1*0901	Dw23
DR10	DQ5(1)	DRβ1*1001	D'SHY'

5

A) HLA PHENOTYPE

Following the recommendations of the World Health Organization (WHO) Committee on Nomenclature, the HLA phenotype of an individual is written with the products of the individual loci grouped together. For example:

HLA- A1,3; B7(Bw6), 8(Bw6); Cw1,2; DR3

If only one of two alleles are identified at a locus, the phenotype may include it twice only if homozygosity is proven by family studies. For example, the phenotype may be either:

A1,1; B7(Bw6),8(Bw6); Cw1,2; DR 2,3
or
A1,-; B7(Bw6),8(Bw6); Cw1,2; DR 2,3

Table 5.6. Examples of allele frequencies in diverse populations

HLA Specificity	Linkage Disequilibrium	Antigen Frequency (%)		
		Caucasian	Black	Japanese
A31		4.4	3.8	**14.8**
B13		**5.9**	1.6	3.4
B53		1.6	**22.6**	0.2
A26	B38-DR4	7.3	3.2	**20.4**
A24	B61-DR2	6.8	8.8	**58.1**
A34	B44	1.2	**9.8**	0.2
A1	B37-Cw6-DR10	**4.4**	2.2	1.6
A33	B58-Cw3	2	**13.7**	1.4

Table 5.7. Restriction of certain alleles in different populations

HLA Specificity	Linkage Disequilibrium	Antigen Frequency (%)		
		Caucasian	Black	Japanese
B5102		0	1.8	0.6
B45		1.2	7.1	0
B67		0	1.6	3.2
A28	B70	8.8	20.8	0
A36	B53	0.8	5.3	0
A1	B8-Cw7	18.1	6.3	0
A30	B42	0	10.5	0
A2	B46-Cw1	0	0	8.8
A3	B47-Cw6-DR6	1.2	0	0.2

Table 5.8. Restriction of haplotypes to certain populations

Haplotype			Possible Origin
A	B	DR	
26	38	4	Jews, Turks, Arabs
3	14	1	Moors
30	18	3	Visigoths
1	8	3	Goths, Germans
3	7	2	Vikings
29	12	7	Basques
2	44	4	Celts

5

B) HLA GENOTYPE

The paternal and maternal contributions to an individual are clearly identified by following WHO nomenclature recommendations for genotype. An individual's HLA genotype is:

a	A1 B7 Cw1 DR2	Paternal
c	A1 B8 Cw2 DR3	Maternal

or

HLA-A1, B7, Cw1, DR2 / A1, B8, Cw2, DR3

VII) PROBLEMS IN IDENTIFYING HLA GENOTYPICALLY IDENTICAL SIBLINGS

When phenotyping the family including parents, it is usually easy to determine the genotypes, especially when each parental locus codes for two different alleles and all alleles are expressed in the siblings (See Fig. 5.4). However, parental HLA homozygosity, gene recombination, or failure to identify class II alleles often result in errors in assigning the correct genotype if phenotypes are determined only for siblings omitting parental typings.

A) PARENTAL HLA HOMOZYGOSITY

Patient: A2,25;B44,18;DR4,7
Sibling 1: A2,25;B44,18;DR4,7
Sibling 2: A2,25;B44,18;DR4,7

When mixed lymphocyte culture was performed with the patient and siblings, a positive response was found for Sibling 1 but no response was found with Sibling 2, indicating that the patient is compatible with Sibling 2.

Reason for apparent incompatibility with Sibling 1:

Father's typing:	a	A2,B44,DR4
	b	A2,B18,DR4
Mother's typing:	c	A25,B18,DR7
	d	A25,B18,DR7
Patient inherited:	a/d	A2,B44,DR4/A25,B18,DR7
Sibling 1 inherited:	a/c	A2,B44,DR4/A25,B18,DR7
Sibling 2 inherited:	a/d	A2,B44,DR4/A25,B18,DR7

Even though the HLA loci tested appear to be identical, haplotype c contains different alleles than haplotype d. For example, perhaps there was a

Table 5.9. HLA polymorphism

Locus	Number of Alleles	Number of Serological Specificities
A	41	27
B	61	59
Cw	18	10
DRβ1	60	20
DRβ3	4	1
DRβ4	1	1
DRβ5	4	1
DQβ1	19	9
DPβ1	38	6

difference in the DQ antigens, DP antigens or minor histocompatibility antigens; differences at these loci can be responsible for MLC reactivity. Therefore, Siblings 1 and 2 are only haplo-identical. The patient is HLA identical with
Sibling 2 and haplo-identical with Sibling 1.

B) GENE RECOMBINATION

Gene recombination occurs in about 2% of siblings. During chromosomal
replication, replicated chromosomes often overlay each other, forming
X-shaped chiasmata. When the chromosomes are pulled apart during meiotic
division, chromosomal breaks can occur at the crossover site, resulting in a
complementary exchange of genetic material. The frequency of recombina-

Table 5.10. Complete listing of recognized HLA specificities

A	B	B	C	D	DR	DQ	DP
A1B5	B50(21)	Cw1	Dw1	DR1	DQ1	DPw1	
A2B7	B51(5)	Cw2	Dw2	DR103	DQ2	DPw2	
A203	B703	B5102	Cw3	Dw3	DR2	DQ3	DPw3
A210	B8	B5103	Cw4	Dw4	DR3	DQ4	DPw4
A3B12	B52(5)	Cw5	Dw5	DR4	DQ5(1)	DPw5	
A9B13	B53	C26	Dw6	DR5	DQ6(1)	DPw6	
A10	B14	B54(22)	Cw7	Dw7	DR6	DQ7(3)	
A11	B15	B55(22)	Cw8	Dw8	DR7	DQ8(3)	
A19	B16	B56(22)	Cw9(w3)	Dw9	DR8	DQ9(3)	
A23(9)	B17	B57(17)	Cw10(w3)	Dw10	DR9		
A24(9)	B18	B58(17)		Dw11(w7)	DR10		
A2403	B21	B59		Dw12	DR11(5)		
A25(10)	B22	B60(40)		Dw13	DR12(5)		
A26(10)	B27	B61(40)		Dw14	DR13(6)		
A28	B35	B62(15)		Dw15	DR14(6)		
A29(19)	B37	B63(15)		Dw16	DR1403		
A30(19)	B38(16)	B64(14)		Dw17(w7)	DR1404		
A31(19)	B39(16)	B65(14)		Dw18(w6)	DR15(2)		
A32(19)	B3901	B67		Dw19(w6)	DR16(2)		
A33(19)	B3902	B70		Dw20	DR17(3)		
A34(10)	B40	B71(70)		Dw21	DR18(3)		
A36	B4005	B72(70)		Dw22			
A43	B41	B73		Dw23	DR51		
A66(10)	B42	B75(15)					
A68(28)	B44(12)	B76(15)		Dw24	DR52		
A69(28)	B45(12)	B77(15)		Dw25			
A74(19)	B46	B7801		Dw26	DR53		
B47							
B48	Bw4						
B49(21)	Bw6						

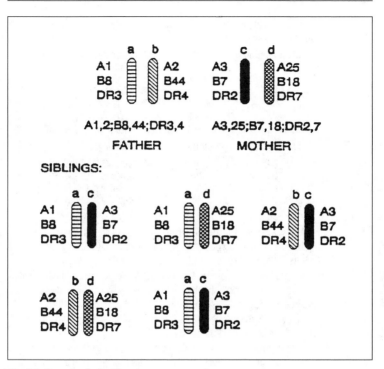

Fig. 5.4. Example of a family typing

tion between two HLA loci is directly related to the distance between the loci (see Fig. 5.1). For example, recombination between HLA-A and HLA-DP is common, but recombination between HLA-DQ and HLA-DR is rare.

In the situation above, it would not have been unusual for recombination to occur between the A and B loci:

Father's typing:	a	A2,Cw7,**B44,DR4**
	b	**A2,Cw5**,B18,DR4
Sibling 1:	ba	A2,Cw5,B44,DR4
	d	A25,B18,DR7

Since there are as many as 10 subtypes of HLA-A2, it is possible to express two different A2 antigens which may not necessarily be detected as different subtypes in routine HLA typing. By typing the parents with the siblings, the more subtle differences in reactivity between the two A2 alleles may be noticed, whereas if the typings are done at different times and by different laboratories (especially if parental typings are not done), this subtlety would not be noticed. Defining the antigens at the C locus can often help to interpret the results. Although the C locus is generally not considered significant in causing GVHD, it can help explain apparent discrepancies in HLA typing results.

C) FAILURE TO IDENTIFY CLASS II ALLELES

Sometimes the incompatibility is caused by failure to adequately identify class II alleles as well as by failure to properly interpret typing data as in the following case.

Results sent to the transplant center by Laboratory X. Class I typings only on patient and siblings.

Identification	Relation	HLA Phenotype
CJ	Patient	A2,28;Cw3;B45
DJ	Sibling	A2,28;Cw3;B45
RJ	Sibling	A2,28;Cw;B45,17
OJ	Sibling	A2,30;Cw5;B45,44
JJ	Sibling	A3;Cw4,5;B45,44

Based on this data, Laboratory X assigned the following genotypes:

Identification	Relation	HLA Phenotype
CJ	**Patient**	**A2,Cw3,B45/A28,Cw-,B-**
DJ	**Sibling**	**A2,Cw3,B45/A28,Cw-,B-**
RJ	Sibling	A2,Cw-,B45/A28,Cw-,B17
OJ	Sibling	A2,Cw-,B45/A30,Cw5,B44
JJ	Sibling	A3,Cw5,B45/A-,Cw4,B44

It looks like there is an identical match between CJ and DJ. But there are five to six possible haplotypes (A2,Cw3,B45; A28,Cw-,B-; A28,Cw-,B17; A30, Cw5,B44; A3,Cw5,B45; A-,Cw4,B44) when there should only be four (two haplotypes from each parent). It is possible that some of the siblings are only half-sibs or it is possible that technical problems made it difficult to detect certain alleles (e.g., C locus). In particular, B17 in RJ and A30 in OJ are difficult to explain. One might suspect the B blanks on the A28 haplotype may be a result of homozygosity at the B locus (B45).

The transplant center requested repeat typings of the sibs including class II typings plus complete class I and class II typings on the parents.

Results of additional typing performed by Laboratory X:

ID	Relation	HLA Phenotype
NJ	Father	A28;Cw2;B17;DR2,6;DQ1;DR52
EJ	Mother	A2,3;Cw3,4;B45;DR4,7;DQ2;DR52,53
CJ	**Patient**	**A2,28;Cw3;B45;DR6,7;DQ1,2;DR52,53**
DJ	Sibling	A2,28;Cw3;B45; no class II typing
RJ	Sibling	A2,28;Cw;B45,17;DR6,7;DQ1,2;DR52,53
OJ	Sibling	A2,30;Cw5;B45,44;DR6,7;DQ1,2;DR52,53
JJ	Sibling	A3;Cw4,5;B45,44;DR5,7;DQ1,2;DR52,53

HLA genotypes were assigned by Laboratory X as follows:

ID	Relation	Haplotype	HLA Genotype
NJ	Father	a	A28,Cw-,*B17*,DR6,DQ1,DR52
		b	A-,Cw2,B-,DR2,DQ1,DR52
EJ	Mother	c	A2,Cw3,B45,DR7,DQ2,*DR52*
		d	A3,Cw4,B45,DR4,DQ2,DR53
CJ	**Patient**	a	**A28,Cw-,*B-*,DR6,DQ1,DR52**
		c	**A2,Cw3,B45,DR7,DQ2,DR53**
DJ	Sibling	a	A28,Cw-,*B-*
		c	A2,Cw3,B45
RJ	Sibling	a	A28,Cw-,*B17*,DR6,DQ1,DR52
		c	A2,Cw3,B45,DR7,DQ2,DR53
OJ	Half-sibling	y	A30,Cw5,B44,DR6,DQ1,DR52
		c	A2,Cw3,B45,DR7,DQ2,DR53
JJ	Half-sibling	z	A-,Cw5,B44,DR5,DQ1,DR52
	recombinant	dc	A3,Cw4,B45,*DR7*,DQ2,DR53

The only informative sibling is RJ. The DR52 identified in the mother does not fit with DR7 (see Table 5.2). Of the four sibs that supposedly inherited the father's **a** haplotype, only one sib expresses the B17 (RJ). None of the sibs inherited the father's second haplotype so it is difficult to feel confident about the assignment of genotypes in this case.

The transplant center then requested samples on all family members and repeated the typings. The findings of the transplant laboratory are as follows:

ID	Relation	HLA Genotype
NJ	Father	A28;Cw2;B58,72;DR6;DQ1;DR52
EJ	Mother	A2,3;Cw4;B44,45;DR7;DQ2;DR53
CJ	**Patient**	**A2,28;Cw3;B45,71;DR7,8;DQ2,3?;DR52?,53**
DJ	Sib	A2,28;Cw2;B45,72;DR6,7;DQ1,2;DR52,53
RJ	Sib	A2,28;Cw ;B45,58;DR6?,7 (Typed elsewhere)
OJ	Half-sib	A2,30;Cw ;B45,x;DR6,7;DQ1,2;DR52,53
JJ	Half-sib	A3;Cw4;B45,44;DR5,7;DQ1,2;DR52,53

The genotypes were assigned as follows:

ID	Relation	Haplotype	HLA Genotype
NJ	Father	a	A28,Cw2,B72,DR6,DQ1,DR52
		b	A28,Cw-,B58,DR-,D1-,DR-
EJ	Mother	c	A2,Cw-,B45,DR7,DQ2,DR53
		d	A3,Cw4,B44,DR7,DQ2,DR53
CJ	**Patient**	**x**	**A28,Cw3,B71,DR8,DQ3?,DR52?**
		c	**A2,Cw-,B45,6,DR7,DQ2,DR53**
DJ	Sibling	a	A28,Cw2,B72,DR6,DQ1,DR52
		c	A2,Cw-,B45,DR7,DQ2,DR53
RJ	Sibling	b	A28,Cw-,B58,DR6?
		c	A2,Cw-,B45,DR7,DQ2,DR53
OJ	Half-sibling	y	A30,Cw-,BX,DR6,DQ1,DR52
		c	A2,Cw-,B45,DR7,DQ2,DR53
JJ	Half-sibling	z	A-,Cw-,B45,DR5,DQ1,DR52
		d	A3,Cw4,B44,DR7,DQ2,DR53

The results of the transplant laboratory are quite different from the initial HLA typing report sent from Laboratory X. In fact, the patient's stated father is not his biological father. What looked like a patient with an identical sibling turned out to be a patient with no identical match but haplo-identical with three sibs.

D) INTER-LABORATORY VARIABILITY

When typing families expressing antigens with significant variability or with poorly characterized antisera, it is sometimes difficult to compare data between laboratories unless they use typing sera from the same sources and use similar testing techniques. It is not unusual to find discrepancies in HLA typing reports. However, even though the antigen specificity may be different, the state of identity should still be maintained. This means that even though one lab detects A24 and another lab identifies it as A2403, this antigenic specificity should be the same in each family member that possesses it (all those with A24 also are the same ones identified as A2403 in the other laboratory). The ability to detect certain antigens requires that the tissue typing laboratory have the sera to test for the presence of the antigen. Sera for some of the more recently characterized specificities are rare and not available to all laboratories.

That is why one lab may be able to identify B17 while another lab assigns B58 (a split of B17). ASHI standards require that all laboratories must be able to type for all HLA-A, HLA-B and HLA-DR specificities officially recognized by the WHO and for which sera are readily available. However, typing for the C locus, DQ locus and DP locus antigens is not mandatory. Keep in mind that although the HLA antigens at each locus may be identical, it is probable that the two siblings are matched for the loci in between, but one should not assume this to be the case.

VIII) IMPLICATIONS OF HLA MATCHING FOR BMT

A) EFFECT OF HLA ON ENGRAFTMENT

The degree of compatibility between donor and recipient has a significant effect on the incidence of graft rejections or failure:

Donor	Incidence
HLA genotypically identical	< 1%
HLA phenotypically identical	5%
One antigen mismatch	7-10%
Two to three antigen mismatch	15-25%

It is possible to have a unidirectional antigen mismatch which would not affect engraftment if the mismatch is one that would not be recognized by the patient; e.g. if the patient is A1,2 and the donor is A1,1 and matched for all other antigens.

B) EFFECT OF HLA ON GRAFT-VERSUS-HOST DISEASE (GVHD)

For unmodified bone marrow grafts, the probability of acute GVHD, Grades II-IV, is

Donor	Incidence
HLA genotypically identical	40%
HLA phenotypically identical	50%
One locus mismatch	75%
Two locus mismatch	80%
Three locus mismatch	90%

It has recently been recognized that the degree of severity of GVHD is dependent on the locus which is discrepant, i.e. HLA-D > HLA-A > HLA-B. There is evidence that a mismatch for HLA-D is not permissible where as mismatches at the A and B loci can sometimes be tolerated. It is also possible to have a unidirectional mismatch which would not increase the risk of GVHD if the mismatch is such that the donor lymphocytes would not recognize the host as foreign; e.g. if the patient is A1,1 and the donor is A1,2 and matched for all other antigens.

SUGGESTED READING

1. Dupont, B, ed. Immunobiology of HLA (2nd vol.) Berlin: Springer Verlag, 1989.
2. Rodey GE. HLA Beyond Tears, Atlanta: De Novo, Inc, 1991.
3. Bodmer JG, Marsh SGE, Albert ED, Bodmer W, Dupont B, Erlich HA, Mach B, Mayr WR, Parham P, Sasazuki T, Schreuder GMTh, Strominger JL, Svejgaard A, Terasaki PI. Nomenclature for factors of the HLA system: Tissue Antigens 1992: 161-173.
4. Standards for Histocompatibility Testing. American Society for Histocompatibility and Immunogenetics. Lenexa: 1994.
5. HLA 1991, Proceedings of the eleventh workshop and conference held in Yokohama, Japan, 6-13 November, 1991, (2nd vol.) New York: Oxford Press, 1992.

6.0 Diseases

6

6

6

6

6

Aplastic Anemia

Richard K. Burt, H. Joachim Deeg

I) ETIOLOGY

Aplastic anemia (AA) arises from failure of marrow to produce in sufficient quantities, all hematopoietic lineages including platelets, neutrophils and red blood cells (RBC). In most cases, AA is due to hematopoietic stem cell damage or a lack of stem cells. Frequently, there is an immunologic component resulting in suppression of hematopoiesis. Patients may present at any age, although it is often a disease of young people (median age 20-25 years). The annual European and North American incidence of AA is roughly 2 per 10^6 individuals. AA may be either acquired or genetically determined (Table 6.1.1). If a drug etiology is suspected and exposure to several drugs has occurred, the most likely culprit would have been taken 6-8 weeks before onset of symptoms.

II) DIAGNOSIS

Symptoms and signs of fatigue, pallor, petechiae and ecchymosis are related to anemia and thrombocytopenia. Patients may also present with infections if severe granulocytopenia ($< 200/\mu l$) has developed. Peripheral blood smear is occasionally unrevealing but generally shows reticulocytopenia, and a paucity of platelets along with a reduced number of granulocytes and possibly prominent monocytes. Marrow biopsy is hypocellular or empty. However, "hot pockets" of hematopoiesis may exist. There may be lymphoid aggregates and, in fact, the pattern of lymphoid infiltration has been used in the classification of AA. Cytogenetic analysis is usually normal. It should be noted, however, that clonal hematopoiesis may be present in patients with AA.

A) DIFFERENTIAL DIAGNOSIS

The differential of peripheral blood pancytopenia is broad, and besides AA, includes hypersplenism, systemic lupus erythromatosis, rheumatoid arthritis, megaloblastic anemias, sepsis, paroxysmal nocturnal hemoglobinuria, myelodysplastic syndromes, leukemia, lymphoma, hairy cell leukemia, myelofibrosis, tuberculosis, fungal infection, lipid storage diseases and carcinomas that have infiltrated the marrow. The presence of splenomegaly, weight loss or lymphadenopathy is unusual for AA and another diagnoses (e.g. hairy cell leukemia, lymphoma) should be considered. The occurrence of chromosomal abnormalities suggests a hypoplastic myelodysplastic syndrome (MDS). However, distinct separation of aplastic anemia with coexistent cytogenetic abnormalities from hypocellular MDS is not always possible. A bright image in marrow spaces (e.g. vertebral bodies) due to increased hydrophobic fat is present on T1 weighted MRI in AA, and in difficult cases may help to distinguish AA from hypoplastic MDS.

III) WORK UP OF APLASTIC ANEMIA

A marrow biopsy is necessary to exclude the possibility of peripheral pancytopenia coexistent with a normal or hypercellular marrow, as may be seen with hypersplenism, megaloblastic anemias, myelodysplasia, leukemia, lymphoma, hairy cell leukemia or carcinomas. An adequate aspirate is essential since some elderly patients with acute leukemia may present with pancytopenia, an empty marrow biopsy and a dry (acellular) iliac crest aspirate. In that event, a successful sternal

Bone Marrow Transplantation, edited by Richard K. Burt, H. Joachim Deeg, Scott Thomas Lothian, George W. Santos. © 1996 R.G. Landes Company.

aspirate may reveal increased blasts and prevent misdiagnosis of leukemia as aplastic anemia. In addition, at least for patients up to 45-50 years of age, blood should be sent for HLA typing.

Evaluation of patients with suspected AA should also include acidified serum lysis (Ham) test to exclude paroxysmal nocturnal hemoglobinuria (PNH), cytogenetic analysis to exclude MDS with a hypocellular marrow, and in young patients or individuals with malformations, diepoxybutane (DEB) breakage studies to exclude Fanconi anemia.

IV) PROGNOSIS

AA is not a malignant disease, however, without appropriate therapy prognosis is guarded because of risk from potentially fatal hemorrhage and infection. Severe aplastic anemia (Table 6.1.2) represents patients with the worst prognosis and 1 year survival of less than 20% with only supportive treatment. Additional patients will show the same overall picture but pancytopenia, in particular granulocytopenia, may be less severe (> 500/μl). In these patients, the risk of fatal complications is lower, and hence, the prognosis is better.

V) TREATMENT CONSIDERATIONS

Patients with severe AA due to a worse prognosis are treated more aggressively and expeditiously usually with BMT. Whereas patients with less severe forms have

Table 6.1.1. Etiology of aplastic anemia

Acquired	Genetic
Idiopathic	Fanconi's anemia +/– physical abnormalities
Toxins—benzene, petroleum products, insecticides, etc.	Dyskeratosis congenita—hyperpigmentation, dystrophic nails, mucosal leukoplakia, pancytopenia
Drugs—nonsteroidal, anticonvulsants, antithyroid, gold, D-penicillamine, quinacrine, sulfonamides, furosamide, phenothiazines, allopurinol, chloramphenicol, etc.	Shwachman Diamond syndrome—pancreatic exocrine insufficiency, neutropenia/pancytopenia
Viruses—hepatitis, infectious mononucleosis	
Radiation	
Chemotherapy (e.g. busulfan)	
Pregnancy	

6.1

Table 6.1.2. Severe aplastic anemia (SAA)

Marrow biopsy < 25% cellularity
And any two of the following:
Absolute neutrophil count < 0.5 x 10⁹/L
Platelets < 20,000 x 10⁹/L
Absolute reticulocyte count (reticulocyte count x RBC) < 60,000

often been given immunosuppressive therapy (IST). While no randomized prospective trial has been done comparing IST versus BMT, the results of both IST and BMT for AA are steadily improving. In some recent studies, the survival after IST is roughly comparable to BMT. Outcome after BMT is best in young patients with an HLA-matched sibling. In most studies, results of immunosuppression are worse in children and patients with an ANC < 200/μl.

Recommendation of BMT is dependent on recipient age and availability of an HLA-matched sibling donor. Some investigators have also suggested that the approach to treatment (BMT vs IST) should depend on severity of disease. However, most patients treated with IST generally do not achieve normal hematopoiesis. After IST, peripheral blood counts, especially platelet counts, remain subnormal, and megaloblastic anemia is common. In patients given IST, the risk of long-term complications such as PNH, MDS or leukemia is increased and has been reported in the range of 5-40%. The relapse rate after IST is high (20-35%) and patients who are not considered for BMT until they have failed IST have a worse prognosis (due to alloimmunization from multiple transfusions, older age and infections). Therefore, most transplantation centers believe that the treatment of choice for all patients less than 45-50 years of age who have an HLA-identical sibling donor is BMT. For all other patients, IST is currently the preferred therapy. BMT from an alternative donor (unrelated or related HLA-mismatched) should only be carried out if IST fails.

VI) IMMUNOSUPPRESSIVE THERAPY

Response to immunosuppression is slow and 1-4 months of therapy may be required before counts improve. IST consisting of either ATG or cyclosporine A (CsA) achieves hematologic response in 30-50% and a survival of 40-60%. Recent trials with a combination of CsA and ATG have improved early hematologic response (65-75%) and survival (65-80%) (Table 6.1.3). Therefore, the current trend in IST is treatment with both CsA and anti-lymphocyte antibodies, either ATG (anti-thymocyte globulin) or ALG (anti-lymphocyte globulin) (Table 6.1.4). Clinically available antibodies directed against human lymphocytes may be of either horse or rabbit origin. The terms ATG and ALG are often used in the literature, sometimes interchangeably, although differences exist in regards to the preparation. Prior to treatment, patients must be skin tested for a wheel and flare response (> 5 mm erythema) indicative of an allergy to horse (or rabbit) proteins. Antihistamines and β-blockers may mask a hypersensitivity reaction and should be avoided for 48 hours prior to skin testing. A wheel and flare response to subcutaneous or intradermal ATG or ALG generally prevents treatment unless the patient is desensitized (Table 6.1.5). Methylprednisolone is included in ATG regimens to decrease the probability of serum sickness. CsA dosing is often adjusted to maintain a therapeutic plasma level (200-400 ng/ml) and avoid organ, particularly renal toxicity. It has recently been proposed to add granulocyte-colony stimulating factor (GCS-F) to ATG plus CsA (or ATG plus CsA plus methylprednisolone) to accelerate hematologic response.

VII) SYNGENEIC BONE MARROW TRANSPLANTATION

If a monozygotic (identical) twin is available as donor, there is theoretically no risk of GVHD, and BMT represents the treatment of choice regardless of patient age. It is of note, however, that without a preparative regimen for immune suppression approximately 50% of syngeneic marrow recipients will fail to engraft. Although some patients have been prepared with irradiation-containing regimens, due to less toxicity we recommend cyclophosphamide alone, generally at a dose

Table 6.1.3. Bone marrow transplantation versus immunosuppressive therapy for severe aplastic anemia

HLA Identical BMT for Severe Aplastic Anemia

Study	Age-Years (median)	Preparative Regimen	GVHD Regimen	% Rejection	% Acute GVHD	% Chronic GVHD	% Survival (median follow-up)
Blood 1991, 78, #9, 2451	5-46 (19)	CY+TAI	MTX, CsA, CsA+MTX	3	32	55	68 (3.8 years)
Blood 1992, 79, #1, 269	1-40	CY+TLI, TAI, TBI	MTX, CsA, CsA+MTX	10	40	45	63 (5 years)
Transplantation, 1990, 49, #4, 720	1-41 (19)	CY+TLI	CsA+MTX	23	22	–	78 (2 years)
Blood 1992, 80, 170a	2-46 (24)	CY+ATG	CsA+MTX	3	15	30	93 (3 years)
BMT 1993, 11, 459	4-53 (21)	CY	CsA	23	5	–	79

Combined ATG and CsA Immunosuppression for Aplastic Anemia

Study	Age-Years (median)	% Response All Patients	% Relapse	% Survival All Patients	% Subnormal Peripheral Blood Counts	Median Follow-Up
NEJM 1991, 324, 1297	7-80 (32)	70	19	64	61	41 months
Blood 1995, 85, #11, 3058	4-79 (28)	78% at 1 year	25% at 1 year, 36% at 2 years	86% at 1 year, 62% at 3 years	NA	33 months

ATG = anti-thymocyte globulin, CsA=cyclosporine, CY= cyclophosphamide, MTX= methotrexate, TAI= total abdominal irradiation, TBI= total body irradiation, TLI= total lymphoid irradiation

6.1

of 200 mg/kg (50 mg/kg day x 4 days). Overall survival is greater than 90%, and some patients have been followed now for more than 30 years.

VIII) HLA-GENOTYPICALLY IDENTICAL SIBLING BMT

For patients less than 45-50 years old with an HLA-identical sibling, BMT is generally considered the treatment of choice with survival ranging from 65-93% (Fig. 6.1.1 and Table 6.1.3). Good prognostic factors are younger age, no prior transfusions, short interval from diagnosis to transplantation (less than 1 month)

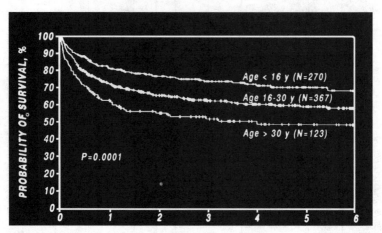

Fig. 6.1.1. Probability of survival after HLA-identical sibling BMT for aplastic anemia, 1987-1994. Reprinted with permission from IBMTR/ABMTR 1995; 2:7.

6.1

Table 6.1.4. Immunosuppressive regimens for SAA

Regimen	Drug	Dose
Frickenhoff (Germany) NEJM 1991; 324:1297	Anti-lymphocyte globulin (ALG)	0.75 ml/kg daily IV over 8-12 hours for 8 consecutive days on day 1-8
	Cyclosporine	12 mg/kg/day (adult) 500 mg/m²/day (child) PO divided BID for 3-6 months (begin day 1)
	Methylprednisolone	5 mg/kg/day PO or IV on day 1-8; 1 mg/kg/day on day 9-14; taper off by day 29
Rosenfield (NIH–USA) Blood 1995; 85(11):3058	Anti-thymocyte globulin (ATG)	40 mg/kg/day IV over 4-6 hours on days 1-4
	Cyclosporine	12 mg/kg/day (adult) 15 mg/kg/day (child) PO divided BID for 6 months (begin day 1), after day 14 dose adjusted by levels
	Methylprednisolone	1 mg/kg/day IV days 1-10 then taper over 2 weeks if no serum sickness
Bacigalupo (Italy) Blood 1995; 85(5):1348	Anti-lymphocyte globulin (ALG)	15 mg/kg/day IV over 6-8 hours, day 1-5
	Cyclosporine G-CSF Methylprednisolone	5 mg/kg/day PO day 1-180 5 μg/kg/day SQ day 1-90 2 mg/kg/day IV day 1-5; 1 mg/kg/day IV day 6-10; then taper off by day 30

Table 6.1.5. ATG (anti-thymocyte globulin) desensitization protocol

ATG is a horse protein and may cause anaphylaxis. Seventy-two hours before skin testing short acting antihistamines (e.g. benadryl) and β-blockers should be discontinued. Long acting antihistamines (e.g. astemazole) may have to be discontinued several weeks before testing. A positive skin test is an immediate (< 5 minutes) wheel and flare greater than 5 mm diameter. Skin testing is done with 0.1 ml of 50 μg/ml ATG. Positive and negative controls are 0.1 ml histamine (1 mg/ml) and 0.1 ml 50% glycerin, respectively. The histamine control must be positive. If not, medications may be interfering with the test and the procedure is not valid.

If the skin test is positive with a negative control, desensitization in an intensive care setting must be done. Due to risk of laryngeal edema, experienced personnel (e.g. anesthesiologists) trained in emergency tracheotomy or cricothyrotomy should also be available. The patient should be NPO for at least 8 hours before the procedure. An intravenous line is necessary and patency maintained with D_5NS. Benadryl, epinephrine and methylprednisolone should be at the bedside. IV epinephrine should only be administered with extreme caution due to risk of arrhythmia. In contrast, SQ epinephrine is safe and rapidly absorbed. In event of anaphylaxis the dose of each medication is:

Drug	Concentration	Adult Dose	Pediatric Dose
Epinephrine	1:1000 (1.0 mg/ml) x 3 doses	0.1-0.5 ml SQ or IM every 10-15 min	0.01 ml/kg SQ or IM every 15 min x 3 doses
Diphenhydramine	50 mg/ml	1 ml in 50 ml IV fluid over 5 min	0.02 ml/kg in 25 ml IV fluids over 5 min
Methyl-prednisolone	125 mg/2 ml	2 ml (125 mg) in 50 ml IV fluids over 5 min	0.03 ml/kg in 25 ml IV fluids over 5 min

Desensitization
Desensitization may be done with either intradermal (standard method) or intravenous (rapid method) injections. The greatest danger is a cavalier attitude arising from a false sense of security due to prior uncomplicated desensitizations. Remember, the general rule for attempting desensitization is "go low and go slow." If the physician has any suspicion of a potentially troublesome desensitization, begin with injections 1-6 of standard desensitization and then proceed to injections 1-5 of the rapid desensitization protocol. Therapeutic ATG is started by continuous infusion immediately after completing desensitization.

6.1

Standard Desensitization
Desensitization is performed on a limb where a tourniquet, if it becomes necessary, can be placed proximal to the injection site. Injections are at 15 min intervals. If an injection produces a large local reaction, the same or a lower dose is repeated before proceeding to the next larger dose. Vital signs should be monitored continuously.

Inj[1]	Method	Dilution of 50 mg/ml ATG Solution	Volume	ATG Dose
1	Intradermal	1/1000	0.02 ml	1 μg
2	Intradermal	1/100	0.02 ml	10 μg
3	Intradermal	1/10	0.02 ml	100 μg
4	Intradermal	Neat (undiluted)	0.02 ml	1000 μg
5	Subcutaneous	Neat	0.1 ml	5000 μg
6	Subcutaneous	Neat	0.2 ml	10, 000 μg
7	Intravenous	40 mg in 500 cc NS or D_5W	2.0 ml per min for 10 min	160 μg/min (1600 μg)

Rapid Desensitization

The bolus intravenous dose interval is 10-15 minutes. Record vital signs continuously. If any symptoms occur (throat tightness, pruritis, hives) either abort, or if desensitization is absolutely necessary repeat at next lower dose.

Inj[1]	Method	mg/ml ATG	Vol/Rate	Total Dose
1	Bolus-IV	0.001 mg/ml	0.1 ml	0.1 μg
2	Bolus-IV	0.01 mg/ml	0.1 ml	1.0 μg
3	Bolus-IV	0.1 mg/ml	0.1 ml	10.0 μg
4	Bolus-IV	0.1 mg/ml	1.0 ml	100.0 μg
5	Continuous IV infusion	40 mg in 500 cc NS or D_5W	2.0 ml per min for 10 min	160 μg/min (1600 μg)

Inj[1] = injection number

and no evidence of infection in the week prior to BMT. Major complications of BMT for AA are graft failure and GVHD.

A) ENGRAFTMENT

Engraftment depends on the conditioning regimen, GVHD prophylaxis regimen, number of donor marrow cells infused and alloimmunization of the patient. Graft failure may be primary (no evidence of engraftment) or delayed (declining counts after transient evidence of engraftment). Treatment of graft failure is a second BMT usually from the same donor and is successful in 5-20% of primary and 30-75% of delayed graft failure patients.

1) Effect of Conditioning Regimen on Engraftment

All patients must be prepared with an immunosuppressive conditioning regimen. Successful regimens include cyclophosphamide either alone or combined with ATG, 30 mg/kg/day for 3 doses or with radiation such as total body irradiation (TBI, 300 cGy) or total lymphoid irradiation (TLI) or thoracoabdominal irradiation (TAI) at doses of 500-750 cGy. Conditioning with cyclophosphamide alone has a high (10-35%) incidence of graft failure (Table 6.1.3). The addition of TLI at 300 cGy was insufficient to improve engraftment. Increasing TLI or TAI to 600 or 750 cGy improves engraftment. However, radiation increases interstitial pneumonitis, GVHD and risk of late tumors.

The preparative regimen we recommend for HLA-matched BMT of AA is cyclophosphamide plus ATG (cyclophosphamide 50 mg/kg/day IV for 4 days on days -5, -4, -3, -2 and ATG 30 mg/kg/day IV for 3 days infused over 10-12 hours on days -5, -4 and -3). Patients with aplastic anemia and chromosomal abnormalities are at increased risk of AML after BMT and should be treated similar to hypoplastic MDS with cyclophosphamide (120-200 mg/kg) and TBI (1200 cGy) or busulfan (16 mg/kg) and cyclophosphamide (120-200 mg/kg) to irradicate the dysplastic clone.

2) Effect of GVHD Prophylaxis on Engraftment

Regimens which are designed to inhibit donor lymphocytes from attacking the host also inhibit host lymphocytes that survived the conditioning regimen from rejecting the donor marrow. Delayed graft failure has been reported to occur when CsA is withdrawn. Therefore, even with successful engraftment and no evidence of GVHD, some physicians wait more than 12 months after marrow infusion before tapering CsA.

3) Effect of Number of Donor Cells on Engraftment

Donor bone marrow cell dose of less than 2.0-3.0 x 10^8 mononuclear cells/kg recipient body weight has been associated with increased probability of graft rejection. Infusion of donor peripheral blood buffy coat cells on days 1-5 after marrow infusion also decreases the probability of graft rejection but increases the incidence and severity of chronic GVHD, and consequently this approach has generally been abandoned.

4) Effect of Alloimmunization on Engraftment

Transfusion of blood products prior to transplant increases the risk of graft rejection due to alloimmunization. The incidence of rejection increases after any transfusion and is substantially increased after more than 20 RBC units or 20-40 units of platelets. However, these observations were made before the use of leukocyte depleted blood products, and it is unclear how many blood products aggressively depleted of leukocytes may be transfused without risk of alloimmunization. If a patient with AA must be transfused before BMT, blood products should be irradiated and leukocyte depleted (including a PALL microfilter). All blood transfusions from relatives should be strictly avoided. Ideally platelets should also be from a single random donor to prevent exposure to multiple alloantigens.

B) GVHD

The risk of acute GVHD (grade II-IV) after HLA-matched sibling BMT is roughly 15-30%. In addition, 30-45% of patients develop chronic GVHD (Table 6.1.3). The most important risk factor for GVHD in HLA-matched siblings is recipient age. The incidence is lower in younger patients. The risk of GVHD is increased with marrow from an alloimmunized donor, in particular a multiparous female. In addition, recipient/donor gender mismatch (especially with female donor to male recipient) increases GVHD in some studies due to H-Y minor antigens encoded on the Y chromosome.

Several GVHD prophylaxis regimens are available (e.g. single agent CsA or combined CsA plus methotrexate). Common regimens, schedules and doses are discussed in the chapter on GVHD.

IX) BMT FROM UNRELATED AND RELATED MISMATCHED DONORS

Patients with severe AA who do not have an HLA-matched sibling and who fail immunosuppressive therapy are candidates for a transplant from a related HLA-one antigen mismatched donor or from an unrelated phenotypically matched donor. For younger patients (< age 35), even an HLA-minor mismatched unrelated donor may be considered. Success with these transplants is lower than with matched sibling BMT. Approximately 15-30% of patients survive with engraftment. Due to a greater risk of graft rejection from HLA disparity, conditioning is generally with cyclophosphamide and radiation (e.g. 200 mg/kg cyclophosphamide and 1200 cGy TBI).

X) AUTOLOGOUS BMT

Autologous BMT has not, in general, been considered an option for patients with AA. However, it has been shown that peripheral blood CD34+ hematopoietic progenitor cells may be harvested from patients with AA after treatment with immunosuppression and G-CSF. In addition, a few patients who failed to engraft have reconstituted their own hematopoiesis. Since cyclophosphamide and ATG do not damage stem cells, it is possible that a cyclophosphamide conditioning regimen without stem cell infusion could induce remission.

SUGGESTED READING

1. Storb R, Champlin RE. Bone marrow transplantation for severe aplastic anemia. Bone Marrow Transplantation 1991; 8:69-72.
2. Socie G, Henry-Amar M, Bacigalupo A et al. Malignant tumors occurring after treatment of aplastic anemia. NEJM, 1993; 329(16):1152-1157.
3. Storb R, Weiden PL, Sullivan FR et al. Second marrow transplants in patients with aplastic anemia rejecting the first graft: Use of a conditioning regimen including cyclophosphamide and anti-thymocyte globulin. Blood 1987; 70(1):116-121.
4. Frickenhofen N, Kaltwasser JP, Schrezenmeier H et al. Treatment of aplastic anemia with anti-thymocyte globulin and methlyprednisolone with or without cyclosporine. NEJM 1991; 324:1297-1304.
5. Bacigalupo A, Piaggio G, Podesta M et al. Collection of peripheral blood hematopoietic progenitors (PBHP) from patients with severe aplastic anemia (SAA) after prolonged administration of granulocyte colony stimulating factor. Blood 1993; 82(5):1410-1414.

6.1

Acute Myeloid Leukemia—AML

Richard K. Burt, Philip Rowlings, George W. Santos

I) ETIOLOGY

In most cases the etiology of AML is unknown. However, an increased risk of AML has been associated with genetic diseases including Down syndrome and genetic disorders of chromosomal breakage (Fanconi's anemia, Bloom syndrome, ataxia telangiectasia). Environmental agents that cause DNA strand breaks are associated with AML and include radiation, benzene, petroleum products, and some chemotherapy drugs especially alkylating agents and etoposide. The incidence of AML, (3-5 per 100,000 annually) increases with age, but does not show appreciable geographic variation. It accounts for 80-85% of all adult acute leukemias.

II) DIAGNOSIS

Onset of AML is generally abrupt with the patient seeking medical help for signs and symptoms of bone marrow failure (anemia, petechiae, easy bruising, infection) or leukemic infiltration (hepatomegaly, splenomegaly, lymphadenopathy, gingival hyperplasia). AML presenting with meningeal carcinomatosis, meningitis, leucostasis or frank hemorrhage may occur but is unusual. The diagnosis of acute leukemia, in most cases depends on a marrow blast count equal to or exceeding 30% of the total nucleated bone marrow cells. However, other criteria are sufficient in acute promyelocytic (M3) or erythroleukemia (M6) (Table 6.2.1). Blasts in the peripheral blood are absent in approximately 10% of patients with AML and are not required for a diagnosis. The blast morphology is categorized according to the French-American-British (FAB) system. The original FAB classification was morphologic and not designed to be of prognostic value. Correlation of FAB with immunophenotype is listed in Table 6.2.2.

A) DIFFERENTIAL DIAGNOSIS

The differential usually is in distinguishing AML from ALL. In AML at least 3% or more of the blasts should be myeloperoxidase positive. Other myeloid/monocytoid lineage-specific stains are Sudan black and esterase's (napthol chloroacetate esterase and alpha naphthyl acetate esterase). Auer rods may occur in myeloblasts or monoblasts and if present are pathognomonic for AML. For difficult cases, immunophenotyping may be useful (Table 6.2.2). Common markers of myeloid lineage are CD11, CD13, CD14, CD15 and CD33.

III) PRETREATMENT EVALUATION

For diagnosis, a bone marrow biopsy and aspirate are mandatory. Routine prechemotherapy evaluation includes: CBC, electrolytes, liver function test (albumin, bilirubin, alkaline phosphatase, transaminases), creatinine, chest x-ray, uric acid, prothrombin time, partial thromboplastin time, resting radionucleotide angiography to noninvasively assess cardiac ejection fraction and HIV serology. If under age 55, the patient, siblings and parents are HLA typed. Refer to the chapter on BMT patient evaluation for pre-BMT work up.

IV) PROGNOSTIC FACTORS

A) PROGNOSTIC FACTORS AT TIME OF PRESENTATION

The three most important factors are age, karyotype and secondary AML due to chemotherapy or following myelodysplastic syndrome.

6.2

Bone Marrow Transplantation, edited by Richard K. Burt, H. Joachim Deeg, Scott Thomas Lothian, George W. Santos. © 1996 R.G. Landes Company.

Table 6.2.1. FAB classification of acute leukemias

FAB Classification	% Blasts in Marrow	Comment
M1—acute myeloblastic leukemia without maturation	30% or more	3% or more of blasts myeloperoxidase positive; auer rods may be present
M2—acute myeloblastic leukemia with maturation	30% or more	10% or more of cells are differentiating along myeloid lineage (promyelocytes, metamyelocytes, myelocytes, bands, neutrophils); auer rods common
M3—acute promyelocytic leukemia	The combination of promyelocytes and blasts is 30% or more	Cytoplasm usually packed with large pink, red or purple granules; auer rods common; in microgranular variant granules only present by electron microscopy
M4—acute myelo-monocytic leukemia	30% or more	Marrow 20-80% monocytic cells or greater than 5×10^9/L peripheral blood monocytes; auer rods may be present; in eosinophilic variant (M4Eo) 1-30% of marrow is abnormal eosinophils
M5—acute monocytic leukemia	M5a > 80% monocytic; monoblasts > 80% of monocytic cells	M5b > 80% monocytic cells; monoblasts < 80% of monocytic cells
M6—acute nucleated erythroleukemia	> 50% of nucleated cells are erythroid; 30% or more of nonerythroid cells are blasts	Immature erythroid cells may have PAS positive cytoplasmic vacuoles; auer rods may be present in nonerthyroid cells
M7—acute megakaryocytic leukemia	30% or more	Blasts have budding, platelet peroxidase by electron microscopy, or platelet surface glycoproteins, e.g. CD42 (glycoprotein Ib), CD41 (glycoprotein IIb)

6.2

Table 6.2.2. Correlation of French-American-British (FAB) subtype of acute myelogenous leukemia with immunophenotype

Antigen	M1	M2	M3	M4	M5	M6	M7
CD13	+++	+++	+++	+++	+++	+++	+
CD33	+++	+++	+++	+++	+++	+++	+
CD15	+++	+++	+++	+++	+++		
CD14		+		+++	+++		
CD11b		+		+++	+++		
CD36				+++	+++	+++	+
CD71						+++	
Glycophorin						+++	
CD41				+	+		+++
CD61				+	+		+++
CD34	++	++		++	++		++
HLA-DR	+++	+++		+++	+++	+++	+

+++ = Virtually all cases positive
++ = A majority but not all cases positive
+ = Occasional or rare case positive

1) **Age**
   Leukemia-free survival decreases with age independent of therapy.

2) **Karyotype**
   t(8;21) or inv(16) are associated with a greater than 90% complete remission (CR) and 60% 5 year survival following standard induction and consolidation chemotherapy. Translocation 15, 17 is almost invariably associated with acute promyelocytic leukemia (FAB M3) and has an intermediate prognosis. Poor prognostic karyotype abnormalities are abnormalities of multiple chromosomes, deletions of 5 or 7, or abnormalities of 12, 8 or 11q23, which are associated with < 40% CR and median survival of approximately 7 months.

3) **Preleukemic or Myelodysplastic Syndrome (MDS)**
   MDS is manifest clinically by peripheral blood cytopenias prior to development of AML and is a poor prognostic factor.

4) **Other**
   Several other factors of questionable significance are: sex (female better), WBC count (high worse), FAB (M3, M4EO better; M5, M6 worse), RAS mutation (present better), immunophenotype (CD13, CD14, CD34 worse; CD2, CD19 better), auer rods (present better), peripheral blast count (low better), hemoglobin (> 12 g/dl better) and platelet count (low worse).

B) **PROGNOSTIC FACTORS AT TIME OF RELAPSE**
   If a patient relapses, the most important indicator of subsequent prognosis is duration of first remission. If initial remission is under 2 months, less than 20% of patients will enter a second CR. If the initial remission is greater than 12 months, 60% of patients will enter a second remission. In general, patients with remissions less than 6 months have a poor prognosis.

C) **PROGNOSTIC FACTORS AT TIME OF BMT**
   The most important prognostic factors for AML patients treated by BMT are stage of disease (induction failure, first remission, relapse, 2nd or greater remission) and recipient age (Table 6.2.3).

V) **TREATMENT PHILOSOPHY**
   AML has an untreated survival of weeks to months and patients at presentation may have multiple complications from cytopenias (infection, bleeding) or leuke-

6.2

Table 6.2.3. Percent leukemia-free survival (LFS) and relapse (R) in AML

Disease Stage	Chemotherapy		Autologous BMT				Allogeneic BMT*	
			Unpurged		Purged			
	LFS%	R%	LFS%	R%	LFS%	R%	LFS%	R%
Failed induction	0	100					10-30	50-60
first relapse	< 10	> 80	< 10	> 80	< 10	> 80	20-30	30-40
second remission	5-20	> 60	20-30	30-65	20-50	30-70	20-30	30-40
first remission	15-45	50-80	35-50	50-60	40-60	35-50	35-65	10-35

first remission by age:	Chemotherapy LFS%	Autologous (purged and unpurged) LFS%	Allogeneic LFS%
< 20 years old	20-45	60	60-65
20-40 years old	15-35	50	45-50
40-55 years old	15-35	40	35-40

* Genotypically matched sibling transplants

mic cell burden (leucostasis, leptomeningeal disease). Therefore, induction chemotherapy is generally initiated semi-emergently. Thus, the treatment of AML has two goals, 1) induction of a CR and 2) postremission therapy to prolong remission and prevent relapse by using either further chemotherapy or BMT (allogeneic, autologous or syngeneic). Since some patients with AML may be cured by either standard chemotherapy (induction and consolidation/intensification) or BMT, the choice of therapy after CR is controversial.

Patient selection bias has also contributed to difficulty in comparing BMT to chemotherapy. While it is not possible to identify an individual patient who will relapse, the statistical probability of relapse depends on good, intermediate or bad risk factors. One should note, however, that 20% of relapses occur within 2 months, 50% within 6 months and 80% within 12 months. If BMT is delayed for several months, the result is a time censoring bias, because patients who remain in CR for several months are less likely to relapse independent of the type of postremission therapy.

VI) CHEMOTHERAPY

Induction is the same independent of FAB classification (Table 6.2.4) except for acute promyelocytic leukemia, M3 with t(15;17), which may be with either chemotherapy or all-trans retinoic acid (ATRA).

A) INDUCTION FOR ALL FAB CLASSIFICATIONS

Upon diagnosis of AML, intravenous hydration with alkalization of the urine (D_5 $1/2$NS with 2 amps $NaHCO_3$ at 200 ml/hour) and allopurinol (200-300 mg/d PO) is initiated to prevent tumor lysis syndrome. Induction chemotherapy with a standard regimen such as "7 + 3" (cytarabine 100 mg/m^2/d continuous infusion for 7 days and daunorubicin 45 mg/m^2/d

6.2

Table 6.2.4. AML regimens

Induction 5 + 2:
Cytarabine 100 mg/m^2/d continuous infusion days 1-5
Daunorubicin 45 mg/m^2/d days 1 and 2

Induction 7 + 3:
Cytarabine 100 mg/m^2/d continuous infusion days 1-7
Daunorubicin 45 mg/m^2/d days 1, 2 and 3

Induction HIDAC:
Cytarabine 1-3 g/m^2 IV over 2-3 hours q 12 hours x 12 doses days 1-6

Induction IC:
Idarubicin 12 mg/m^2/d IV days 1-3
Cytarabine 100 mg/m^2 continuous infusion days 1-7

Consolidation HIDAC:
Cytarabine 1-3 g/m^2 IV over 2-3 hours q 12 hours x 12 doses days 1-6
or
Cytarabine 1-3 g/m^2 IV over 2-3 hours q 12 hours x 6 doses days 1, 3, and 5

Salvage MC:
Mitoxantrone 12 mg/m^2/d days 1-3
Cytarabine 100-200 mg/m^2/d continuous infusion days 1-7

Salvage MV:
Mitoxantrone 10 mg/m^2/d IV days 1-5
Etoposide 100 mg/m^2/d IV days 1-5

for 3 days) results in 60-80% of patients entering complete remission (CR). A 7 day course of cytarabine is optimal since 5 days of cytarabine are inferior and 10 days are equivalent. Intermediate-dose cytarabine (500 mg/m^2) gives the same CR rate as low-dose cytarabine (100 mg/m^2). High-dose cytarabine (1-3 g/m^2) has been reported to result in a higher CR rate but also has greater toxicity and morbidity.

Daunorubicin may be infused on either the first or last 3 days of cytarabine. Daunorubicin is preferred to doxorubicin due to less mucositis. However, in the "7+3" regimen, other equally effective anthracyclines such as idarubicin (12 mg/m^2/d for 3 days) or mitoxantrone (12 mg/m^2/d for 3 days) may substitute for daunorubicin. Mitoxantrone may be less cardiotoxic and possibly preferable in older patients. The acridine dye amsacrine (100-200 mg/m^2/d for 5 days) has also been substituted for daunorubicin with equal results.

The addition of an epidophyllotoxin (e.g. etoposide) to cytarabine and daunorubicin may improve remission duration. Etoposide is given as 75 mg/m^2/day on days 1-7 of the standard "7 + 3" induction regimen.

One or two courses of induction therapy may be necessary for complete remission. In general, if the first course of induction therapy has not emptied the marrow of blasts by day 14, the induction course with the same drugs and the same dose is repeated. An exception is M3 in which the presence of blasts in the marrow by day 10-14 is not necessarily an indication of induction failure.

A complete remission must be greater than 30 days duration and is defined as less than 5% blasts in the marrow, no circulating blasts in the peripheral blood and no extramedullary disease.

B) INDUCTION FOR ACUTE PROMYELOCYTIC LEUKEMIA, FAB-M3

For patients with M3, induction may be with either chemotherapy such as "7+3" or all-trans retinoic acid (ATRA) at a dose of 50 mg/m^2/d PO divided twice a day for 90 days or until CR. In contrast to chemotherapy for M3, ATRA does not trigger DIC and bleeding. However, ATRA may transiently increase peripheral blasts and promyelocytes, causing organ damage from leucostasis (interstitial infiltrate, dyspnea, stroke, myocardial infarction or renal failure). If the combined blast and promyelocyte count rises to greater than 10,000/µl, some institutions use single agent chemotherapy (e.g. idarubicin, mitoxantrone or etoposide) to temporarily control counts and prevent leucostasis.

VII) ALLOGENEIC BMT

Relapse after allogeneic BMT is lower than after standard consolidation chemotherapy or autologous BMT. However, regimen toxicity and morbidity is greater with allogeneic BMT than either chemotherapy or autologous BMT. Allogeneic HLA-matched sibling BMT has a 20-30% early (within the 1st 100 days) mortality predominantly due to conditioning regimen related toxicity, infection and GVHD. Toxicity, including severity of GVHD, increases with recipient age. Therefore, allogeneic BMT is generally limited to patients less than 55 years old.

A) TIMING OF ALLOGENEIC BMT

Allogeneic BMT is usually done after obtaining a complete remission. However, it is controversial whether allogeneic BMT should be done after induction or following induction and 1-2 cycles of consolidation.

B) CONDITIONING

Conditioning regimens are described in the chapter on preparative regimens. Prior treatments (e.g. radiation), toxicities and institutional preference deter-

6.2

mine the regimen. Few randomized trials are available, and in nonrandomized trials, leukemia-free survival is remarkably similar. Common conditioning therapies are: cyclophosphamide and TBI; or etoposide and TBI; or cytarabine and TBI; or etoposide and cyclophosphamide and TBI; or busulfan and cyclophosphamide (refer to chapter on conditioning regimens).

C) GVHD PROPHYLAXIS

GVHD prophylaxis is discussed in the chapter on GVHD. For AML, GVHD pharmacology prophylaxis regimens include cyclosporine or methotrexate combined with cyclosporine or triple immunosuppressive therapy with cyclosporine, methotrexate and prednisone. They are described in detail in the chapter on GVHD.

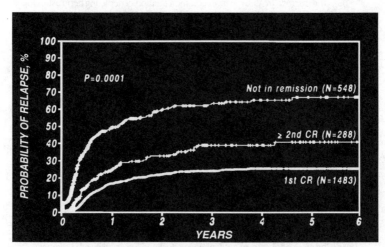

Fig. 6.2.1. *Probabliity of relapse after HLA-identical sibling transplants for acute myelogenous leukemia, 1987-1994. Reprinted with permission from IBMTR/ABMTR 1995; 2:6.*

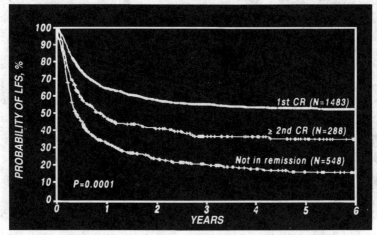

Fig. 6.2.2. *Probablilty of leukemia-free survival after HLA-identical sibling transplants for acute myelogenous leukemia, 1987-1994.Reprinted with permission from IBMTR/ABMTR 1995; 2:7.*

6.2

VIII) AUTOLOGOUS BMT

Compared to allogeneic, autologous BMT has a lower regimen-related mortality (under 10-15%) due to absence of GVHD. However, relapse rates are higher (Table 6.2.3). Conditioning regimen dose intensity is limited by extramedullary toxicity. Decreasing relapse after autologous BMT may depend on purging the marrow of neoplastic cells and/or posttransplant immune modulation.

A) PURGING

The role of purging remains controversial. Gene marker studies have documented that relapse can occur from leukemia cells reinfused in the graft. However, it is debatable whether most relapses arise from residual disease remaining in the body following myeloablation or from leukemic cells re-infused with the marrow. In AML the most common method of depleting neoplastic clones has been with ex vivo chemical purging using agents such as mafosphamide or 4-hydroperoxycyclophosphamide (4-HC). A side effect of chemical purging seen predominantly in AML is prolonged engraftment from damage to hematopoietic progenitor cells. Despite these caveats, nonrandomized studies suggest that relapse is lower after purging. Purging appears most beneficial in patients who undergo BMT within less than 6 months of entering remission or patients who required more than 40 days to achieve a complete clinical remission.

B) IMMUNE MODULATION

IL-2 with or without LAK cells has been used as immunotherapy beginning 20-90 days after autologous BMT. LAK cells caused significant toxicity. IL-2 alone had toxicities of capillary leak with pulmonary edema, fever, nausea, diarrhea and rash. The optimal timing or dose of IL-2 is unknown. It is too early to determine if IL-2 or other cytokines after autologous BMT will decrease relapse. After autologous BMT, cyclosporine A is capable of inducing an autoimmune-like disease with rash similar to mild GVHD. However, whether this will be associated with an antileukemia effect and lower relapse is unknown.

6.2

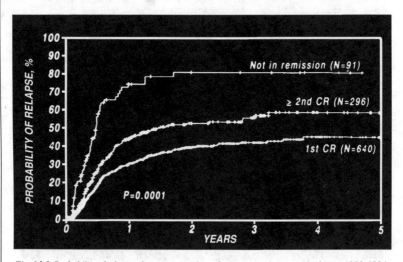

Fig. 6.2.3. Probability of relapse after autotransplants for acute myelogenous leukemia 1989-1994. Reprinted with permission from IBMTR/ABMTR 1995; 2:6.

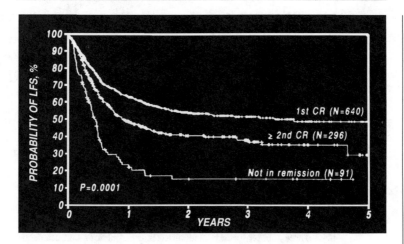

Fig. 6.2.4. Probability of leukemia-free survival after auto-transplants for acute myelogenous leukemia, 1989-1994. Reprinted with permission from IBMTR/ABMTR 1995; 2:6.

C) TIMING OF AUTOLOGOUS BMT HARVEST

Autologous marrow is harvested when the marrow blast count is less than 5%, the peripheral platelet count has recovered to greater than 100,000/µl, and ANC greater than 1500/µl. At some institutions autologous marrow is harvested at time of first remission and not used until leukemia recurs. At other institutions autologous marrow is harvested, processed and reinfused in the same remission.

D) TIMING OF AUTOLOGOUS MARROW REINFUSION

Autologous BMT may be done after induction or after 1-2 cycles of consolidation. Autologous BMT done within 6 months of diagnosis results in lower leukemia-free survival than after 6 months. This may be due to time censoring bias, since more aggressive leukemias relapse early. The median time from diagnosis to autologous BMT in AML is 5 months, suggesting that most patients receive 1-2 cycles of consolidation. However, another approach is induction and then autologous BMT in place of consolidation with chemotherapy.

E) CONDITIONING REGIMENS

The same TBI-containing regimens used in allogeneic BMT of AML may be used for autologous BMT. Since immunosuppression is not necessary for engraftment of autologous marrow, non-TBI regimens are more common with autologous BMT. Examples of such regimens include busulfan and cyclophosphamide (BU/CY); or cyclophosphamide and carmustine and etoposide (CBV); or carmustine and amsacrine and etoposide and cytarabine (BAVC) (refer to chapter on conditioning regimens for dose and schedule).

IX) SYNGENEIC BMT

If an identical (monozygotic) twin is available, syngeneic BMT in first remission may be done with a 3 year leukemia-free survival and relapse of approximately 42% and 52% respectively. Syngeneic BMT has a higher relapse rate than HLA identical sibling BMT. However, due to absence of GVHD and less treatment related mortality, survival after syngeneic BMT for AML in first CR is comparable to HLA identical sibling BMT.

X) UNRELATED OR MISMATCHED BMT

The toxicity of unrelated or HLA-mismatched BMT is too high for patients who may be curable by less morbid therapy. However, in advanced leukemias (relapse, second remission or later), unrelated BMT gives a leukemia-free survival of approximately 30% for patients transplanted in remission and 12% for patients transplanted in relapse. Unrelated BMT, due to complications of chronic GVHD, may diminish the quality of life for roughly 40% of surviving patients.

XI) TREATMENT OF AML BY DISEASE STAGE

A) FIRST REMISSION

Most patients (60-80%) achieve first remission with induction therapy. Further therapy is necessary or almost all patients (> 90%) will relapse in a median of 4 months. Even promyelocytic leukemia (M3) in CR after all-trans retinoic acid will relapse in almost all cases unless postremission chemotherapy is given. There is no consensus on postremission therapy. Options include 2-3 cycles of consolidation/intensification chemotherapy or BMT (autologous or allogeneic). Three arm, prospective, randomized British, American, European and Italian trials (UKMRC, ECOG, EORTC and GIMEMA, respectively) comparing chemotherapy versus autologous versus allogeneic BMT during first remission for AML are in progress. The analysis of leukemia-free survival, relapse and overall survival from these trials may help to better define therapy for AML in first remission.

1) Chemotherapy in First Remission

Postremission chemotherapy in which the same drugs used in induction are repeated is called consolidation. Intensification is the use of drugs that are non-cross resistant to the induction drugs. The optimal number of intense postremission chemotherapy cycles is 2 to 3. In AML, unlike ALL, there is no role for maintenance chemotherapy. Postremission therapy varies by institution. An example of postremission therapy is one cycle of high-dose cytarabine($3 g/m^2$ q12 hours for 8 doses) and daunorubicin ($45 mg/m^2/d$ for 3 days) followed in 1 month by a second cycle of either low-dose cytarabine ($100-200 mg/m^2/d$ for 7 days) and daunorubicin ($45 mg/m^2/d$ for 3 days), or etoposide ($100 mg/m^2/d$ for 5 days) and mitoxantrone ($10 mg/m^2/d$ for 3-5 days).

2) Allogeneic BMT in First Remission

Although not always significant, all trials comparing postremission chemotherapy to allogeneic BMT have demonstrated superior leukemia-free survival for allogeneic BMT. However, with increasing age (> 45), overall survival between chemotherapy and allogeneic BMT becomes comparable. Patients with t(8,21) or inversion 16 karyotype have a good prognosis with chemotherapy (> 60% 5 year leukemia-free survival). Thus, in patients who are older than 45 or have t(8,21) or inversion 16 karyotype, it is reasonable to reserve allogeneic BMT until relapse or second remission. Otherwise, if an HLA-matched sibling is available, allogeneic BMT is generally regarded as the treatment of choice for patients in first remission. Leukemia-free survival after allogeneic BMT is age dependent (Table 6.2.3).

3) Autologous BMT in First Remission

In comparison to chemotherapy, autologous BMT results in at least similar and perhaps 10-15% higher leukemia-free survival (Table 6.2.3). Indeed,

6.2

in nonrandomized studies, the results of purged autologous BMT appear to rival the results from allogeneic BMT (Table 6.2.3).

B) FAILED INDUCTION

If a patient fails to obtain complete remission with 2 cycles of induction therapy, by definition the leukemia is drug resistant. If induction caused a partial remission(marrow blasts 5-20% or < 5% for less than 30 days), a CR may occur with high dose cytarabine (2-3 g/m² q12 hours for 4-12 doses) alone or in combination with an anthracycline (e.g. daunarubicin 45-60 mg/m²/d for 3 days). The combination of etoposide (100 mg/m²/d for 5 days) and mitoxantrone (10 mg/m²/d for 5 days) has been reported to give a 60% CR in patients who achieved only a partial remission (PR) with prior induction therapy. If induction therapy did not even succeed in inducing a partial remission, the only possibility of cure is allogeneic BMT which results in a 10-20% leukemia-free survival (Table 6.2.3).

C) RELAPSE OR SECOND OR GREATER REMISSION

Autologous BMT or allogeneic BMT from a genotypically matched or mismatched sibling or phenotypically unrelated donor is the treatment of choice.

1) Chemotherapy in Relapse or Second Remission

Chemotherapy induces a second remission in 60% of patients but second or greater remissions are generally brief and of shorter duration than the previous remission. Leukemia-free survival from chemotherapy in second remission is less than 20%.

2) Allogeneic BMT in Relapse or Second Remission

Results from the Fred Hutchinson Cancer Research Center indicate that leukemia-free survival for AML is similar for allogeneic BMT in untreated relapse (28%) and second remission (30%). This supports doing allogeneic BMT in untreated relapse instead of attempting a second CR prior to BMT. However, this remains controversial, and some institutions perform allogeneic BMT in untreated early (< 30% blasts in the marrow) relapse but induce a second remission prior to allogeneic BMT for fulminant (> 30% blasts) relapse.

3) Autologous BMT in Relapse or Second Remission

In untreated relapse, autologous BMT using marrow purged and stored in first remission has been reported to give poor results (> 90% relapse). If a second remission can be obtained with chemotherapy, purged autologous BMT gives a leukemia-free survival of 20-50% and median remission of greater than 400 days. At some institutions leukemia-free survival after autologous purged BMT is similar in first and second remission (Table 6.2.3). Even if cure is not achieved, autologous BMT is capable, in the majority of patients, of causing an inversion. This term denotes a remission which is longer in duration than the previous remission.

XII) CNS TREATMENT

A) LUMBAR PUNCTURE (LP)

A diagnostic LP is not routinely recommended for patients with AML regardless of treatment with chemotherapy or BMT. An LP is recommended for headache, neurologic signs or asymptomatic patients with monoblast (M4, M5) morphology. The platelet count should be greater than 50,000/μl at the time of LP.

6.2

B) CNS PROPHYLAXIS

CNS prophylaxis is not recommended for AML patients regardless of treatment with chemotherapy or BMT. After allogeneic BMT with a TBI conditioning regimen, CNS relapse is 2% whether or not there is a history of prior CNS leukemia. Prophylactic intrathecal methotrexate after allogeneic BMT with TBI does not change the rate of CNS relapse for AML.

C) CNS DISEASE

Treatment for AML involving the CNS is the same as for ALL and is discussed in the chapter on ALL.

SUGGESTED READING

1. Foon KA, Gale RP. Therapy of acute myelogenous leukemia. Blood Reviews 1992; 6:15-25.
2. Gorin NC et al. Autologous bone marrow transplantation for acute myelogenous leukemia in Europe: Further evidence of the role of marrow purging by mafosfamide. Leukemia 1991; 5(10):896-904.
3. McCauley DL. Therapy Review, Treatment of adult acute leukemia. Clinical Pharmacology 1992; 11:767-796.
4. Santos GW. Marrow transplantation in acute nonlymphocytic leukemia. Blood 1989; 74:901-908.
5. Zittoun RA et al. Autologous or allogeneic bone marrow transplantation compared with intensive chemotherapy in acute myelogenous leukemia. NEJM 1995; 332:217-223.

6.2

Adult Acute Lymphoblastic Leukemia (age >15 years old)

Richard K. Burt, Philip Rowlings, George W. Santos

I) ETIOLOGY

For most patients the cause of ALL is unknown. Risk factors for ALL are similar to AML and incidence is increased by congenital or environmental conditions that predispose to DNA damage. ALL is predominately a disease of childhood with a peak incidence at age 5-6. In adults, incidence increases with age and is highest over age 60. ALL accounts for 15-20% of adult acute leukemias. Like all leukemias, ALL shows little geographic variation.

II) DIAGNOSIS

ALL similar to AML presents abruptly with the patient complaining of pallor, fatigue, petechiae, bleeding or fever. Bone pain and arthalgias are more common with ALL than AML and may be the initial symptoms. Similar to AML, cytopenias (thrombocytopenia, anemia, granulocytopenia) are common. In contrast to AML, ALL more commonly presents with extramedullary involvement: lymphadenopathy, splenomegaly, hepatomegaly, CNS disease (meningeal lymphomatosis) and testicular enlargement due to infiltration with lymphoblasts. A diagnosis of acute leukemia depends on greater than 30% blasts in the marrow. In ALL, the blasts are of lymphoid lineage. Distinguishing lymphoid from myeloid blasts can usually be done with morphology, cytochemical stains and immunophenotyping (Table 6.3.1). Morphologic classification of ALL is based on the FAB system (Table 6.3.2). The original FAB classification was based on morphology and not designed to have prognostic significance. However, FAB L1 has a better prognosis than FAB L2. FAB L3 has a distinctive cell type (Burkitt's cells) and has an invariable immunophenotype of mature B cells with surface membrane immunoglobulin. FAB L3 has the worst prognosis of the three subtypes. More recent classifications have attempted to incorporate immunophenotype and cytogenetic data.

A) DIFFENTIAL DIAGNOSIS

ALL may be confused with osteomyelitis or juvenile rheumatoid arthritis due to presentation with bone pain or arthalgias. Thrombocytopenia may be confused for ITP unless a bone marrow is performed. Due to pancytopenia, ALL may be confused with aplastic anemia. An empty marrow in aplastic anemia in comparison to a marrow packed with blasts in ALL prevents misdiagnosis. However, on occasion in children, ALL may be present with pancytopenia and an "empty" or hypocellular bone marrow. In this case, with time, increased lymphoblasts will occur in the marrow. Due to similar presentation, ALL has also been confused with neuroblastoma and nonmalignant causes of lymphocytosis such as mononucleosis.

III) CHILDHOOD VERSUS ADULT ALL

For the purpose of this chapter, adult ALL is defined as ALL in any patient 15 years of age or older. Childhood ALL is discussed under BMT for childhood neoplasms. ALL is separated into adult and childhood forms for many reasons (Table 6.3.3).

6.3

Bone Marrow Transplantation, edited by Richard K. Burt, H. Joachim Deeg, Scott Thomas Lothian, George W. Santos. © 1996 R.G. Landes Company.

Table 6.3.1. Distinguishing features of lymphoid (ALL) and myeloid (AML) blasts

Feature	AML	ALL
Morphology		
nuclear/cytoplasmic ratio	low	high
nucleoli	2-5	0-2
chromatin	open	clumped
granules	+	+/–
auer rods	+/–	–
Cytochemical		
myeloperoxidase	+	–
sudan black	+	–
esterases	+/–	–
periodic acid schiff	–	+/–
TdT	–	+/–
Immunophenotype		
myeloid		
CD11, CD13, CD14, CD15, CD33	+/–	–
lymphoid- T cell		
CD1, CD2, CD3, CD4, CD5, CD7, CD8	–	+/–
lymphoid-B cell		
CD19, CD20, CD21, CD24,	–	+/–
lymphoid-CALLA		
CD10	–	+/–

6.3

Table 6.3.2. FAB histology of ALL

Feature	L1	L2	L3
Size	Small	Large, heterogeneous	Large, homogeneous
Nuclear Shape	Smooth, occasional indentation	Irregular, indentation common	Smooth
Nucleoli	Absent, small	1 or more, large	1 or more, prominent
Cytoplasm	Scanty	Variable	Abundant
Basophilia	Slight	Variable	Prominent
Vacuoles	Variable	Variable	Prominent

Scoring System to Differentiate L1 and L2

Criteria	Score
High nuclear/cytoplasmic ratio in > 75% of cells	+1
Low nuclear/cytoplasmic ratio in > 25% of cells	–1
0 to 1 small nucleoli in > 75% of cells	+1
1 or more prominent nucleoli in > 25% of cells	–1
Irregular nucleus in > 25% of cells	–1
Large cells > 50% of cells	–1

When 200 cells are counted a score 0 to +2 = L1; A score of –1 to–4 = L2

Table 6.3.3. Childhood Versus Adult ALL

Feature	Childhood ALL	Adult ALL
Response to chemotherapy		
CR	> 90%	70-85%
LFS	50-75%	25-35%
Immunophenotype		
B cell[1]	1-2%	2-7%
T cell[2]	10-15%	9-25%
common (CALLA)[3]	75%	50%
null[4]	5-10%	10-30%
Cytogenetics		
hyperdiploid	30%	< 1%
Philadelphia chromosome, t(9,22) or bcr/abl positive by PCR	5%	20-40%
WBC at diagnosis 10^6/µl		
< 10	50-55%	40-45%
10-50	25-35%	25-35%
> 50	10-20%	20-30%
FAB classification		
L1	85%	30%
L2	14%	60%
L3	1%	5-10%

[1] B cell (mature B cell) is defined as presence of surface immunoglobulin
[2] T cell is defined by presence of T cell receptor (CD3) on surface membrane
[3] Common ALL antigen (CALLA) is defined by surface antigen CD10 which is present on most immature cells of B lymphocyte lineage.
[4] Null cell is a lymphoblast negative for CALLA, surface immunoglobulin or T cell receptor, but with B lineage markers such as CD19 or CD20
CR= complete remission, LFS= leukemia-free survival

6.3

IV) PRETREATMENT EVALUATION

For diagnosis, a bone marrow biopsy and aspirate with cytogenetic analysis is mandatory. Routine prechemotherapy evaluation includes: CBC, electrolytes, liver function test (bilirubin, alkaline phosphatase, transaminases), creatinine, chest x-ray, uric acid, prothrombin time, partial thromboplastin time, lactate dehydrogenase, resting radionucleotide angiography to noninvasively assess cardiac ejection fraction and HIV serology. A lumbar puncture with cytology should be done to rule out CNS disease. If under age 55, the patient, siblings and parents are HLA typed. Refer to the chapter on BMT patient evaluation for pre-BMT work up.

V) PROGNOSIS

Adult ALL is generally thought of as having a worse prognosis than childhood ALL. However, 20-30% of adult patients are without adverse prognostic features and have a prognosis similar to childhood ALL.

A) PROGNOSIS AT PRESENTATION

Prognosis in adult ALL is defined by age, WBC count on presentation, time required to achieve complete remission (CR), immunophenotype and cytogenetics. Good prognosis adult ALL has a leukemia-free survival with chemotherapy of 50-60%. In general, bad prognosis adult ALL has a leukemia-free survival with chemotherapy of 20-40%. Philadelphia (Ph) chromosome positive (translocation 9, 22) ALL has the worst prognosis with leukemia-free survival after chemotherapy of 0-15%.

1) Age

There is an incremental decrease in remission duration and survival with increased age. Poor prognosis cut off is between 25-35 years old. The Ulm, Frankfurt, Munich (UFM) ALL group has reported that for patients under age 35, the median duration of chemotherapy remission is 31 months and the 5 year leukemia-free survival is 43%. Over age 35, the median duration of chemotherapy remission is 15 months and the 5 year leukemia-free survival is 21%.

2) WBC

An incremental decrease in remission duration and survival occurs with increasing WBC count on presentation. Poor prognosis cut off is between 25,000-35,000/μl. The German UFM ALL group has reported that patients with leukocyte counts under 30,000/μl have a 32 month chemotherapy remission duration and 5 year leukemia-free survival of 42%. Patients with a leukocyte count over 30,000/μl have a chemotherapy remission duration of 15 months and 5 year leukemia-free survival of 32%.

3) Time to Response

Duration of remission and survival decrease when greater than 4-8 weeks are required to achieve complete remission. According to the German UFM group, if complete remission can be obtained within 4 weeks, median duration of remission is 31 months. If greater than 4 weeks are necessary to obtain CR, median duration of remission is 12 months.

4) Immunophenotype

B cells are defined as having surface immunoglobulin. T cells are defined as having on their surface membranes components of the T cell receptor (e.g. CD3). Originally T cells were defined as cells that formed rosettes with sheep erythrocytes (E-rosette positive). The majority of lymphoblasts from patients with ALL do not express immunoglobulin on their surface nor form E-rosettes, but express the common ALL antigen (CALLA or CD10), and were subsequently termed CALLA positive cells. Almost all CALLA positive cells are of early B cell lineage. B cell lineage is divided into:

    a)   pro-B cell (germ cell immunoglobulin genes)
    b)   pre-pre-B cell (rearranged immunoglobulin genes)
    c)   pre-B cell (cytoplasmic immunoglobulin present)
    d)   mature B cell (surface immunoglobulin present)
    e)   plasma cell (no surface immunoglobulin)

In adults, T cell ALL has the best prognosis while B cell ALL (L3) has the worst. Null cell and common ALL (CALLA positive) have an intermediate prognosis.

5) Cytogenetics

Ph chromosome positive or BCR-ABL polymerase chain positive ALL has an especially poor prognosis with the lowest CR rates (< 65%) and poorest chemotherapy leukemia-free survival (0-15%).

6) Other

Other factors inconsistently reported to affect ability to achieve or maintain complete remission are: CNS leukemia (present is worse), sex (female better), hepatosplenomegaly (present is worse), and β_2-microglobulin (> 4.0 mg/L worse).

B) PROGNOSIS AT RELAPSE

Patient age and the interval between remission and relapse are the most important prognostic factors predictive of subsequent treatment outcome, independent of therapy. Older age and patients who relapse early (< 12-18 months after CR) or relapse while on therapy have a worse prognosis.

VI) TREATMENT PHILOSOPHY

Treatment of adult ALL was modeled after successful therapy of childhood ALL, which involves induction to achieve a remission, various schedules of consolidation, prolonged maintenance therapy and CNS prophylaxis. The total duration of therapy is usually 2-3 years. Remission occurs in 60-85% of adult ALL patients treated with induction chemotherapy that includes vincristine, prednisone and an anthracycline. After obtaining a CR, further treatment is mandatory. However, the choice of treatment (chemotherapy versus allogeneic or autologous BMT) is controversial. Randomized trials of chemotherapy versus allogeneic or autologous BMT have generally combined children with adults and good with poor prognosis patients. Due to the rarity of ALL in adults, few if any, institutions have enough patients for randomized trials properly analyzed according to risk factors.

VII) CHEMOTHERAPY

There is no standard ALL regimen. Several are found in the literature. In order to illustrate the principles of therapy, four different ALL chemotherapy regimens are listed in Tables 6.3.4a, 6.3.4b, 6.3.4c and 6.3.4d.

A) INDUCTION

In general, the highest rates of remission are obtained with a combination of vincristine, prednisone and an anthracycline with or without L-asparaginase. Although L-asparaginase does not increase the CR rate, it is believed to prolong remission duration. For patients who cannot tolerate an anthracycline (e.g. low ejection fraction), the MOAD regimen with a CR rate of 76% is an alternative (Table 6.3.4c).

B) CONSOLIDATION

In general, consolidation involves use cyclophosphamide, cytarabine and methotrexate with repetition of some or all of the induction drugs.

C) MAINTENANCE

In childhood ALL, maintenance therapy will prolong the remission and improve leukemia-free survival. Despite its frequent use, it is controversial and unknown if maintenance therapy is necessary for all or some subset of adult ALL. Short course intense adult ALL regimens such as the VAAP protocol (Table 6.3.4d) do not use maintenance therapy.

D) CNS PROPHYLAXIS

Prophylaxis of the CNS to prevent relapse is standard in ALL. Various methods exist including intrathecal (IT) drugs, CNS irradiation or high-dose intravenous methotrexate or cytarabine. The only form of CNS prophylaxis in the MOAD protocol is high-dose intravenous methotrexate.

VIII) ALLOGENEIC BMT

At any stage of disease, allogeneic BMT from a matched sibling results in a lower risk of relapse than chemotherapy or autologous BMT. However, allogeneic BMT has the highest early mortality (20-30% in the first 100 days) due to the conditioning regimen and acute GVHD, and the highest late morbidity predominately due to chronic GVHD. Therefore, the optimal timing of allogeneic BMT depends on patient

6.3

age, disease stage and risk of relapse for different subgroups of ALL patients (see treatment by disease stage). Due predominately to the increased risk of GVHD, allogeneic BMT is restricted to patients under age 55.

A) TIMING OF ALLOGENEIC BMT

Allogeneic BMT is usually done after obtaining a CR with induction therapy. It is controversial whether allogenous BMT is done after induction or induction followed by consolidation. CNS prophylaxis is not given until the time of BMT and generally combines preparative regimen TBI and IT methotrexate (see section in this chapter on CNS disease).

B) GVHD

Prevention of GVHD is similar for all types of leukemia and is discussed in the chapter on GVHD. Methotrexate and corticosteroids which are used to prevent GVHD also decrease relapse after BMT for ALL. These agents decrease relapse probably because they are cytotoxic to lymphoblasts. Cyclosporine alone or T cell depleted marrow increases the relapse rate. Commonly used GVHD prophylactic regimens are methotrexate combined with cyclosporine, or metho-

6.3

Table 6.3.4a. Hoelzer adult ALL regimen (Blood 1988; 71 (1):123-131)

Drug	Dose	Route/Day
Induction		
Phase 1 (4 weeks)		
prednisone	60 mg/m²	PO day 1-28
vincristine	1.5 mg/m² (max 2.0 mg)	IV day 1, 8, 15, 22
daunorubicin	25 mg/m²	IV day 1, 8, 15, 22
L-asparaginase	5000 U/m²	IV day 1-14
Phase 2 (4 weeks)		
cyclophosphamide	650 mg/m²	IV day 29, 43, 57
cytarabine	75 mg/m²	IV day 31-34, 38-41, 45-48, 52-55
6-mercaptopurine	60 mg/m²	PO day 29-57
methotrexate	10 mg/m² (max 15 mg)	IT day 31, 38 45, 52
1st maintenance		
(8 weeks)		
6-mercaptopurine	60 mg/m²	PO daily weeks 10-18
methotrexate	20 mg/m²	PO or IV weekly for weeks 10-18
Consolidation		
(begins week 20)		
Phase 1 (4 weeks)		
dexamethasone	10 mg/m²	PO day 1-28
vincristine	1.5 mg/m² (max 2.0 mg)	IV day 1, 8, 15, 22
doxorubicin	25 mg/m²	IV day 1, 8, 15, 22
Phase 2 (2 weeks)		
cyclophosphamide	650 mg/m² (max 1000 mg)	IV day 29
cytarabine	75 mg/m²	IV day 31-34, 38-41
thioguanine	60 mg/m²	PO day 29-42
2nd maintenance		
(2 years)		
6-mercaptopurine	60 mg/m²	PO daily weeks 29-130
methotrexate	20 mg/m²	PO or IV weekly for weeks 29-130

CNS prophylaxis—If CR obtained in phase I, CNS prophylaxis starts during phase II. If CR not obtained by completion of phase II, prophylaxis begun when CR obtained. Prophylaxis consists of cranial radiation (2400 cGy) and IT methotrexate (10 mg/m²) once a week for 4 weeks.

trexate combined with cyclosporine and corticosteroids, or cyclosporine combined with corticosteroids (doses and schedule in chapter on GVHD).

C) CONDITIONING

Conditioning regimens for ALL generally use TBI since lymphocytes are relatively radiation sensitive and TBI provides CNS prophylaxis. Common regimens

Table 6.3.4b. Linker adult ALL regimen (Blood 1991; 7(11):2814-2822)

Drug	Dose	Route/ Day
Induction		
daunorubicin	50 mg/m²	IV day 1, 2, 3
vincristine	2 mg	IV day 1, 8, 15, 22
prednisone	60 mg/m²	PO days 1-28
L-asparaginase	6000 U/m²	IM days 17-28
If day 14 marrow > 5% blasts		
daunorubicin	50 mg/m²	IV day 15
If day 28 marrow > 5% blasts		
daunorubicin	50 mg/m²	IV day 29, 30
vincristine	2 mg	IV day 29, 36
prednisone	60 mg/m²	PO days 29-42
L-asparaginase	6000 U/m²	IM days 29-35
Consolidation		
Treatment A (cycles 1, 3, 5 and 7 alternate monthly with treatment B)		
daunorubicin	50 mg/m²	IV day 1, 2
vincristine	2 mg	IV day 1, 8
prednisone	60 mg/m²	PO days 1-14
L-asparaginase	12,000 U/m²	IM days 2, 4, 7, 9, 11, 14
Treatment B (cycles 2, 4, 6 and 8 alternate monthly with treatment A)		
teniposide	165 mg/m²	IV day 1, 4, 8, 11
cytarabine	300 mg/m²	IV day 1, 4, 8, 11
Treatment C (cycle 9 begins 1 month after starting cycle 8)		
methotrexate	690 mg/m²	IV over 42 hours
leucovorin	15 mg/m²	IV every 6 hours x 12 doses beginning at 42 hours
Maintenance (begins after cycle 9 and ends after 30 months of continuous CR)		
methotrexate	20 mg/m²	PO weekly
6-mercaptopurine	75 mg/m²	PO daily

CNS prophylaxis begun within 1 week of CR and consists or 1800 cGy (in 10 fractions over 12-14 days) cranial radiation and concurrent IT methotrexate 12 mg every week for 6 weeks. If CNS disease present at diagnosis, cranial radiation increased to 2800 cGy and IT methotrexate started with induction and given every week for 10 weeks then every month for 1 year.

6.3

Table 6.3.4c. Adult ALL MOAD regimen (Leukemia 1993; 7(8):1236-1241)

Drug	Dose	Route / Day
Induction—6 continuous 10 day cycles		
methotrexate	100 mg/m^2–escalated by 50% each cycle to 225 mg/m^2, then by 25% to minimal toxicity	IV day 1
vincristine	2 mg/m^2	IV day 2
L-asparaginase	500 U/kg	IV or IM day 2
dexamethasone	6 mg/m^2	PO day 1-10
Consolidation—6 cycles repeated every 10 days		
methotrexate	100 mg/m^2- escalated by 50% each cycle to 225 mg/m^2, then by 25% to minimal toxicity	IV day 1
L-asparaginase	500 U/kg	IV or IM day 2
Cytoreduction—12 monthly cycles		
vincristine	2 mg	IV day 1
dexamethasone	6 mg/m^2	PO day 2-6
methotrexate	100 mg/kg—increase by 25% each cycle to minimal toxicity	IV infusion over 6 hours day 1
leukovorin	5% of methotrexate dose divided over 3 days	IV infusion begin 2 hours after end of methotrexate
Maintenance—monthly for a total of 3 years of therapy		
vincristine	2 mg	IV day 1
dexamethasone	6 mg/m^2	PO day 1-5
methatrexate	15 mg/m^2	PO weekly
6-mercaptopurine	100 mg/m^2	PO daily

CNS prophylaxis—no radiation, no IT drugs, only high-dose IV methotrexate as described in protocol

6.3

Table 6.3.4d. VAAP adult ALL regimen—short intensive treatment (Eur J Haematol 1988; 41: 489-495)

Drug	Dose	Route/Day
Induction		
vincristine	1.4 mg/m^2	IV day 1 and 7
cytarabine	1000 mg/m^2	2 hour infusion q12 hour day 1-6
doxorubicin	25 mg/m^2	IV day 1 and 7
prednisone	60 mg/m^2	PO day 1-7
methotrexate	15 mg	IT day 1
Consolidation—1 or 2 cycles—6 week interval between induction and each consolidation		
vincristine	1.4 mg/m^2	IV day 1 and 5
doxorubicin	25 mg/m^2	IV day 1 and 5
cytarabine	3000 mg/m^2	2 hour infusion q12 hour day 1-4
prednisone	60 mg/m^2	PO day 1-5
methotrexate	15 mg	IT day 1

CNS prophylaxis-2400 cGy cranial irradiation, beginning after consolidation, IT methotrexate on day 1 of induction and consolidation, high-dose cytarabine in consolidation

are: cyclophosphamide and TBI, etoposide and TBI cytarabine and TBI, melphalan and TBI, or etoposide and cyclophosphomide and TBI. The combination of busulfan and cyclophosphamide has also been used and may be as effective as TBI regimens. Conditioning agent dose and schedule are found in the chapter on conditioning regimens.

IX) AUTOLOGOUS BMT

The role of autologous BMT in ALL remains controversial. Morbidity of autologous BMT is low (< 5-10%). However, relapse after autologous BMT is high (> 50%). The future of autologous BMT in adult ALL hinges on decreasing relapse.

A) TIMING OF AUTOLOGOUS BMT

Marrow is usually harvested in the same remission as BMT. Therefore, the patient must be in complete remission at the time of harvest. Some centers

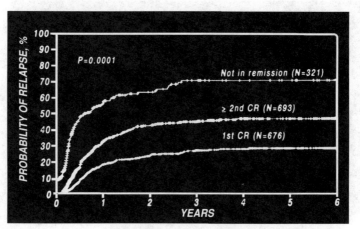

Fig. 6.3.1. Probability of relapse after HLA-identical sibling transplants for acute lymphoblastic leukemia, 1987-1994. Reprinted with permission from IBMTR/ABMTR 1995; 2:5.

6.3

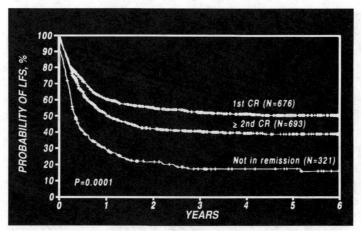

Fig. 6.3.2. Probability of survival leukemia-free after HLA-identical sibling transplants for acute lymphoblastic leukemia, 1987-1994. Reprinted with permission from IBMTR/ABMTR 1995; 2:7.

have done autologous BMT following months or even years of induction, consolidation, intensification and maintenance therapy. This reserves BMT for patients already likely to be cured since more aggressive disease will relapse early (time bias censoring). Most centers do autologous BMT after induction and 1-2 cycles of intense consolidation.

B) CONDITIONING

Autologous BMT ALL protocols generally use the same preparative regimens as allogeneic BMT (see above).

C) PREVENTION OF RELAPSE

Prevention of relapse postautologous BMT has focused on purging, maintenance chemotherapy and immune modulation.

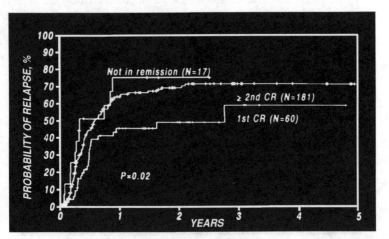

Fig. 6.3.3. Probability of relapse after autotransplants for acute lymphoblastic leukemia, 1989-1994. Reprinted with permission from IBMTR/ABMTR 1995; 2:5.

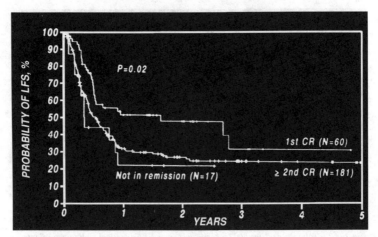

Fig. 6.3.4. Probability of leukemia-free survival after autotransplants for acute lymphoblastic leukemia, 1989-1994. Reprinted with permission from IBMTR/ABMTR 1995; 2:5.

1) **Purging**

Unlike AML where most purging is done with chemicals, most experience with purging in ALL has been with antibodies to lymphocyte antigens. In ALL, as in AML, it remains unclear if purging decreases relapse.

2) **Maintenance**

In childhood ALL, maintenance chemotherapy prolongs remission and decreases relapse. Maintenance chemotherapy is a standard part of most adult ALL protocols even though its value in adults is unproved. After autologous BMT, maintenance with oral chemotherapy (6-mercaptopurine and methotrexate) has been reported to decrease risk of relapse. However, a formal trial needs to be done for confirmation.

3) **Immune Modulation**

Attempts to decrease relapse with cytotoxic lymphocytes or cytokines such as IL-2 are ongoing. It is unknown if this approach will be successful.

X) **SYNGENEIC BMT**

If an identical twin is available, syngeneic BMT is an option. In an IBMTR analysis of data combining childhood and adult ALL in first remission, syngeneic BMT had a relapse rate of 36% and leukemia-free survival of 57%. This was similar to allogeneic BMT in first remission. Due to lack of GVHD, if an identical twin is available, syngeneic BMT may be considered as an option for all ALL patients in remission.

XI) **UNRELATED BMT**

Due to the increased morbidity from GVHD, an unrelated BMT is generally reserved for second or greater remission in patients who do not have a matched sibling. Unrelated BMT in second remission results in a 20-25% leukemia-free survival. Studies are ongoing to decrease acute GVHD in unrelated BMT using T cell depleted (TCD) marrow. Unlike the experience of TCD in matched sibling BMT, depletion of T cells from unrelated marrow may decrease acute GVHD but may not effect relapse.

6.3

XII) **TREATMENT OF ADULT ALL BY DISEASE STAGE**

A) FIRST REMISSION

For patients in first remission, good risk adult ALL is treated with chemotherapy. Ph+ adult ALL is treated with allogeneic BMT if a matched sibling is available. Ph- but otherwise poor risk adult ALL may be offered either allogeneic BMT from a matched sibling or chemotherapy. The role of autologous BMT in first remission is not yet defined. However, in a protocol setting, it is reasonable to treat poor risk adult ALL in first remission with autologous BMT. Alternatively, consideration may be given to harvesting marrow while the patient is in first remission; to be used in the event of relapse.

1) **Chemotherapy in First Remission**

Chemotherapy for good risk adult ALL (age < 35, WBC at presentation < 30,000/μl, time to remission < 4 weeks, absence of *bcr/abl* gene rearrangement) has a leukemia-free survival of 50-60%. Good risk adult ALL should be treated with chemotherapy in first remission.

Patients with Ph+ ALL have a survival of 0-16% with chemotherapy and should be treated with allogeneic BMT if a matched donor is available.

Many single institution studies of adult patients have, in general, lumped together poor risk features including Philadelphia chromosome positive and negative patients. These studies have reported a leukemia-free survival of 20-40% with chemotherapy. However, the IBMTR has reported in a retrospective multicenter comparison that the leukemia-free survival of Ph negative but otherwise poor prognosis adult ALL is similar in first remission between patients treated with either chemotherapy (38%) or allogeneic BMT (44%). Therefore, the preferred treatment (allogeneic BMT vs chemotherapy) for Ph negative but poor risk adult ALL is controversial.

2) Allogeneic BMT in First Remission

Leukemia-free survival after allogeneic BMT in adult ALL is age dependent. The leukemia-free survival is 60% for ages 15-19, 50% for ages 20-29, < 40% for age 30-35, and 20-35% for age > 35. Allogeneic BMT from a matched sibling is the treatment of choice in Ph+ ALL. In single institution nonrandomized studies, leukemia-free survival after allogeneic BMT for Ph+ ALL is 30-40% compared to 0-16% for chemotherapy.

3) Autologous BMT in First Remission

The role of autologous BMT in adult ALL in first remission is controversial and remains to be defined. Since clonal proliferation may occur at any stage of lymphoid differentiation, ALL is heterogeneous in terms of its common precursor neoplastic cell. Those derived from stem cells such as Ph+ ALL are, similar to CML, unlikely to be cured by chemotherapy or autologous BMT. For poor prognosis ALL, which is believed to arise from more committed precursors, autologous BMT is acceptable in a protocol setting.

B) FAILED INDUCTION

For patients who fail induction therapy, options include salvage chemotherapy or immediate allogeneic BMT. Salvage chemotherapy usually combines high-dose cytarabine and either an anthracycline or epipodophyllotoxin or high-dose methotrexate and L-asparaginase (Table 6.3.5). If salvage chemotherapy induces a remission, the prognosis remains poor, and early BMT should be considered. If salvage chemotherapy is unsuccessful, the only curative treatment is allogeneic BMT with 10-20% leukemia-free survival.

C) RELAPSE

Unlike AML, there is no data available for ALL comparing allogeneic BMT at untreated relapse versus second remission. In practice, an attempt to achieve a second remission is usually done prior to proceeding with BMT.

D) SECOND OR GREATER REMISSION

Long term prognosis depends on length of first remission. If relapse occurrs within 18 months, leukemia-free survival with chemotherapy is < 5%. If relapse occurred after 18 months, leukemia-free survival with chemotherapy is 10%. Because of the poor prognosis with chemotherapy, ALL in second or greater remission is treated with BMT(autologous, sibling matched or unrelated). Interpreting survival rates for BMT in second remission must take into account that these patients are very select, having survived to a second remission.

1) Allogeneic Sibling Matched BMT

Results of allogeneic BMT in second remission are 25-35% leukemia-free survival.

2) Autologous BMT

For adult ALL in second remission, autologous BMT results in a 20% leukemia-free survival.

3) **Unrelated BMT**

For adult ALL in second remission, unrelated phenotypically matched BMT results in a 20-25% leukemia-free survival.

XIII) **PROPHYLAXIS OF THE CNS IN ADULT ALL**

Due to the risk of relapse, in both non-BMT and BMT patients, some form of CNS prophylaxis is routine for ALL. Treatment options include intrathecal (IT) drugs, intraventicular drugs, cranial radiation, or high-dose intravenous cytarabine or methotrexate. The traditional CNS prophylactic regimen is combined IT drugs and cranial irradiation.

A) **INTRATHECAL (IT) DRUGS**

Few effective antileukemic drugs may be safely injected into the intrathecal space. Most experience is with methotrexate and cytarabine given alone or both drugs mixed together with hydrocortisone as a single injection (triple intrathecal chemotherapy). Some institutions adjust the intrathecal dose of methotrexate (10 mg/m^2) and cytarabine (30-100 mg/m^2) according to body surface area. However, the CNS space is constant after age 3. Since body surface area does not correlate with CNS volume, methotrexate and cytarabine dosing is given at some institutions according to age not body dimensions (Table 6.3.6).

B) **CRANIAL IRRADIATION AND IT THERAPY**

Most ALL meningeal leukemia prophylaxis regimens use a combination of cranial irradiation (1800-2400 rads) and IT methotrexate (usually 4-6 weekly doses). Due to direction of CSF flow out of the ventricles, intralumbar chemotherapy has variable distribution into the ventricles. Therefore, cranial radiation is combined with intralumbar chemotherapy to provide adequate

6.3

Table 6.3.5. Examples of salvage chemotherapy for adult ALL

Regimen	Drug	Dose	Route/Day
Cytarabine Mitoxantrone			
	Cytarabine	2 g/m^2	IV q12 hour x 8 doses
	Mitoxantrone	12 g/m^2	CI daily x 2 days
Cytarabine Idarubicin			
	Cytarabine	2 g/m^2	IV over 3 hours daily x 3 days
	Idarubicin	12 mg/m^2	IV daily x 3 days
Cytarabine Amsacrine			
	Cytarabine	3 g/m^2	IV over 3 hours daily x 3 days
	Amsacrine	200 mg/m^2	IV daily x 3 days
Cytarabine Teniposide			
	Cytarabine	3 g/m^2	IV q12 hours x 12
	Teniposide	100 mg/m^2	IV day 7 and 14
Methotrexate Asparaginase Repeat course every 7-14 days	Methotrexate	50-80 mg/m^2	IV day 1, increase MTX 50 mg/m^2 with each course
	L-asparaginase	20,000 U/m^2	IV 3 hours after MTX

Table 6.3.6. Intralumbar dose of chemotherapy by patient age

1) Intralumbar Antileukemia Agent—given as single drug

Age (years)	IT Methotrexate Dose (mg)	IT Cytarabine Dose (mg)
< 1	6	–
1-2	8	30
2-3	10	50
3 or older	12	70

2) Intralumbar Triple Intrathecal Chemotherapy (TIC)—mixed together in single injection

Age (years)	IT Methotrexate Dose (mg)	IT Cytarabine Dose (mg)	IT Hydrocortisone Dose (mg)
< 1	8	16	8
1-2	10	20	10
3-8	12	24	12
8 or older	15	30	15

6.3

craniospinal prophylaxis. Craniospinal irradiation can substitute for the combination of intralumbar drugs and cranial XRT. However, craniospinal radiation compromises marrow reserve in the vertebra. Another substitute for cranial radiation and intralumbar chemotherapy is intraventricular chemotherapy.

C) INTRAVENTRICULAR CHEMOTHERAPY

Surgical placement of an Ommaya reservoir allows instillation of chemotherapy directly into the ventricles and circumvents the problem of CSF flow which hinders distribution of intralumbar chemotherapy into the ventricles. Due to neurosurgical placement of the reservoir, this procedure is usually reserved for actual meningeal leukemia which relapses after cranial radiation.

D) HIGH-DOSE INTRAVENOUS CHEMOTHERAPY

Antileukemic drug penetration into the CSF is generally low. The CNS to plasma drug concentration ratio for methotrexate is 0.03. The CSF to plasma ratio for cytarabine is 0.1-0.25. However, therapeutic CSF levels may be achieved by using high-dose intravenous cytarabine (2-3 g/m^2 q12hour x 6-12 doses) or high-dose intravenous methotrexate (5 g/m^2) with leucovorin rescue. Adequate prophylaxis of the CNS has been reported using high-dose intravenous therapy without IT drugs or radiation.

E) BMT CNS PROPHYLAXIS

Unlike AML, prophylaxis of the CNS in ALL patients undergoing BMT decreases meningeal relapse. Prior cranial radiation due to previous CNS leukemia is not an absolute contraindication to TBI conditioning. In patients receiving TBI conditioning, Fred Hutchinson Cancer Center has given prophylactic IT methotrexate twice before marrow infusion then every other week from day 32 to day 102. If the patient had no prior history of CNS disease, the risk of meningeal relapse without prophylaxis was 19% versus 4% with IT methotrexate. For ALL patients with active CNS disease or a history of prior CNS leukemia, the risk of CNS relapse after BMT was 52% without IT methotrexate versus 17% with IT methotrexate.

XIV) **TREATMENT OF ACTIVE CNS DISEASE**

A) **TREATMENT OF CNS LEUKEMIA IN THE NON-BMT PATIENT**

CNS leukemia (blasts in the CSF) is usually treated with IT methotrexate once or twice a week until the CNS is clear. Remission in the CSF is consolidated with maintenance intrathecal chemotherapy (e.g. once a week for 4 weeks then once a month for 12 months) alone or combined with cranial irradiation (2400 cGy).

B) **TREATMENT OF CNS LEUKEMIA IN THE BMT PATIENT**

Treatment of CNS leukemia at the time of BMT is similar to prophylaxis in BMT patients without CNS leukemia. One approach is TBI conditioning and IT methotrexate twice before marrow infusion, repeated every other week from day 32 to day 102, then every other month for a total of 8 doses.

XV) **TREATMENT OF THE TESTES**

Testicular relapse has occurred after BMT for ALL. Therefore, unlike AML, some institutions boost the testes with an additional 400-500 cGy in patients undergoing BMT for ALL. This causes permanent sterility while after TBI alone fertility may occasionally be preserved.

SUGGESTED READING

1. Preti A, Kantarjian H M. Management of adult acute lymphoblastic leukemia: present issues and key challenges. Journal Clinical Oncology 1994; 12(12):1312-1322.
2. Hoelzer D. Treatment of acute lymphoblastic leukemia. Seminars in Hematology 1994; 31(1):1-15.
3. Horowitz MM et al. Chemotherapy compared with bone marrow transplantation for adults with acute lymphoblastic leukemia in first remission. Annals of Internal Medicine 1991; 115:13-18.

6.3

Chronic Myeloid Leukemia—CML

Richard K. Burt, Philip Rowlings, George W. Santos

I) ETIOLOGY

Chronic myeloid leukemia (CML) is a clonal disorder of hematopoietic precursors that results in excessive accumulation of white blood cells (WBC). CML is separated into a chronic phase dominated by increased mature and intermediate myeloid forms and a transforming phase (accelerated and blast crises) dominated by cells of immature myeloid or less commonly lymphoid lineage. Besides increasingly immature cells, the transforming phase is characterized by enlarging spleen size, additional cytogenetic abnormalities and difficulty controlling the WBC count with therapy. The etiology of CML is unknown, but similar to ALL and AML, it is associated with exposure to radiation. Over 90% of patients with CML have the Philadelphia (Ph) chromosome, a translocation between chromosome 9 and 22. Approximately 25-30% of CML patients without the Ph chromosome are positive by polymerase chain reaction for the Ph translocation. This translocation gives rise to an altered tyrosine kinase suspected of playing a pivotal role in the pathogenesis of CML. The American and European annual incidence is approximately 1.5 per 10^5 people. The median age at presentation is 50-60.

II) DIAGNOSIS

In developing countries or for patients without routine access to health care, CML usually presents as fatigue, low grade fever and minimal to massive splenomegaly causing left upper quadrant fullness. Elevated platelet and white blood cell count may lead to leukostasis causing headache, tinnitus, priapism, stroke, myocardial or splenic infarction and peripheral thrombosis. Abnormal platelet function may rarely cause a hemorrhagic diathesis despite an elevated platelet count. In developed countries, the majority of patients are asymptomatic and come to medical attention by an elevated WBC count on routine blood tests obtained for unrelated reasons. The peripheral blood neutrophils may be only minimally elevated, or markedly increased (white blood cell count > 1 million/ml) with immature myeloid forms including myelocytes, metamyelocytes, promyelocytes and myeloblasts. Platelet count may be low, normal or high. Basophilia is common. Anemia and nucleated red cells may be present. The bone marrow is hypercellular with a left shifted M:E ratio often > 10:1 (normal 2.5:1) and increased megakaryocytes. Myelofibrosis and glycolipid swollen histiocytes (sea blue histiocytes) may be present.

A) DIFFERENTIAL DIAGNOSIS

The differential of an elevated WBC count is broad and includes infections, autoimmune diseases, drugs (corticosteroids, epinephrine, lithium, etc.), neoplasms, trauma, ketoacidosis, thyrotoxicosis, transfusion reactions, splenectomy and myeloproliferative disorders other than CML (agnogenic myeloid metaplasia, essential thrombocytosis and polycythemia vera). Features that help distinguish CML from other causes of neutrophilia are presence of splenomegaly, basophilia, low leukocyte alkaline phosphatase and the Ph chromosome.

6.4

Bone Marrow Transplantation, edited by Richard K. Burt, H. Joachim Deeg, Scott Thomas Lothian, George W. Santos. © 1996 R.G. Landes Company.

Occasionally a patient with CML may present clinically with only an elevated platelet count suggesting essential thrombocytosis, or predominant marrow fibrosis suggesting agnogenic myeloid metaplasia or an elevated RBC count suggesting polycythemia vera. In these cases the diagnosis of CML is established by identification of the Ph chromosome. Therefore, any patient with a myeloproliferative disorder should have marrow or peripheral blood sent for cytogenetic analysis.

III) WORK UP OF CML

All patients should have a CBC with differential, electrolyte and hepatic panel, coagulation screen, bone marrow biopsy and aspirate with chromosomal analysis by karyotype, and if less than 55 years of age, HLA typing of patient, siblings and parents. Pre-BMT work up is covered in the chapter on BMT patient evaluation.

IV) PROGNOSIS

CML usually begins in an indolent manner with an increase in mature myeloid cells. As CML progresses, it enters a poorly defined accelerated phase (AP) in which the disease becomes less responsive to treatment with more immature cells in the peripheral blood. The blast phase may be either myeloid or lymphoid and is defined similarly as any acute leukemia with > 30% blasts in the marrow. Median survival for patients in chronic phase (CP) CML has been reported to be between 3-4 years but with earlier diagnosis perhaps a more accurate estimate is 5-6 years. Survival during CP CML has been subdivided into low, intermediate and high risk groups based on sex, spleen size, hematocrit, platelet count and percentage of circulating blasts. However, probably the best predictor of survival is a cytogenetic response to therapy, i.e. a reduction or disappearance of the Ph chromosome following therapy with interferon or after BMT. Rare patients have been reported to survive more than 20 years in CP. CML in transformation has a worse prognosis. Median survival is 12-16 months for accelerated phase and 3-12 months for blast crises.

V) TREATMENT PHILOSOPHY

The only proven curative therapy for CML is allogeneic BMT. Therefore, the treatment of choice for younger patients (less than 55 years old) in CP who have a matched sibling donor is early referral for allogeneic BMT. Although not curative, interferon is capable of suppressing Ph+ hematopoiesis in some patients and possibly prolonging survival. Thus, all other CP patients should be started on interferon α. If a major cytogenetic response (< 35% Ph+ cells) occurs over 12-18 months, interferon should be continued and autologous marrow stored in the event of disease progression. If interferon is ineffective or not tolerated, the disease may be controlled with hydroxyurea and the patient considered for autologous, unrelated, or related HLA-mismatched BMT. CML in transformation (AP or blast crises) is treated with cytotoxic chemotherapy and if possible allogeneic BMT from a related or unrelated donor.

VI) HYDROXYUREA, BUSULFAN

Traditional therapy of CML with agents such as hydroxyurea or busulfan will control leukocyte counts but, in general, does not change the natural history of the disease in terms of progression to a blast crisis. Side effects of busulfan include variable and prolonged myelosuppression, gonadal failure, pulmonary and marrow fibrosis and an Addisonian-like picture. In addition, prior exposure to busulfan may increase BMT mortality. Therefore, for potential future BMT candidates, hydroxyurea is generally preferred over busulfan. Hydroxyurea is adjusted according

6.4

to counts and is usually given orally between 0.5-2.0 g/day. Side effects are cytopenias, rashes, fever, diarrhea, nausea and vomiting.

VII) INTERFERON α

Treatment with interferon α gives a hematologic remission (i.e. normalization of leukocyte count) in 70-80% of patients. Unlike hydroxyurea or busulfan, interferon may result in a cytogenetic response. A complete cytogenetic response (0% Ph+ cells) has been reported in the range of 10-30%. A major cytogenetic response (< 35% Ph+ cells) occurs in approximately 30-40% of patients. Maximal cytogenetic response may require 12-18 months of treatment. The dose of interferon α is approximately 5 x 10^6 U/m^2 SQ daily but is adjusted according to toxicity and cytopenias. Cytogenetic responses are higher if, toxicity permitting, the dose can be adjusted to maintain the white blood cell count between 2000-3000 cells/ml. Toxicities include fever, chills, anorexia, fatigue, insomnia, depression, thrombocytopenia and a variety of immune-mediated disorders including hypothyroidism, nephrotic syndrome, systemic lupus, rheumatoid arthritis, arrhythmias and congestive heart failure. For patients who obtain a major cytogenetic response, interferon increases survival by greater than 2 years for CP CML. Nevertheless, interferon is not considered curative, and even with a complete cytogenetic response, patients are kept on maintenance therapy. Prior interferon treatment does not adversely affect BMT outcome, however, interferon is a radiosensitizer and should be discontinued a minimum of 2 weeks prior to TBI or severe, even life-threatening, mucositis may develop.

VIII) GENOTYPICALLY MATCHED SIBLING BMT

If a matched sibling is available, syngeneic or allogeneic BMT is the treatment of choice for all patients less than 55 years old. LFS is 50-60% in CP, 30-35% in AP and 10-20% in blast crises. For a patient in CP, transplant within 1 year of diagnosis results in higher survival (LFS > 80%) than those transplanted later. Therefore, patients should be referred early rather than later for allogeneic BMT.

Patients undergoing BMT in second CP have a survival of 30-40%. Since remission can be achieved in the majority of patients in lymphoblastic crises, it is reasonable to attempt a second chronic phase prior to BMT. For blast crises of myeloid lineage, a second remission is difficult to obtain and of short duration. For

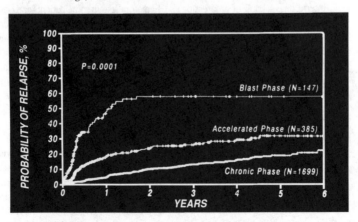

Fig. 6.4.1. Probability of relapse after HLA-identical sibling BMT for chronic myelogenous leukemia, 1987-1994. Reprinted with permission from IBMTR/ABMTR 1995; 2:7.

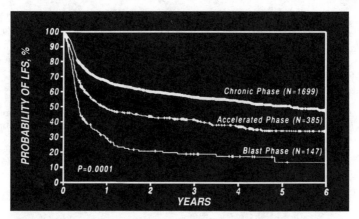

Fig. 6.4.2. Probability of leukemia-free survival after HLA-identical sibling BMT for chronic myelogenous leukemia, 1987-1994. Reprinted with permission from IBMTR/ ABMTR 1995; 2:7.

these patients, it is arguable whether to proceed directly to BMT or to first attempt inducing a second chronic phase.

Preparative regimen dose and schedule are discussed in the chapter on conditioning regimens. The two most common CML conditioning regimens are cyclophosphamide(120 mg/kg) and TBI(1200 cGy, fractionated as 200 cGy BID) or busulfan (16 mg/kg) and cyclophosphamide(120-200 mg/kg). GVHD prophylaxis regimens are presented in the chapter on GVHD. Common regimens are: 1) cyclosporine A (CsA) and methotrexate (MTX); 2) CsA and steroids; 3) CsA and MTX and steroids.

6.4

IX) AUTOLOGOUS BMT

CML patients without an HLA-matched sibling may be considered for autologous BMT. Autografts for CML in transformation (AP or blast crises) have given disappointing results with a second CP of short duration. Pending development of better purging methods, autologous BMT for CML in transformation has generally been abandoned. Early results suggest that autologous BMT for CML in chronic phase may prolong survival especially if some Ph- cells are present after autografting. At present, there are no data to suggest an advantage of peripheral blood stem cells (PBSC) over marrow or use of purged versus unmanipulated marrow. Some centers performing PBSC autografts for CML, initially cytoreduce with chemotherapy (e.g. ICE- idarubicin, cytarabine, etoposide) and perform cytogenetic analysis when the WBC count is rebounding. If there is a decrease in the number of Ph+ cells with cytoreduction, then a second course of chemotherapy is given to mobilize stem cells for an autograft. If improvement in the number of cytogenetically positive cells cannot be demonstrated with initial cytoreduction, autologous BMT is abandoned. Although reasonable, currently, there are no data to suggest an advantage to this approach or between autografting in early versus late CP. It is important to stress that, currently, autografting for CML in CP is not curative, but may prolong survival.

X) UNRELATED BMT

Patients who do not have an HLA-matched sibling may be considered candidates for phenotypically matched unrelated or HLA-mismatched sibling BMT, especially if they do not obtain a major cytogenetic response with interferon.

Unrelated BMT for CML has a high incidence of graft failure (approximately 10%), grade II-IV acute GVHD (> 50%) and extensive chronic GVHD (50-60%). Leukemia-free survival for unrelated BMT is 35-40% in CP, 20-25% for AP and 0-5% for blast crises.

SUGGESTED READING

1. Kantarjian HM, Deisseroth A, Kurzrock R, Estrov Z, Talpaz M. Chronic myelogenous leukemia: A concise update, Blood 1993; 82(3):691-703.
2. McGlave P, Bartsch G, Anasetti C, Ash R, Beatty P, Gajewski J, Kernan N. Unrelated donor marrow transplantation for chronic myelogenous leukemia: Initial experience of the National Marrow Donor Program. Blood 1993; 81(2):543-550.
3. Hoyle C, Gray R, Goldman J. Autografting for CML in chronic phase: an update. British Journal of Haematology, 1994; 86:76-81.

6.4

Chronic Lymphocytic Leukemia

Issa Khouri, Richard Champlin

I) INCIDENCE AND ETIOLOGY

Chronic lymphocytic leukemia (CLL) is characterized by clonal expansion of abnormal lymphocytes of B cell lineage in more than 95% of cases and T cell lineage in less than 5% of cases. It is the most prevalent adult leukemia in the Western hemisphere, accounting for 30% of leukemias diagnosed in the United States. The incidence is similar by race but increases with advancing age. Fewer than 20% of patients are younger than age 50. It is more prevalent in males than females (ratio of 2:1). The cause of CLL is unknown; it is not associated with radiation or exposure to occupational hazards or alkylating drugs. Among leukemias, CLL has the strongest tendency for family aggregation.

II) CYTOGENETICS AND PATHOLOGY

Cytogenetic abnormalities occur in about 50% of patients with CLL, trisomy 12 being most common, followed by aberrations involving the long arm of chromosome 14 or 13. The detection of trisomy 12 increases with the use of fluorescence in situ hybridization. The absence of a chromosomal abnormality is associated with a better prognosis. No association with oncogene activation has been identified. Unlike chronic myelogenous leukemia, which may acquire new cytogenetic changes with the progression of the disease, cytogenetic evolution does not generally occur in CLL.

The bone marrow histology can be classified into nodular, interstitial, mixed and diffuse pattern. Diffuse histology is more common in advanced stages. B lymphocyte CLL cells express surface immunoglobulin. The cells express typical B cell antigens such as CD19, CD20, CD23 and CD24 and the T cell antigen CD5. They do not react with other antibodies that identify T cell antigens such as CD2, CD3, CD4 or CD8. Heavy or light chain immunoglobulin gene rearrangement is usually present.

6.5

III) DIAGNOSIS

The requirements for the diagnosis of CLL include the presence of a sustained lymphocytosis in the peripheral blood of $10 \times 10^3/\mu l$ or higher, with the documentation of either a B-CLL phenotype or 30% involvement of the nucleated cells of the bone marrow. Peripheral blood lymphocytosis greater than $5 \times 10^3/\mu l$ may suffice to make a diagnosis of CLL if both marrow involvement and a B cell phenotype criteria are met.

IV) STAGING

Staging in CLL is based on clinical assessment of tumor burden and features of bone marrow failure that include anemia and thrombocytopenia. Rai developed the first staging system (Table 6.5.1), which was later revised into three (rather than five) groups that differ with respect to prognosis (Table 6.5.2). The Binet system proposed in 1977 is of similar utility (Table 6.5.3). Although the International Workshop on CLL recommended an integrated system (Table 6.5.4), the Rai and Binet staging systems continue to be more commonly used in patient stratification, and their prognostic value has been validated by several investigations.

Bone Marrow Transplantation, edited by Richard K. Burt, H. Joachim Deeg, Scott Thomas Lothian, George W. Santos. © 1996 R.G. Landes Company.

V) PROGNOSIS

Several characteristics have been associated with prognosis, this includes age, sex, lactate dehydrogenase, serum albumin and lymphocyte count. The most important prognostic features, however include 1) advanced clinical staging (Table 6.5.2), 2) diffuse pattern of bone marrow involvement, 3) lymphocyte doubling time < 12 months, 4) elevated beta-2 microglobulin and 5) abnormal karyotype.

VI) CLINICAL COURSE

Patients with CLL are often asymptomatic at the time of diagnosis. The disease is usually identified incidentally on a routine blood count examination. Patients can be broadly divided into two groups, indolent or aggressive, depending on the course of the disease. Patients with indolent disease have low tumor burden

Table 6.5.1. Staging for CLL—Rai system

Stage	Characteristics
0	Lymphocytosis only
I	Lymphocytosis and lymphadenopathy
II	Lymphocytosis and lymphadenopathy and/or splenomegaly (with or without lymphadenopathy)
III	Lymphocytosis plus anemia (hemoglobin < 11 g/dl)
IV	Lymphocytosis plus thrombocytopenia (platelets count < 100 x 10^9/L)

Table 6.5.2. Staging for CLL—the revised Rai system

Stage	Risk Category	Median Survival (years)
0	Low	> 10
I/II	Intermediate	6
III/IV	High	2

Table 6.5.3. Staging for CLL—Binet system

Stage	Characteristics
A	< 3 lymphoid sites involved and hemoglobin > 10 mg/dl, platelets > 100 x 10^9/L
B	> 3 lymphoid sites involved and hemoglobin > 10 mg/dl, platelets > 100 x 10^9/L
C	Hemoglobin < 10 mg/dl and/or platelets < 100 x 10^9/L, independent of lymph sites involvement

*Areas of involvement include the cervical, axillary and inguinal lymph nodes, spleen and liver.

Table 6.5.4. Staging for CLL—integrated system

Stage	Characteristics	Corresponding Binet/Rai Stage
A	< 3 lymphoid sites involved	A/0, A/I, A/II
B	> 3 lymphoid sites involved	B/I, B/II
C	Anemia and/or thrombocytopenia	C/III, C/IV

6.5

with lymphocyte doubling time of more than 1 year and enjoy an excellent quality of life.

A) INFECTION

The most common complication and cause of death in CLL is infection. It occurs mainly in patients with advanced stages, and encapsulated bacteria are the most frequent cause. A new spectrum of infections is emerging in the group of patients treated with purine nucleoside analogs (2-CDA, fludarabine); this includes pneumocystis, listeria and nocardia. Susceptibility to infections results from hypogammaglobulinemia, associated B cell impairment, T cell dysfunction and neutropenia.

B) CYTOPENIAS

Anemia and thrombocytopenia may result from decreased production secondary to massive bone marrow infiltration, excessive pooling in the spleen, immune destruction of red blood cells (RBC) and platelets. Pure RBC aplasia and rarely, immune neutropenia can be observed.

C) TRANSFORMATION

CLL may transform into an aggressive lymphoma, Richter's syndrome (diffuse large cell lymphoma), which occurs in 3-10% of cases and is usually poorly responsive to chemotherapy. Whether this transformation represents clonal evolution of CLL or an independent disorder is unsettled. Other less common types of transformation include "prolymphocytoid" transformation, acute lymphoblastic leukemia of the L2 type and multiple myeloma.

Patients with CLL have an increased risk of nonhematologic neoplasms, mostly skin, colon and lung. Whether this is due to the effect of treatment or an intrinsic susceptibility of these patients, is not yet settled.

6.5

VII) DIFFERENTIAL DIAGNOSIS

Within the mature B cell malignancies, prolymphocytic leukemia, hairy cell leukemia and its variants, leukemic phase of follicular or diffuse small cleaved cell lymphoma, splenic lymphoma with villous lymphocytes and Waldenstrom's macroglobulinemia can sometimes be confused with CLL. The integration of cytology and membrane immunophenotyping can differentiate these entities from B cell CLL (Table 6.5.5).

VIII) INDICATIONS FOR INITIATING THERAPY

It is generally accepted that therapy is indicated in patients with advanced stage (Binet C, Rai III or IV) CLL. For patients without hematologic evidence of bone marrow impairment, however, the decision is less clear. For such patients an analysis of the pattern of bone marrow involvement, cytogenetics and beta-2 microglobulin should be done. Often a period of observation is correlated to determine the lymphocyte doubling time in the peripheral blood. Patients with normal cytogenetics, normal beta-2 microglobulin, a nondiffuse marrow and lymphocyte count doubling time greater than 1 year usually have indolent disease and may be followed on a watch and wait basis.

IX) CRITERIA FOR REMISSION IN CLL

The International Workshop on CLL (IWCLL) and the National Cancer Institute Sponsored Working Group (NCISWG) developed two recommendations of objective criteria to standardize the analysis of response to treatment. The criteria for complete remission (CR) and partial remission (PR) for the two groups are summarized in Table 6.5.6. Both allow for nodules and focal lymphocytosis to be present in the bone marrow biopsy of CR patients. This issue needs to be addressed, since it has been shown that patients who have persistent nodules but are otherwise in

Table 6.5.5. Mature B cell leukemia

Mature B Cell Leukemia	Immunophenotype	Cytology
Chronic lymphocytic leukemia (CLL)	CD19+, CD20+(dim), CD23+, surface Ig+(dim), CD5+, CD22-/+, CD11C-/+, CD25-/+, CD10-	Small mature lymphocytes, occasional prolymphocyte < 10%
Prolymphocytic leukemia (PLL)	CD19+, CD20+, CD22+, surface Ig+, CD5-/+, CD23-	Predominance of prolymphocytes— large cells, round nuclei, single prominent nucleolus
Hairy cell leukemia	CD19+, CD20+, CD25+, surface Ig+, CD5-	Abundant cytoplasm with villous (hairy) projections, slightly immature nucleus may be indented or reniform
Mantle cell lymphoma –leukemic phase	CD19+, CD20+, CD5+, CD22+, surface Ig+, CD23-	Small to intermediate-sized lymphocyte with nuclear clefting
Follicular small cleaved cell lymphoma –leukemic phase	CD19+, CD20+, surface Ig+, CD10+/-, CD5-	Small lymphoid cell with prominent nuclear cleft ("buttock cell")
Splenic lymphoma	CD19+, CD20+, surface Ig+, CD5-	Usually from splenic marginal zone, circulating cells may mimic hairy cell leukemia
Waldenstrom's macroglobulinemia	CD19+, CD20+, surface Ig+, CD5-,	Plasmacytoid cells
Plasma cell leukemia	Negative for surface lymphoid markers, CD38+, surface Ig- but cytoplasmic Ig+	Eccentric nucleus, basophilic cytoplasm

6.5

Table 6.5.6. Remission criteria for CLL—comparison of the International Workshop on CLL (IWCLL) and the National Cancer Institute Sponsored Working Group for CLL (NCISWG)

	IWCLL	NCISWG	IWCLL	NCISWG
Criteria	CR	CR	PR	PR
Lymphadenopathy /organomegaly	None	None	Decrease in Binet stage	> 50% decrease
Neutrophils	> 1.5 x 10⁹/L	> 1.5 x 10⁹/L		> 1.5 x 10⁹/L or 50% increase
Platelets	> 100 x 10⁹/L	> 100 x 10⁹/L		> 100 x 10⁹/L or > 50% increase
Hemoglobin		> 11 g/dl		> 11 g/dl or > 50% increase
Lymphocytes	< 4 x 10⁹/L	< 4 x 10⁹/L		> 50% decrease
Bone marrow	Normal or focal lymphoid infiltrate	< 30% lymphocytes		

CR, have shorter time to progression than those patients with no remaining nodules. Furthermore, sensitive methods are required to detect residual malignant cells. Flow cytometry to detect coexpression of CD5 and CD19 or CD20 markers and analysis of gene rearrangement by Southern blot and by the polymerase chain reaction ("immunoglobulin fingerprinting") are under study in this setting.

X) TREATMENT OPTIONS

A) CONVENTIONAL THERAPY (TABLE 6.5.7)

This involves repeated administration of an alkylating agent such as chlorambucil or cyclophosphamide, usually combined with prednisone. Response rates as high as 70% have been reported, however, few of these patients achieve a CR. Combination regimens such as CVP (cyclophosphamide, vincristine and prednisone) with or without doxorubicin, CAP (cyclophosphamide, doxorubicin and cisplatin), POACH (cyclophosphamide, doxorubicin, cytarabine, vincristine and prednisone) are more toxic and not clearly superior to the combination of chlorambucil and prednisone.

B) NEW CHEMOTHERAPEUTIC DRUGS (TABLE 6.5.7)

Recently three purine nucleoside analogs, fludarabine, chlorodeoxyadenosine (2-CDA) and deoxycoformycin, have been studied in the treatment of CLL. Therapy with these agents is frequently complicated by opportunistic infections (listeria monocytogenes, pneumocystis carina), myelosuppression and thrombocytopenia.

1) Fludarabine

Studies with fludarabine at the dosage of 25-30 mg/m^2/d IV for 5 consecutive days every 4 weeks can induce a CR rate of 70% in previously untreated patients. Complete response at the level of detection by Southern blotting for immunoglobulin gene rearrangement and two-parameter flow cytometry has been demonstrated with this modality of treatment. These data suggest that fludarabine may be the single most active agent in

6.5

Table 6.5.7. *Examples of chemotherapy for chronic lymphocytic leukemia*

I) Primary Chemotherapy
CHL +P
Chlorambucil 0.1–0.4 mg/kg PO day 1
Prednisone 100 mg/d PO for 2 days
Repeat cycle every other week, increase chlorambucil by 0.1 mg/kg increments every 2 weeks until control of disease or toxicity

Fludarabine
Fludarabine 25-30 mg/m^2/d IV for 5 consecutive days
Repeat cycle every 4 weeks x 4-6

2-CDA (2-chlorodeoxyadenosine)
2-chlorodeoxyadenosine 0.09 mg/kg/day IV continuous infusion over 24 hours for 5-7 consecutive days
Repeat every 4 weeks for 4-6 cycles

2-deoxycoformycin (pentostatin)
Pentostatin 4 mg/m^2 IV
Repeat weekly or every other week for 4-6 cycles

II) BMT Conditioning Regimen
Cyclophospamide 60 mg/kg/day IV for 2 consecutive days, followed by total body irradiation 1000-1400 cGy in fractionated doses of 200 cGy BID.

CLL. However, despite the high response rates, all patients' relapse and survival is not substantially prolonged.

2) 2-CDA and deoxycoformycin

2-CDA or deoxycoformycin can be effective in CLL with response rate of 30% and 45% respectively in previously treated patients.

C) ALLOGENEIC BONE MARROW TRANSPLANTATION

Currently, the only therapy considered curative for CLL is allogeneic BMT, but it is limited to patients less than 55 years old who have an HLA-identical sibling. More than 100 patients with CLL have received genotypically matched sibling transplantation worldwide. The most common preparative regimen used was cyclophosphamide (120 mg/kg) and fractionated total body irradiation. The graft-versus-host disease prophylaxis varied; most patients received a combination of cyclosporine and methotrexate. Although the follow-up on these patients is of short duration, important results have been observed, in that complete remission with eradication of minimal residual disease could be obtained in patients who have been refractory to conventional alkylator therapy and to fludarabine. Greater than 60% of these patients who otherwise would have had a median survival of 7 months, are still in remission more than 3 years postallogeneic transplant. Unlike the best responders to conventional therapy, patients who achieved CR with transplantation have shown normalization of their immunoglobulin deficit.

D) AUTOLOGOUS BONE MARROW TRANSPLANTATION

Autografting for CLL has recently been evaluated for patients who do not have an HLA-identical sibling or who are older than 55 years of age.

Most patients had their marrow harvested when minimally involved with disease. Autologous transplantation has proven to be effective for patients who had minimal tumor burden at the time of the transplant. Approximately 75% of patients with advanced disease achieve CR. Unlike the case with allogeneic transplantation, patients who have extensive disease, and are refractory to fludarabine have a remission duration of less than a year. The difference in disease-free survival between allogeneic and autologous transplants is likely due to the allogeneic graft-versus-leukemia effect and the probability of disease involvement of the autologous graft. Investigational studies are undertaken to evaluate innovative approaches for marrow purging in autotransplantation.

For patients who had either an allogeneic or autologous graft, clonal peripheral blood lymphocytosis can take many weeks to disappear, and disease in the bone marrow and lymph nodes responds slowly over several months. Polyclonal lymphocytic proliferation can occur in the bone marrow and peripheral blood after transplant, and this should not be confused with residual disease. Recurrence of autoimmune cytopenias has been observed in patients with CR after autotransplantation, suggesting that this phenomenon may be produced by remnant normal B lymphocytes rather than by CD5+ CLL cells.

SUGGESTED READINGS

1. O'Brien S, del Giglio A, Keating M. Advances in the biology and treatment of B-cell chronic lymphocytic leukemia. Blood 1995; 85:307-318.
2. Khouri I, Keating MJ, Vriesendorp HM et al. Autologous and allogeneic bone marrow transplantation for chronic lymphocytic leukemia: preliminary results. J Clin Oncol 1994; 12:748-758.
3. Rabinowe SN, Soiffer RJ, Gribben JG et al. Autologous and allogeneic bone marrow transplantation for poor prognosis patients with B-cell chronic lymphocytic leukemia. Blood 1993; 82: 1366-1376.

6.5

Multiple Myeloma

David H. Vesole, Ann E. Traynor, Sundar Jagannath, Guido Tricot, Bart Barlogie

I) ETIOLOGY

Multiple myeloma is a malignancy of terminally differentiated B lymphocytes, which is incurable with standard chemotherapy. The etiology of multiple myeloma remains unknown. Risk factors are similar to those described for other malignancies of hematopoietic bone marrow cells, and include ionizing radiation, environmental and industrial exposures. A genetic element may have a role in some cases.

II) EPIDEMIOLOGY

The median age at diagnosis of multiple myeloma is 65 years. Approximately 13,000 cases are diagnosed yearly, accounting for 1% of all types of malignant disease and slightly more than 10% of all hematologic malignancies. The cumulative incidence of myeloma in African American men is 10 per 1,000, roughly twice that of Caucasian men and African American women, and three times that of Caucasian American women. Multiple myeloma is characterized by a neoplastic proliferation of a single clone of plasma cells engaged in the production of a monoclonal protein. The disease may be localized or disseminated, indolent or progressive. The malignant plasma cells in multiple myeloma are inherently resistant at the time of diagnosis, which is reflected in the low complete remission rates ($< 10\%$) to standard therapy. This resistance may result from the many genetic aberrations incurred during a prolonged subclinical course of 15-20 years.

III) DIAGNOSIS (TABLE 6.6.1)

A) SIGNS AND SYMPTOMS

1) Pain

Bone pain, particularly in weight bearing bones such as the back, pelvis or hips, is present in two-thirds of patients at diagnosis. Lytic lesions are appreciated by plain film and are commonly seen in the skull, vertebrae, ribs, pelvis and proximal long bones. Tumor necrosis factor-β (TNF-β) and IL-1, produced by the myeloma cells, both appear to enhance osteoclast activation. Their activity, in turn, is stimulated by the myeloma paracrine growth factor, IL-6. This osteoclast activity, induced by neoplastic plasma cells, results in osteolytic lesions and demineralization.

2) Hypercalcemia

Hypercalcemia develops in approximately 30% of patients, and often contributes to renal insufficiency. Hypercalcemia develops because of increased bone resorption and may cause lethargy, nausea and constipation as well as renal insufficiency.

3) Anemia

Hematopoiesis is often impaired, probably due to cytokine dysregulation, and results in anemia in 60% of patients. However, leukopenia and thrombocytopenia are uncommon. Weakness and fatigue are often associated with the anemia.

4) Renal Insufficiency

Renal insufficiency is due predominantly to the development of the "myeloma kidney" in which distal convoluted tubules and collecting tubules

6.6

Bone Marrow Transplantation, edited by Richard K. Burt, H. Joachim Deeg, Scott Thomas Lothian, George W. Santos. © 1996 R.G. Landes Company.

become obstructed by large laminated casts consisting mainly of Bence-Jones (light chain) protein. Insufficiency occurs in 30-40% of all patients. Hypercalcemia exacerbates the toxicity of the Bence-Jones protein, and sites of calcium precipitation coincide with areas of light chain deposition. Light chain deposits occur not only intraluminally, but interstitially, in the tubular cells themselves, and in the mesangium. All of these deposits may be accompanied by calcium deposits. Lambda light chain deposits are more commonly associated with renal failure. Kappa light chain deposits are more commonly associated with proximal tubular dysfunction and Fanconi's syndrome. Nephrotic range proteinuria is predominantly albuminuria and occurs as a result of light chain deposition in the vasa recta, the tubular basement membrane and the glomerular mesangial or endothelial matrix. The light chain subtype is more commonly lambda when nephrotic range proteinuria occurs. Systemic amyloidosis (AL) may be present, and occurs in up to 15% of myeloma patients. Dehydration, hyperuricemia, sepsis, direct infiltration of myeloma cells, cryoglobulinemia, hyperviscosity, exposure to nonsteroidal anti-inflammatory drugs, aminoglycosides, amphotercin and intravenous contrast media each may contribute to acute renal insufficiency in the setting of myeloma.

5) **Infections**

Susceptibility to bacterial infections, particularly pneumococcal pneumonia and other infections with encapsulated organisms, is enhanced in the setting of multiple myeloma, in part due to diminished production of normal immunoglobulins. Fever is rarely ascribable to the neoplastic proliferation itself, and 12% of patients present with fever at diagnosis, indicating the frequency of infection as a reason for seeking medical attention.

6.6

Table 6.6.1. Diagnosis of multiple myeloma

Major Criteria for Diagnosis of Multiple Myeloma
I Plasmacytoma on tissue biopsy
II Bone marrow plasmacytosis > 30%
III Monoclonal spike on serum electrophoresis
> 3.5 g % for IgG
> 2.0 g % for IgA
> 1.0 g/24 hours for Bence-Jones proteinuria

Minor Criteria for Diagnosis of Multiple Myeloma
a Bone marrow plasmacytosis 10-30%
b Monoclonal globin spike
< 3.5 g % for IgG
< 2.0 g % for IgA
< 1.0 g/24 hours for Bence-Jones proteinuria
c Lytic bone lesions
d IgM < 50 mg%, IgA < 100 mg%, or IgG < 600 mg%

Diagnosis of Multiple Myeloma
In a symptomatic patient requires either one major and one minor criteria:
I+b, I+c, I+d (I+a not sufficient)
II+b, II+c, II+d
III+a, III+c, III+d
or three minor criteria which include a+b:
a+b+c or a+b+d

The incidence of gram negative bacterial infections and infections with herpes zoster are also increased in the setting of myeloma, as a result of B and T lymphocyte deficiencies.

6) Abnormalities of Hemostasis

Thrombocytopenia is uncommon early in the course of myeloma, but is a common complication of the terminal stage of disease, in association with diffuse infiltration of marrow. Secondary myelodysplasia may be a late feature of the disease as well. Acquired von Willebrand's disease has been described. In addition, monoclonal proteins may complex with fibrin monomers, impairing fibrin polymerization. Patients with amyloidosis experience amyloid deposits in the vessel wall which impair vascular integrity. Amyloid can also cause precipitation of factor X and factor VII. Circulating anticoagulants which have been described in myeloma include a heparin-like substance, antibody to factor VIII, antibody to factor X and antibody to factor VII.

7) Neurologic Dysfunction

Neurologic dysfunction is present in 2% of patients at diagnosis and develops in 20-30% of patients during the course of disease. Spinal cord compression is associated with upper motor neuron signs. Hyperviscosity can be manifested by headache, blurring of vision, progressive obtundation, vertigo, ataxia, hemiparesis or seizure. A demyelinating, degenerative, predominantly sensory or sensorimotor, distal, symmetric polyneuropathy can be seen. Amyloid infiltration commonly causes carpal tunnel syndrome and distal dysesthesias, but may also be associated with autonomic neuropathy.

B) LABORATORY FEATURES

1) Hematology

A normocytic, normochromic anemia occurs in nearly every patient with multiple myeloma. Bone marrow aspirate and biopsy reveal increased numbers of plasma cells of various stages of differentiation from mature to markedly anaplastic appearance.

2) Monoclonal Gammopathy

Serum protein electrophoresis (SPEP) separates serum proteins by electrical charge. Most immunoglobulins migrate in either the γ or less frequently the β region. The monoclonal band is appreciated as a peak on densitomitry tracing of the electrophoretic gel. SPEP shows a monoclonal spike (M spike) in 90% of myeloma patients. Urine electrophoresis shows a globulin spike in 75% of patients. Immunoelectrophoresis or immunofixation utilizes antiserum to the individual heavy or light chains to identify the isotype of monoclonal protein present, of which 60% are IgG, 20% are IgA, 1% are IgD and 10% are light chain only. Immunoelectrophoresis is based on diffusion (Ouchterlony plate) of the M protein against anti-isotype antibody to form a precipitant arc. Immunofixation uses electrophoresis to separate serum proteins and subsequent immunoprecipitation with monospecific antibodies to identify the heavy or light chain isotype. Immunofixation may detect a monoclonal protein which was not apparent on routine electrophoresis (SPEP) or immunoelectrophoresis.

3) Serum Creatinine

Serum creatinine level is elevated (> 2.0 mg/dl) in 30-40% of patients.

6.6

4) **Serum Calcium**

Hypercalcemia is present in roughly 30% of patients.

5) **Skeletal Abnormalities**

Conventional roentgenograms reveal punched out "lytic" lesions, osteoporosis and/or fractures in about 80% of patients. The vertebrae, skull, thoracic cage, pelvis and proximal humeri and femurs are the most frequent sites of involvement.

C) **DIFFERENTIAL DIAGNOSIS**

Bone pain, anemia and renal insufficiency constitute a triad that is strongly suggestive of multiple myeloma. The diagnosis depends on the demonstration of increased numbers of plasma cells in the bone marrow. The differential diagnosis of monoclonal protein production and monoclonal gammopathy includes monoclonal gammopathy of undetermined significance (MGUS), macroglobulinemia (Waldenstroms), chronic lymphocytic leukemia (CLL), lymphoma (lymphoplasmacytoid lymphoma), amyloidosis and solitary or extramedullary plasmacytoma.

IV) **STAGING SYSTEM AND PROGNOSTIC FACTORS**

A) **DURIE-SALMON STAGING SYSTEM**

Distinguishing patients with low, intermediate and high volumes of tumor mass before institution of therapy is useful for prognosis (Tables 6.6.2 and 6.6.3). The three stages are divided into substages A (serum creatinine < 2 mg/dl) and B (serum creatinine ≥ 2 mg/dl).

B) **PROGNOSTIC FACTORS** (TABLE 6.6.3)

Myeloma is an incurable malignancy when treated with conventional chemotherapy. Syngeneic and allogeneic bone marrow transplantation have been associated with the only reported cures. The median survival for patients treated with standard therapy is approximately 36 months and ranges from a median of 15 months for those presenting with stage IIIB disease to 60 months for

6.6

Table 6.6.2. Assessment of tumor mass (Durie-Salmon)

High Tumor Mass (stage III) (> 1.2×10^{12} tumor cells/m²)
One of the following abnormalities must be present:
Hemoglobin < 8.5 g%, hematocrit < 25 volume %
Serum calcium > 12 mg %
Very high myeloma protein production
IgG peak > 7 g%
IgA peak > 5 g %
Bence-Jones protein > 12 g/24 hour
> 3 lytic lesions on bone survey (bone scan not acceptable)
Low Tumor Mass (stage I) (< 0.6×10^{12} malignant cells /m²)
All of the following must be present:
Hemoglobin > 10.5 g % or hematocrit > 32 volume %
Serum calcium normal
Low myeloma protein production
IgG peak < 5 g%
IgA peak < 3 g %
Bence-Jones protein < 4 g/24 hour
No bone lesions
Intermediate Tumor Mass (stage II) (0.6-1.2×10^{12} malignant cells/m²)
All patients who do not qualify for high or low tumor mass are considered to have intermediate tumor mass

Table 6.6.3. Prognostic factors and survival for multiple myeloma

Durie-Salmon Stage	Median Survival
Stage I	> 60 months
Stage II	36-48 months
Stage III	6-24 months
β2M and CRP levels	**Median Survival**
Both β2M and CRP < 6 mg/dl	54 months
Either β2M or CRP > 6 mg/dl	27 months
Both β2M and CRP > 6 mg/dl	6 months

β2M = beta -2-microglobulin, CPR = C–reactive protein

patients presenting with stage IA disease. The serum beta-2 microglobulin (β2M) level represents the single most important prognostic variable that is readily obtained in a quantitative and reproducible fashion. In addition, the plasma cell labeling index, the serum C-reactive protein (reflecting in vivo interleukin-6 activity) and blood soluble interleukin-6 receptor levels have been recognized as independently helpful parameters. Plasmablastic morphology portends a rapidly fatal course, associated with high tumor cell labeling index and elevation of serum lactic dehydrogenase (LDH).

V) **PRETREATMENT EVALUATION**

A) INITIAL EVALUATION

The standard evaluation of a patient with suspected multiple myeloma includes a complete blood count and differential, a blood chemistry profile, a radiographic skeletal survey and a bone marrow aspiration and biopsy. The aspirate may also be evaluated for plasma cell labeling index, for cytogenetic analysis and for cytometric DNA analysis where these tests are available. Serum and urine electrophoresis and immunofixation of serum and urine identify the monoclonal protein produced, and densitometry of these electrophoreses quantitate the protein. A 24 hour urine collection is performed to quantitate the total protein excretion as well as the monoclonal light chain excretion. Quantitative serum immunoglobulin analysis is used to assess the distribution and adequacy of normal immunoglobulins. Magnetic resonance imaging (MRI) of the spine and pelvis are not routine evaluations, but are particularly useful in detection of a suspected soft tissue mass and in ruling out spinal cord compression. MRI is also useful to differentiate multiple myeloma from a solitary plasmacytoma. Prognostic indicators should be evaluated at diagnosis, including a β2M, C-reactive protein and, when available, plasma cell labeling index. Finally, if the patient is under the age of 60 years, the patient and siblings should undergo HLA typing, anticipating possible allogeneic bone marrow transplantation.

B) PRETRANSPLANT EVALUATION

The initial evaluation above should be repeated prior to high-dose therapy. Refer to the chapter on BMT patient evaluation for pre-BMT evaluation.

VI) **TREATMENT OF NEWLY DIAGNOSED MYELOMA**

A) STANDARD CHEMOTHERAPY

The mainstays of treatment for myeloma include glucocorticoids, which are most effective in high doses, alkylating agents and local radiation. A meta-analysis of the results of 18 trials, involving 3,814 myeloma patients, comparing

6.6

treatment with melphalan and prednisone to combination chemotherapy revealed no significant difference in the survival of patients receiving either of these two forms of therapy. Most individual trials comparing a single alkylating agent to multiple agents have shown no significant survival benefit. High-dose dexamethasone is the most active single agent for inducing responses in refractory myeloma patients and in the previously untreated patient. Nevertheless, glucocorticoids have never been found to enhance survival in patients randomized to glucocorticoids vs placebo or melphalan vs melphalan and prednisone. Cumulative exposure to melphalan is associated with increasing risk of marrow toxicity, including myelodysplasia and impaired stem cell production.

B) AUTOLOGOUS TRANSPLANTATION FOR NEWLY DIAGNOSED MYELOMA

Based upon the preliminary results of the University of Arkansas "Total Therapy" program, utilizing high-dose therapy and stem cell transplant early in the course of myeloma, and the randomized French myeloma group trial demonstrating a relapse-free survival and overall survival advantage in patients undergoing high-dose therapy with autologous transplantation compared to standard therapy, a similar trial sponsored by the National Cancer Institute was initiated in the United States in early 1994. Newly diagnosed untreated patients should be referred to participating centers for consideration of this protocol. This randomized trial will offer a valid assessment of the impact of early high-dose therapy and stem cell infusion when compared to more conventional therapy.

VII) THERAPY FOR PREVIOUSLY TREATED MYELOMA

6.6

A) SALVAGE CHEMOTHERAPY

Salvage therapy is needed for patients who have relapsed after initial induction therapy or who have developed resistance to first-line treatment, either due to primary refractoriness or as a result of resistant relapse after remission. The scope of available salvage regimens narrows with the number of regimens to which resistance has been established. In the presence of resistance restricted to alkylating agents, DEX alone induces responses in about 30% of patients regardless of whether a response had been obtained previously. VAD (Table 6.6.4) benefits about 50% of relapsing patients and about 30% of those with primary unresponsive disease. EDAP (Table 6.6.4) chemotherapy is an effective regimen in patients with plasmablastic myeloma.

B) AUTOLOGOUS TRANSPLANTS

With additional resistance to DEX or VAD, tumor control can be established with dose-intensive therapy using autologous transplants. With combination chemotherapy-radiotherapy regimens, given either in single or tandem cycles, almost 80% of patients achieve greater than 75% tumor cytoreduction, including true complete response rates of 15-20% with median relapse-free survival and overall survival on the order of 2-3 years posttransplant, respectively. The frequency and duration of antitumor effect declines with the duration of previous therapy (more than 12 months), increased β2M and non-IgA isotype. Because autologous transplantation is not curative using currently available methods, subsequent DEX pulsing and/or interferon-α maintenance may be useful in further extending the duration of disease control.

Table 6.6.4. Treatment regimens in use for myeloma

Regimen	Drug	Dose	Repeat cycle
EDAP	Etoposide	100 mg/m²/d CI days 1-4	q 35-42 days
	Dexamethasone	40 mg/m²/d PO days 1-5	
	Cytarabine	1 g/m² IV day 5	
	Cisplatin	20-25 mg/m²/d CI days 1-4	
IFN/DEX	Interferon	3-5 million units SQ QD or TIW	q 35 days
	Dexamethasone	40 mg/d PO days 1-4, 9-12, 17-20	
MP	Melphalan	8 mg/m²/d PO days 1-4	q 28 days
	Prednisone	40 mg/m²/d PO days 1-7	
VAD	Vincristine	0.4 -0.5 mg/d CI days 1-4	q 35 days
	Doxorubicin	10 mg/m²/d CI days 1-4	
	Dexamethasone	40 mg/d PO days 1-4, 9-12, 17-20	
VBAP	Vincristine	1.0 mg IV day 1	q 28-42 days
	Carmustine	30 mg/m² IV day 1	
	Doxorubicin	30 mg/m² IV day 1	
	Prednisone	60 mg/m²/d PO days 1-7	
VBMCP	Vincristine	1.2 mg/m² IV day 1	q 35-42 days
	Carmustine	20 mg/m² IV day 1	
	Melphalan	8 mg /m²/d PO day 1-4	
	Cyclophosphamide	400 mg/m² IV day 1	
	Prednisone	40 mg/m²/d PO days 1-7	
VMCP	Vincristine	1.0 mg IV day 1	q 28-42 days
	Melphalan	6 mg/m² PO days 1-4	
	Cyclophosphamide	125 mg/m²/d IV days 1-4	
	Prednisone	60 mg/m²/d PO days 1-4	

6.6

VIII) TRANSPLANTATION

A) AUTOLOGOUS TRANSPLANTS

Transplant-related mortality is less than 5%. However, even in the setting of consolidation therapy for first remission, the frequency of true complete response usually does not exceed 50%. Depending on prognostic factors, event-free and overall survival durations are on the order of 3 and 5 years, respectively, from transplantation. No plateau has been realized in the survival curves following autologous transplantation and at this time autologous transplantation may not be considered curative.

1) Timing

Individuals up to age 70 years, lacking other significant health problems, should remain candidates for autologous transplant trials. Optimal timing for autologous BMT (i.e. newly diagnosed or after relapse) has not been established. However, when considering autologous bone marrow transplantation or peripheral blood stem cell (PBSC) transplant, hematopoietic progenitor cells should be collected prior to initiation of a chronic alkylating agent or interferon therapy, because of the potential for subsequent stem cell impairment. Patients who had received more than 14 courses of melphalan in a Toronto series never achieved adequate peripheral stem cell harvest and those who had received 10-14 courses obtained adequate harvest less than half the time in spite of a high number of attempts. When compared to alkylating agents, the preservation of adequate hematopoietic stem cell function is likely with dexamethasone (DEX) administered alone or in combination with vincristine and doxorubicin (VAD) (Table 6.6.4).

2) Conditioning

The optimal conditioning regimen has yet to be established for multiple myeloma. The more common preparative regimens are outlined in Table 6.6.5. Melphalan alone or in combination with total body irradiation is an active myeloablative regimen in the treatment of myeloma.

3) Source of Hematopoietic Stem Cells

Various sources of hematopoietic stem cells are currently utilized for hematopoietic reconstitution following myeloablative therapy, including bone

---

Table 6.6.5. Transplant preparative regimens in multiple myeloma

Autologous
MEL 140-200 mg/m^2, single dose or divided over 2 days
MEL 140 mg/m^2 + TBI 850-1200 cGy
BU + CY + TT or MEL
TBI + others

Allogeneic
BU + CY
BU + MEL
CY + TBI
MEL + TBI
TBI + others

Abbreviations : BU = busulfan; CY = cyclophosphamide; MEL =melphalan;
TBI = total body irradiation (850-1375 cGy); TT =thiotepa

6.6

marrow, PBSC or a combination of the two. However, in terms of long-term prognosis, PBSC, especially when "mobilized" by high-dose chemotherapy (e.g. cyclophosphamide, 4 g/m^2) and/or hematopoietic growth factors (e.g. G-CSF), shorten the duration of bone marrow aplasia compared to unprimed bone marrow as the sole source of stem cells. With the utilization of PBSC and hematopoietic growth factors, the duration of bone marrow aplasia is shortened markedly with severe neutropenia (< 500/μl) and thrombocytopenia (< 50,000/μl) usually lasting no longer than 1 week. The advantage of one modality over another has not been reported in single institution or in randomized trials. The most crucial determinant for adequate stem cell function is the duration of prior alkylating agent therapy, particularly with melphalan or nitrosoureas. Significant compromise of hematopoietic function is seen in more than 50% of patients who have had more than 1 year of prior therapy with such drugs.

Tumor cells are present in both bone marrow and peripheral blood stem cells, even in newly diagnosed patients and those patients in complete remission. The presence of bone marrow plasmacytosis of up to 30% has not affected relapse-free or overall survival. The etiology of relapse following myeloablative therapy remains unknown, whether due to re-infusion of contaminating tumor cells or inadequate tumor cytoreduction. Tumor cell removal strategies include the use of monoclonal antibodies, immunotoxins and chemotherapy. Positive selection of normal hematopoietic cells using CD34 antigen expression, either by cell separation or ex vivo expansion, is currently being evaluated.

4) Hematopoietic Growth Factors

Hematopoietic growth factors, such as GM-CSF or G-CSF, either alone or in combination with chemotherapy such as high-dose cyclophosphamide can be utilized for collecting PBSC, and reduce marrow aplasia posttransplant. Newer hematopoietic growth factors, such as PIXY 321 (GM-CSF/IL-3), IL-6 and IL-11 to enhance peripheral blood stem cell mobilization and neutrophil and platelet counts posttransplant are in the investigative phase.

5) Maintenance Therapy

Interferon-α (IFN) is the only immunomodulatory agent evaluated in randomized trials for maintenance therapy. Trials with thrice weekly or daily administration of 2-5 million units subcutaneously have been explored during remission induction with standard MP or combination regimens. Collectively, these data show longer durations of disease control in patients receiving IFN but overall survival is usually not extended. Improved relapse-free and overall survival with IFN maintenance posttransplantation have been reported in a single randomized trial.

B) ALLOGENEIC TRANSPLANTS

The potential advantages of allogeneic transplants for myeloma are the absence of tumor cell re-infusion that occurs with autologous transplant and the benefit of a graft-versus-myeloma effect. Disease-free survival has been observed as long as 14 and 7 years post syngeneic and allogeneic transplantation, respectively. Due to age limitation (< 60 years) and limited availability of histocompatible sibling donors, less than 10% of myeloma patients are candidates for allotransplant. The treatment-related early mortality (< 100 days) for

6.6

allotransplants in myeloma is approximately 25-40%. Fifty percent of patients survive at 1 year. Improved outcome has been observed in patients who enter myeloablative therapy after less extensive prior therapy (< 2 years), and among patients with low pretransplant β2M and attainment of a complete response with myeloablative therapy. Late relapses have been observed, particularly when allotransplants have been used in the salvage setting and when complete remissions have failed to be attained with myeloablative therapy. Overall, 25-30% of patients survive beyond 5 years after allogeneic transplant for myeloma, and a significant percentage of these survivors may be cured of their disease.

1) Conditioning

The optimal allogeneic conditioning regimen has not been defined. However, the combination of either melphalan and TBI or busulfan and cyclophosphamide are the two most common regimens. A dose intensification study of chemotherapy in 20 myeloma patients receiving allogeneic transplant at the University of Washington varied doses of busulfan and cyclophosphamide. With a combination of busulfan administered at 14 mg/kg and cyclophosphamide administered at 120 mg/kg the complete response rate was 75%, and the transplant related death rate was 0/4 patients. Two of four patients remained in remission at long-term follow up. There were no deaths ascribable to veno-occlusive disease. The continued CR rate for this entire trial was 35% at 190-1271 days of follow up, but only half of the patients in the trial had had responsive disease at the time of their transplant. The rate of complete remission achieved with busulfan and cyclophosphamide preparation for allogeneic transplantation in another study completed at Arkansas was 50% for patients who had received less than 24 months of prior therapy and 18% for patients with greater than 24 months of prior therapy.

2) Graft-vs-Myeloma Effect

Few syngeneic bone marrow grafts have been carried out to treat myeloma, because of the low probability that appropriate patients will have an identical twin. It is therefore impossible to compare syngeneic to allogeneic transplantation to try to derive an appreciation of graft-vs-myeloma effect. The largest experience with syngeneic transplant is in Seattle, where two out of seven patients remain disease-free at 8 and 14 years post-bone marrow transplant. Retrospective analysis of the allogeneic transplant experience of the European Bone Marrow Transplant Registry indicates that GVHD has varied in severity and frequency in relation to use of T cell depletion or drug-based GVHD prophylaxis, but GVHD incidence and severity have seemed to have little bearing on disease relapse. Within 79 months of follow up by the EBMT, 43 of 90 allogeneically transplanted patients had died, 8 of them from progressive disease and the rest from transplant-related mortality. Historically, high transplant- related mortality and low frequency of allogeneic transplant for this disease have made appreciation of an immunologic effect on remission maintenance less likely.

A number of immunologic interventions to enhance autologous and allogeneic immune response to myeloma are now under investigation. In order to test the hypothesis that tumor antigen-specific immunity could be transferred from transplant donor to recipient, a healthy sibling donor was immunized pretransplant with immunoglobulin G, which had been puri-

6.6

fied from the plasma of the recipient and conjugated to an immunogenic carrier protein and emulsified in an adjuvant. A myeloma idiotype-specific T cell response was successfully transferred to the recipient, and led to the recovery of a recipient CD4+ T cell line with unique specificity for the myeloma idiotype. A pilot clinical trial designed to improve the potency and duration of the transferred idiotype-specific response is now in progress.

The transduction of donor lymphocytes with herpes simplex virus-thymidine kinase (HSV-TK) gene is designed to make them vulnerable to the effects of systemically administered gancyclovir, so that adverse effects of lymphocyte infusion, such as severe graft-vs-host disease, can be controlled by gancyclovir administration. Two clinical trials are now becoming activated utilizing donor buffy coat infusions of HSV-TK-transduced lymphocytes, one in conjunction with T cell depleted allografts for myeloma, and one for the treatment of patients who relapse after allogeneic transplant.

SUGGESTED READING

1. Barlogie B et al. Autologous and allogeneic transplants for multiple myeloma. Sem Hematol 1995; 32:31-44.
2. Bensinger WI et al. Phase I study of busulfan and cyclophosphamide in preparation for allogeneic marrow transplant for patients with multiple myeloma. J Clin Oncol 1992; 10:1492-1497.
3. Hoover RG et al. Autoregulatory circuits in myeloma. Tumor cell cytotoxicity mediated by soluble CD16. J Clinical Invest 1995; 95:241-247.
4. Kwak LW et al. Transfer of myeloma idiotypic-specific immunity from an actively immunized marrow donor. Lancet 1995; 345:1016-1024.
5. Morbacher A, Anderson KC. Bone marrow transplantation in multiple myeloma. In: Malpas JS, Bergsagel DE, Kyle RA eds. Myeloma, Biology and Management. Oxford: Oxford University Press, 1995.
6. Stevenson FD et al. Immunologic variable region genes as DNA vaccines for myeloma. Abst presented at Vth International Workshop on Multiple myeloma.
7. Vesole DH et al. High-dose therapy for refractory multiple myeloma: Improved prognosis with better supportive care and double transplants. Blood 1994; 84(3):950-956.
8. Gahrton G et al. Allogeneic bone marrow transplantation in multiple myeloma. N Engl J Med 1992; 325:1267-1273.
9. Jagannath S et al. Hemopoietic stem cell transplants for multiple myeloma, Oncology 1994; 11:89-103.

6.6

Bone Marrow Transplantation in Lymphomas

S. Martin-Algarra, Philip J. Bierman, James O. Armitage

I) INTRODUCTION

Chemotherapy alone or combined with radiation therapy has been the standard treatment for Hodgkin's disease (HD) and non-Hodgkin's lymphomas (NHL) during the last 30 years, but a considerable number of patients cannot be cured with these approaches and may be considered candidates for bone marrow transplantation (BMT). In this chapter unless otherwise stated, bone marrow transplant (BMT) refers to autologous BMT (not allogeneic).

A) NHL

1. Nearly all patients with advanced stage low grade NHL die from their disease (median survival 5-10 years).
2. A substantial number (approximately 50%) of intermediate and high grade NHL patients relapse, despite an initial response, and die of malignancy.

B) HD

1. Approximately 50% of patients with advanced HD fail to achieve a lasting response.
2. There is a small, but still considerable number of patients with advanced HD (20-30%) who do not attain a complete response after standard treatments.

II) ETIOLOGY

A) NHL

The etiology of NHL is unknown, but different viral, immunological, environmental and hereditary factors have been implicated (Table 6.7.1). Characterizations of nonrandom chromosomal translocation have been instrumental in defining oncogenes relevant to lymphoma (*c-myc, bcl-1, bcl-2, bcl-6* and others). Rearranged oncogenes, as well as immunoglobulin and T cell receptor gene rearrangements are used as markers of clonality or for characterization and subclassification of lymphomas. Some of the more consistent karyotypic abnormalities are:

1. t(8;14) associated with 80% of Burkitt lymphomas, and with T cell lymphomas;
2. t(2;8) associated with 10% of Burkitt lymphomas;
3. t(8;22) associated with 10% of Burkitt lymphomas;
4. t(14;18) associated with 85% of follicular lymphomas (small cleaved cell lymphomas);
5. t(11;14) associated with anaplastic large cell lymphomas;
6. t(2;5) associated with Mantle cell lymphomas and with T cell lymphomas;
7. t(3q27) present in about one-third of diffuse B cell large cell lymphomas.

B) HD

The diagnose of HD relies on the presence of the Reed-Sternberg cell or its mononuclear variants, in an appropriate cellular background. Nevertheless,

6.7

Table 6.7.1. Possible etiologic factors for NHL

Immunodeficiency
  Congenital
      ataxia-telangiectasia
      severe combined immunodeficiency
      Wiskott-Aldrich syndrome
      X-linked lymphoproliferative syndrome
  Acquired
      autoimmune diseases
         rheumatoid arthritis
         Sjogren syndrome
      iatrogenic immunodeficiencies
         organ transplantation
         azathioprine, cyclosporine or anti-CD3 treatments
      HIV infection/AIDS
Occupational/Environmental
  Hair dyes
  Ionizing radiation
  Pesticides
      phenoxy herbicides
      triazine herbicides
      organophosphate insecticides
  Organic solvents
  Ultraviolet light
Viral
  Epstein-Barr virus
  HTLV-I

6.7

Reed-Sternberg-like cells are not pathognomonic of HD and can be found in reactive lymphoid hyperplasia, such as infectious mononucleosis, NHL and nonlymphoid malignancies like sarcomas and carcinomas. The origin of the Reed-Sternberg cell is unknown, but laboratory evidence supports both T and B lymphocyte as well as a macrophage cell of origin. In contrast to NHL, no specific genomic abnormalities (e.g. translocation) have been documented in HD. However, the Epstein-Barr virus (EBV) may be involved in the pathogenesis of HD.

III) EPIDEMIOLOGY
A) NHL

The incidence of NHL is increasing steadily. Data from the NCI's Surveillance, Epidemiology and End Results Program demonstrate a 4% per year increase in men and a 3% per year increase in women since 1970. This finding seems to be predominantly due to an increase in diffuse aggressive lymphoma (nearly 30% of all NHL), and cannot be completely attributed to the AIDS epidemic or to improvements in diagnosis methods. In 1995 it is estimated that there will be 52,700 new cases of NHL, accounting for 23,300 deaths. The incidence of NHL increases with age and is most common after the fourth decade of life.

B) HD

The age-specific incidence rates of HD has a bimodal distribution: the peak incidence is between ages 15-34 years old, with a second peak after 55 years of age. Contrary to NHL, the age-adjusted incidence of HD has declined slightly

over the same time period, specially among the elderly. The estimated incidence
and mortality in the United States for HD in 1996 is 7500 and 1510, respectively.

IV) CLASSIFICATION

A) NHL

Several schemes of classification of NHL have been used during the last 20
years. A correlation among three common classifications is shown in
Table 6.7.2. Although the classification systems may appear complicated and
overwhelming, for clinical purposes most NHL fall into one of three categories: low grade, intermediate grade, and high grade, grouped by natural history (Table 6.7.3). Three common classification systems are:

1) Working Formulation for Clinical Usage (WF)

Since its introduction in 1982, the WF has been the classification most
used in the United States. This is due to its simplicity and its predictive
value for survival. Nevertheless, the WF was initially designed as a method
of translating between the different pre-existing schemes, rather than a
separate system.

2) Kiel Classification

Updated in 1988, the Kiel classification is based on the postulated relationship of neoplastic lymphoid cells to their normal counterparts in the
immune system. Morphological analysis and immunologic studies are used
in this classification to define specific disease entities.

3) Revised European-American Lymphoma Classification (REAL)

Recently introduced, this new scheme includes several subtypes of lymphoma not previously well classified, and uses molecular biology techniques to better characterize and differentiate among entities.

B) HD

Unlike NHL, general agreement exists regarding classification of HD. The
Rye classification relates histopathologic subtypes to clinical behavior and prognosis (Table 6.7.4).

V) STAGING AND PROGNOSTIC FACTORS AT DIAGNOSIS

A) NHL

Although the Ann Arbor staging system (see HD staging below) is routinely
used in NHL, several other characteristics, besides anatomical stage and the
presence of systemic symptoms, have been shown to be important prognostic
predictors of response and disease-free survival. Accordingly, an International
Prognostic Factor Index for aggressive NHL has been developed (Table 6.7.5).

B) HD

The Ann Arbor staging system (Table 6.7.6) is based on anatomical stage and
systemic symptoms. It is currently used in both HD and NHL, but its prognostic implications are not as important in NHL since most of these patients at the
time of presentation already have disseminated and extranodal disease. This
staging system is based on anatomical sites of involvement and the presence
or absence of these systemic symptoms, also called B symptoms: unexplained
fever > 38°C, 10% weight loss in the 6 months prior to diagnosis and night
sweats. Pruritus, ranging from localized and mild to generalized and severe, is
a common manifestation in HD, but its pathogenesis and significance is unclear, and it is not considered a B symptom. In 1989 the Cotswald modifications were introduced for the staging of HD (Table 6.7.6).

6.7

Table 6.7.2. Comparison of the REAL classification with the Kiel classification and working formulation

Kiel Classification	Revised European-American Lymphoma Classification	Working Formulation
B-lymphoblastic	Precursor B-lymphoblastic lymphoma/leukemia	Lymphoblastic
B-lymphocytic CLL B-lymphocytic, prolymphocytic leukemia	B cell chronic lymphocytic leukemia/ prolymphocytic leukemia/small lymphocytic lymphoma	Small lymphocytic consistent with CLL
Lymphoplasmacytoid immunocytoma	Lymphoplasmacytoid lymphoma	Small lymphocytic, plasmacytoid
Centrocytic	Mantle cell lymphoma	Diffuse, small cleaved cell
Centroblastic-centrocytic, follicular	Follicular center lymphoma, follicular	
	Grade I	Follicular, predominantly small cleaved cell
	Grade II	Follicular, mixed small and large cell
Centroblastic follicular	Follicular center lymphoma, follicular Grade III	Follicular, predominantly large cell
Centroblastic-centrocytic, diffuse	Follicular center lymphoma, diffuse, small cell (provisional)	Diffuse, mixed small and large cell
Monocytoid, including marginal zone	Extranodal marginal zone B cell lymphoma (low grade B cell lymphoma of MALT type)	Small lymphocytic Diffuse small cleaved cell Diffuse mixed small and large cell
	Nodal marginal zone B cell lymphoma (provisional)	
	Splenic marginal zone B cell lymphoma (provisonal)	

6.7

Table 6.7.2. Comparison of the REAL classification with the Kiel classification and working formulation (cont'd.)

Kiel Classification	Revised European-American Lymphoma Classification	Working Formulation
Hairy cell leukemia	Hairy cell leukemia	
Plasmacytic	Plasmacytoma/myeloma	Extramedullary plasmacytoma
Centroblastic (monomorphic, polymorphic and multilobated subtypes)	Diffuse large B cell lymphoma	Diffuse, large cell
B-Immunoblastic		Large cell immunoblastic
B-large cell anaplastic (Ki-1)	Primary mediastinal large B cell lymphoma Large cell anaplastic B cell	Diffuse, large cell Large cell immunoblastic
Burkitt's lymphoma	Burkitt's lymphoma	Small noncleaved cell, Burkitt's
	High grade B cell lymphoma, Burkitt-like (provisional)	Small noncleaved cell, non-Burkitt's
T-lymphoblastic	Precursor T-lymphoblastic lymphoma/leukemia	Lymphoblastic
T-lymphocytic, CLL type T-lymphocytic, prolymphocytic leukemia	T cell chronic lymphocytic leukemia/ prolymphocytic leukemia	Small lymphocytic Diffuse small cleaved cell
T-lymphocytic, CLL type	Large granular lymphocytic leukemia -T cell type -NK cell type	Small lymphocytic Diffuse small cleaved cell

6.7

Small cell cerebriform (mycosis fungoides, Sezary syndrome)	Mycosis fungoides/Sezary syndrome	Mycosis fungoides
T-zone Lymphoepithelioid Pleomorphic, small T cell Pleomorphic, medium-sized and large T cell	Peripheral T cell lymphomas, unspecified including provisional subtype: subcutaneous panniculitic T cell lymphoma	Diffuse small cleaved cell Diffuse mixed small and large cell Diffuse large cell
T-immunoblastic	Hepatosplenic T cell lymphoma (provisional)	Large cell immunoblastic
Angioimmunoblastic (AILD, LgX)	Angioimmunoblastic T cell lymphoma	
	Angiocentric lymphoma	Diffuse small cleaved cell Diffuse mixed small and large cell Diffuse large cell Large cell immunoblastic
Pleomorphic small T cell, HTLV1	Adult T cell lymphoma/leukemia	
Pleomorphic medium-sized and large T cell, HTLV1		Diffuse, mixed small and large cell Diffuse, large cell
T large cell anaplastic (Ki-1)	Anaplastic large cell lymphoma, T and null cell types	Large cell immunoblastic

Modified from Blood. 1994; 84(5): 1361-1392

6.7

6.7

Table 6.7.3. Characteristics of the subtypes of NHL, according to the WF

Subtype	%	Growth	Cell Type	Immunophenotype % B cell	Immunophenotype % T cell	Potentially Curable	5 yr Survival Rate
Low grade							
Small lymphocytic	4	Diffuse	Small round cells	99	1	Unproved	58%
Follicular small cleaved cell	23	Follicular	Small cleaved cells	100	0	Unproved	70%
Follicular mixed cell	8	Follicular	Small cleaved cells intermediate number of large cells	100	0	Controversial	69%
Intermediate grade							
Follicular large cell	4	Follicular	Many large cleaved and noncleaved cells	100	0	Controversial	59%
Diffuse small cleaved cell	7	Diffuse	Small cleaved or small irregular	90	10	Controversial	43%
Diffuse mixed cell	7	Diffuse	Mixed small and large cells	60	40	Yes	47%
Diffuse Large cell	20	Diffuse	Predominantly large cleaved or noncleaved cells	90	10	Yes	43%
High grade							
Immunoblastic	8	Diffuse	Predominantly immunoblasts (plasmacytoid, clear or pleomorfic)	80	20	Yes	33%
Lymphoblastic	4	Diffuse	Round or convoluted lymphoblasts	5	95	Yes	39%
Small noncleaved cell	5	Diffuse	Intermediate-sized round cells	100	0	Yes	38%

(*) Modified from: Armitage J.O. Treatment of Non-Hodgkin's lymphoma. New Eng J Med 1993, 328: (14) 1023-1030.

VI) NATURAL HISTORY, STANDARD THERAPY AND RESULTS WITH BMT

A) NHL (TABLES 6.7.3, 6.7.7a-b AND 6.7.8)

1) Low Grade Lymphomas (small lymphocytic, follicular small cleaved cell, follicular mixed cell)

Low grade NHL accounts for approximately one-third of all cases of NHL in adults. All follicular lymphomas are B cell in origin. These lymphomas are usually extensive at diagnosis (80% stage III-IV; < 10% stage I), and most often have bone marrow involvement. The clinical course is usually indolent, and patients have a relatively long survival (5-10 years), with or without aggressive therapy.

a) Early-Stage (Ann Arbor I-II) Low Grade NHL

These patients are not candidates for BMT. Radiation therapy is the treatment of choice. Involved field, extended field and total nodal radiation have been used with good results, and long-term survival in patients 40 years old or younger is over 80%. However, in older

Table 6.7.4. Rye classification of Hodgkin's disease and clinical/pathologic correlations

Rye Classification	Incidence	Stage at Diagnosis	Prognosis
Lymphocyte predominant -diffuse pattern -nodular pattern	15%	Usually presents in neck or inguinal nodes as localized stage I disease	Very good
Nodular sclerosis	70%	Usually presents above the diaphragm as localized disease, stage I or II. Mediastinum is commonly involved	Good
Mixed cellularity	10%	Often present with abdominal disease	Intermediate
Lymphocyte depleted	5%	Usually stage III or IV at diagnosis	Poor

6.7

Table 6.7.5. International prognostic factors index for NHL

Risk Factors
Age > 60
Ann Arbor Stage III or IV
> 1 extranodal site
Performance status 2, 3 or 4
Lactic dehydrogenase above normal levels

Risk Category	Number of Risk Factors
Low	0-1
Low-intermediate	2
High-intermediate	3
High	4-5

From: New Eng J Med 1993; 329(14):987-994.

patients long-term survival is significantly reduced, and there is no clear evidence of cure in this particular subgroup of patients. For all patients, freedom from relapse is over 40% at 15 years. The role of chemotherapy, alone or in combination with radiotherapy, in these patients remains to be proved.

b) *Advanced-Stage (Ann Arbor III-IV) Low Grade NHL*

The choice of initial therapy is controversial. Asymptomatic patients may be managed conservatively, with observation alone. Single alkylating agents or combination chemotherapy induce a high rate of response, but relapses are the rule, and there is no definite evidence that aggressive therapy is better than conservative management. Interferon α and other biological response modifiers have been used with promising results and are now the object of prospective studies. Performance status, age, gender, LDH, hemoglobin, histologic subtype, marrow involvement, number of extranodal sites, nodal size, tumor burden, cytogenetics and immunophenotype have all been reported to influence survival. Some centers are currently treating patients in first complete remission with high-dose therapy and autologous stem cell rescue, but it is too early to assess the efficacy of these approaches.

c) *Relapsed Low Grade NHL*

BMT is usually attempted in low grade disseminated NHL after first or subsequent relapses. Because of the high frequency of bone marrow involvement in these lymphomas, most studies included some methods of purging the marrow in an attempt to eliminate the risk of infusing contaminating tumor cells. Preliminary reports indicate a longer time to treatment failure in patients treated with au-

---

Table 6.7.6. Ann Arbor staging classification and Cotswald modifications*

Stage I: Involvement of a single lymph node region or lymphoid structure.
Stage II: Involvement of two or more lymph node regions on the same side of the diaphragm.
Stage III: Involvement of lymph node regions or structures on both sides of the diaphragm.
Stage IV: Involvement of extranodal site(s) beyond that designated "E".
For All Stages:
   A. No symptoms
   B. Fever (> 38° C), sweats, weight loss (> 10% body weight over 6 months)
For Stages I to III:
   E: Involvement of a single, extranodal site contiguous or proximal to known nodal site.

Cotswald Modifications
Subscript "**X**" to be used if bulky disease is present.
The number of anatomical regions involved should be indicated by a subscript (e.g., II_3).
Stage III may be subdivided into:
   III1: with or without splenic, hilar, celiac or portal nodes
   III2: with para-aortic, iliac, mesenteric nodes
Staging should be identified as clinical stage (**CS**) or pathologic stage (**PS**). PS at a given site will be
   denoted by a subscript (ie, M=bone marrow; H=liver; L=lung; O=bone; P=pleura; D=skin).
Unconfirmed/uncertain complete remission (**CR[u]**) if persistent radiologic abnormalities of uncertain
   significance.

(*) Modified from Weinshel E.L., Peterson B.A., Hodgkin Disease, CA Cancer J. Clin. 1993; 43:327-346.

tologous BMT, but there is no definite evidence of a survival advantage using high-dose therapy and autologous stem cell rescue.

2) Intermediate Grade Lymphomas (follicular large cell, diffuse small cleaved cell, diffuse mixed cell, diffuse large cell and immunoblastic lymphomas)

These NHL are more common in adults, but present in all age groups. Most cases are composed of transformed B (85%) or T Lymphocytes (15%). Nodal (65%) and extranodal (35%) sites can be affected. Immunoblastic lymphoma is usually classified as a high grade lymphoma since survival without therapy is measured in weeks, similar to lymphoblastic lymphoma and small noncleaved cell lymphoma (Table 6.7.3) However, response to chemotherapy is alike between diffuse large cell and immunoblastic lymphomas and, therefore, from a practical point of view can be considered under intermediate grade NHL.

a) Early-Stage (Ann Arbor I-II) Intermediate Grade NHL

Localized intermediate grade lymphomas are reported to occur only in 10-25% of cases. These patients are generally treated with 3-4 cycles of a chemotherapy regimen such as CHOP and involved field

Table 6.7.7a. Single and combination agents used to treat low grade NHL

Acronym/Agents	Dose	Route	Days	Repeat Cycle
Single Agents				
chlorambucil	0.08-0.12 mg/kg	PO	daily	
	or 0.4-1.0 mg/kg	PO	1	day 28
cyclophosphamide	50-100 mg/m²	PO	daily	
	or 300 mg/m²/day	PO	1-5	day 28
fludarabine	25 mg/m²/day	IV	1-5	day 28
pentostatin	4 mg/m²	IV	1	day 14
cladribine	0.09 mg/kg/day	CI*	1-7	day 28
	or 0.14 mg/kg/day	IV (2 h)	1-5	day 28
Combination Therapies				
CVP				day 21
cylophosphamide	400 mg/m²	PO	1-5	
vincristine#	1.4 mg/m²	IV	1	
prednisone	100 mg/m²	PO	1-5	
COPP				day 28
cyclophosphamide	400-650 mg/m²	IV	1 and 8	
vincristine#	1.4 mg/m²	IV	1 and 8	
procarbazine	100 mg/m²	PO	1-14	
prednisone	40 mg/m²	PO	1-14	
CHOP				day 21
cyclophosphamide	750 mg/m²	IV	1	
doxorubicin	50 mg/m²	IV	1	
vincristine#	1.4 mg/m²	IV	1	
prednisone	100 mg/m²	PO	1-5	

6.7

* (continuous infusion).
# 2 mg maximum dose.
Modified from: Foon KA, Fisher RI. Lymphomas. In: Beutler E, Lichtman MA, Coller BS, Kipps TJ, eds. Williams Hematology. New York: McGraw-Hill, Inc., 1995:1082.

Table 6.7.7b. Combination chemotherapy for intermediate and high grade NHL

Acronym/Agents	Dose	Route	Days	Repeat Cycle
CHOP (see Table 6.7.7A)				
COP-BLAM				day 21
cyclophosphamide	400 mg/m²	IV	1	
doxorubicin	40 mg/m²	IV	1	
vincristine #	1 mg/m²	IV	1	
procarbazine	l00 mg/m²	PO	1-10	
prednisone	40 mg/m²	PO	1-10	
bleomycin	15 U	IV	14	
ProMACE/CytaBOM				day 21
clophosphamide	650 mg/m²	IV	1	
doxorubicin	25 mg/m²	IV	1	
etoposide	120 mg/m²	IV	1	
cytarabine	300 mg/m²	IV	8	
bleomycin	5 U/m²	IV	8	
vincristine #	1.4 mg/m²	IV	8	
methotrexate	120 mg/m²	IV	8	
leucovorin	25 mg/m²	PO	9 (q 6 h x 4)	
prednisone	60 mg/m²	PO	1-14	
MACOP-B				one 12-week cycle
methotrexate	400 mg/m²	IV	8, 36, 64	
leucovorin	15 mg	PO (q 6 h x 6)	9, 37, 65	
doxorubicin	50 mg/m²	IV	1, 15, 29, 43, 57, 71	
vincristine #	1.4 mg/m²	IV	8, 22, 36, 50, 64, 78	
bleomycin	10 U/m²	IV	22, 50, 78	
prednisone	75 mg	PO	1-84 (tapered 70-84)	
cyclophosphamide	350 mg/m²	IV	1, 15, 29, 43, 57, 71	
m-BACOD				day 21
methotrexate	200 mg/m²	IV	8, 15	
leucovorin	10 mg/m²	PO (q 6 h x 6)	9, 16	
bleomycin	4 U/m²	IV	1	
doxorubicin	45 mg/m²	IV	1	
cyclophosphamide	600 mg/m²	IV	1	
vincristine	1 mg/m²	IV	1	
dexamethasone	6 mg/m²	PO	1-5	
ESHAP		-		day 21
Etoposide	40-60 mg/m²	IV/2 h	1-4	
Methylprednisolone	500 mg/m²	IV	1-4	
Cytarabine	2 g/m²	IV/3 h	5	
Cisplatin	25 mg/m²	CI*	1-4	
DHAP				day 21
Dexamethasone	40 mg	PO or IV	1-4	
Cisplatin	100 mg/m²	CI*	1	
Cytarabine	2 g/m²	IV (12 h x 2)	2	

2 mg maximum dose
* continuous infusion
Modified from: Foon KA, Fisher RI. Lymphomas. In: Beutler E, Lichtman MA, Coller BS, Kipps TJ, eds. Williams Hematology. New York: McGraw-Hill, Inc., 1995:1084.

6.7

Table 6.7.7c. Combination chemotherapy regimens for Hodgkin's disease

Acronym/Agents	Dose	Route	Days	Repeat Cycle
MOPP				day 28
nitrogen mustard	6 mg/m²	IV	1, 8	
vincristine	1.4 mg/m²	IV	1, 8	
procarbazine	100 mg/m²	PO	1-14	
prednisone*	40 mg/m²	PO	1-14	
MVPP				day 28
nitrogen mustard	6 mg/m²	IV	1, 8	
vinblastine	6 mg/m²	V	1, 8	
procarbazine	100 mg/m²	PO	1-14	
prednisone	40 mg/m²	PO	1-14	
ChIVPP				day 28
chlorambucil†	6 mg/m²	PO	1-14	
vinblastine†	6 mg/m²	IV	1, 8	
procarbazine+	100 mg/m²	PO	1-14	
prednisone	40 mg/m²	PO	1-14	
ABVD				day 28
doxorubicin	25 mg/m²	IV	1, 15	
bleomycin	10 U/m²	IV	1, 15	
vinblastine	6 mg/m²	IV	1, 15	
dacarbazine	375 mg/m²	IV	1, 15	
MOPP/ABVD (alternate cycles of MOPP and ABVD)				
MOPP/ABV hybrid				day 28
nitrogen mustard	6 mg/m²	IV	1	
vincristine#	1.4 mg/m²	IV	1	
procarbazine	100 mg/m²	PO	1-7	
prednisone	40 mg/m²	PO	1-14	
doxorubicin	35 mg/m²	IV	8	
bleomycin	10 U/m²	IV	8	
vinblastine	6 mg/m²	IV	8	
MOPP/ABVD hybrid				day 28
nitrogen mustard	6 mg/m²	IV	1	
vincristine#	1.4 mg/m²	IV	1	
procarbazine	100 mg/m²	PO	1-7	
prednisone	40 mg/m²	PO	1-7	
doxorubicin	25 mg/m²	IV	15	
bleomycin	10 U/m²	IV	15	
vinblastine	6 mg/m²	IV	15	
dacarbazine	375 mg/m²	IV	15	

6.7

*In the original report Prednisone was given only on cycles 1 and 4. It now is commonly given with every cycle. † 10 mg maximum dose; + 150 mg maximum dose; # 2 mg maximum dose. Modified from: Horning S.J., Hodgkin Disease. In: Beutler E, Lichtman MA, Coller BS, Kipps TJ, eds. Hematology. New York: McGraw-Hill, Inc., 1995:1066.

Table 6.7.8. Failure-free survival (FFS) in NHL patients after autologous BMT

Histology	Conditioning	FFS (year)		Value
Low grade NHL				
relapsed	CY/TBI	30-40%	(3)	+/++
Intermediate grade NHL				
poor risk	BEAM/CBV CY/VP/TBI	67-85%	(3)	++/+++
1st PR NHL		76%	(2)	+++
refractory NHL		< 15%	(2)	+
relapsed		20-40%	(2)	++++
High grade NHL				
lymphoblastic NHL (poor risk)	CY/TBI	60%	(2)	+++
Burkitt lymphoma (PR, 1st or 2nd sensitive relapse)	BEAM	40-50%	(2)	+++
HD				
high risk HD	BEAM/CBV	70-80%	(2)	+
1st PR HD	CBV/BEAM	73%	(48mo)	+++
refractory HD	BEAM/CBV	22%	(3)	++
relapsed HD	BEAM	53%	(3)	++++
	CBV±P	40-64%	(84% if > 1y from CR)	

Key: +   : Little evidence of improving survival
     ++   : Some evidence of improving survival
     +++ : Good evidence of improving survival
     ++++ : Strong evidence of improving survival
PR: Partial Remission
CR: Complete Remission

6.7

Table 6.7.9. Conventional approach to treatment of HD and 10 years (y) disease-free surival (DFS)

Stage	Treatment	10y DFS
IA peripheral*	Mini-mantle/inverted Y	> 90%
I-IIA supradiaphragmatic*	Mantle irradiation ±para-aortic fields	85-90%
I-IIA infradiaphragmatic	Inverted Y ±para-aortic fields	85-90%
I-II B supradiaphragm*	Subtotal nodal irradiation	70-75%
I-II B infradiaphragmatic	Combined modality	~80%
I-II bulky, mediastinal	Combined modality	~80%
III A	Chemotherapy†	75-80%
III B-IV	Chemotherapy†	55-65%

* With staging laparotomy. Spleen included in para-aortic field if no laparotomy done.
† Some patients might benefit from involved field radiation.

radiation (4000-4500 cGy). Some centers advocate 6 cycles of chemotherapy without irradiation. Disease-free survival with either approach is greater than 80%. Results with radiation therapy alone are worse than with chemotherapy or combined modalities. Localized extranodal intermediate grade lymphomas are usually treated with surgery or radiotherapy followed by chemotherapy. Bone marrow transplantation is generally not considered in patients with low stages at diagnosis. However, in some centers, patients with bulky mass (> 10 cm), three or more sites of disease or other poor prognostic factors are considered for high-dose therapy and stem cell rescue.

b) *Advanced-Stage (Ann Arbor III-IV) Intermediate Grade NHL*

If untreated, prognosis is poor. However, after treatment with combination chemotherapy 50-80% of patients can achieve complete remission, and 5 year disease-free survival rates are 30-50%. Patients with disseminated intermediate grade NHL are generally evaluated for autologous BMT if they fail to enter complete remission with initial therapy, or following relapse from a front line chemotherapy regimen. However, as in acute leukemia, the best results with BMT for NHL have involved treatment of patients in first complete remission. It has been estimated that autologous BMT for all patients with intermediate NHL in first complete remission will increase cure in 13%, but will result in 70% unnecessary BMT treatments. Therefore, consolidation in first remission with BMT should be reserved for patients with a high risk of relapse; although even this statement is controversial. Data from City of Hope using autologous BMT as consolidation in high risk NHL (high LDH, bulky disease, extranodal disease, advanced stage) showed a 54 month disease-free survival of 84%. Additional pilot studies have confirmed that very high cure rates can be achieved in NHL when autologous BMT is performed as a part of primary therapy. However, the influence of selection bias is uncertain. A multi-institutional randomized trial of early transplantation versus consolidative sequential chemotherapy for patients with aggressive NHL, recently published by French investigators, did not show significant differences in 3 year survival or disease-free survival (59% vs 52% respectively). However, in the subset of patients with poor prognosis features, a trend toward improved disease survival was found. New studies, as well as longer follow up are needed before any conclusion can be obtained.

c) *Relapsed Intermediate Grade NHL*

For patients who relapse after complete remission, prognosis with conventional chemotherapy is poor. With salvage therapy at conventional doses approximately 5-10% of patients survive. The most important prognostic factors in these patients are duration of first remission and whether another complete remission can be achieved prior to BMT. There is increasing evidence that outcome of relapsed intermediate grade lymphomas is improved with high-dose chemotherapy followed by BMT. In unselected groups of patients, the cure rate after autologous BMT appears to be approximately 30%, with

6.7

rates of 30-50% in those patients whose tumors remain sensitive to standard dose chemotherapy. An ongoing international prospective randomized trial is comparing transplantation and conventional salvage chemotherapy in patients with sensitive disease, but final results are still pending.

d) Failure to Achieve Complete Remission for Intermediate Grade NHL

In patients with advanced NHL who have a partial response after induction chemotherapy, a 76% disease-free survival 2 years after BMT and 12% early mortality have been described by researchers from the University of British Columbia. For patients who do not achieve at least a partial response (overt chemotherapy-resistance) results with autologous BMT are not encouraging with 2 year disease-free survival less than 15%. Alternative salvage approaches for those patients include the use of new agents (fludarabine, pentostatin, 2-chlorodeoxyadenosine), allogeneic BMT or other experimental approaches like monoclonal antibodies and interleukin-2 (IL-2).

3) High Grade Lymphomas (lymphoblastic lymphoma, small noncleaved cell lymphoma)

a) Lymphoblastic Lymphoma

Lymphoblastic lymphoma is a malignancy of immature T cells that primarily affects male adolescents and young adults. It is frequently diagnosed with anterior mediastinal masses, superior vena cava syndrome, pleural effusion and pericardial effusion. Bone marrow and central nervous system involvement are common. If untreated, it usually evolves into a picture resembling T cell acute lymphocytic leukemia. Aggressive chemotherapy, similar to acute lymphoblastic leukemia regimens, including central nervous system prophylaxis cures over 60-70%. However patients with extensive bone marrow, central nervous system involvement or high LDH do poorly, with less than 20% long-term survival. The outcome of poor risk patients can be improved to around 60% disease-free survival with either autologous or allogeneic BMT.

b) Small Noncleaved Cell Lymphoma (Burkitt's Lymphoma)

Burkitt's lymphoma is the most frequent NHL in children, however it is rare in adults (5-7%). Cells are of B lineage, expressing monoclonal IgM surface immunoglobulin. The sporadic form is rarely associated with EBV. Sites of tumor involvement usually are abdomen, gastrointestinal tract and bone marrow. There is a high risk of tumor lysis syndrome with initial chemotherapy, but more than 70% of patients are cured after first line chemotherapy with current protocols. Most of the relapses occur in the first 8 months. After this interval, most patients in remission are likely to be cured. Approximately 40-50% of those patients not cured with conventional chemotherapy (partial remission after induction or sensitive first or second relapse) can be cured with autologous or allogeneic BMT. Limited irradiation to sites of bulky disease is sometimes used as adjunctive treatment but may not be necessary.

6.7

B) HODGKIN'S DISEASE (TABLES 6.7.6, 6.7.7C AND 6.7.8; FIG. 6.7.1)

Adequate staging, supervoltage radiation therapy techniques and combination chemotherapy have dramatically improved the prognosis of HD patients. Long-term disease-free survival with standard chemotherapy, radiation or combined chemotherapy and radiation ranges from 40-90% depending on stage and histological subtype. Therefore, high-dose chemotherapy and stem cell rescue have been traditionally considered only in advanced stage HD, mainly after failure to achieve a complete remission, after relapse and, sporadically, as a consolidative treatment in high risk patients.

1) Early Transplantation in HD

BMT in first complete remission has not been frequently used in HD due to the difficulty with identifying patients with high risk of relapse after conventional therapy. However, investigators from Italy reported 87% disease-free survival 3 years after transplantation in a series of patients with stage IVB at diagnosis who achieve complete remission with the alternating regimen MOPP/ABVD, compared with 33% of historical patients with similar adverse prognostic factors and response to MOPP/ABVD. An international randomized trial on early transplantation is now being carried out for high risk HD.

2) Failure to Achieve Complete Remission in HD

Prognosis of HD patients who fail to enter complete remission is poor (0% disease-free survival and median survival of 16 months). With autologous BMT, 20-40% long-term disease-free survival rates have been described. Autologous BMT must be considered in advanced HD as soon as it is evident that the patient is not achieving a prompt complete remission.

3) Relapsed HD

In patients with untreated relapse, a 6 year event-free survival of 62% has been described by Vancouver investigators using high-dose chemotherapy and autologous BMT. However, the use of conventional chemotherapy before conditioning, in an attempt to reduce tumor burden, is the standard approach. A recent report from the British National Lymphoma Investigation on patients with relapsed or refractory HD randomized to autologous BMT with high-dose chemotherapy (BEAM) versus conventional low-dose chemotherapy (mini-BEAM) has shown an actuarial 3 year event-free survival significantly better in the patients who received autologous BMT (53% versus 10%). This report indicates that high-dose chemotherapy and autologous BMT are superior to conventional-dose salvage therapy in relapsed or refractory HD. Failure-free survival after BMT is higher in patients whose first remission lasts more than 1 year (60-80%), than in those with shorter remissions (approximately 40% 4 year failure-free survival).

VII) PRETRANSPLANT STAGING

Prior to admission for BMT, all patients need to be restaged and re-evaluated in order to define accurately the extent of the disease and assess their physical and psychological ability to tolerate high-dose chemotherapy or chemoradiotherapy (Table 6.7.10).

VIII) CONDITIONING REGIMENS (TABLE 6.7.11)

A) NHL

Most high-dose preparative regimens for NHL include one or more alkylating agents, etoposide, occasionally cytarabine and sometimes total body

6.7

Table 6.7.10. Patient data and work up studies needed before autologous BMT for lymphoma

Diagnosis and initial staging
 diagnosis (histologic subtype)
 date of diagnosis
 prior biopsy results (lymph node, skin, bone marrow, etc.)
 performance status at diagnosis
 staging at diagnosis (with specification of diagnostic tools used, bone marrow biopsy result and presence or absence of B symptoms):
  Ann Arbor/Cotswald in HD
  Ann Arbor/International Prognostic Index in NHL
Front-line Therapy and Response
 primary treatment:
  protocol name
  number of cycles and dates
  total dose administered and dose per m^2 of body surface area of doxorubicin and bleomycin
  radiation therapy doses, fields and dates
 tolerance to treatment (specifying major toxicities, need for hospitalization and hematopoietic growth factors administered)
 remission achieved (e.g. early vs delayed; complete vs partial) and dates
Relapse and Salvage Therapy
 duration of the first remission and date of relapse
 diagnosis at relapse
 site and extension of relapse (with specification of diagnostic tools used for evaluation)
 performance status at relapse
 salvage treatment, names, doses, fields (if radiation therapy was administered), and dates
 response to salvage treatment
Status at Transplantation
 duration of response after salvage treatment
 time span from last treatment
 disease status prior to transplantation
Rescue Product Characteristics
 type of rescue product planned to be used
 clonogenicity of rescue product (CFU-GM, BFU-E, CD34+ cell number and others)
 mobilization maneuvers (in peripheral blood stem cells products)
 minimal residual disease detection studies (PCR, long-term cultures, etc.)
 ex vivo manipulation (purging, expansion, T cell depletion etc.)
Work-Up Studies:
 bone marrow biopsy
 laboratory:
  CBC with differential and platelet count
  electrolytes, BUN, creatinine
  LDH, liver function tests
  beta-2 microglobulin
  HIV/HSV/CMV/hepatitis serology
  HLA type of patient and donor (in allogeneic transplantation)
 radiology:
  chest x-rays PA and L
  CT scan of the chest, abdomen and pelvis
 nuclear medicine studies:
  gallium scan
  radionuclide ventriculography (see cardiac assesment)
 pulmonary function tests

6.7

Table 6.7.10. Patient data and work up studies needed before autologous BMT for lymphoma (cont'd)

cardiac assessment:
  ECG
  two-dimensional echocardiography and/or radionuclide ventriculography
social worker assessment
if required, due to disease characteristics:
  MRI of the brain
  spinal tap
  upper GI and bowel series and/or endoscopy
  bronchoscopy
  bone scan
  psychologic/Psychiatric evaluation
  others

---

irradiation (TBI). There is no agreement about which regimen is the most active or the better tolerated, and well designed prospective randomized studies are needed before any conclusion can be obtained. Most publications are of small series of patients in different clinical situations treated with several protocols, and comparisons are extremely difficult. The optimal dose of any particular agent in terms of efficacy remains unknown. The role of TBI is equally unclear. TBI is widely used in low grade and lymphoblastic lymphomas. However, the majority of patients with intermediate lymphomas do not receive TBI. There have not been prospective studies comparing TBI-transplant regimens for lymphoma, but investigators from Memorial Sloan-Kettering have suggested that etoposide, cyclophosphamide and TBI yielded results superior to cyclophosphamide and TBI alone.

6.7

B) HD

As in NHL, most of the regimens used in HD are derivations of the BACT (carmustine, cytarabine, cyclophosphamide and 6-thioguanine) regimen. Other agents frequently used are etoposide and melphalan. The most popular combination regimen is CBV, which has been used by different investigators in varying doses and schedules. This regimen is attractive due to the low extramedullary toxicity profile, and because most HD patients have not been previously exposed to these agents. Treatment-related mortality with this regimen in experienced centers is less than 5%, but higher mortality rates mainly due to sepsis or interstitial pneumonia, have been described in patients with low performance status treated with increased doses of CBV. Cytarabine in conventional doses is not considered to be an active agent in HD but used in combination with carmustine, melphalan and etoposide (BEAM) in patients with relapsed or refractory HD, induces 50% complete remission and 45% 2 year survival. TBI is not often used in HD because many patients have already received prior radiation.

IX) STEM CELL SOURCE

A) AUTOLOGOUS BONE MARROW

One of the most important points to consider when autologous bone marrow is used as a rescue product after myeloablative therapy, is that of tumor cell contamination. Histologic evidence of absence of tumor contamination has usually been required prior to transplant. However, several reports have

Table 6.7.11. Most used preparative regimen for autologous BMT in HD and NHL *

Regimen	Drugs	Doses	Most Used in
CBV	cyclophosphamide	4.8 - 7.2 g/m^2	HD
	carmustine	300-600 mg/m^2	intermediate grade NHL
	etoposide	750-2400 mg/m^2	
BEAC	carmustine	300 mg/m^2	HD
	etoposide	600-800 mg/m^2	intermediate grade NHL
	cytarabine	800 mg/m^2	Burkitt lymphoma
	cyclophosphamide	140 mg/kg or 6 g/m^2	
BEAM	carmustine	300 mg/m^2	HD
	etoposide	400-800 mg/m^2	intermediate grade NHL
	cytarabine	800-1600 mg/m^2	Burkitt lymphoma
	melphalan	140 mg/m^2	
CY/TBI	cyclophosphamide	120-200 mg/kg	low grade NHL
	total body irradiation	800-1320 cGy	intermediate grade NHL
VP16/CY/TBI	etoposide	60 mg/kg or 750 mg/m^2	intermediate grade NHL
	cyclophosphamide	100-120 mg/kg	high grade NHL
	total body irradiation	1200-1375 cGy	

*Modified from: Bierman PJ, Armitage JO. Autologous Bone Marrow Transplantation for Non-Hodgkin Lymphoma. In: Forman SJ, Blume KG, Thomas ED,eds. Bone Marrow Transplantation. Cambridge: Blackwell Scientific Publications 1994:683-695.

6.7

recently emphasized that despite absence of histologic contamination of the harvested rescue product, the presence of contaminating tumor cells is so frequent that occult involvement may be universal. Different techniques have been used to remove clonogenic tumor cells from autologous bone marrow of patients with lymphoma, including chemotherapy agents, immunological and mechanical maneuvers (see chapter on purging). However, a moderate decrease in the repopulation capability of the rescue product after in vitro purging maneuvers is common. Also, there are long-term survivors after transplantation with overtly contaminated products, and in the majority of patients, relapses occur at sites of prior disease, which suggests that inadequate conditioning causes the relapse. For these reasons, the advantages of using purged marrow in autologous transplantation for lymphoma are still uncertain.

B) PERIPHERAL BLOOD STEM CELLS (PBSC)

Circulating PBSC are an alternative to autologous bone marrow, particularly useful in patients with prior pelvic irradiation, tumor contaminated bone marrow or inability to tolerate bone marrow harvesting procedures. The additional advantage of more rapid hematopoietic reconstitution has also been consistently described with PBSC products. However, contamination of the apheresis product with tumor cells is not uncommon (see chapter on hematopoietic stem cell source). Innovative approaches directed to avoid reinfusion of tumor cells and expansion of hemopoietic precursors, including CD34+ isolation, and ex vivo co-culturing with different cytokines, are currently being studied in several institutions around the world. Preliminary results are promising.

C) ALLOGENEIC BONE MARROW

The vast majority of transplants for lymphoma (NHL and HD) have been autologous and little data is available on allogeneic transplants. However, the number of allogeneic transplants being done in lymphomas is increasing. For low grade lymphomas, autologous BMT cannot currently be considered curative, and the only potential curative therapy is allogeneic BMT. For intermediate and high grade lymphomas despite marked improvement in disease-free survival, tumor recurrence remains a dominant cause of treatment failure after autologous transplantation (approximately 50% relapse rate). Allogeneic bone

6.7

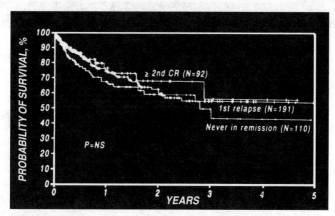

Fig. 6.7.1. Probability of survival after autotransplants for Hodgkin's disease, 1989-1994. Reprinted with permission from IBMTR/ABMTR 1995; 2:7.

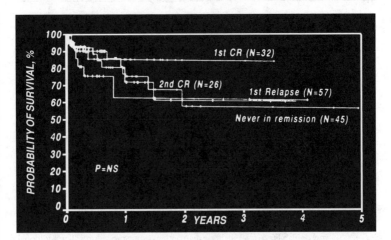

Fig. 6.7.2. Probability of survival after autotransplants for low grade non-Hodgkin's lymphoma, 1989-1994. Reprinted with permission from IBMTR/ABMTR 1995; 2:8.

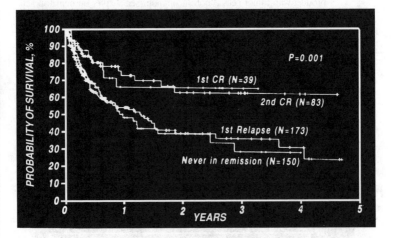

Fig. 6.7.3. Probability of survival after autotransplants for intermediate grade or immunoblastic non-Hodgkin's lymphoma, 1989-1994. Reprinted with permission from IBMTR/ABMTR 1995; 2:8.

marrow has the advantage of being a tumor-free product, and donor graft-versus-lymphoma effects may play an important role in tumor eradication. Allogeneic BMT relapse rate is approximately 25-30% less than autologous BMT, but this benefit is offset by graft-versus-host disease related mortality (see chapter on graft-versus-tumor effect). Anecdotal reports using allogeneic PBSC in NHL have recently been published.

X) GROWTH FACTORS

Prolonged severe neutropenia is a major cause of acute morbidity and mortality after dose intensification and stem cell rescue. Results of four randomized trials have shown that time to recovery is consistently reduced by 4-13 days when growth factors are used. Also the number of febrile days, incidence of positive cultures, requirement for intravenous antibiotics and duration of hospitalization are usually shortened but results have not always been consistent (Table 6.7.12) (see chapter on growth factors).

XI) **THERAPY AFTER TRANSPLANTATION**

A) CONSOLIDATIVE RADIATION THERAPY

Peritransplant radiotherapy is a commonly used tactic, especially in patients with bulky disease. Although its value has not been proven in randomized studies, this approach is justified because patients tend to relapse in sites of bulky disease, and radiation therapy is an effective tool when given after conventional therapy. The putative toxicity of 15-30 Gy is acceptable if there is no prior TBI. However extensive locoregional radiation to marrow-bearing areas may produce considerable myelosuppression after transplant.

B) ADJUVANT IMMUNOTHERAPY

Relapse of disease is one of the greatest shortfalls following autologous transplantation. In most cases this problem is thought to be the result of failure of the conditioning regimen to completely eradicate the tumor. These findings have led to active attempts at modulating the immune function following transplantation as a therapeutic strategy to decrease relapse rates. Methods of immunomodulation include low dose IL-2, interferons or other cytokines.

XII) **TREATMENT OF LYMPHOMA RELAPSE AFTER BMT**

In early relapses (< 1 year), the proximity of the BMT procedure makes any aggressive maneuver difficult, but some patients tolerate further chemotherapy. Radiation therapy should be considered in localized relapses. In late relapses (> 1 year), a second transplantation with autologous or even allogeneic products can be considered, mainly in patients with good performance status.

XIII) **FOLLOW UP AND LATE COMPLICATIONS**

A) MYELODYSPLASTIC SYNDROME (MDS) AND SECONDARY MALIGNANCY

Researchers from Dana-Farber Cancer Institute have noted that 50% of hematologically normal patients tested postautologous BMT, harbor clonal karyotypic abnormalities, but in most of these patients, no adverse events occurred even several years after such an ominous finding. MDS is being reported with increasing frequency after autologous BMT for lymphoma (7-14.5%, 5 year incidence). This incidence is similar to those patients exposed to prolonged alkylating agents and/or low dose radiation. Most of the features associated with developing MDS have been related to prior treatments, although one recent study suggests that patients with NHL who are age 40 or older at the time of transplant and who received total body irradiation as conditioning regimen are at greater risk of developing MDS. Myelodysplasia rarely develops after allogeneic transplantation.

B) PULMONARY

Delayed pulmonary damage is a well recognized complication of allogeneic BMT, but its incidence is much lower after autologous BMT. The most frequent complications are shortness of breath and lung fibrosis, and have been described mainly in patients with prior irradiation BCNU, or bleomycin exposure.

C) CARDIAC

Subclinical cardiac alterations are frequently detected in long-term follow up of lymphoma patients treated with BMT, but life-threatening cardiac toxicity

6.7

Table 6.7.12. Randomized studies comparing the use of growth factors versus placebo after bone marrow transplantation in lymphomas (+)

Growth Factor	Dose/Duration	Reduction in:					
		d < 500ANC	Hosp.	+Cultures	Antib.	Fever	Mort.
rhG-CSF	10-20 μg/kg/d SQ or CI d1-1000 ANC	11.5d p < .0005	3d NSD	– NSD	4.5d p = .052	8.5d p < .0001	– NSD
rhGM-CSF	250 μg/m²/d SQ or CI d1– < d 30	6d p < .001	1d NSD	– NSD	1 NSD	– NSD	– NSD
rhGM-CSF	250 μg/m²/d 4-6h IV d1– < d 30	4d p < .02	– NSD	p = .04	– NSD	– NSD	– NSD
rhGM-CSF	250 μg/m²/d SQ or CI d1– < d 30	13d p < .0001	1d NSD	– NSD	NSD	NSD	NSD
rhGM-CSF	250 μg/m²/d 2h IV d1– < d 21	7d p < .001	6d p < .001	– NSD	3d p < .009	NSD	– NSD

Hosp. = Reduction in days of hospitalization.
Antib.= Reduction in days with antibiotics.
Mort. = Reduction in mortality

d = days
CI = continuous intravenous infusion.
NSD = Non-statistical Differences

SQ = subcutaneous

(+) Stahel RA, Jost LM, Cerny T et al. J Clin Oncol 1994; 12(9):1931-8; Khwaja A, Linch DC, Goldstone AH et al. Br J Haematol 1992; 82:317-23; Nemunaitis J, Rabinowe SN, Singer JW et al. N Engl J Med 1991; 324:1773-8; Advani R, Chao NJ, Horning SJ et al. Ann Intern Med 1992; 116:183-9; Link H, Boogaerts MA, Carella AM et al. Blood 1992; 80(9):2188-95.

6.7

is uncommon. Late cardiac effects after 4000 cGy mantle field radiation have been described in HD patients as long as 30 years after treatment. Usually a left ventricular block is placed to minimize whole organ exposure (< 3200 cGy), but if pericardial lymph nodes or direct pericardial involvement is suspected, a higher dose may have been delivered. Therefore, follow up is required for patients who received mantle radiation prior to BMT.

D) GONADAL FUNCTION AND FERTILITY

Several studies have shown that gonadal function following standard chemotherapy regimens is impaired. Long-term testicular damage in men, and high incidence of premature menopause in older women have been consistently reported, although, it has been described that regimens with reduced amount of alkylating agents, like ABVD, cause less gonadal damage in HD patients. In transplanted patients, most of the published studies referred to high-dose cyclophosphamide with or without total body irradiation. After BMT, luteinizing hormone (LH) and follicle-stimulating hormone (FSH) levels are usually elevated in women older than 26 years of age. Some younger patients had normal menstrual cycles as early as 6 months after transplantation. Most men have normal testosterone, LH and FSH levels after BMT. The effects of conditioning regimens other than cyclophosphamide and/or total body irradiation are difficult to ascertain, but pregnancy and normal childbirth have been sporadically observed after high-dose chemotherapy and autologous BMT.

SUGGESTED READING

NON-HODGKIN'S LYMPHOMAS

1. Armitage JO. Bone marrow transplantation for indolent lymphoma. Seminars in Oncology 1993; 20(5):136-142.
2. Armitage JO. Treatment of Non-Hodgkin's lymphoma. New Engl J Med 1993; 20:1023-1030.
3. Haioun C, Lepage E, Gisselbrecht C, Coiffier B et al. Comparison of autologous bone marrow transplantation with sequential chemotherapy for intermediate-grade and high-grade nonHodgkin's lymphoma in first complete remission: a study of 464 patients. J Clin Oncol 1994; 12(12):2543-2551.
4. Horning SJ. Treatment approaches to the low grade lymphomas. Blood 1994; 83(4):881-884.
5. Nademanee A, Schmidt GM, O'Donnell MR et al. High-dose chemoradiotherapy followed by autologous bone marrow transplantation as consolidation therapy during first complete remission in adult patients with poor-risk aggressive lymphoma: a pilot study. Blood 1992; 80:1130-1134.
6. Petersen FB, Appelbaum FR, Hill R et al. Autologous marrow transplantation for malignant lymphoma: a report of 101 cases from Seattle. J Clin Oncol 1990; 4(4):638-647.

HODGKIN'S DISEASE

1. Bierman PJ, Bagin RG, Jagannath S et al. High-dose chemotherapy followed by autologous hematopoietic rescue in Hodgkin's Disease: Long-term follow-up in 128 patients. Ann Oncol 1993; 4:767-773.
2. Bierman PJ, Vose JM, Armitage JO. Autologous transplantation for Hodgkin's disease: coming of age? Blood 1994; 83(5):1161-1164.
3. DeVita VT, Hubbard SM. Hodgkin's disease. New Eng J Med 1993; 328 (8):560-565.
4. Linch DC, Winfield D, Golstone AH et al. Dose intensification with autologous bone marrow transplantation in relapsed and resistant Hodgkin's disease: Results of a BNLI randomized trial. Lancet 1993; 341:1051-1055.
5. Phillips GL. Transplantation for Hodgkin disease. In: Forman S J, Blume KG, Thomas ED, eds. Bone Marrow Transplantation. Cambridge: Blackwell Scientific Publications 1994:696-708.
6. Reece DE, Connors JM, Spinelli JJ et al. Intensive therapy with cyclophosphamide, carmustine, etoposide, cisplatin and autologous bone marrow transplantation for Hodgkin's disease in first relapse after combination chemotherapy. Blood 1994; 83:1193-1199.

6.7

MISCELLANEOUS

1. Armitage JO. Bone marrow transplantation. New Eng J Med 1994; 330:827-838.
2. Coiffier B, Phillip T, Burnett AK, Symann ML. Consensus conference on intensive chemotherapy plus hematopoietic stem-cell transplantation in malignancies: Lyon, France June 4-6, 1993. J Clin Oncol 1994; 12(1):226-231.
3. Harris NL, Jaffe ES, Stein H et al. A revised European-American classification of lymphoid neoplasm: a proposal from the international Lymphoma Study Group. Blood 1994; 84(5):1361-1392.
4. Stone RM. Myelodysplastic syndrome after autologous transplantation for lymphoma: the prize of progress? Blood 1994; 83(12):3437-3440.
5. The International Non-Hodgkin's Lymphoma Prognostic Factors Project. A predictive model for aggressive non-Hodgkin's lymphoma. New Eng J Med 1993; 329(14):987-994.
6. The Non-Hodgkin's Lymphoma Pathologic Classification Project. National Cancer Institute sponsored study of classification of non-Hodgkin's lymphomas: summary and description of a working formulation for clinical usage. Cancer 1982; 49:2112-2135.
7. Vose JM, Kennedy BC, Bierman, PJ et al. Long-term sequelae of autologous bone marrow or peripheral stem-cell transplantation for lymphoid malignancies. Cancer 1992; 69:784-789.

6.7

Adult Solid Tumors

Mary C. Territo, Michael C. Lill

I) INTRODUCTION

In comparison to hematologic malignancies, there is no convincing evidence of a graft-versus-tumor effect, and the extent of gross marrow involvement for most of the solid tumors is low. There is, therefore, little experience with allogeneic transplantation in solid tumors. Consequently, this discussion is confined to autologous BMT.

II) BREAST CANCER

The largest experience with autologous transplantation in solid tumors is in breast cancer. Breast cancer is one of the leading causes of cancer mortality in women in the United States. The disease is known to be sensitive to chemotherapy, and a significant dose-response relationship has been demonstrated in vitro and in clinical trials.

A) INCIDENCE

There are more than 114, 000 new cases of breast cancer each year in the United States and 37,000 deaths. It is estimated that the lifetime risk of development of breast cancer for a woman born in the United States is 1 in 9.

B) RISK FACTORS

Risk factors for the development of breast cancer include societal influences, probably dietary in nature in the Western world, a positive family history of breast cancer, particularly if the onset was premenopausal and especially if it affects more than one first degree female relative, a previous history of breast cancer and a prolonged period of unopposed menstruation, e.g. early menarche, late first pregnancy, late menopause. Most patients with breast cancer have no definable risk factors apart from being female and living a Western lifestyle.

C) PROGNOSIS

Poor prognostic factors include positive nodes, large tumor size (> 5 cm), aneuploidy, high grade, high S phase, ER/PR receptor negative, Her-2/neu (c-erb) positive.

D) THERAPY BY STAGE

Definitions for conventional treatment modalities are listed in Table 6.8.1a; staging and survival by stage in Table 6.8.1b. Local regional control is either modified radical mastectomy or treatment aimed at breast conservation (lumpectomy and radiation or quadrantectomy and radiation). The breast conservation approach involves addition of radiation which is generally limited to only the involved breast. However, involved breast and regional nodal irradiation is given if greater than 4 nodes are positive, extracapsular extension is present, surgical margins are positive or the primary tumor is greater than 5 cm. Breast conservation therapy is contraindicated in patients with large primary tumors (> 4-5 cm), multicentric disease, diffuse malignant appearing microcalcifications, pregnancy (due to fetal exposure to radiation) or history of irradiation to the chest, e.g. mantle field.

6.8

Table 6.8.1a. Breast cancer treatment terminology

Treatment	Definition	Indications
Radical mastectomy	En bloc removal of breast, axillary lymph nodes, pectoralis major and pectoralis minor	Rarely indicated
Modified radical mastectomy	**Patey**- total mastectomy, axillary dissection, excision of pectoralis minor (preservation of pectoralis major) **Auchinclass**–total mastectomy, axillary dissection (preservation of pectoralis major and minor)	Standard surgical approach for all stages
Breast conservative therapy–i.e. lumpectomy and radiation, quadranectomy and radiation	Total excision of tumor with normal margin of surrounding tissue, followed by radiation treatment to residual breast	Stage I and II disease if tumor < 5 cm, and negative or only focally positive margins. **Contraindicated** in multifocal or multicentric disease, diffuse microcalcifications or presence of extensive intraductal component with positive margins
Total mastectomy	Removal of entire breast including nipple and areola (no axillary dissection, preservation of pectoralis major and minor)	Noninvasive ductal carcinoma (ductal carcinoma in situ–DCIS)
Regional nodal irradiation	Irradiation to axillary nodes and supraclavicular nodes +/– internal mammary nodes	> 3-4 lymph nodes positive, > 5 cm primary, presence of gross extracapsular extension. **Not recommended** if complete (level I, II and II) lymph node dissection even if nodes positive
Adjuvant chemotherapy	Systemic chemotherapy/hormonal therapy to eradicate subclinical micrometastatic disease	Recommended for node positive pre- and postmenopausal and node negative patient with tumor > 2 cm, ER-, nuclear grade 3, high S phase, aneuploidy or high cathepsin D. May consider if tumor 1-2 cm, ER+/-, nuclear grade 2, aneuploid with low S phase, low cathepsin D level. **Not recommended** for ER+, diploid, low S phase, tumor < 1 cm, mammographically detected disease only, DCIS
Neoadjuvant chemotherapy	Systemic chemotherapy given before local therapy (surgery or surgery/radiation)	Preoperative management for T4d (inflammatory) or noninflammatory locally advanced technically unresectable tumors. Being evaluated for locally advanced operable tumors
Autologous bone marrow transplantation	High-dose myeloablative chemotherapy with autologous marrow rescue	Chemosensitive metastatic disease (best results if patient in complete re-mission before transplant) and local disease with high risk of relapse (i.e. > 10 nodes +). Being evaluated for patients with more than 4 nodes positive

6.8

1) Local Disease (Stage I and II)

a) Stage I (T1N0)

Treatment is modified radical mastectomy or breast conservation treatment. Adjuvant therapy depends on menopausal status. These patients are not candidates for BMT.

i) Premenopausal

Adjuvant therapy (CMF, AC or CAF for 4-6 cycles) (Table 6.8.1c) is given to premenopausal patients with poor prognostic factors including size (> 1 cm), aneuploidy, high grade, high S phase or ER/PR receptor negative.

Table 6.8.1b. Breast cancer TNM staging and survival

Tumor Size (T)
TX—primary tumor cannot be assessed
T0—no evidence of a primary tumor
Tis—carcinoma in situ
T1—tumor 2 cm or less in greatest dimension
T2—tumor between 2 and 5 cm in greatest dimension
T3—tumor more than 5 cm in dimension
T4—tumor of any size with direct extension to chest wall or skin
T4a—extension to chest wall (chest wall includes ribs, intercostal muscle, serratus anterior but not the pectoralis major or minor muscles)
T4b—edema, peau d'orange, or ulceration of skin or satellite skin nodules
T4c—presence of both T4a and T4b
T4d—inflammatory carcinoma (tumor embolization to dermal lymphatics causing induration of skin with an erythematous edge)

Regional lymph Nodes (N) (internal mammary or ipsilateral axillary) (other nodes such as supraclavicular, cervical, contralateral axillary or contralateral internal mammary are M1)
NX—regional nodes cannot be assessed
N0—no regional nodes involved
N1—metastasis to mobile ipsilateral axillary nodes
N2—metastasis to fixed ipsilateral nodes
N3—metastasis to ipsilateral internal mammary nodes

Staging (stage IV = M1, metastatic disease)

	T1	T2	T3	T4
N0	I	IIa	IIb	IIIb
N1	IIa	IIb	IIIa	IIIb
N2	IIIa	IIIa	IIIa	IIIb
N3	IIIb	IIIb	IIIb	IIIb

Survival with Standard Chemotherapy and Local Control (surgery, irradiation)
1) Early stage disease

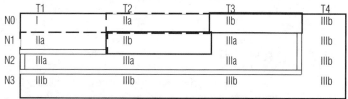

tumor size		10 year survival
T1	< 2 cm	70-80%
T2	2-5 cm	50-65%
T3	> 5 cm	40-45%

number of positive axillary nodes	10 year survival
0	75%
1-3	40%
4 or more	25%

2) Metastatic disease (M1) 10 year survival < 1-5%

6.8

ii) Postmenopausal
   a) ER+ Tamoxifen 10-20 mg PO BID for 5 years
   b) ER- Adjuvant therapy controversial; most centers offer no adjuvant therapy

  b) Stage II Node Negative, Less than 5 cm (T2NO)
      These patients are not candidates for BMT.

  i) Premenopausal
      Modified radical surgery or breast conservation therapy followed by adjuvant chemotherapy with CMF, CA or CAF (Table 6.8.1c).

  ii) Postmenopausal
      a) ER+ Modified radical surgery or breast conservation therapy followed by tamoxifen
      b) ER- Modified radical therapy or breast conservation therapy followed by adjuvant chemotherapy with CMF, AC or CAF

  c) Stage II Node Negative, Large Primary Size (T3N0)
      Patients with large primary tumors are not candidates for breast conservation. Treatment is modified radical mastectomy and adjuvant chemotherapy with CMF or CAF. These patients are candidates for BMT after standard therapy.

  d) Stage II Node Positive(T1N1 and T2N1)
      If less than 4 nodes are positive, the patient is not a candidate for BMT. The majority of patients with more than 4-10 positive nodes will relapse and therefore are candidates for BMT after standard therapy.

  i) Premenopausal
      Modified radical mastectomy or breast conservation therapy followed by adjuvant chemotherapy with CMF or CAF.

  ii) Postmenopausal
      a) ER+ Modified radical mastectomy or breast conservation therapy followed by tamoxifen
      b) ER- Modified radical mastectomy or breast conservation therapy followed by adjuvant chemotherapy with CMF or CAF

2) Locally Advanced Disease (Stage III)
   These patients are candidates for BMT after standard therapy.

  a) Stage IIIA
      Treatment is modified radical mastectomy, chest wall and regional nodal radiation and chemotherapy (CAF, CMF, etc.).

  b) Stage IIIB
      If disease is surgically resectable, treatment is modified radical mastectomy, postoperative chest wall and regional lymph node radiation and chemotherapy (CMF, CAF, etc.). Inoperable disease (chest wall fixation, inflammatory) is treated with neoadjuvant chemotherapy. If after chemotherapy, the tumor is resectable (mobile, no inflammatory changes), a modified radical mastectomy is done followed by chest wall and regional lymph node radiation.

3) Metastatic Disease (Stage IV)
   Metastatic breast cancer is essentially incurable with standard doses of chemotherapy. Patients with stage IV disease at presentation do not require additional biopsies unless there are unusual features. ER+ bone only

6.8

disease generally has a better prognosis. Otherwise, there is currently insufficient data to determine whether sites of metastatic disease have any impact on disease-free survival. Patients responsive to chemotherapy are candidates for BMT.

4) Relapsed Disease

Patients with relapsed disease should have the diagnosis confirmed histologically. Except for ER+ bone only disease or isolated nodal relapse with a long disease-free interval, these patients have a short survival and are candidates for BMT.

E) SUMMARY OF BMT FOR BREAST CANCER (FIG. 6.8.1)

1) Local Disease

Transplants in the adjuvant setting, in general have focused on women with greater than 10 lymph nodes positive or stage III disease. However, the results of standard therapy for women with greater than 4 nodes positive is also poor (Table 6.8.1b), and a number of centers are transplanting these patients. Before BMT, patients should have 2-6 courses of adjuvant chemotherapy such as CAF or CMF (Table 6.8.1c). High-risk local disease (T3, inflammatory or greater than 10 nodes positive) treated with BMT has a 75% 4 year disease-free survival. This compares very favorably with historical controls of 25% using standard therapy.

2) Metastatic (Stage IV)

Patients should have up to 6 cycles of chemotherapy with CAF or CMF or similar combinations to reduce tumor bulk and demonstrate chemosensitivity. Patients who fail to demonstrate chemosensitivity do not benefit from transplantation. Ideally the marrow should be free of disease after 6 cycles of chemotherapy. Survival after PBSC transplant for patients with marrow metastasis is lower (approximately 10-20%) than for patients without marrow metastasis (approximately 20-30%). It is unclear what level of marrow contamination by malignant cells is acceptable or whether purging of PBSCs or marrow will decrease relapse in breast cancer. Overall, patients with chemotherapy-sensitive metastatic disease who undergo transplantation have a 10-20% chance of longterm disease-free survival. For patients with metastatic disease who are in a complete remission after conventional chemotherapy, the long-term disease-free survival with BMT is 30-40%. This compares very favorably with historical controls of less than 5% using standard therapy.

F) CONDITIONING REGIMENS FOR BREAST CANCER

Chemotherapy combinations including cyclophosphamide, etoposide, cisplatin, carboplatin, thiotepa or BCNU have been utilized. Two common regimens are CTCb or BCC (Table 6.8.1c).

G) SOURCE OF HEMATOPOIETIC STEM CELLS FOR BREAST CANCER BMT

Most centers now use mobilized autologous PBSC as the primary transplant material. The need to collect bone marrow to provide a backup is generally considered unnecessary. At UCLA, most patients engraft with a neutrophil count $> 0.5 \times 10^9$/L on day +10 following transplantation and engraft with a platelet count $> 50,000/\mu l$ by day +13. PBSC collections have also been shown to be less contaminated with tumor cells than bone marrow. Some provocative early data also raise the possibility of a lower risk of relapse with the use of peripheral blood progenitor cells. Selection for CD34 positive cells from peripheral blood

6.8

progenitor cells has been demonstrated to significantly purge breast cancer cells from the transplant product. The significance of reduction in transplanted tumor burden is unclear.

1) **Mobilizing PBSC**

There are a variety of methods which can be utilized to mobilize PBSC. These include 1) chemotherapy (which needs to produce significant neutropenia) and/or 2) cytokines (the mechanisms by which cytokines mobilize PBPC are not known). Most studies have used either G-CSF

6.8

Table 6.8.1c. Examples of breast cancer chemotherapy regimens

Chemotherapy Regimens

CAF
cyclophosphamide 600 mg/m^2 IV day 1
adriamycin 60 mg/m^2 IV day 1
5-fluorouracil 600 mg/m^2 IV day 1 and 8
Repeat cycle every 28 days for 4 cycles.
(Proc Am Soc Clin Oncol 1992; 11:51)

CAF
cyclophosphamide 500 mg/m^2 IV day 1
adriamycin 50 mg/m^2 IV day 1
5-fluorouracil 500 mg/m^2 IV day 1
Repeat cycle every 21 days for 6 cycles.
(Cancer 1977; 40: 625-632)

CMF
cyclophosphamide 100 mg/m^2/day PO
on days 1-14
methotrexate 30-40 mg/m^2 IV
day 1 and 8
5-fluorouracil 400-600 mg/m^2 IV
day 1 and 8
Repeat cycle every 28 days for 4-6 cycles.
(NEJM 1976; 294:405-410)

AC
adriamycin 40 mg/m^2 IV day 1
cyclophosphamide 200 mg/m^2 PO days 3-6
Repeat cycle every 21 days.
(Cancer 1975; 36:90-97)

CMF
cyclophosphamide 600 mg/m^2 IV
day 1 and 8
methotrexate 40 mg/m^2 IV day 1 and 8
5-fluorouracil 600 mg/m^2 IV day 1 and 8
Repeat cycle every 28 days for 4-6 cycles.
(WB Saunders, Philadelphia, 1990; 169-173)

FAC
5-fluorouracil 500 mg/m^2 IV day 1 and 8
adriamycin 50 mg/m^2 IV day 1
cyclophosphamide 500 mg/m^2 IV day 1
Repeat every 21 days.
(Cancer 1979; 43:1225-1233)

II BMT Conditioning Regimens

CTCb
carboplatin (800 mg/m^2 total dose) by continuous intravenous infusion over 96 hours day -7 to -3.
cyclophosphamide (total dose of 6 g/m^2) given at a fixed dose of 1.5 g/m^2/d by continuous infusion over 96 hours days -7 to -3.
thiotepa (500 mg/m^2 total dose) given at 125 mg/m^2/daily by continuous intravenous infusion over 96 hours day -7 to -3.
(J Clin Oncol 1990; 8:1239-45)

BCC
cyclophosphamide (total dose 5625 mg/m^2) given as 1875 mg/m^2/d in 300 ml D$_5$W IV over 1 hr x 3 days on day -6,-5 and -4.
cisplatin (total dose 165 mg/m^2) given as 55 mg/m^2/d in 1000 ml NS IV as a continuous infusion x 3 days on day -6,-5 and -4. Infusion begins at the same time as the cyclophosphamide infusion, but is administered through a separate lumen.
BCNU 600 mg/m^2 in 500 ml D$_5$W IV on day -3. Infusion rate is 5 mg/m^2/min, unless hypotension (systolic blood pressure < 80 mm Hg) not responsive to fluid challenge occurs, at which time the dose rate is reduced to 2.5 mg/m^2/min. If hypotension persists, dopamine is begun and titrated to maintain a systolic blood pressure > 80 mm Hg.
(J Clin Oncol 1993; 11:1132-43)

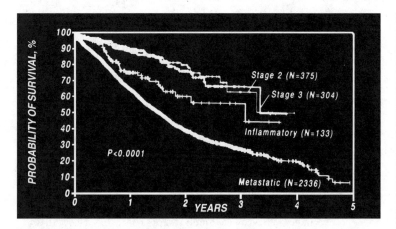

Fig. 6.8.1. Probability of survival after autotransplants for breast cancer, 1989-1994. Reprinted with permission from IBMTR/ABMTR 1995; 2:8.

3-10 μg/kg/day SQ or IV or GM-CSF 7 μg/kg/day. Studies are in progress to examine novel cytokines or combinations of cytokines or combinations of chemotherapy and cytokines. The principal advantage of using cytokines alone for mobilization is the predictability of timing of leukapheresis. With G-CSF priming, the maximal number of CD34 cells are harvested on days 5-7 of G-CSF. First leukapheresis after chemotherapy induced neutropenia is usually initiated when the absolute neutrophil count reaches 10^9 cells/L.

At the time of maximal mobilization of PBSC the patient usually undergoes 3-6 sequential daily leukaphereses in order to harvest an adequate number of stem cells (> 4-8 x 10^8 mononuclear cells/kg). Leukapheresis requires a continuous flow of 50-70 ml/min blood through the machine. We achieve this by the use of a large bore, tunneled central venous catheter, which is then used for central venous access throughout the transplant period. Most patients with breast cancer utilizing our current strategy attain this number of cells after 2 or 3 leukaphereses. In the future, we are likely to set a target of > 5-7 x 10^6 CD34 cells/kg. After harvest, the cells are then cryopreserved, usually in DMSO. Some centers select for CD34+ cells prior to cryopreservation in order to purge breast cancer cells and decrease the volume of DMSO which will be reinfused.

H) POSTTRANSPLANT CARE

During the pancytopenic phase posttransplantation, patients receive appropriate red cell and platelet transfusion support as necessary. We usually transfuse filtered irradiated PRBC for Hb < 9/dl and filtered irradiated single donor platelets for platelet counts < 10 x 10^9/L. Patients who develop fevers or signs of infection receive broad spectrum antibiotic coverage. Many will not develop neutropenic fevers during the transplant.

1) Growth Factors

Patients receive G-CSF 10 μg/kg/d IV or SQ or GM-CSF 7 μg/kg/d SQ starting day 0 and continued until the absolute neutrophil count is greater than 5000/μl for 3 consecutive days, or greater than 10,000/μl for 1 day.

The addition of cytokines posttransplant offers a modest advantage in terms of rate of engraftment.

2) Post-BMT Radiation

Patients with a history of locally extensive breast cancer (stage II with > 4 nodes positive, IIIA and B and inflammatory) who have not undergone chest wall radiotherapy have been reported to have increased local relapse after BMT. These patients should receive irradiation with 45-50 Gy to the chest wall and regional lymph nodes (supraclavicular and axillary) and a 10-15 Gy scar boost, once full hematopoietic recovery has occurred (usually 6-7 weeks posttransplant). Post-BMT radiotherapy may cause a fall in peripheral blood counts necessitating a delay or interruption of radiation.

3) Post BMT Tamoxifen

Patients whose primary tumors were estrogen receptor positive can be treated with tamoxifen 10 mg PO twice daily for 5 years, although the role of hormone prophylaxis posttransplantation is not clear.

III) TESTICULAR CANCER AND OTHER GERM CELL TUMORS

Germ cell tumors are those which arise from the pluripotential cells of embryonic primitive streak. These cells normally populate the gonads, but occasionally some are left behind during migration from the yolk sac. Over 95% of germ cell tumors originate in the testes, but 1-5% originate in the ovaries, or in extragonadal midline locations as a result of malignant transformation of residual germinal elements in the mediastinum, peritoneum or other sites.

A) HISTOLOGY

Histologically the tumors can be classified as seminomas or nonseminoma (embryonal, teratomas, choriocarcinomas) or a mixed pattern of 2 or more elements.

B) TUMOR MARKERS

Alpha fetal protein (AFP) has a half life of 5 days. It is not present in a pure seminoma. The beta subunit of human chorionic gonadotropin (β-HCG) has a half life of 24 hours. β-HCG is elevated in nonseminomas and 15% of seminomas. Most nonseminomas (80-90%) have an elevated β-HCG or AFP. Placental alkaline phosphatase is elevated in 90% of seminomas.

C) INCIDENCE

Germ cells tumors account for 1% of all cancers, but are the most common carcinoma among men age 15-35 years. Three percent are bilateral and may occur synchronously or metachronously. The risk of developing testicular cancer is increased 5-50 times in patients with a history of cryptorchidism or atrophic testis as seen following mumps orchitis.

D) PROGNOSIS

Most patients with disseminated testicular cancer are cured of their disease through the use of standard-dose chemotherapy with or without surgery. However, presenting factors which are associated with poor prognosis include:

1) Histology

Nonseminoma germ cell tumors have a worse prognosis. However, this is due to a more advanced stage at presentation. When corrected stage for stage, the prognosis is similar.

6.8

2) **β-HCG**

Human chorionic gonadotropin levels > 10,000 IU/ml indicate greater tumor burden and worse prognosis.

3) **Bulky Disease**

Large tumor mass have a worse prognosis, e.g. number of lung metastases > 10 cm, abdominal metastases > 5 cm.

4) **Response to Initial Therapy**

The most important criteria determining long-term outcome is failure to achieve a complete response with initial induction therapy, or slow clearance of tumor markers (HCG > 3 day half-life, α-fetoprotein (AFP) > 7 days half-life).

5) **Relapse**

Once patients relapse the majority will go on to die of their disease.

E) GERM CELL TUMOR STAGING AND THERAPY (TABLE 6.8.2)

1) **Local Disease (Stage I)**

Stage I disease is limited to the testes and can be treated with surgery alone or surgery and adjuvant radiation for seminomas. However, tumors that show evidence of vascular invasion and embryonal (undifferentiated) histology have an increased risk of relapse (about 50%) and may benefit from adjuvant chemotherapy.

Table 6.8.2a. Germ cell cancer TNM and Royal Marsden Hospital staging

6.8

Tumor Size (T)
TX—primary tumor cannot be assessed (no orchiectomy done)
T0—no evidence of primary tumor (no tumor found at orchiectomy)
Tis—intratubular, preinvasive tumor
T1—tumor limited to testis
T2—tumor invades beyond tunica albuginea or into epididymis
T3—tumor invades spermatic cord
T4—tumor invades scrotum
Regional Lymph Nodes (N) (aortic, paraaortic, external iliac, paracaval, intrapelvic, inguinal)
NX—nodes cannot be assessed
N0—no node metastasis
N1—metastasis to single node 2 cm or less in greatest dimension
N2—metastasis to a single node 2-5 cm in greatest dimension or; multiple nodes, none more than 5 cm in greatest dimension
N3—metastasis in a lymph node more than 5 cm in dimension
Stage

Royal Marsden Stage	TNM (AJC) Stage
I = limited to testis	I= any T, N0
IIA = < 2 cm diameter retroperitoneal nodes IIB = 2-5 cm diameter retroperitoneal nodes IIC = 5-10 cm diameter retroperitoneal nodes IID = > 10 cm diameter retroperitoneal nodes	II= any T, N1, N2, or N3
III = supra- and infradiaphragmatic nodes	III = any M
IV = hematogenous metastasis	

2) Disseminated Disease (Stage II, III and IV)
Patients with disease outside the testes have diverse prognoses with tumor bulk or extent of disease being the most important parameter (Table 6.8.2b). These patients are treated with platinum-based chemotherapy regimens such as BEP or EP (Table 6.8.2c).

3) Failure to Achieve CR
Failure to achieve a complete response following initial induction therapy, or slow clearance of tumor markers (HCG > 3 day half life, AFP > 7 day half life) imparts a poor prognosis.

Table 6.8.2b. Germ cell cancer survival with standard therapy

1) Survival by Royal Marsden stage

Testicular cancer Royal Marsden stage	5 year survival with standard therapy
I	99%
IIA	95%
IIB, C, D	75%
III, IV	66%

2) Survival for subsets of stage II, III and IV disseminated testicular cancer, according to extent of disease (University of Indiana criteria)

Extent of disseminated disease	Criteria	5 year survival with standard therapy
Minimal	1) elevated serum markers only 2) unresectable, nonpalpable retroperitoneal nodes 3) fewer than 5 pulmonary metastases per lung field and none larger than 2 cm 4) cervical nodes + or — nonpalpable retroperitoneal nodes	90%
Moderate	1) palpable abdominal mass, no supradiaphragmatic disease 2) 5-10 pulmonary metastases per lung field and largest < 3 cm, or any larger than 2 cm	70%
Advanced	1) primary mediastinal nonseminoma germ cell, or > 10 metastasis per lung field, or largest pulmonary metastases > 3 cm 2) palpable abdominal mass plus supradiaphragmatic disease 3) liver, bone or CNS metastasis	50%

6.8

a) *Slow Clearance of Tumor Markers (HCG, AFP)*

Suboptimal decline or plateau in serologic tumor markers predicts treatment failure and treatment should be switched to a salvage chemotherapy regimen such as VeIP (Table 6.8.2c).

b) *Residual Mass*

A residual benign teratoma may also be confused with failure to achieve a complete response. Following chemotherapy, if tumor markers (β-HCG, AFP) have normalized but residual radiographic abnormalities persist, the mass should be resected. In over 90% of these patients the resected specimen will be a benign teratoma or fibrosis and the patients need only be closely followed. Those who have residual carcinoma should be treated with two additional courses of chemotherapy and close follow up.

Table 6.8.2c. Examples of chemotherapy regimens for germ cell cancer

Disseminated Disease (stage II-IV)

BEP
bleomycin 30 units IV on days 2, 9 and 16
etoposide 100 mg/m^2/day IV on days 1 to 5
cisplatin 20 mg/m^2/ day IV on days 1 to 5
Repeat cycle at 21 day intervals for 4 cycles.
(NEJM 1987; 316:1435-1440)

EP
Etoposide 100 mg/m^2/day IV on days 1 to 5
Cisplatin 20 mg/m^2/day IV on days 1 to 5
Repeat cycles every 3-4 weeks for 4 cycles. The EP regimen eliminates the risk of pulmonary toxicity from bleomycin, and 4 cycles of EP may be used in place of 3 cycles of BEP for good risk patients. EP is not recommended for poor risk patients.

Salvage Regimen
VeIP
vinblastine 0.11 mg/kg/day IV on day 1 and 2
ifosfamide 1.2 g/m^2/day IV on days 1 to 5
cisplatin 20 mg/m^2/day IV on days 1 to 5
Cycle is repeated at 21 day intervals.
(Ann Intern Med 1988; 109, 540-546)

VIP
etoposide 75 mg/m^2/d IV on days 1 to 5
ifosfamide 1200 mg/m^2/day on days 1 to 5
cisplatin 20 mg/m^2/d IV on days 1 to 5
Repeat cycle every 21 days.
(Ann Intern Med 1988; 109, 540-546)

BMT Conditioning Regimen
carboplatin 500 mg/m^2 IV over 45 min x 3 days (day-8,-6,-4)
etoposide 400 mg/m^2 IV over 90 min x 3 days (day-8,-6,-4)
cyclophosphamide 50 mg/kg IV over 45-60 min x 3 days (day-8, -6,-4)
(Cancer Research 1993; 53:3730-35)

6.8

4) Relapse

When evaluating patients for relapse, radiologic and serologic data can be misleading. False positive elevations of serum β-HCG can occur due to cross-reactivity with luteinizing hormone and may be secondary to hypogonadism or marijuana use. Administration of testosterone will suppress β-HCG in the case of hypogonadism but not in the case of recurrent disease. Occasionally subpleural pulmonary nodules due to bleomycin may be confused with recurrent disease. Patients with relapsed germ cell tumor are treated with a salvage regimen such as VeIP (Table 6.8.2c).

F) BMT FOR GERM CELL CANCER

Germ cell tumors are chemosensitive, afflict young people, and the bone marrow is rarely involved by tumor, all of which make autologous BMT a reasonable therapeutic option. However, there is no universally accepted criteria as to which patients should be considered for transplantation. Studies have been undertaken in 3 different groups.

1) BMT for Advanced Disseminated Disease

Patients with advanced disease on presentation (Table 6.8.2b; i.e. > 10 lung metastasis, > 5 cm abdominal metastasis, AFP > 1000 Ku/L β-HCG > 10,000 IU/L, extragonadal primary, liver or CNS metastasis) have a 50% 5 year survival with standard chemotherapy. The use of BMT as initial consolidation in these patients is controversial but under investigation.

2) Suboptimal Decline or Plateau in Serologic Tumor Markers

This predicts for treatment failure, and in some centers therapy is switched to a salvage regimen given for 2-3 cycles pending arrangements for BMT. Whether BMT will improve long-term survival is currently unknown.

3) BMT for Failure to Achieve Complete Remission

Patients who fail to achieve a complete response to initial platinum-based standard chemotherapy have a poor prognosis, long-term disease-free survival < 5%. Long-term disease-free survival after BMT is 5-20%.

4) BMT for Relapse

Approximately 10% of patients who enter a complete remission will relapse. Salvage chemotherapy such as VeIP has a 60% response rate but only 20-30% long-term disease-free survival. BMT appears to give superior results with 30-50% long-term disease-free survival for patients with responsive relapse. For resistant relapse, long-term disease-free survival after BMT is 5-20%.

G) CONDITIONING REGIMEN FOR GERM CELL TUMORS

The high-dose conditioning chemotherapy which has been utilized for BMT include: carboplatin (900-1500 mg/m^2) or cisplatin (180 mg/m^2), and etoposide (1200-1700 mg/m^2); with or without cytoxan (50-150 mg/kg) or ifosfamide (6-12.5 g/m^2). The role of multiple sequential courses of high-dose chemotherapy (double autologous transplants) with hematopoietic stem cell rescue is under investigation.

H) SOURCE OF HEMATOPOIETIC STEM CELLS FOR GERM CELL CANCER BMT

Almost all reported series of transplant for germ cell cancers have utilized bone marrow as the stem cell source. Testicular cancers are unlikely to involve the marrow and purging has not been utilized. At least one study has utilized peripheral blood stem cells with good results.

6.8

I) **POSTTRANSPLANT CARE OF PATIENTS WITH GERM CELL TUMORS**

Response is monitored with physical exam, chest x-ray, serum tumor markers and other radiologic studies such as CT scan depending on sites of disease. Any residual masses are removed by surgery.

IV) **EPITHELIAL OVARIAN CANCER**

A) **INCIDENCE**

Epithelial carcinoma of the ovary is the fourth most frequent cause of cancer death in women, with 20,000 women per year in the United States developing the tumor, and 12,500 dying annually. The peak incidence is between ages 40-60 years.

B) **RISK FACTORS**

Risk is increased in women with early menarche, late menopause, high dietary fats and use of talc as a perineal dusting powder. There is also a genetic link in families with a high incidence of breast and ovarian cancers. Risk is decreased in women who have used oral contraceptives, had multiple pregnancies and breast fed.

C) **HISTOLOGY**

The large majority of ovarian cancers are epithelial carcinomas with less than 5% being germ cell or sex cord-stromal tumors. Germ cell tumors of the ovary should be approached similarly to testicular germ cell tumors. Epithelial carcinoma of the ovary actually arise from surface epithelial cells that cover the ovary which then invaginate into the parenchyma of the ovary. The tumor may remain confined to the ovary for some time or spread by exfoliation into the peritoneal cavity or via lymphatics. Hematologic spread to parenchymal organs or bone may occur in more advanced stages. The degree of cellular differentiation is an independent prognostic factor and relates to response to treatment and survival in early stages. The appearance of epithelial tumors of the ovary may be cystadenocarcinoma (42%), mucinous cystadenocarcinoma (12%), endometrial carcinoma (15%), undifferentiated carcinoma (17%) or clear cell carcinoma (6%).

6.8

D) **OVARIAN CANCER STAGING**

Except for metastatic disease, staging requires surgery with removal of primary tumor, multiple blind biopsies, peritoneal cytology and a meticulous and thorough intraoperative search for disease throughout the pelvis and abdomen. Table 6.8.3a lists the International Federation of Gynecology and Obstetrics (FIGO) staging classification for ovarian cancer. Serologic markers of disease include CA-125, CEA and occasionally β-HCG. A preoperative mammogram is necessary because patients with ovarian cancer are at increased risk of breast cancer, and an ovarian mass can occasionally represent metastatic breast cancer. If gastrointestinal signs or symptoms are present, radiologic and endoscopic investigation should be done prior to staging surgery.

E) **THERAPY AND SURVIVAL BY STAGE**

There are no accurate methods for early detection. This disease is often asymptomatic until late stages; about 70% of patients present with advanced disease confined to the abdomen.

1) **Stage IA and IB**

Surgery is the mainstay of treatment. Low grade histology is treated with surgery alone with a 90% long-term survival. Postoperative adjuvant chemotherapy is given for moderate or high grade stage IA or IB disease.

2) Stage IC, II, III and IV

These stages are treated with surgical debulking followed by chemotherapy (Table 6.8.3b). Ovarian tumors are sensitive to cyclophosphamide, ifosfamide, carboplatin, cisplatin, adriamycin, mitoxantrone, hexamethylamine and taxol. ^{32}P radioisotope (^{32}P-chromic phospate) may be used intraperitoneally only for microscopic disease since penetration is less than 5 mm. Although controversial, patients with a clinical complete remission (negative physical exam, no radiograph or serologic evidence of disease) generally undergo a second look laparotomy. About half of the patients who have what appears to be a complete response at second look operations will relapse. Patients with negative second looks but advanced stage (III or IV) on presentation, high histology grade and suboptimal debulking at the initial surgery are most likely to relapse.

3) Relapse

Once relapse has occurred all patients succumb to disease with a median survival of 12-18 months. Initial response to salvage therapies is 30-60%. Disease refractory to platinum-based regimens show a 25% response to taxol. The role of second debulking is controversial.

F) OVARIAN CANCER PATIENT SELECTION FOR BMT

It is not clear which patient population will best benefit from high-dose chemotherapy with hematopoietic stem cell reconstitution. There are no large

---

Table 6.8.3a. Ovarian cancer FIGO stage and survival

6.8

Stage
I—Limited to ovaries
    IA—one ovary involved, no ascites, capsule intact
    IB—both ovaries involved, no ascites, capsule intact
    IC—IA or IB and tumor on surface of ovary or rupture capsule or malignant cells in peritoneal
        wash or ascites present with malignant cells
II—Pelvic extension
    IIA—pelvic wall extension
    IIB—extension to uterus or tubes
    IIC—IIA or IIB and tumor on surface of ovary or rupture capsule or malignant cells in
        peritoneal wash or ascites present with malignant cells
III—Peritoneal implants or positive retroperitoneal or inguinal nodes
    (liver capsule metastasis = stage III)
    IIIA—microscopic seeding of abdominal peritoneal surface
    IIIB—abdominal peritoneal implants not exceeding 2 cm
    IIIC—abdominal peritoneal implants exceeding 2 cm or positive retroperitoneal or inguinal
        nodes
IV—Distant metastasis (liver parenchymal metastasis = stage IV; pleural effusion must have malignant cells to be stage IV)

Survival with Standard Therapy
Survival depends on stage, optimal initial debulking and findings of second look laparotomy

FIGO Stage	5 year survival		Results of second look laparotomy	3 year survival
I	75-80%		no residual disease	80%
II	60-65%		< 2 cm disease	50%
III	20-30%		> 2 cm disease	30%
IV	< 20%			

trials which have defined the role of transplantation and no randomized studies are yet available in this disease. BMT has been utilized in the following groups:

1) BMT as Part of Initial Treatment

Most patients with macroscopic residual disease after initial cytoreductive surgery will die of their tumors. BMT can be considered in these patients as part of their initial therapy. These patients could receive several cycles of chemotherapy followed by high-dose chemotherapy and autologous BMT.

2) BMT for Persistent Disease at Second Look

Patients who have microscopic disease at second look operation have a 50-60% 2 year survival but limited long-term survival and may benefit from BMT. Patients with macroscopic disease at second look operation have a very poor prognosis with a median survival of about 15 months and could be considered candidates for BMT. There is no information about the role of second debulking pre-BMT.

3) BMT for Relapsed or Refractory Disease

BMT has been shown to induce responses in relapsed and refractory disease, but effect on long-term survival has yet to be defined. The role of surgical debulking pre-BMT is unknown.

G) CONDITIONING REGIMEN FOR OVARIAN CANCER

High-dose chemotherapy combinations with approaches similar to those used in germ cell tumors and lung cancer can be used in ovarian cancer patients. These tumors are chemosensitive to agents such as cyclophosphamide, ifosfamide, carboplatin, cisplatin, mitoxantrone and taxol. The regimen MCC

6.8

Table 6.8.3b Examples of chemotherapy regimens for ovarian cancer

Primary Chemotherapy
PT
cisplatin 75 mg/m^2 IV on day 2
taxol 135 mg/m^2 IV over 24 hours on day 1
Repeat every 21 days for 6 courses. Taxol clearance is decreased with prolonged neutropenia if cisplatin given first. Therefore, taxol precedes cisplatin. (J Clin Oncol 1991; 9:1692-1703)

Cisp/Carbo/CPA
cisplatin 50 mg/m^2/day IV day 1
cyclophosphamide 600 mg/m^2 IV day 1
carboplatin 280 mg/m^2 IV day 1
Repeat every 28 days for 6 courses. (J Clin Oncol 1991; 9:1793-1800)

CAP
cyclophosphamide 650 mg/m^2 IV day 1
adriamycin 50 mg/m^2 IV day 1
cisplatin 50 mg/m^2 IV day 1
Repeat cycle every 28 days for 6 courses. (Lancet 1987; 2:353-)

BMT Conditioning Regimen
MCC
cyclophosphamide 40 mg/kg/day on days -8, -6 and -4
carboplatin 300 mg/m^2/day for 5 days on days -8 to -3
mitoxantrone 25 mg/ m^2/day on days -8, -6 and -4
Hydrate D5NS with 20 mg KCl/l at 3000 ml/m^2/day during and for 72 hours after carboplatin therapy. (J Clin Oncol 1994; 12:176-83)

(Table 6.8.3b) has been reported to induce responses in patients with ovarian cancers which were refractory to standard therapy.

H) CONDITIONING REGIMEN TOXICITY

Toxicity varies according to conditioning regimen. Nonhematologic toxicity in the 25 patients reported with the MCC regimen include: nausea, vomiting, diarrhea, anorexia and alopecia in everyone; and partial transient deafness 2-3 weeks following the chemotherapy in four patients, renal failure in two patients and congestive heart failure in one patient.

I) SOURCE OF HEMATOPOIETIC STEM CELLS FOR OVARIAN BMT

Most studies of BMT in ovarian cancer have utilized bone marrow as the stem cell source. Ovarian cancers are unlikely to involve the marrow and purging has not been utilized. At least one study has utilized peripheral blood stem cells with good results.

V) LUNG CANCER

Carcinoma of the lung is the leading cause of cancer-related mortality in men and women with an estimated 120,000 deaths yearly in the United States. Non-small cell lung cancer (NSCLC) (squamous cell, adenocarcinoma and large cell carcinoma) has been treated by autologous BMT, but responses were incomplete or of short duration. Therefore, BMT for NSCLC has generally been abandoned pending better chemotherapy regimens. About 20% of lung cancers are histologically of the small cell type which frequently presents with metastasis, has a rapid growth rate, and has a dramatic response to chemotherapy, but most patients die of recurrent disease within 2 years. Therefore, BMT is a therapeutic option for SCLC.

6.8

A) SCLC STAGE AND SURVIVAL (TABLE 6.8.4a)

SCLC is staged by the extent and area of tumor involvement into limited stage (disease confined to one hemithorax and regional nodes, i.e. disease that can be encompased by one radiation port) or extensive stage disease (involvement of any organ or structure not included in the definition of limited stage).

Table 6.8.4a. Small cell lung cancer (SCLC) staging and survival

Staging
Limited stage. Disease which can be encompassed by a radiation port; i.e. hemithorax, ipsilateral mediastinal and supraclavicular nodes. Some studies include malignant pleural effusion, contralateral supraclavicular, and ipsilateral and contralateral cervical nodes as limited stage.
Extensive stage. Involvement of any organ/structure not included in definition of limited stage.

Survival of SCLC with Standard Therapy

	Limited Stage	Extensive Stage
Response rate	90%	60%
Complete response rate	40-60%	10-20%
Median survival	12-18 months	8 months
5 year survival	5-20%	< 1%

Table 6.8.4b. Examples of chemotherapy for small cell lung cancer

Primary Chemotherapy
EP
cisplatin 20 mg/m²/day IV on days 1-5
etoposide 80 mg/m²/day IV on days 1-5
Repeat cycle every 21 days. (J Clin Oncol 1992, 10; 282-291)

CAV
cyclophosphamide 1000 mg/m² IV on day 1
adriamycin 40 mg/m² IV on day 1
vincristine 1.0 mg/m² (max 2 mg) IV on day 1
Repeat cycle every 3 weeks for 6 cycles. (J Clin Oncol 1992, 10; 282-291)

BMT Conditioning Regimen
1)
cyclophosphamide 1875 mg/m²/d IV over 2 hrs x 3 days (Day -7,-6,-5)
cisplatin 55 mg/m²/d IV by continuous infusion over 72 hrs (Day -7 through -5)
BCNU 60 mg/m²/dose IV over 1 hr twice daily x 4 days (Day -7,-6,-5,-4)
Consolidative radiation begins after patients recover from the acute morbidity of BMT, i.e.
normalization of blood counts and stabilization of pulmonary function tests. Patients received
chest radiotherapy:
50-56 Gy in 25-30 fractions given over 5-6 weeks, and **cranial irradiation** 30 Gy in 15 fractions
given over 3 weeks. (J Natl Cancer Inst 1993, 85; 559-66)

2) VICE
etoposide (1500 mg/m²) 500 mg/m²/day on days -4,-3,-2
ifosfamide (12 g/m²) 4g/m²/day on days -4,-3,-2
carboplatin (750 mg/m²) 250 mg/m²/day on days -4,-3,-2
epirubicin (150 mg/m²) 50 mg/m²/day on days -4,-3,-2
After transplant all patients receive **chest irradiation** 50 cGy in 25 fractions, and for those
in complete remission, **cranial irradiation** 30 cGy in 15 fractions. (Seminars in Oncol 1995;
vol. 22;(1) supp 2:3-8)

6.8

B) THERAPY FOR SCLC

The primary treatment modality regardless of stage is chemotherapy (Table 6.8.4b). Chemotherapy combinations include cyclophosphamide, doxorubicin, vincristine, etoposide and cisplatin.

1) **Limited Stage**

Limited stage SCLC is treated with chemotherapy (e.g. cisplatin and etoposide—Table 6.8.4b) and concurrent thoracic radiotherapy. The sequencing of radiation and chemotherapy is controversial, but early data supports concurrent treatment over sequential. Chest radiotherapy decreases local relapse and in one meta-analysis improved overall survival. Patients who obtain a CR are candidates for prophylactic cranial radiation. CNS relapse is decreased by cranial irradiation but does not change survival. Initial response rates in limited stage SCLC are high (80-90%), remission duration is only 9-16 wks, and median survival is short (12-18 mo), with 2 year survivals of 10-20% and 5 year survival of only 5-20%.

2) **Extensive Stage**

Extensive Stage is treated with chemotherapy. Radiotherapy does not effect survival in extensive stage and is reserved for patients that have a good response to chemotherapy or require symptomatic palliation of local disease. Long-term survival with extensive stage is poor (< 1%).

C) BMT FOR SCLC

Considering the large number of individuals with lung cancers, it is surprising that the number of trials investigating the use of high-dose chemotherapy with transplantation is relatively small, and the role which this approach plays in the treatment of SCLC remains unclear. The most encouraging study so far involves SCLC patients who demonstrated complete or partial response to initial conventional dose induction chemotherapy. Responsive patients were treated with high-dose cyclophosphamide, cisplatin and BCNU followed by marrow infusion, and after recovery from the transplant, chest and cranial radiation therapy. With this approach, actuarial 2 year disease-free survival for limited stage and extensive stage disease has been reported to be 57% and 35%, respectively. These results appear markedly superior to standard therapy (Table 6.8.4b). Nevertheless, convincing evidence proving a significant advantage of BMT over conventional-dose chemotherapy does not yet exist.

D) CONDITIONING REGIMEN FOR SCLC

Intensive chemotherapy regimens have included various combinations of cyclophosphamide, etoposide, cisplatin, carboplatin and BCNU followed by marrow transplantation. When local radiotherapy was not included, a large number of the recurrences occurred in the lungs or CNS. An example of a BMT conditioning regimen for SCLC is shown on Table 6.8.4b.

E) SOURCE OF STEM CELLS FOR SCLC

Concern for the use of marrow transplantation in patients with SCLC relates to the high rate of marrow involvement in these individuals and the chance that the cryopreserved marrow product may be contaminated with tumor cells, thus contributing to the high rate of relapse posttransplant in most series. It is not clear that peripheral blood stem cells are less likely than marrow to be contaminated by tumor. It has not yet been demonstrated that purging the marrow preparation of tumor cells (using immunological or other methods), or the positive selection of CD34 positive cells can impact on the relapse rate.

6.8

Bone Marrow Transplantation for Pediatric Malignancy

Robert A. Krance, Helen E. Heslop, Malcolm K. Brenner

I) INTRODUCTION

Bone marrow transplantation has defined at times controversial indications in hematologic malignancies depending on several factors including type of malignancy and stage (see specific diseases below). The utility of bone marrow transplantation in pediatric solid tumors by and large remains undefined. That most of the childhood tumors manifest dose-response behavior would portend a favorable role for this therapy. Yet now, after more than a decade of clinical trials, its actual benefit is questioned. Clinical reports show that virtually every pediatric solid tumor has been treated with intensive chemotherapy and bone marrow rescue; however, the broadest application has been in neuroblastoma, rhabdomyosarcoma, Ewing sarcoma, and high grade glioma. Bone marrow transplantation (BMT) for malignant disease in children has many distinct advantages and disadvantages compared to transplants in adults.

A) DISADVANTAGES OF BMT IN CHILDREN

1) Conditioning Regimen

Intensive conditioning procedures may have a much greater impact on a developing individual than on a mature person. The effect of total body irradiation on the infant brain is probably the most egregious of these. Adverse effects on growth and endocrine function are almost inevitable with all conditioning regimens used, and are addressed more fully in the section on long-term effects of BMT.

2) Most Childhood Tumors Are Rare

Most childhood cancers (with the exception of ALL) are rare compared to common adult tumors. Consequently, it has proved difficult to design randomized studies of sufficient power to show conclusively that BMT is the procedure of choice for treating childhood malignancy. Most of the childhood tumors manifest dose response behavior, suggesting a favorable role for BMT. However, without support from randomized studies, it is only in patients with CML, or with the most advanced leukemia that BMT is unequivocally indicated, since in these patients, BMT may eradicate otherwise incurable malignancies. In all other cases, BMT is an experimental procedure, and should be presented to the family as such; the physician hopes, but does not know, that the dose intensification and GVL effect allowed by BMT will serve to cure more individuals.

3) Pharmacokinetics

The pharmacokinetics of many drugs in infants and young children are different from those in adults. Busulfan is the best studied example, but there are many other instances in which drug doses and schedules that are appropriate for adults may be inappropriate for the very young. This may lead to suboptimal results, either in terms of an excessive relapse rate or excessive toxicity.

6.9

B) ADVANTAGES OF BMT IN CHILDREN

Fortunately, BMT in children also has several advantages over the procedure in adults. For example, the incidence of severe GVHD is lower, and tolerance of mismatched or unrelated donor marrow grafts is higher. Since the resilience of other major organ systems is somewhat greater than in adults, deaths from almost all procedure-related complications are lower in children.

II) CHILDHOOD ACUTE LYMPHOBLASTIC LEUKEMIA (ALL)

A) INCIDENCE

ALL is the single most common cause of childhood cancer with an annual incidence of 29.4 per million population for children under the age of 15, with a peak age incidence between 2 and 6. The etiology is unknown but it has been hypothesized that an abnormal immune response to viral antigens may be involved in the pathogenesis. The incidence of ALL is increased in some inherited conditions such as Down's syndrome.

B) CLINICAL FEATURES

Children usually present with symptoms such as pallor, infection and bleeding that are a consequence of low counts but may present with atypical symptoms such as joint pain. Diagnosis is made by bone marrow examination including immunophenotyping and cytogenetics (see chapter on ALL).

C) PROGNOSIS

Features associated with a poor prognosis in childhood ALL include:

1) Cytogenetic

Abnormalities including the t(4;11) and t(9;22) are associated with a poor prognosis and may justify the earlier use of BMT. The t(1;19) was previously associated with a poor prognosis, but this is no longer so if an intensive chemotherapy regimen is used.

2) Age

Children less than 1 year who have an 11q23 chromosomal abnormality have a poor prognosis.

3) Sex

Males have a worse prognosis.

4) White Blood Cell Count

High WBC count at presentation is associated with worse prognosis.

5) Response

Delayed response to treatment with persistent circulating blasts at day 7 or persistent marrow blasts at day 14 imparts a worse prognosis.

6) Immunophenotype

The majority of children have an early pre-B phenotype and are CD19+ CD10+. Patients with CD10- and T cell ALL have previously had a poorer prognosis but do not if they receive more intensive regimens. Surface Ig+ B ALL responds poorly to conventional ALL chemotherapy but has a good prognosis if short term intensive therapy is employed (see below).

7) DNA Index

A DNA index of < 1.16 confers a poor prognosis.

D) TREATMENT

A variety of chemotherapy regimens have been employed. All usually consist of induction therapy with prednisone, vincristine and asparaginase in conjunction with additional drugs including etoposide, cytarabine, daunomycin or adriamycin and methotrexate followed by maintenance therapy with 6-mer-

6.9

captopurine and methotrexate interspersed with consolidation therapy. CNS prophylaxis is an important component of therapy and includes intrathecal chemotherapy with methotrexate (sometimes in conjunction with cytarabine and hydrocortisone) in association with systemic methotrexate. High risk patients may also receive cranial irradiation.

With intensive chemotherapy regimens, the cure rate with chemotherapy is around 70%. Because some survivors may develop regimen-related toxicity, such as secondary AML in patients who receive etoposide and cardiotoxicity in patients receiving anthracyclines, most studies are currently aiming to target therapy to high and low risk groups. Therefore a low risk patient may receive therapy with anti-metabolites alone, whereas early BMT may be considered for a patient with many high risk features. Patients with surface Ig positive ALL require different therapy and have a 80-90% cure rate if treated with short term intensive chemotherapy including cyclophosphamide, methotrexate and cytarabine. Patients with this diagnosis also require more intensive CNS prophylaxis.

E) **INDICATIONS FOR BMT**

With modern intensive chemotherapy the likelihood of cure after "front-line" therapy in childhood ALL is around 70%. Therefore, BMT in first remission is only undertaken in patients who have cytogenetic abnormalities such as the t(9:22) or t(4:11) which confer a poor prognosis. Transplantation of infants (children under the age of 1 year) in first remission is controversial. Although infant leukemia responds poorly to chemotherapy, the results of BMT are unsatisfactory as well, with a high level of regimen-related toxicity and a high relapse rate. The clearest indication for transplantation is in children who relapse whilst they are receiving chemotherapy, or who suffer a disease recurrence within 1 year of completing treatment. For these patients, the chance of cure with further chemotherapy alone is very low, and BMT should be offered. For children who relapse more than 1 year off therapy, opinion is divided. Some centers recommend early transplantation, particularly if an MHC identical sibling is available. Other groups prefer to retreat patients with salvage chemotherapy, performing a BMT only if a second treatment failure occurs.

6.9

1) **Type of BMT**

Most studies show a high risk of relapse when autologous BMT is used for ALL, and an allogeneic BMT is generally preferred. If the patient does not have a matched sibling, a mismatched family member or closely matched unrelated donor should be sought.

2) **Conditioning Regimen**

Most conditioning regimens for ALL employ total body irradiation (TBI) in conjunction with combinations of cyclophosphamide, cytarabine and etoposide. Alternatively busulfan and cyclophosphamide may be given. In mismatched or unrelated donor BMT additional immunosuppression with anti-thymocyte globulin may be added. In infants, irradiation is not normally used, because of the likelihood of severe developmental delay.

3) **Specific Toxicities**

Children who receive their transplant after relapse have already received intensive primary therapy followed by salvage therapy. Predictably, therefore, they have a high risk of regimen-related toxicity, dominated by veno-occlusive disease and posttransplant nephritis. Children who have previously received cranio-spinal irradiation at a young age have a risk of developmental delay and reduced growth. BMT from a mismatched family member or unrelated donor is associated with a higher risk of GVHD and regimen-related toxicity, but the risk of relapse after such transplants may be reduced.

4) Outcome

Long-term data on survival after transplant for high risk patients in first remission is not yet available. When BMT is used in second or subsequent remission, the long-term disease-free survival ranges from 10-50%. Survival is highest in children who receive an HLA-matched sibling donor transplant in second remission of disease which has relapsed off therapy. It is lowest in patients receiving marrow from a mismatched family member or matched unrelated donor for treatment of disease in third or later remission which has relapsed on therapy. In any patient group, infants fare worse than older children. When BMT is used in patients with active disease the results range from poor to dismal, and few centers now offer unrelated or mismatched donor BMT to patients with resistant ALL.

5) Late Effects

Growth hormone deficiency may occur, especially if the patient has received prior cranial irradiation. Around 50% of patients will develop abnormal thyroid function tests, and around 10% will require replacement therapy. Most patients will also develop gonadal failure, and females require ovarian replacement therapy.

6.9

III) CHILDHOOD ACUTE MYELOID LEUKEMIA (AML)

A) ETIOLOGY

Some inherited conditions such as Down's syndrome as well as conditions with an increased susceptibility to chromosomal breakage such as Bloom's syndrome or Fanconi's anemia have an increased risk of AML. In most cases however, the etiology is unknown.

B) DIAGNOSIS AND TREATMENT

Most patients present with signs and symptoms of bone marrow failure. The diagnosis, morphology and cytogenetics are reviewed in the adult AML chapter. Chemotherapy regimens used in pediatric AML are similar to those used in adult AML. In most chemotherapy studies the long-term survival is 20-30%. If a patient undergoes autologous BMT or allogeneic BMT from a matched sibling donor in first remission the cure rate rises to 40-70%. Most centers, therefore, recommend BMT in first remission. Because more procedure-related morbidity is associated with BMT from a matched unrelated donor or partially matched family member, these procedures are only undertaken in patients with refractory or relapsed AML or with high risk features. Conditioning regimens and late effects are essentially identical to those in ALL.

IV) JCML/MYELODYSPLASIA

As the results with chemotherapy are very poor for juvenile chronic myeloid leukemia or myelodysplasia occurring in the pediatric age group, allogeneic bone marrow transplantation is recommended for all such patients. However, there is dispute as to the need to give courses of ablative chemotherapy prior to the conditioning regimen.

V) CHRONIC MYELOID LEUKEMIA (CML)

Adult Ph+ CML is rare under the age of 20 and the etiology, diagnosis and therapy are as described in the chapter on CML. Allogeneic BMT is the only curative modality of treatment for CML and is the preferred treatment for pediatric patients with a matched sibling donor. The 5 year disease-free survival is 80-90% if BMT is undertaken in chronic phase. If a matched sibling is not available, and a matched unrelated donor or mismatched family member is used the survival falls to 50-60%. In these patients it may therefore be reasonable to undertake a trial of γ-interferon, as around 20% of patients will have a cytogenetic remission on this therapy. Autologous transplantation is still an experimental approach in this disease and should be confined to centers participating in clinical trials.

VI) NEUROBLASTOMA

A) BACKGROUND

After brain tumors, neuroblastoma is the most common nonlymphohematopoietic tumor of childhood comprising approximately 10% of pediatric malignancies. Pathologically neuroblastoma is classified as a tumor of postganglionic sympathetic neuroblasts, tissue derived from primitive neural crest.

B) DIAGNOSIS

Neuroblastoma is a tumor of infants and young children (5 years and less), and while an abdominal primary is most common, classically arising in the adrenal gland, any site in the sympathetic nervous system can be primary. As evidence of the developmental relationship of these tumors to the sympathetic nervous system, most patients will be noted to have increased catecholamine secretion. At diagnosis the tumor may be limited to a single organ, locally and/or regionally invasive, and/or widely disseminated. Bone, bone marrow, liver and skin are among the most common metastatic sites. Although classical light and electron microscopic findings of neuroblastoma are recognized, the diagnosis of neuroblastoma by these methods is sometimes problematic because its histopathologic appearance can be ambiguous and indistinguishable from other "small blue round cell tumors" of childhood, e.g. Ewing sarcoma, rhabdomyosarcoma, non-Hodgkin lymphoma and primitive neuroectodermal tumor. Furthermore symptoms and signs of bone marrow involvement by neuroblastoma may suggest acute leukemia. Awareness of its protean manifestations provides the necessary index of suspicion to consider the diagnosis of neuroblastoma.

6.9

1) Studies Supportive of the Diagnosis

a) Catecholamines

Elevated catecholamines (urine catecholamine metabolites, homovanillic acid (HVA) and vanillylmandelic acid (VMA), or urine or serum dopamine) are present in 90% of neuroblastoma.

b) Cytogenetics

Abnormal cytogenetic findings, e.g. deletion 1p, homogeneously staining regions (HSR) and double minute chromatin bodies (dmins), increase in copy number and expression of the proto-oncogene N-myc; tissue staining by immunologic reagents, e.g. antibodies to neuron-specific enolase, synaptophysin and the ganglioside GD2.

2) Definitive Diagnosis

By the criteria of the International Neuroblastoma Staging System, the diagnosis of neuroblastoma is established if: unequivocal pathologic diagnosis is made from tumor tissue by light microscopy (with or without immunohistology, electron microscopy, increased urine or serum catecholamine or metabolites); or bone marrow aspirate or trephine biopsy contains unequivocal tumor cells (e.g. syncytia or immunocytologically positive clumps of cells) and increased urine or serum catecholamines or metabolites.

C) PROGNOSIS

Age and disease stage at diagnosis are the most important clinical predictors of outcome in neuroblastoma.

1) Infants < 12 Months

Infants (< 12 months) have an intermediate prognosis even with advanced disease. However, neuroblastoma stage 4S (Table 6.9.1) is usually limited to infants < 12 months, and by definition is widely disseminated yet carries a good prognosis and is associated with "hyperdiploid" DNA content and low N-myc copy number.

2) Children Beyond 12 Months

Children beyond a year of age at diagnosis with advanced disease have a dismal prognosis with conventional chemotherapy.

3) Localized Disease

Most patients with localized, nonmetastatic tumors have an excellent outcome regardless of age.

4) Tumor Cell Biology

Tumors with "diploid" DNA content (DNA index, (DI = 1)) are associated with advanced disease whereas "hyperdiploid" tumors (DI > 1) are typically lower stage. The finding of homogeneous staining regions (HSR) or double minutes (dmins) which arise from increase copy number and expression of *n-myc* is usually associated with advanced stage disease. Deletion 1p (perhaps containing the locus of a tumor suppressor gene) is highly concordant with "diploid" tumors and is associated with advanced stage neuroblastoma. In general but especially for infants < 12 months, there is concordance of "diploid" DNA content (DI = 1), *n-myc* amplification, advanced stage and poor response to treatment. Likewise "hyperdiploid" DNA content (DI > 1) and absence of *n-myc* amplification predict favorable treatment response regardless of disease extent.

D) PRETREATMENT EVALUATION

1) Radiologic Imaging

Radiologic studies should include images of the primary tumor (e.g. MRI or CT), examination of chest and abdomen and other sites as clinically indicated by CT, MRI, and/or bone scan (MIBG or technetium scan), and plain bone radiographs.

2) Bilateral Iliac Crest Bone Marrow Aspirates and Biopsies

Marrow is a common metastatic site.

3) Biopsy of Enlarged Nodes

4) Serologic Markers

Elevated serum levels of neuron specific enolase, ganglioside GD2 and ferritin are associated with advanced stage.

5) DNA Content of Tumor Cells

DNA index (DI), measured by flow cytometry, correlates well with chromosome number.

6) **Gene Amplification**

7) **Karyotype**

E) STAGING (TABLE 6.9.1)

As the International Neuroblastoma Staging System (INSS) staging definition is recent in origin, data regarding treatment based upon it are few. Extrapolating from older staging systems, some important inferences can be made.

1) **Stage 1**

Surgery is adequate treatment for stage 1 neuroblastoma—90% disease-free survival (DFS).

2) **Stage 2 and 3**

DFS following surgery and chemotherapy in children with stage 2A and infants with stage 2B/3 is approximately 80%.

3) **Stage 4S**

Disagreement over the indications for and the method of treatment for infants with stage 4S neuroblastoma persists.

4) **Stage 4**

Advanced stage disease, stage 4, while often responsive to aggressive chemotherapy has a high probability of recurrence—DFS 20%, and it is for this cohort, encompassing greater than 60% of patients over 1 year of age, for whom bone marrow transplantation has been extensively applied.

F) NONTRANSPLANT TREATMENT

As for many malignancies, optimal treatment depends on a multi-disciplinary approach.

6.9

Table 6.9.1. International neuroblastoma staging system

Stage	Definition
1	Localized tumor with complete gross excision, with or without microscopic residual disease; representative ipsilateral lymph nodes negative for tumor microscopically (nodes attached to and removed with the primary tumor may be positive).
2A	Localized tumor with incomplete gross excision; representative ipsilateral nonadherent lymph nodes negative for tumor microscopically.
2B	Localized tumor with or without complete gross excision, with ipsilateral nonadherent lymph nodes positive for tumor. Enlarged contralateral lymph nodes must be negative microscopically.
3	Unresectable unilateral tumor infiltrating across the midline,* with or without regional lymph node involvement; or localized unilateral tumor with contralateral regional lymph node involvement; or midline tumor with bilateral extension by infiltration (unresectable) or by lymph node involvement.
4	Any primary tumor with dissemination to distant lymph nodes, bone, bone marrow, liver, skin and/or other organs (except as defined for stage 4S).
4S	Localized primary tumor (as defined for stage 1, 2A or 2B), with dissemination limited to skin, liver and/or bone marrow (limited to infants <1 year of age).

*The midline is defined as the vertebral column. Tumors originating on one side and crossing the midline must infiltrate to or beyond the opposite side of the vertebral column.
Marrow involvement in stage 4S should be minimal, i.e. <10% of total nucleated cells identified as malignant on bone marrow biopsy or on marrow aspirate. More extensive marrow involvement would be considered to be stage 4. The MIBG scan (if performed) should be negative in the marrow.

1) **Surgery**

Aggressive surgical intervention is determined by considerations of resectability and awareness of the high probability of favorable response to chemotherapy. Regional lymph nodes remote from the tumor should be sampled and, for patients with abdominal primaries, the liver biopsied whenever possible. Attempts at complete surgical resection following chemotherapy treatment are associated with fewer complications.

2) **Chemotherapy**

Combination therapy that may include vincristine, imidazole carboxamide, cyclophosphamide, doxorubicin, teniposide, etoposide, carboplatin and/or cisplatin is most commonly used to treat neuroblastoma; cyclophosphamide and doxorubicin are sufficient chemotherapy for low stage/low risk cases (nonmetastatic unresectable tumors and infants with hyperdiploid tumors), but for higher risk patients optimal therapy is not defined.

3) **Radiotherapy**

Neuroblastoma is a radiosensitive tumor (tumorcidal dose 15-30 Gy). Response rates and disease-free survival (DFS) among INSS stage 2B and 3 are better with the addition of radiotherapy to doxorubicin and cyclophosphamide—DFS approximately 60%.

G) **PRETRANSPLANT THERAPY**

Although the optimal treatment prior to transplantation is not established, wide consensus holds that transplantation should be reserved until maximum tumor reduction. In most instances, neuroblastoma responds to chemotherapy (see agents above), and many patients may be rendered disease-free by chemotherapy alone. In those circumstances where response is incomplete, it is common practice to perform second and even third surgical procedures to obtain complete tumor resection and/or to administer local irradiation if feasible to sites of residual disease before proceeding to transplant. Results are likely to be superior if patients undergo transplant before relapse or progression of disease.

H) **AUTOGRAFT VERSUS ALLOGRAFT**

Bone marrow is a common site of metastases in advanced neuroblastoma (80% of stage 4 patients), and the concern for residual but undetectable marrow disease initially favored the use of allogeneic bone marrow transplantation for these patients. Subsequent experience with both autologous and allogeneic bone marrow transplant has produced no evidence to support the superiority of allogeneic over autologous transplantation. Several reports suggest that the morbidity inherent in allogeneic transplantation outweighs any theoretical benefit. Thus the majority of transplants performed for patients with neuroblastoma have been autologous.

I) **MARROW PURGING**

Perhaps the major issue to be resolved in the application of autologous transplantation is the role of bone marrow purging. Using sensitive techniques such as immunocytology or reverse transcriptase PCR to detect tumor cells, up to 40% of clinically tumor-free autologous bone marrow may be contaminated by neuroblastoma cells even when all other indicators are negative for tumor. Gene marking of autologous marrow has unequivocally established that harvested marrow can contribute to relapse posttransplant. However it is not certain that every instance of contamination invariably causes relapse. Treating harvested marrow with antibodies specific for a particular tumor-associated antigen and removing the antibody coated tumor cells from the marrow (e.g. by the use of immunomagnetic beads which bind the antibody-neuroblastoma cell complex) may be one way to circumvent the problem of tumor cell contamination.

J) TRANSPLANT PREPARATORY REGIMEN

To date no bone marrow transplant regimen is clearly superior. Regimens have included chemotherapy without radiation (etoposide and carboplatin; or BCNU; or teniposide; or melphalan) or combined drug therapy (e.g. melphalan and vincristine; or cisplatin and etoposide and melphalan; or melphalan and cisplatin and teniposide and doxorubicin) with fractionated total body irradiation (dose range 1000-1500 cGy).

K) BMT OUTCOME

For patients undergoing transplant in complete remission or very good partial remission (> 90% primary tumor reduction, no metastatic disease except for bone, > 90% decrease in catecholamine excretion), disease-free or progression-free survival is 30-40% at 2 years. In the most recent trials, approximately 10-15% of patients die from transplant-related morbidity and the remainder relapse. While these results appear superior to historical data, adequate follow up will be needed to confirm these impressions. Indeed longer follow up from earlier transplant studies indicate that relapses occur beyond 2 years. Further, improvements in the traditional management of neuroblastoma with chemotherapy, irradiation, surgery and hematopoietic growth factors require randomized trials comparing these two approaches.

VII) RHABDOMYOSARCOMA

6.9

A) BACKGROUND

The most common childhood soft tissue sarcoma, rhabdomyosarcoma, comprises 5% of pediatric cancers, about 250 cases per year. This tumor, like the other soft tissue sarcomas, arises from primitive mesenchymal cells. Rhabdomyosarcoma may occur at virtually any body site; the primary anatomic distribution of rhabdomyosarcoma is: orbit—10%; parameningeal (includes nasopharynx, middle ear/mastoid, paranasal sinuses and infratemporal fossa)—15%; other head and neck—10%; genitourinary excluding special pelvic primary tumors—10%; special pelvic primary tumors (bladder, prostate, vagina, uterus)—15%; extremity—20%; trunk and other sites—20%.

B) PRESENTATION

Painless swelling without cause is a common finding at presentation but signs and symptoms largely depend on the anatomic location of the tumor. Diagnosis may be delayed because of the apparent benignity of the mass lesion.

C) HISTOPATHOLOGIC FEATURES

Rhabdomyosarcomas share some features of fetal skeletal muscle. Routine histopathology examination and ancillary studies include:

1) **Immunochemistry**

Rhabdomyosarcoma may be distinguished by antibodies to muscle-specific actin, to myoglobin and desmin, and most recently monoclonal antibodies to MyoD (protein family specific to muscle).

2) **Electron Microscopy**

Ultrastructural evidence for muscle actin-myosin or Z-bands are present in rhabdomyosarcoma.

3) Histology

Based upon light microscopic findings, rhabdomyosarcoma is classified:
a) embryonal—60% (includes botryoid variant)
b) alveolar—20%
c) pleomorphic/undifferentiated—20%
Distinguishing between these variants is important (see prognosis) and can be difficult. Several new findings are relevant.

  a) Translocation

  t(2;13) detected by reverse transcriptase PCR or by FISH is commonly found in alveolar rhabdomyosarcoma, the resulting fusion gene product presumably functions inappropriately as a transcription regulatory protein.

  b) DNA Index

  Embryonal rhabdomyosarcoma often has hyperdiploid DNA content, while alveolar rhabdomyosarcoma is associated with tetraploid DNA content.

D) DIFFERENTIAL DIAGNOSIS

The rhabdomyosarcoma differential diagnosis may include any of the other pediatric "small round blue cell tumors" including Ewing sarcoma and primitive neuroectodermal tumor as well as the other soft tissue sarcomas including fibrosarcoma, alveolar soft part sarcoma and synovial sarcoma.

E) PROGNOSIS

Primary site, postoperative extent of tumor and treatment are most important prognostic variables in RMS. Regarding the latter item, the Intergroup Rhabdomyosarcoma Study Clinical Grouping System (Table 6.9.2) is most commonly used and is the system for which associated outcome data are most available; however a modification of the tumor-node-metastasis (TNM) staging system has been adopted for rhabdomyosarcoma classification. Adequate staging requires extensive imaging of the primary tumor, both for surgery and radiotherapy planning. Some staging studies depend on the primary site. For example, parameningeal and paraspinal tumors may invade the nervous system (up to 50% cases). Lymphatic extension to lymph nodes is frequent in genitourinary, extremity and head and neck primaries. Other metastatic sites are lung, liver, bone and bone marrow.

1) Histology

Rhabdomyosarcoma with alveolar or undifferentiated histology compared to embryonal histology treated in the same fashion is more likely to fail treatment, but to some extent these differences can be overcome by intensive multi-modality therapy.

2) Site of Disease

Patients with tumors in the unfavorable sites, extremity, parameningeal and trunk fare worse than orbit, head and neck and genitourinary sites.

3) Intergroup Rhabdomyosarcoma Study (IRS) Groups (Table 6.9.2)

IRS group I and II comprise slightly more than one third of all patients, group III slightly less than one half. The outcome for group IV patients has not improved with additional therapy and better treatment is needed.

F) TREATMENT

Treatment requires the combined approach of surgery and chemotherapy; radiotherapy is necessary in all but group I tumors. Overall 65-70% of patients

6.9

Table 6.9.2. Clinical grouping system used by the IRS Committee for Rhabdomyosarcoma

Group I	Localized disease, completely resected Confirmed to organ or muscle of origin Infiltration outside organ or muscle of origin; regional nodes not involved
Group II	Compromised or regional resection including Grossly resected tumors with "microscopic" residual tumor Regional disease, completely resected, with nodes involved and/or tumor extension into an adjacent organ Regional disease with involved nodes, grossly resected, but with evidence of microscopic residual tumor
Group III	Incomplete resection or biopsy with gross residual disease remaining
Group IV	Distant metastases present at onset

are cured, this reflects cure of 90% of favorable histology and site patients but less than 30% of group IV patients.

1) IRS Group I

IRS group I have > 95% DFS with surgery followed by vincristine and actinomycin-D.

2) IRS Group II

Group II patients with favorable site and nonalveolar histology have 90% DFS with vincristine and actinomycin-D and local radiotherapy.

3) IRS Group III

Group III patients fare better with the use of additional chemotherapy agents, e.g. vincristine and actinomycin-D combined with cyclophosphamide. DFS is 60-70%. The impact of unfavorable histology on outcome has been reduced by the addition of intensive therapy.

4) IRS Group IV

The DFS of group IV patients has not demonstrably improved over the span of the IRS studies. These patients are candidates for BMT.

G) BONE MARROW TRANSPLANT

Patients with rhabdomyosarcoma who present with metastases, disease at unfavorable sites or fail following standard chemotherapy are at high risk of failure with conventional treatment. BMT conditioning regimens and results are similar to Ewing sarcoma (see below).

VIII) EWING SARCOMA

A) BACKGROUND

Originally defining a "small round blue cell" tumor of bone, Ewing sarcoma now denotes a tumor that may be osseous or extraosseous; moreover the t(11;22), the characteristic translocation of Ewing sarcoma, is also present in primitive neuroectodermal tumor (PNET) and a variant small cell osteosarcoma, suggesting a common genetic step in tumorigenesis for these entities. Although PNET is considered to possess neural cell pedigree, the origin of Ewing sarcoma is less certain, possibly neural, but most certainly not bone. Current opinion conceptualizes these tumors as variants within the family, Ewing sarcoma.

B) DIAGNOSIS

Ewing sarcoma is most common during the second decade (two-thirds) with another quarter occurring during the first decade. It is so uncommon among

6.9

black patients that the diagnosis should be questioned. In this tumor family, Ewing sarcoma is the most common diagnosis, five times more common than PNET. Primary sites are:

1) **Primaries for Osseous Ewing Sarcoma**

Half originate in extremity long bone (equal distribution between proximal and distal primary sites), and half along central axis (in descending order of frequency: pelvis, chest wall, spine or paraspinal, head or neck).

2) **Primaries for Extraosseous Ewing Sarcoma and PNET**

These are far more common in the trunk than in the extremities. PNET most frequently involves chest wall and may be synonymous with Askin tumor.

C) PRESENTING SIGNS AND SYMPTOMS

These are related to the primary site; pain, palpable mass, pathologic fracture and fever are common but nonspecific manifestations. Although all patients are assumed to have microscopic metastases at diagnosis, 20-30% of patients will have overt metastases to lung, bone and bone marrow. Metastases to the vertebrae are common and back pain at diagnosis or at any time along the course of treatment must be thoroughly evaluated less spinal cord compression supervene. For purposes of treatment planning and evaluation, imaging studies must adequately define the extent of primary tumor as well as identify gross metastatic disease.

D) DIFFERENTIAL DIAGNOSIS

The differential diagnosis for Ewing sarcoma includes both benign and malignant disorders. Among the latter are the "small round blue cell" tumors such as non-Hodgkin lymphoma, neuroblastoma and rhabdomyosarcoma. Although classic histiotypes are recognized, distinction between the tumors of the Ewing sarcoma group, as well as segregating them from other "small round blue cell" tumors on the basis of light microscopy, can be difficult; consequently both electronmicroscopy and immunocytochemistry are crucial adjuncts in making the diagnosis.

1) **Translocation**

t(11;22) is characteristic of these tumors; RT-PCR and FISH are useful techniques for demonstrating its presence.

2) **Cell Surface Antigen**

An antibody raised against a cell surface antigen encoded by the MIC2 gene is closely associated with Ewing sarcoma and rare in other "small round blue cell" tumors.

3) **Neuronal Differentiation**

Within the Ewing sarcoma group evidence of neuronal differentiation is characteristic of PNET.

E) PROGNOSIS

For the present, no single factor predicts the probability of 5 year DFS more accurately than the presence or absence of clinically evident metastatic disease at diagnosis. The DFS for patients without metastases approaches 70% with multi-agent dose intense chemotherapy, but for patients with metastases it is about 20-30%. Tumor size and site are important, large tumors and pelvic primaries generally fare worse. Multi-agent dose intense chemotherapy may diminish the importance of these factors. It is uncertain whether PNET and

Ewing sarcoma carry the same prognosis. Elevated LDH at diagnosis is associated with poorer DFS.

F) TREATMENT

As is true for virtually all pediatric solid tumors, the best treatment for Ewing sarcoma requires the judicious and ordered combination of surgery, radiotherapy and chemotherapy. Although local control and systemic treatment cannot be separated conceptually, the majority of patients who present without clinical evident metastasis fail systemically (65% systemic, 25% local, 10% combined).

1) Chemotherapy

All patients, regardless of the completeness of resection, require chemotherapy and current evidence favors dose intense cyclophosphamide and doxorubicin in combination with vincristine, actinomycin-D, ifosphamide and etoposide. Under investigation, chemotherapy may convert inoperable tumors to operable.

2) Radiotherapy

Following complete tumor resection radiotherapy is unnecessary. Radiotherapy (50-55 Gy) is essential for unresectable primaries but the optimal approach, i.e. radiotherapy, surgery or both, for the majority of Ewing sarcoma is uncertain.

G) BONE MARROW TRANSPLANTATION FOR RHABDOMYOSARCOMA AND EWING SARCOMA

Patients with rhabdomyosarcoma or Ewing sarcoma who present with metastases, disease at unfavorable sites or fail following standard chemotherapy are at high risk of failure with conventional treatment. New approaches to treatment for these high risk patients are justified; unfortunately there is as of yet no clear benefit to transplantation.

6.9

1) Conditioning Regimen

Conditioning regimens use high-dose chemotherapy alone (e.g. melphalan) or in combination (e.g. melphalan and busulfan; vincristine, doxorubicin and cyclophosphamide; melphalan, etoposide and carboplatin) combined with local and/or total body irradiation (8-12 Gy) followed by autologous marrow or stem cell transplant.

2) Disease-free Survival with BMT

a) High Risk Localized Disease

BMT has resulted in 40-50% 5 year DFS for high risk patients with localized disease.

b) Metastatic Disease

BMT in patients presenting with metastases results in DFS less than 20-45%. While these results may be viewed as encouraging, the number of patients undergoing transplant is small, and patient selection does not always permit comparison to patients treated by conventional methods.

c) Chemotherapy Resistant Disease

In those circumstances where the tumor is unresponsive to chemotherapy or has recurred, intensive high-dose combination chemotherapy with autologous marrow rescue may induce partial response, but salvage rates rarely exceed 10% for patients with refractory disease.

IX) GLIOBLASTOMA MULTIFORME

Glioblastoma multiforme is histopathologically and clinically the most malignant form of astrocytoma. For most patients, surgery and radiotherapy with or without chemotherapy seldom are curative; 5 year survival rates are less than 10%. The role of chemotherapy in the treatment of glioblastoma multiforme is uncertain primarily because no clinical data have demonstrated improved survival for chemotherapy treated patients. However, the observation that half of patients with tumor respond to chemotherapy (e.g. CCNU, vincristine and prednisone or cyclophosphamide and vincristine), at least partially, has led to the use of dose intense chemotherapy, followed by autologous marrow or pharesis infusion.

A) CONDITIONING REGIMEN

Regimens have included marrow ablative doses of chemotherapy (e.g. thiotepa, etoposide and nitrosurea; thiotepa and cyclophosphamide), autologous marrow infusion and brain irradiation.

B) DISEASE-FREE SURVIVAL WITH BMT

Thirty to sixty percent of treated patients have shown complete or partial response. Follow up is insufficient, and the number of glioblastoma patients treated is small; therefore, the lasting benefits of this approach remain unproven. Nevertheless, survival after surgery and/or radiotherapy is so dismal that these preliminary findings deserve further evaluation.

X) OTHER BRAIN TUMORS

There are case reports of patients with recurrent brain tumor (medulloblastoma, ependymoma and primitive neuroectodermal tumor) who have been salvaged with high-dose chemotherapy and autologous marrow reinfusion. The ultimate utility for this approach will depend upon the availability of effective chemotherapeutic agents. To be effective these agents must be able to penetrate the blood-brain and blood-tumor barrier in concentrations that are tumoricidal. The scarcity of effective agents is problematic for the entire domain of neuro-oncology.

XI) OTHER TUMORS

Patients with recurrent or refractory germ cell tumor, retinoblastoma or other pediatric soft tissue sarcomas may be candidates for salvage therapy with ablative chemotherapy and autologous bone marrow transplantation. Fortunately, initial therapy for these entities is usually effective and only rare patients fail treatment and are in need of this approach. At the same time, the limited experience in treating failing patients hampers the development of effective ablative regimens, and the outcome following transplantation is poor, regardless of the conditioning regimen used.

6.9

SUGGESTED READING

1. International Bone Marrow Transplant Registry. Transplant or chemotherapy in acute myelogenous leukemia. Lancet 1989; i:1119-1122.
2. Rivera G, Pinkel D, Simone JV, Hancock ML, Crist WM. Curing children of acute lymphoblastic leukemia: 30 years of "Total Therapy" at St Jude Children's Research Hospital. New England Journal of Medicine 1993; 329:1289-1295.
3. Pui C, Crist WM, Look AT. Biology and clinical significance of cytogenetic abnormalities in childhood acute lymphoblastic leukemia. Blood 1990; 76:1449-1463.
4. Shuster JJ, Cantor AB, McWilliams N, Graham-Pole J, Castleberry RP, Marcus R, Pick T, Smith EI, Hayes FA. The prognostic significance of autologous bone marrow transplant in advanced neuroblastoma. Journal of Clinical Oncology 1991; 9:1045-1049.
5. Horowitz ME, Kinsella TJ, Wexler LH, Belasco J, Triche T, Tsokos M, Steinberg SM, McClure L, Longo DL, Steis RG et al. Total-body-irradiation and autologous bone marrow transplant in the treatment of high risk Ewing's sarcoma and rhabdomyosarcoma. Journal of Clinical Oncology 1993; 11:1911-1918.

Hereditary Disorders of the Immune System Treatable by Bone Marrow Transplantation

Richard E. Harris, Cynthia DeLaat, Lisa Filipovich

I) INTRODUCTION

Hereditary disorders of the immune system include not only the primary lymphocyte immune deficiencies but also disorders of neutrophil and monocyte production and function. This chapter will cover those hereditary immune disorders which are reasonable candidates for treatment with marrow transplantation, including severe combined immune deficiency and its several variants, Wiskott-Aldrich syndrome, reticular dysgenesis, hemophagocytic lymphohistiocytosis, X-linked lymphoproliferative syndrome, Kostmann's agranulocytosis, Chediak-Higashi syndrome, chronic granulomatous disease and CD11/18 deficiency. These disorders are all believed to be due to genetic mutations. Some of the affected genes have now been identified and sequenced, opening the door to gene therapy for at least some of the primary immune deficiencies in the decades to come. Until this eventuality, the only curative therapy for the majority of these disorders is stem cell transplantation, principally bone marrow transplantation. The first successful marrow transplant was performed in a child with severe combined immune deficiency at the University of Minnesota in 1968. This first marrow recipient is still alive and well today, cured of his otherwise fatal immune deficiency.

A prematurely fatal primary immune deficiency is suspected in any child who presents early in life with severe life-threatening infections, unusual infections (*Pneumocystis pneumonia*), persistent infections with common organisms (chronic oral thrush or *Candida* diaper rash) or failure to thrive (FTT) with chronic diarrhea. Many of the immune deficiencies are also associated with abnormal physical characteristics such oculocutaneous pseudoalbinism as in Chediak-Higashi syndrome, or abnormalities of other cell lines as with the microthrombocytopenia seen in Wiskott-Aldrich syndrome. The precise diagnosis can often be suspected on the basis of an exam and history and a few simple laboratory tests. However, others require more sophisticated immune or gene studies to make a clear diagnosis.

II) SEVERE COMBINED IMMUNE DEFICIENCY AND VARIANTS

SCID is a general term to describe several immune deficiencies which have in common significant defects in both B and T cell immunity. All three of the common SCID types can present with early onset of rashes, malabsorption and diarrhea, failure to thrive, chronic oral candidiasis, and life-threatening opportunistic infections such as *Pneumocystis carinii* and cytomegaloviral pneumonitis. Most children, if untreated, die within the first 2 years of life.

A) SCID SUBTYPES

1) SCID with B cells

The most common form of SCID in North America is termed SCID with B cells or X-linked SCID, representing about half of SCID patients. These

6.10

patients have few or no circulating T cells, but do have circulating, though usually nonfunctional, B cells. This disorder results from a defect in the IL2R-γ chain (or "common gamma chain") on the X-chromosome.

2) SCID Without B or T cells

A widely recognized form of SCID includes SCID without B or T cells, also termed the "classical" or Swiss form of SCID and is probably due to genetic defects in RAG (recombination activating genes).

3) ADA-Deficient SCID

Patients with adenosine deaminase (ADA) deficiency lack the enzyme adenosine deaminase, which leads to elevated intracellular levels of deoxyadenosine and its phosphorylated metabolites which are directly toxic to T and B cells.

4) SCID Variants

a) Omenn's Syndrome

A SCID variant, known as Omenn's syndrome, presents with a scaly rash, diarrhea, leukocytosis with eosinophilia, hepatosplenomegaly and lymphadenopathy with lymphoid replacement by Langerhans and reticulum cells. Elevated IgE, even in the face of hypogamma-globulinemia, is typical. These patients otherwise present in a similar fashion with early onset of severe opportunistic infections.

b) Cartilage Hair Hypoplasia

Another uncommon variant, cartilage hair hypoplasia, is characterized by short-limbed dwarfism and ectodermal dysplasia, which also is associated with a progressive and severe immune deficiency.

c) Purine Nucleoside Phosphorylase Deficiency

Purine nucleoside phosphorylase (PNP) deficiency, another rare variant of SCID, is associated with progressive immune dysfunction, especially of T cell function. Children are normal at birth, then by age 3-5 years develop severe viral, fungal and bacterial infections. Disseminated varicella is a special risk in this subgroup.

d) Reticular Dysgenesis

The very rare entity known as reticular dysgenesis is characterized by severe combined immune deficiency in association with neutropenia and lymphoid hypoplasia, but generally with normal production of red cells and platelets. These infants usually die shortly after birth unless the infant is placed into a protected environment and an immediate marrow transplant is performed within the first month or so of life.

e) Other

ZAP 70 (zeta-associated protein-70) deficiency, SCID with NK cells and other unusual forms are largely restricted to inbred populations.

B) DIAGNOSIS

The primary diagnosis of SCID is based on finding marked decreases of T lymphocyte function (depressed response to mitogens) usually in combination with an absence or nonfunction of B cells, characteristically, hypogammaglobulinemia. Phenotypic marker and mitogen response studies can be carried out in many laboratories across the country, but more sophisticated immune function and gene studies are generally only performed at a few specialized laboratories.

6.10

C) PRE-BMT CONSIDERATIONS

1) Infections

Underlying infections should be carefully diagnosed and treated aggressively. Patients should be placed on prophylaxis for *Pneumocystis carinii* pneumonia and thrush and should receive replacement intravenous immune globulin (IVGG) 400 mg/kg every 3-4 weeks. Levels of IgG should be tested to maintain a trough in excess of the normal lower limit for age (generally > 400 ng/ml) and the dose or interval adjusted to assure adequate levels at all times. Patients who are IgA deficient should receive IVGG products devoid of IgA. Some centers place the newly diagnosed SCID patient into laminar air flow isolation. Studies have not proven the efficacy of this approach. These patients at the very least, though, should be protected from exposure to individuals with any evidence of an infection, viral or otherwise. The air handling system for the patient's hospital room should provide positive pressure in the room and should be HEPA filtered to reduce the chance of aspergillus infection. In some cases of SCID, patients are better "isolated" in their home to decrease the risk of nosocomial infection.

2) Blood Products

All blood products must be irradiated and should be leukofiltered to prevent infusion of viable lymphocytes capable of engrafting and causing GVHD. Filtration also reduces the risk of transmission of viral infectious agents such as cytomegalovirus. CMV-negative blood products are preferred for patients with SCID to further reduce the risk of transmission of CMV to the patient.

6.10

3) Nutrition

Most patients with SCID have failure to thrive with malabsorption and diarrhea. They usually require hyperalimentation or elemental feeding by an NJ tube to deliver adequate nutrition.

D) BONE MARROW TRANSPLANTATION

Bone marrow transplantation should be performed as soon as possible in all patients with SCID, if feasible. Of special note is the observation that SCID patients often develop marked inflammatory responses as immune reconstitution begins. These patients must be closely monitored for organ toxicity during the first few months of immune reconstitution. Steroids are sometimes administered during this phase to modulate the severity of the immune response and to limit organ toxicity, especially interstitial pneumonitis.

1) Matched Sibling

If the patient does perchance have a matched sibling donor available, the infusion of non-T cell depleted marrow is recommended and usually results in lymphoid engraftment with less than a 20% incidence of significant GVHD and a survival of about 90% in recent years. Many patients who have nonfunctional B cells do not engraft with donor B cells, but their host B cells often are found to function normally once HLA-matched donor T cells engraft. Rarely is myeloablution (i.e. conditioning regimen) required for those fortunate patients with SCID and a matched sibling donor.

2) Alternate Donors

For patients without matched sibling donors use of haploidentical T cell depleted related marrow or cord blood from closely matched unrelated donors (with or without T cell depletion) can be considered. The optimal "alternative donor" transplant will depend on the current projected survival rate with different approaches, the SCID phenotype and the urgency of the situation. In related haploidentical transplants most centers give a T cell depleted graft without conditioning. If lymphocyte engraftment is not documented, a second transplant from the same donor is attempted following myeloablation with cyclophosphamide and busulfan. For a phenotypically matched unrelated transplant, myeloablation with cyclophosphamide and busulfan is done prior to graft infusion.

E) GENE THERAPY

With identification of the gene responsible for ADA deficiency, gene therapy studies have become a reality. Three patients with ADA deficiency diagnosed at birth at the Children's Hospital of Los Angeles have received therapy with gene corrected lymphocytes and continued therapy with ADA enzyme infusions. These patients are being followed closely for permanent reconstitution with ADA-replete lymphocytes, potentially allowing discontinuance of the enzyme infusions in the future.

III) WISKOTT-ALDRICH SYNDROME (WAS)

WAS is an X-linked recessive immune disorder characterized by progressive T cell dysfunction and impaired production of antibodies to polysaccharide antigens, eczema and thrombocytopenia associated with small platelet size. The T cell immune dysfunction is variable and progressively deteriorates over time, and the B cell function is characterized by low serum IgM and low or absent isohemagglutinins. The specific genetic defect in WAS has been recently localized to the *wasp* gene on the X-chromosome, though little is known to date about the function of this gene.

Without transplant, patients with WAS generally die during the first or second decade of life from bleeding associated with the thrombocytopenia or from infectious complications. Late in the course of the disease, patients also develop autoimmune complications such as autoimmune hemolytic anemia. In addition, patients with WAS may develop lymphoid or other malignancies as their terminal event.

Matched sibling donor marrow transplantation for WAS is highly successful with a 90% predicted cure rate. Not only is the immune deficiency and thrombocytopenia cured, but the propensity toward autoimmune disease and lymphoid malignancies is eliminated. While T cell depleted haploidentical transplantation has not proven highly successful in WAS, use of unrelated donor transplantation for young patients appears to be a promising alternative. Most centers give myeloablative conditioning with cyclophosphamide and busulfan with or without etoposide to ensure engraftment.

IV) HEMOPHAGOCYTIC LYMPHOHISTIOCYTOSIS (HLH)

HLH, previously termed familial erythrophagocytic lymphohistiocytosis (FEL), is a rapidly fatal disorder of the immune system inherited in an autosomal recessive manner. It is characterized by the massive infiltration of the marrow, liver, spleen and lymph nodes by cells of the monocyte/macrophage system as well as by lymphocytes. Hemophagocytosis by macrophages is a prominent feature. Patients generally present in the first 6 months of life with organomegaly, fever, cytopenias, rashes, hypofibrinogenemia, elevated triglyceride levels and coagulopathy. Many patients (in our experience 2/3) present with CNS involve-

6.10

ment as well. Immune studies reveal a consistent lack of natural killer cell activity and often abnormal cytotoxic T cell function. Hypofibrinogenemia, elevated triglyceride levels and coagulopathy are common in HLH. There is no clear evidence to support a malignant etiology in this disease. HLH should be differentiated from virally associated hemophagocytic syndromes (VAHS) by studies to rule out CMV, EBV, HIV, rubella and varicella. Probes for some of these can be used to test biopsied tissues as well. VAHS generally presents later in life, beyond infancy.

A) PRE-BMT CONSIDERATIONS

Upon presentation, children with HLH should be thoroughly evaluated including biopsies of the marrow, liver and lymph nodes and a spinal tap. Abdominal ultrasound and chest x-ray is recommended as well as a brain MRI, even in patients with a negative CSF examination. Immune studies should be performed, especially for natural killer cell function. Peripheral blood counts and triglyceride and fibrinogen levels should be obtained. HLA typing of family members should also be obtained and a search for an unrelated donor should be initiated if no suitable match is found within the family. Once a diagnosis of HLH is established, treatment should be initiated with a combination of steroids and etoposide. During the maintenance phase, cyclosporine may be used, although there is no definitive evidence to strongly support its role. Patients with CNS involvement should additionally receive intrathecal methotrexate. The exact recommended regimen is available through the Histiocyte Society (609 New York Road, Glassboro, NJ 08028, phone 609/881-4911, fax 609/569-6614) which provides detailed protocols of therapy. Physicians are encouraged to enter newly diagnosed patients onto the national trials for treatment of HLH sponsored by the Histiocyte Society. For patients who are unable to undergo a marrow transplant for lack of a donor and who have failed therapy with etoposide, steroids and cyclosporine, immune therapy with ATG and/or cyclosporine should be considered, per the recommendations of the Histiocyte Society.

6.10

B) BONE MARROW TRANSPLANTATION

Patients who achieve an initial response to therapy should be offered marrow transplantation with a matched sibling donor, if available. If not, a search among other family members and for an unrelated donor should be rapidly pursued. If a suitable donor is found, then patients should be offered BMT in first response of the disease. The generally recommended preparative therapy includes a combination of cytoxan, busulfan and etoposide.

C) POST-BMT CONSIDERATIONS

Of interest is that patients resolve all evidence of their disease over a period of several months. Evidence of persistent disease at the day +100 evaluation does not definitely predict failure from the transplant. Such patients should be observed closely for at least a year with repeated evaluations posttransplant before considering further therapy.

V) X-LINKED LYMPHOPROLIFERATIVE SYNDROME (XLP)

XLP, also known as Purtillo's syndrome, is an immune deficiency historically associated with an inability to control EBV infection. Many patients are well until exposed to this virus, when they develop either rapidly progressive primary EBV infection with hepatosplenomegaly and pulmonary and/or CNS involvement, or they develop agammaglobulinemia and/or aplastic anemia. Some patients die from EBV-associated lymphoma. Most patients die by the age of 10 years. Recently, however, it has been recognized that patients with XLP may experience immunologic abnormalities, lymphoma and other infections even prior to EBV exposure.

The diagnosis of XLP is based on a strong family history and evidence of inadequate immune response to EBV infection combined with the development of any of the clinical scenarios noted above. Patients with XLP also exhibit abnormal NK activity, as in HLH.

The only currently available potentially curative therapy is marrow transplantation. Few transplants for this disease have yet been performed, but the outcomes look encouraging. The recommendations for preparative therapy are similar to that for patients with HLH.

VI) CHEDIAK-HIGASHI SYNDROME (CHS)

CHS is an autosomal recessive disorder of neutrophil function characterized by the appearance of large lysosomes in many tissues, often visualized in lymphocytes and neutrophils on a peripheral smear. The neutrophils characteristically contain giant primary granules associated with a severe neutrophil dysfunction of chemotaxis and intracellular bacterial killing resulting in severe and recurrent pyogenic infections. Patients with CHS also have poor T cell and NK cell function. Pseudoalbinism due to the presence of melanosomes with large granules which do not disperse light, giving the skin and hair a very light blond color is a characteristic phenotypic finding. These patients can be diagnosed by examination of their neutrophils on a peripheral blood smear for the presence of the characteristic giant primary granules. Patients with CHS also develop peripheral neuropathies. These patients often die at an early age from infection or later from development of the "accelerated phase" often associated with EBV infection, which is reminiscent of HLH. In the accelerated phase the patient develops massive lymphohistiocytic infiltration of the liver, spleen, bone marrow and CNS. Bone marrow infiltration results in pancytopenia.

The only known curative therapy is marrow transplantation. Several reports of successful transplants for CHS are in the literature. It is recommended that patients with CHS undergo marrow transplant at an early age, before the development of the accelerated phase and evolution to acute myelogenous leukemia. Again, the recommended preparative therapy is that used for HLH and XLP.

VII) KOSTMANN'S AGRANULOCYTOSIS

A few successful marrow transplants have been reported for this severe form of congenital neutropenia caused by a defect in c-kit ligand, but since G-CSF has proven beneficial in most of these patients, marrow transplant is only recommended for those patients who fail G-CSF therapy. The transplant should be done before the development of serious and chronic infections or organ dysfunction from infection or its treatment precludes the likelihood of a successful transplant.

VIII) CHRONIC GRANULOMATOUS DISEASE (CGD)

CGD is a neutrophil functional defect in oxidative metabolism involving the NADPH oxidase enzyme complex and is characterized by the inability of the neutrophil to kill catalase positive organisms following their ingestion. Patients with CGD develop abscesses and granulomas in the subcutaneous tissues, lungs and lymph nodes due to *Staphylococcus aureus* and various gram negative rods as well as various fungi (especially *Aspergillus*). A few successful marrow transplants for CGD have been documented in the literature. However, therapy with gamma-interferon and the use of prophylactic trimethoprim-sulfamethoxazole has markedly improved the outcome of patients with CGD. Few patients with CGD are

6.10

today being offered transplant due to the availability of these alternative therapies. Patients who fail to respond to gamma- interferon and who have ongoing severe infectious complications might be considered for therapy with marrow transplantation. Transplant therapy though should be used early enough in the disease course to minimize the risk of infectious or toxic deaths from the transplant.

IX) CD11/18 DEFICIENCY (LEUKOCYTE ADHESION DEFICIENCY)

This severe neutrophil adhesive protein deficiency (leukocyte adhesion deficiency—LAD) can be identified by a history of delayed umbilical cord separation. It is typified by early and severe infections usually due to *Pseudomonas* and *Staphylococcus epidermidis*, high peripheral neutrophil counts and poor wound healing. Many patients with severe deficiency die at an early age from sepsis or necrotizing enterocolitis. The neutrophils are unable to adhere to the vascular endothelium and enter the tissues to sites of active infection. Most patients have been found to lack the CD18 moiety of the CD11/18 complex. Thus patients cannot form the LFA-1 complex (CD11a/18) or the Mo-1 complex (CD11b/18), both important leukocyte integrins involved in neutrophil adherence via the ICAM-1 adhesive molecules. T cell lymphocyte function is also abnormal in these patients and has been postulated as a reason for the relatively low rate of graft rejection in these patients when donors other than matched siblings have been utilized.

Once a firm diagnosis is established and the patient is found to have a severe deficiency ($< 5\%$ expression of the CD11/18 complexes on the surface of the neutrophils), marrow transplant should be pursued as the preferred mode of therapy. Transplant should be carried out at an early age to avoid the problems of unresolved infections and organ damage from prior therapy. Even a low percentage of engrafted donor cells can restore functional immunity.

6.10

SUGGESTED READING

1. Stiehm ER. Immunologic Disorders of Infants and Children (4th ed.). Philadelphia: W. B. Saunders Company, 1995. (Various chapters in section II deal with the specific hereditary immune deficiencies.)
2. Möller G, ed. Genetic basis of primary immunodeficiencies. Immunological Reviews 1994; 138:221.

Lysosomal and Peroxisomal Storage Diseases

Richard E. Harris, Nancy Leslie, William Krivit

6.11

I) INTRODUCTION

Over 200 marrow transplants have been performed in the past decade in patients with various lysosomal and peroxisomal storage diseases. Marrow transplantation is currently the only method of providing a self-replenishing source of enzyme to these patients. Transplantation of marrow stem cells from a nondeficient donor will provide cells of the monocyte/macrophage system which are of donor origin and therefore competent to produce the target enzyme. These donor derived cells replace the host fixed tissue macrophages in the liver (Kupffer's cells), bone (osteoclasts), lung (pulmonary macrophages), lymph nodes (fixed and wandering histiocytes) and skin (Langerhans cells). At least for some of the lysosomal and peroxisomal storage diseases, enzyme produced in the fixed tissue macrophage system is capable of uptake into recipient cells, where it is targeted to lysosomes. By this process accumulated storage material can be hydrolyzed, resulting in reduction of organomegaly and clinical stabilization or improvement in organ function. Since the marrow continuously replenishes the body with donor fixed tissue macrophages, the effects can be expected to be permanent.

In many lysosomal and peroxisomal storage diseases, CNS pathology causes major morbidity. In these situations, a major goal of therapy would be for enzyme to reach the central nervous system by replacement of recipient cells with donor microglial cells, hopefully resulting in stabilization and prevention of further damage to the central nervous system. Evidence is rapidly accumulating that indeed there is an influx of donor microglial cells into the brain tissues. The microglial cell population represents 10-15% of all cells in the central nervous system. As a result, these patients often do have documentable improvements in their brain MRI and in their level of neurological and neuropsychological functioning. This replacement by donor microglial cells is slow, though, taking 12 months or more. Thus, many patients may continue to have deterioration of the CNS for several months before any slowing or arrest of neurological damage can occur. Alternatively, enzymes produced in the new monocyte-macrophage system may enter the plasma via exocytosis. The number of cells in the monocyte-macrophage system is equivalent to that of hepatocytes in the liver. Production of protein is also equivalent. If present in the plasma at high enough levels, newly produced enzyme might be expected to obtain direct entrance into the central nervous system if there is disruption of the blood-brain barrier.

Improvement with bone marrow transplantation for the various lysosomal and peroxisomal storage diseases may be incomplete. For instance, in Hurler syndrome, the chondrocytes remain unable to clear glycosaminoglycan accumulation, most likely due to their lacunar isolation. Therefore, skeletal defects remain even though complete engraftment is obtained. Similarly, complete correction of the central nervous system manifestations is not achieved. In order to determine whether the expected improvement outweighs the risks and expense of marrow transplantation, there is a need for prospective comparison of patients undergoing marrow

Bone Marrow Transplantation, edited by Richard K. Burt, H. Joachim Deeg, Scott Thomas Lothian, George W. Santos. © 1996 R.G. Landes Company.

transplantation to those patients not undergoing BMT. To satisfy this need, over 25 transplant institutions have joined together in an NIH-funded consortium known as the Storage Disease Collaborative Study Group (Table 6.11.1) in order to rapidly obtain the information needed to determine the efficacy of BMT for these disorders.

II) ETIOLOGY

The lysosomal and peroxisomal storage diseases are for the most part due to single gene defects. In most cases, the enzyme deficiency results from point mutations or deletions in the gene coding for the missing protein. Since the disease state results from deficiency or absence of a normal gene product, these diseases are recessive, either autosomal (e.g. Hurler syndrome) or X-linked (e.g. Hunter syndrome). A wide variety of mutant alleles may be described for any given disease classification. However, certain alleles are associated with a mild phenotype while others may be found only with a more severe phenotype. Therefore, even within a single classification (e.g. type 1 Gaucher disease), phenotypes may vary considerably. This has importance in determining the efficacy of treatment.

III) DIAGNOSIS

The clinical diagnosis in each of these diseases is first suspected by delays in development, attention deficit disorders, inability to walk normally, loss of milestones or by the recognition of abnormal somatic features or organomegaly. A tentative diagnosis may be made by analysis of abnormal substrates, such as mucopolysaccharide or oligosaccharide in urine. For most diseases, direct enzyme assay in serum, leukocytes or cultured fibroblasts will be required for confirmation.

IV) PRETREATMENT EVALUATION

The goals of the pretransplant evaluation are two-fold: 1) to identify organ dysfunction which might affect the outcome of the transplant and 2) to determine the status of the patient's primary disorder for comparison after transplant and with other patients with the same diagnosis. Patients may become ineligible for transplant either due to severe organ dysfunction which makes the transplant too risky or due to severe damage to the CNS or other organs by the primary disease which would prevent the transplant from being of any significant benefit to the patient. Some children with lysosomal and peroxisomal enzymatic deficiencies have a rapid deterioration. Some may die within the first 1-2 years of life. Others may not begin to have clinical manifestations until later and will live into their adult years. Some conditions are associated with progressive loss of cognitive function, whereas in others, normal intelligence is preserved even in the face of impaired organ function. One would expect that affected siblings would exhibit similar phenotypes, but exceptions are found in the literature. Therefore, complete clinical evaluation, complemented by careful family history, enzymatic confirmation and, if possible, analysis of specific mutations all play a role in counseling families about possible outcomes and in analyzing results of attempted therapies.

6.11

Table 6.11.1. Consortium address—Storage Disease Collaborative Study Group

Contact—William Krivit, M.D.
Office—University of Minnesota Hospitals and Clinics, Box 391
516 Delaware Street, Minneapolis, Minnesota 55455
Phone—612-624-6116
Fax—612-624-2682

V) TREATMENT PHILOSOPHY

Physicians following a child with an inherited metabolic disorder should know the up-to-date results of marrow transplantation for that specific disorder and the current recommendations for therapy. Information can be obtained through the Storage Disease Collaborative Study Group offices at the University of Minnesota Hospitals and Clinics (Table 6.11.1).

For many of the lysosomal and peroxisomal storage diseases, the outcome with transplant is not yet known because not enough transplants have been done for that disorder or patients have not been followed long enough posttransplant to determine the ultimate outcome. For some patients, such as in adreno-leukodystrophy, there may be actual improvement in CNS function after transplant. But for other patients, such as those with Hunter syndrome or Sanfilippo syndrome, a marrow transplant may lead to a more drawn out course attendant with more demands upon the family and upon society. Families may demand everything medically possible be done for their child only later to regret the decision because of the long drawn out demands for care of a marginally functional individual.

For some diseases such as Hurler syndrome, enough transplants have been done to know what to expect years after the transplant. Such a transplant results in stabilization of intelligence, disappearance of hepatosplenomegaly, improvement in hearing and prevention of death due to coronary artery stenosis after the first posttransplant year. However, the orthopedic problems seem to progress in spite of the transplant, though at a slower rate. These patients are likely to develop severe kyphoscoliosis and resultant myelopathy and other orthopedic problems which may become severely incapacitating in the years after the transplant.

VI) CONDITIONING REGIMEN AND GVHD PROPHYLAXIS

Various preparative regimens and GVHD prophylaxis have been utilized in these transplants. There is currently no widely accepted best preparative therapy or GVHD prophylaxis regimen. A consortium-sponsored conference on these issues was held in June of 1995 and a summary from this conference is available from the consortium office. In patients undergoing a matched sibling transplant, preparative therapy with cytoxan and busulfan is usually adequate to ensure engraftment and a regimen of cyclosporine and steroids or cyclosporine and methotrexate usually prevents severe GVHD. However, in patients with a relative other than a fully matched sibling donor or patients with an unrelated donor, a more aggressive preparative regimen is generally recommended to prevent graft rejection. The current recommendation from the University of Minnesota is to prepare the patient with a combination of cytoxan, busulfan and moderate dose total body irradiation with partial T cell depletion and GVHD prophylaxis with cyclosporine, anti-thymocyte globulin (ATG), and both pre- and posttransplant methylprednisolone. Other centers are testing a similar approach of partial T cell depletion but with different preparative regimens. At the Children's Hospital Medical Center in Cincinnati a regimen of cytoxan, busulfan and etoposide is being tested in conjunction with partial T cell depletion, anti-thymocyte globulin and cyclosporine. Physicians are encouraged to contact transplant centers experienced in the transplantation of children with lysosomal and peroxisomal storage diseases to obtain the latest recommendations or to refer their patient to such a center.

6.11

VII) HURLER SYNDROME

Hurler syndrome is an autosomal recessive disorder caused by a deficiency of α-L-iduronidase. The disease is characterized by hepatosplenomegaly, persistent rhinitis, coarse facial features, corneal clouding, claw hands, coronary artery stenosis, dysostosis multiplex, kyphoscoliosis, short stature, hydrocephalus and progressive severe mental retardation. Most patients have significant mental retardation by age 3 and die by age 10 years from cardiac failure, pneumonia or complications of hydrocephalus. The consortium has now collected data on over 80 transplants for Hurler syndrome.

A) DIAGNOSIS

The diagnosis of Hurler syndrome is made by the assay of α-L-iduronidase in the peripheral blood leukocytes or fibroblasts, and the diagnosis is supported by increased amounts of glycosaminoglycans (GAG), specifically heparan and dermatan sulfate, in the urine. Genomic analysis on cultured fibroblasts as well as the earlier age of onset of symptoms separates the Hurler from the Hurler-Scheie and Scheie patients.

B) PRETRANSPLANT EVALUATION

To assure accurate diagnosis and for later comparison purposes, prior to the transplant the patient should have samples obtained for genotypic analysis (cultured fibroblasts), enzymatic analysis (peripheral blood leukocytes or cultured fibroblasts) and excretion of heparan and dermatan sulfate (urine). Patients should have a cardiac and abdominal ultrasound performed as well as brain MRI with spectroscopy. A full ophthalmologic evaluation with tonometry to look for evidence of glaucoma and an electroretinogram (ERG) to evaluate for retinal disease are recommended. Audiogram is recommended as well as consultation by the otolaryngology service. Placement of pharyngeal eustachian (PE) tubes is no longer routinely recommended since the marrow transplant will usually stop the excessive mucosal secretion of GAG. All patients should have a thorough neuropsychological evaluation performed. A skeletal survey with 6 foot spine films, and an orthopedic evaluation should be obtained. Routine T&A and tympanostomy are no longer routinely recommended prior to transplantation, since the transplant itself results in resolution of GAG in these tissues and amelioration of symptoms such as sleep apnea and recurrent otitis media.

C) BMT INDICATIONS

Currently the recommendations are that patients under the age of 24 months with a developmental quotient (DQ) greater than 70 should be offered a marrow transplant if they have a matched sibling donor or other HLA-compatible relative or an acceptable unrelated donor.

D) BMT RESULTS

BMT results are superior to nontransplanted control patients followed by the consortium among whom median survival is only 4.6 years of age. Also, the transplanted patients show a higher intelligence quotient compared to the surviving controls. Similar results have been obtained by the European transplant community. Genotyping of the patients in the U.S. consortium confirms that almost all patients have the severe Hurler I-H genotype and phenotype, yet do show definite benefit from marrow transplantation.

6.11

1) **Matched Sibling**

Matched sibling donor transplants have resulted in a survival of 78%.

2) **Related Nonsibling Transplants**

Related nonsibling transplants have a survival of 52%.

3) **Unrelated Donor**

Unrelated matched donor transplants have a survival of 50% at 2 years (35% survival with stable donor engraftment, follow up from 0.3-5.6 years, $N = 40$ patients). The leading cause of death among the unrelated donor transplant patients was GVHD (5 of 20 patients). Thirty-three percent of patients surviving 100 days experienced grade II-IV acute GVHD. Six patients developed extensive chronic GVHD. Only 24 of the 40 patients had documented initial engraftment with donor cells. Of the 20 survivors, 5 did not have full donor engraftment. The best engraftment rate was seen in those receiving both irradiation and T cell depletion (TCD). The lowest rate of engraftment was seen in those receiving neither irradiation nor TCD (10/12). T cell depletion ($N = 17$) was not associated with an increased incidence of graft rejection (36% vs 40%), but was associated with a lower incidence of significant acute GVHD (20% vs 44%).

E) **POSTTRANSPLANT COURSE**

1) **Hearing**

Transplanted patients have reversal of the GAG accumulations in the cochlea, middle ear and ossicles. Audiograms have shown improvement in most patients (8 of 12 tested) such that amplification is often no longer required. Improvements in both conductive and sensorineural testing have been documented.

2) **Vision**

Progressive loss of vision is typical in nontransplanted patients and is due to a combination of corneal clouding, glaucoma and GAG accumulation around the optic nerve with retinal degeneration. After transplant there is usually some improvement in the degree of corneal clouding and the optic nerve and retina may improve as well. Electroretinograms (ERG) initially show improvement after transplantation, but some patients have been shown to later have worsening in their ERG. Glaucoma is generally prevented or reversed by transplantation. Patients must be monitored closely by an ophthalmologist to determine the effect of transplantation on vision.

3) **Cardiac**

Heart failure and death due to coronary artery stenosis has not been seen after the first posttransplant year. Prior to the first year posttransplant, care must be taken to avoid unusual stresses to the heart, especially hypotension. Sudden cardiac failure resulting in death has been precipitated by septic shock, surgery, anesthesia and reactions to phenytoin and other drugs capable of causing hypotension. Even though coronary artery stenosis disappears, mitral and aortic regurgitation may even progress following successful transplantation, though cardiomegaly and heart failure have been infrequent, in contrast to nontransplanted controls.

4) **CNS**

Findings posttransplant demonstrate a positive effect in the central nervous system of Hurler patients. Brain MRI findings often return to normal

6.11

in marked contrast to nontransplanted patients. Accumulation of GAG surrounding the vessels in the Virchow-Robin spaces, termed "Hurler bodies", is prevented or reversed. Hydrocephalus with ventriculomegaly is prevented or arrested. Even patients with borderline elevated CSF pressures had return to normal pressures after transplant, obviating the need for shunting in most eligible patients pretransplant. Neuropsychological studies usually show stabilization at the patient's pretransplant level of testing.

5) **Orthopedic**

Physicians must be aware that the orthopedic problems are not prevented by the transplant, and close long-term orthopedic follow up will be required. Though generally less severe than seen in the nontransplanted control patients, orthopedic problems remain the major ongoing medical concern to successfully engrafted children. Surgical correction of genu valgum, dislocated hips, kyphoscoliosis, claw hand and carpal tunnel syndrome is necessary in most patients. Because of odontoid hypoplasia, the risk of C1-C2 subluxation during intubation for anesthesia must be recognized and avoided to prevent cord compression and paresis. The minimal improvement in the skeleton after transplant is most likely explained by a lack of cell-to-cell contact between the chondrocyte and the donor derived monocyte-macrophages.

6) **Liver/Spleen**

Following successful engraftment, the liver and spleen normalize in size over the first 3-6 months

7) **Other**

The chronic profuse watery rhinitis (which is high in GAG) quickly disappears posttransplant, often being the first sign of successful engraftment. A recently recognized problem in the successfully engrafted patients is that of social acceptance by peers. With maintenance of IQs in the normal to low normal range, as these children reach their middle childhood years, they recognize how they are different from other children their age. Psychological support and counseling will be a necessary part of their routine care posttransplant.

6.11

VIII) **HUNTER SYNDROME**

Hunter syndrome is a sex-linked recessive disorder caused by a deficiency in iduronate sulfatase. Thirteen transplants have been reported to the consortium in patients with Hunter syndrome with uniformly dismal results.

A) SEVERE FORM—EARLY ONSET

The severe form of the disease is due to a gene deletion and is associated with the early development of hepatosplenomegaly, progressive severe mental retardation and skeletal dysplasia. Of the 4 patients with the severe phenotype and with documented full engraftment for several years, all have shown significant progression of neurological deterioration. Some patients, though, have been reported to have improvement in joint mobility and in their degree of hyperactivity and aggressiveness. The consortium recommendations are currently not to offer marrow transplantation to children with the severe form of Hunter syndrome.

B) MILD FORM—ADULT ONSET

The milder adult onset form is compatible with life into adulthood and normal intelligence. Information on 4 patients with the mild form of Hunter syndrome with documented long-term engraftment are contained in the corsortium database. These patients with the mild phenotype appear to have indeed benefited from marrow transplant, though longer follow up is required to be certain.

IX) SANFILIPPO SYNDROME

The four variants of Sanfilippo syndrome are inherited in an autosomal recessive fashion and each varies somewhat in its clinical presentation. As in patients with Hunter syndrome, the results have not been encouraging. All 16 of the 28 transplanted patients who are fully engrafted and have been followed for more than 18 months after transplant have shown a steady downhill course with progressive mental retardation, aggressiveness and hyperactivity after marrow transplant. Patients only slowly and incompletely clear the storage material from the tissues. Continued cardiac deterioration has also been noted in the surviving fully engrafted patients. Therefore, the consortium does not currently recommend transplantation for any of the variants of Sanfilippo syndrome. It is felt that the failure of a positive effect from the transplant is due to an inability to transfer the enzyme from the donor monocyte-macrophages into the deficient recipient tissue cells. Additionally, it appears that little of the enzyme produced in the circulating donor leukocytes is secreted into the plasma.

X) MAROTEAUX-LAMY SYNDROME (MLS)

The first patient transplanted at the University of Minnesota with a storage disease had Maroteaux-Lamy syndrome, a deficiency of arylsulfatase B inherited in an autosomal recessive fashion. These patients present with hepatosplenomegaly and dysostosis multiplex and die in the second or third decade from pulmonary or cardiac insufficiency. A late complication is the development of hydrocephalus, though intelligence usually remains normal. All patients successfully engrafted with MLS have had improvement in all organ systems, and none have yet developed hydrocephalus. Thus BMT is currently strongly recommended in MLS. The gene responsible for MLS has been cloned, and a feline animal model is available for evaluation of both marrow transplantation and gene therapy.

XI) GAUCHER DISEASE

Gaucher disease is an autosomal recessive disorder resulting from a deficiency of the lysosomal enzyme β-glucocerebrosidase (acid-β-glucosidase). Accumulation of the substrate glucosylceramide in macrophage-derived cells leads to a variable degree of hepatosplenomegaly, destructive bone disease and compromise of bone marrow function. Based on the presence or absence of CNS involvement, there are three subtypes of Gaucher disease.

A) TYPE 1 GAUCHER

Type 1 is distinguished by its absence of CNS involvement. Modified glucocerebrosidase and its recombinant form are commercially available as Ceredase® and Cerazyme®, respectively. Enzyme replacement is a safe and effective treatment for the great majority of persons with type 1 Gaucher disease. In the rare individual with severe disease who is unresponsive to enzyme replacement therapy, BMT may be a viable option.

B) TYPE 2 GAUCHER

Type 2 is characterized by early onset of rapidly progressive neurodegeneration and often severe pulmonary pathology. Average age of death is 9 months. No benefit from marrow transplantation after onset of neurodegeneration has been conclusively demonstrated for type 2.

6.11

C) **TYPE 3 GAUCHER**

Type 3 Gaucher disease exhibits variable onset and progression of CNS degeneration with or without pulmonary pathology. Marrow transplant can be offered for persons with type 3. Long-term predicted outcome is improved if transplant is done before presence of significant CNS and/or pulmonary disease.

XII) **METACHROMATIC LEUKODYSTROPHY (MLD)**

MLD is a demyelinating autosomal recessive disorder due to a deficiency of the enzyme arylsulfatase A. There are several forms of MLD.

A) **PSEUDODEFICIENT MLD**

Since pseudodeficient patients can be found among family members of patients affected with the late infantile form of MLD, sulfatide loading studies must be performed in fibroblast cultures to separate out the pseudodeficient patients. Children with the pseudodeficient state are able to normally degrade sulfatides, do not excrete undegraded sulfatides in the urine, do not develop clinically significant symptoms and should not undergo BMT. The presence of sulfatiduria in a deficient patient also is used to separate the truly deficient patients from those with pseudodeficiency. Genomic analysis on cultured fibroblasts can be performed to convincingly separate the truly deficient from the pseudodeficient children.

B) **LATE INFANTILE MLD**

Patients with the late infantile form of MLD present with difficulty in walking and running followed by a disturbance in speech between the second and fifth year of life. Loss of other milestones typically rapidly ensue, ending in a decerebrate state. Patients with true deficiency should have pretransplant nerve conduction velocity measurements and sural nerve biopsies performed in addition to brain MRI. Several transplants have been reported in children with symptomatic late infantile MLD. These patients with clinically evident disease have unfortunately continued to progress in spite of an otherwise successful transplant, though at a slower rate. It is currently recommended that children with clinically evident late infantile MLD not undergo BMT. However, presymptomatic patients are good candidates. This generally requires the candidate be a patient who was diagnosed from family studies after diagnosis of a symptomatic sibling or relative. Six patients with preclinical late infantile MLD have undergone successful transplant. Patients do not show clinical progression and their MRI improves over the several months after transplant, often returning to normal. These results to date suggest benefit from the transplant for this select group of late infantile MLD patients.

6.11

C) **JUVENILE, ADOLESCENT AND ADULT MLD**

Patients with the juvenile, adolescent and adult forms of the disease generally do not rapidly progress to decerebration, but first develop behavioral disturbances and attention deficit disorders during the first or second decade of life. Progression to a decerebrate state often requires years in these milder cases. Thus these patients may benefit from a marrow transplant if done in the early stages of the disease.

XIII) **ADRENOLEUKODYSTROPHY (ALD)**

ALD is a peroxisomal sex-linked recessive storage disease characterized by accumulation of high levels of very long chain fatty acids in the plasma and the CNS. The lack of a peroxisomal membrane transport protein prevents the very long chain fatty acids (VLCFA) from reaching lignoceryl-ligase CoA, the enzyme normally responsible for shortening the chain length. The accumulation of the VLCFA causes destruction of the adrenal glands and subsequently leads to demyelination in the white matter of the brain. However, there is more than just the high levels of VLCFA which leads to the damage, since patients with the mild form of ALD known as adrenomyeloneuropathy (AMN) have similarly high levels of VLCFA but progress only very slowly. Patients with ALD present in the middle childhood years with hyperactivity, shortened attention spans, worsening school performance and progressive visual and hearing losses. Eventually these patients become completely decerebrate, and most die by age 10-12 years. Lorenzo's oil (named for the son of Augusto Odone, an Italian without prior medical training, who discovered and initiated this therapy after his son was diagnosed with ALD) has been shown to normalize the levels of VLCFA in the plasma and to slow the progression of the disease, but does not stop the progression. Thus, though Lorenzo's oil is a useful adjunct to therapy, it should not be used in place of a marrow transplant.

A) ADRENOMYELONEUROPATHY

The milder form of ALD known as adrenomyeloneuropathy affects adults and has a propensity toward the development of damage to the spinal cord. Since these patients survive for decades, BMT is not recommended for this group of patients.

B) ADRENOLEUKODYSTROPHY

The mechanism by which marrow transplantation effects improvement in patients with ALD is not understood; nevertheless, over 20 children with ALD have received a marrow transplant. The results have shown that patients who undergo marrow transplant after significant motoric symptoms develop have not fared well. Most have progressed; some even have acutely worsened following the transplant. In contrast, those patients transplanted early in the course of their disease have had improvements in their neurological functioning and have generally had an arrest of the disease process. Thus, BMT is clearly recommended for ALD by the consortium provided the patient has been diagnosed fairly early in the disease process and does not yet exhibit motoric symptoms. Patients with visual or aural dysfunction but without motoric problems are probably reasonable candidates for BMT. Additionally, since many patients with ALD without motoric problems have had improvements in their IQ measurements following BMT, it might be reasonable to consider even those patients with IQs below 70, provided no motoric deficits are present. There is consensus though that all patients with ALD should be on a diet (Lorenzo's oil) for at least 4 months prior to attempting transplant to normalize the levels of VLCFA and reduce the serum viscosity and its effect on cell membrane functions. If this is not done, there is a significant risk of worsening of the patient's condition shortly after the transplant characterized by a malignant syndrome of hepatic dysfunction, gastrointestinal bleeding and coagulopathy. Additionally, patients with ALD should not receive lipid infusions with their parenteral nutrition during the transplant admission. Following successful transplantation, Lorenzo's oil may be safely discontinued, even if the levels of VLCFA have not yet normalized.

C) ASYMPTOMATIC PATIENT

Debate still exists as to what to do with asymptomatic children diagnosed from screening of relatives of a proband, since some will never develop symptoms of ALD. Some feel the transplant should be delayed until clear cut symp-

6.11

toms develop. Others feel a transplant should be performed before onset of any symptoms, even before abnormalities develop on MRI.

XIV) GLOBOID CELL LEUKODYSTROPHY (GLD)

GLD is an autosomal recessively inherited deficiency of galactocerebrosidase.

A) INFANTILE FORM

The infantile form, termed Krabbe's disease, is characterized by the rapid neurological deterioration in infancy with intractable seizures and decerebration. This disease seems to progress too rapidly to allow time for correction by BMT, thus BMT is not recommended by the consortium, unless perhaps the transplant could be performed within a very few weeks after birth. In these patients, identification at birth would have to be confirmed after suspicion of the diagnosis because of a previously affected sibling.

B) JUVENILE AND ADULT FORM

The juvenile and adult forms of GLD are much milder and more slowly progressive and are associated initially with ataxia, tremor and problems in memory and school work. Three patients with the juvenile and adult forms of GLD have undergone transplant with definite benefit. Thus currently the consortium recommends further transplants be done for the juvenile and adult forms to confirm the positive benefit seen in these patients.

XV) OTHER MISCELLANEOUS SYNDROMES

Patients with several other forms of lysosomal storage disease have undergone marrow transplants. Because of promising results in single patients or in animal models, BMT might be considered for patients in the very early stages of Wolman disease, I-cell disease, fucosidosis, mannosidosis, Sly disease and Batten disease. Physicians are advised to contact the consortium about any patient with any of these diagnoses or others not listed to obtain the current results of and recommendations for transplantation.

6.11

XVI) CORD BLOOD TRANSPLANTATION

Cord blood banks are rapidly accumulating HLA-typed cord bloods for use in unrelated donor transplants. Over 75 cord blood transplants had been performed by mid-1995, including several in children with lysosomal and peroxisomal storage diseases. Of special interest is that the incidence of GVHD seems to be surprisingly low following cord blood transplantation. Some centers will pick one antigen mismatched unrelated cord blood over a fully matched unrelated donor marrow. T cell depletion is probably not needed for successful cord blood transplantation even with 2 antigen mismatched cord blood transplants. Thus, the ability to find a donor for nearly all patients with lysosomal and peroxisomal storage diseases in need of a transplant may in the future become reality, especially since the NIH will be providing large funding support for unrelated cord blood banking in the near future.

XVII) GENE THERAPY

Perhaps in the future some of these diseases will be amenable to enzyme replacement therapy or gene insertion therapy. However, if it is shown that marrow transplantation after full engraftment does not ameliorate the disease done in appropriate clinical settings, then these gene therapy approaches will need to follow different tactics. Thus, gene researchers can guide their efforts based on the results of marrow transplantation.

Table 6.11.2. BMT for lysosomal and peroxisomal storage diseases

Disease	Subtype	Defect	Consortium Recommendations as Potential BMT Candidate
Hurler		Autosomal recessive– α-iduronidase	Yes
Hunter	Severe form— early onset	X-linked– iduronate sulfatase	No
	Mild form— adult onset	X-linked– iduronate sulfatase	Yes
Sanfilippo		Autosomal recessive 4 different enzymatic defects any of which cause accumulation of heparin sulfate	No
Maroteaux-Lamy		Autosomal recessive– arylsulfatase B	Yes
Gaucher	Type 1	Autosomal recessive– β-glucocerebrosidase	No
	Type 2	Autosomal recessive– β-glucocerebrosidase	No
	Type 3	Autosomal recessive– β-glucocerebrosidase	Yes
Metachromatic leukodystrophy	Late infantile	Autosomal recessive– arylsulfatase A	No—if symptomatic Yes—if pre-symptomatic
	Juvenile, ado-lescent, adult	Autosomal recessive– arylsulfatase A	Yes
Adrenoleuko-dystrophy	Adrenomyelo-neuropathy	Perioxosomal X-linked	No
	Adrenoleuko-dystrophy	Perioxosomal X-linked	Yes—if early in course No—if signifi-cant motor symptoms
Globoid cell leukodystrophy	Infantile (Krabbe's disease)	Autosomal recessive– galactocerebrosidase	No—unless diag-nosed within a few weeks of birth
	Juvenile and adult	Autosomal recessive– galactocerebrosidase	Yes

6.11

SUGGESTED READING

1. Krivit W, EG Shapiro. Bone Marrow Transplantation for Storage Diseases. In: Forman, Blume, Thomas eds. Bone Marrow Transplantation Blackwell Scientific Publications, 1994; 66:883-893.
2. Krivit W, Sung JH, Lockman LA, Shapiro EG. Bone Marrow Transplantation for Treatment of Lysosomal and Peroxisomal Storage Diseases: Focus on Central Nervous System Reconstitution, Chapter 119. In: Editors: Rich, Shearer, Fleischer, Schwartz. Textbook of Clinical Immunology. Mosby, 1995.
3. Shapiro EG, Lockman LA, Balthazor M, W Krivit. Neuropsychological outcomes of several storage diseases with and without bone marrow transplantation. J Inherit Metabol Dis 1995; 18:413-429.
4. Krivit W, Lockman LA, Watkins PA, Shapiro EG. The future of bone marrow transplantation as treatment for adrenoleukodystrophy, metachromatic leukodystrophy, globoid cell leukodystrophy, and Hurler syndrome. J Inherit Metabol Dis 1995; 18:398-412.

6.11

The Clinical Management of Thalassemia Patients Undergoing Bone Marrow Transplantation

Emanuele Angelucci, Guido Lucarelli

I) INTRODUCTION

The treatment of transfusion dependent β-thalassemia (Cooley's anemia) has improved during the last 2 decades. In the 1970s the introduction of regular transfusional and chelation therapy dramatically prolonged survival and in the 1980s allogeneic bone marrow transplantation (BMT) provided the possibility, although not without risk, of cure. As of today, transfusion/chelation therapy and marrow transplant are the only available options. Other approaches such as gene therapy or stimulation of fetal hemoglobin with drugs such us 5-azacytidine, hydroxyurea or butyric acid hold great expectation for the future but are currently without routine clinical application. The choice of BMT for a disease in which a prolonged survival is achievable must be carefully considered. Furthermore, thalassemia patients present during the transplant course with problems which are in some instances different from those usually encountered on a BMT ward.

II) DEFINITION OF THALASSEMIA

The thalassemias are a group of inherited disorders of hemoglobin synthesis, characterized by the reduced or absent production of one or more of the globin chains of hemoglobin. They are classified, depending on which globin chain is affected, into α-, β-, δβ- and εγδβ- thalassemia. Because stable accumulation of normal adult hemoglobin requires balanced production of α- and β-globin, the clinically most relevant forms of this disorder are α- and β-thalassemia. The α-thalassemia syndrome occurs because of inheritance of mutations, usually as a result of deletions of one, two, three or four of the α-globin genes. β-thalassemia is rarely caused by structural gene deletion, but most commonly by point mutation (over 80 of which have been described). Homozygous β-thalassemia disease results from the inheritance of two mutant β-globin alleles. In β^0-thalassemia the gene expression is abolished while in β^+ thalassemia, it is reduced. Severe anemia presents in the first year of life as γ-chain synthesis declines. Anemia is caused by intramedullary ineffective erythropoiesis, extravascular hemolysis and reduction of hemoglobin synthesis. Ineffective erythropoiesis, hemolysis and transfusions are responsible for iron overload.

β-thalassemia major is highly variable with most patients requiring regular blood support since the first year of life (Cooley's anemia). β-thalassemia intermedia is a clinical term, with no specific molecular correlate, used to describe patients with two mutant gene alleles who do not require regular blood transfusion for survival (Table 6.12.1 reports the most important form of α- and β-thalassemia).

III) TREATMENT OF β-THALASSEMIA MAJOR

A randomized clinical trial comparing BMT to conventional treatment presents many difficulties, and consequently, has not been performed. Without randomized trials, the approach to treatment (medical therapy versus BMT) is controversial.

6.12

Bone Marrow Transplantation, edited by Richard K. Burt, H. Joachim Deeg, Scott Thomas Lothian, George W. Santos. © 1996 R.G. Landes Company.

In term of cost analysis the only available data refer to the North American situation in which the cost of BMT correspond to 5 years of conventional therapy.

A) MEDICAL THERAPY

Prerequisite for prolonged survival in medically treated patients are availability of safe blood products to treat anemia, chelation drugs (deferoxamine) to prevent severe iron overload and good compliance with therapy. A median survival of 17 years has been reported with low transfusion regimen (regular transfusion to maintain hemoglobin level between 7-8 g/dl) without chelation in a cohort of 71 patients treated in New York. In the same study the median survival increased to 31 years with adequate transfusion (hemoglobin between 10.5-11.5 g/dl) and chelation treatment . Guidelines of a good medical treatment are maintaining a hemoglobin level of 10-12 g/dl (transfusional requirement is usually 12-15 ml/kg of packed red blood cells every 4 weeks) and the subcutaneous or intravenous infusion of deferoxamine at a variable dose of 20-40 mg/kg over 8-12 hours a day for at least 5 days a week for the patients entire life. Deferoxamine dose is usually adjusted on the basis of serum ferritin level. Most centers consider a serum ferritin level < 1300 ng/ml as an indicator of optimal chelation.

1) Disadvantage of Medical Therapy

Toxicity is related to deferoxamine and to iron overload. Iron overload leads to cardiac failure, growth retardation, delayed puberal development, liver disease, diabetes and arthritis. Toxicities of deferoxamine include abnormality of growth, vision, hearing, renal toxicity, thrombocytopenia, musculoskeletal pain, *Yersinia enterocolitica* infection etc. Assessment of height velocity and annual ophthalmology and audiology examination

6.12

Table 6.12.1. Thalassemia

α-Thalassemia	Genetic Defect	Clinical Feature	Therapy
Carrier	– α/αα	Normal	None
Trait	– α/–α, –/αα	Mild hypochromic Anemia	None
Hb H	–/–α	Usually moderate anemia (Hgb 7-10 g/dl), rarely severe (Hgb 3-5 g/dl)	Supportive–folic acid, transfusions generally not required or infrequent
Hydrops fetalis	–/–	death in utero	stillborn, or die shortly after birth

β-Thalassemia	Clinical Feature	Therapy
Minor (trait)	Mild anemia, no transfusion requirement	None
Intermedia	Irregular transfusion requirement	Supportive, judicious transfusions
Major (Cooley's)	Transfusion dependent severe anemia, bone deformity due to expanded marrow, hepatosplenomegaly	Transfusion and chelation versus BMT

should be obtained. This treatment is expensive, tedious and painful. Even more important deferoxamine is not available in all the parts of the world (the introduction of an oral chelator could improve this situation).Transmission of human immunodeficiency virus (HIV) and of hepatitis C virus (HCV) raises concern about the safety of chronic transfusion therapy.

2) Results with Medical Therapy

Recently, a 55% cardiac disease-free survival has been reported by three North American groups after 15 years of chelation therapy in a cohort of 97 patients (mean age at the end of the study 23 years). This percentage was dramatically different in patients responsive to chelation therapy (91%) compared with those who were not responsive (< 40%).

B) MARROW TRANSPLANT

In the largest series reported (222 consecutive patients) the probability of survival and event-free survival of thalassemia patients after BMT plateaus after 1 year at 82% and 75% respectively (Fig. 6.12.1). In this group the incidence of extensive chronic GVHD was 6%. We developed a system for assigning patients undergoing BMT to prognostically useful categories. Three risk factors have been identified: the presence of hepatomegaly (enlargement of more than 2 cm below the costal margin); the presence of liver fibrosis in the pretransplantation liver biopsy and a history of inadequate chelation therapy in the years before transplantation (see section III for definition of regular chelation). These three factors stratify the patients into 3 groups: class 1–none of the risk factors, class 2–one or two of the risk factors and class 3–all three factors. Survival and event-free survival after BMT were, respectively, 94% and 94% for class one, 80% and 77% for class two, 61% and 53% for class three patients. Fig. 6.12.2 reports the event-free survival in these three categories of patients. The data of class 1 patients have been updated in 1993 on 89 consecutive patients with a survival and event-free survival, respectively of 92% and 85% (Fig. 6.12.3).

1) Patients Responding to Medical Therapy Without Organ Damage

The dilemma is BMT with a 8% probability of early death but with the possibility of cure versus continued medical treatment. It is not possible to predict how long patients will be compliant or how long they will be responsive to chelation. At the same time, it is not possible to exclude complications (e.g. viral hepatitis) of medical treatment. Delaying BMT can cause progression from a good to a bad clinical situation, from a low risk to a high risk class.

2) Patients not Responding to Chelation Therapy

For patients not responsive to chelation therapy, BMT should be attempted before the development of organ damage. Unfortunately, in practice this is a theoretical point. For those in which the organ damage has already been documented, it must be discussed if the kind of organ damage and its severity (i.e. cirrhosis) preclude a prolonged survival. Preliminary results from long observation suggest that improvement of clinical conditions and in some cases also *restitutio ad integrum* is possible after BMT.

IV) PRETRANSPLANT EVALUATION

A prerequisite to be considered a BMT candidate is the availability of an HLA-identical donor. Genotypically mismatched transplants demonstrate a low success

6.12

rate and unrelated phenotypically matched transplants are still experimental in this category of patients. Alternate donor transplants must, therefore, be reserved for patients who for any reason cannot benefit from regular transfusional and chelation therapy and do not have an HLA-matched sibling. Patients between 1 and 32 years old have been transplanted. Rather than age, clinical conditions are important for transplant outcome; however it is uncommon that a patient > 15 years old will meet the criteria for class 1. All the possible consequences of chronic transfusional therapy (iron overload, viral infection, etc.) must be evaluated prior to BMT. Consequently, several organ systems that may be affected by hemochromatosis or transfusion related infections must be evaluated.

A) LIVER

The liver represents the main site of iron deposition. Liver biopsy permits an accurate evaluation of liver fibrosis, liver iron overload and chronic hepatitis.

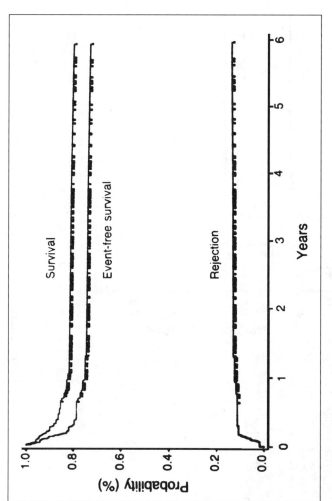

6.12

Fig. 6.12.1. Probability of survival, event-free survival, and rejection in 222 patients under 16 years old with thalassemia treated with allogeneic marrow transplantation. An event was defined as rejection, the recurrence of thalassemia or death. Reprinted by permission of the New England Journal of Medicine, (322; 418, 1990).

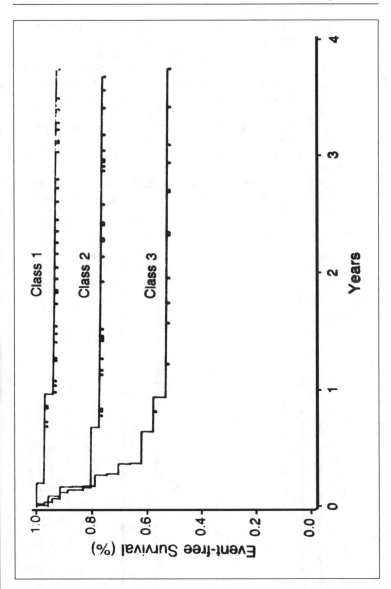

Fig. 6.12.2. Probabilities of event-free survival after transplantation in 99 patients with thalassemia. The patients were assigned to class 1 (n=39), class 2 (n=36) or class 3 (n=24) as described in the test. An event was defined as rejection, the recurrence of thalassemia or death. Reprinted by permission of the New England Journal of Medicine, (322: 420, 1990).

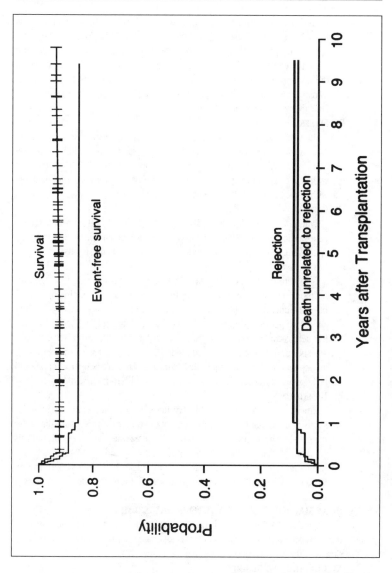

Fig. 6.12.3. Kaplan-Meier analysis of survival, event-free survival, rejection and death from causes unrelated to rejection after marrow transplantation in 89 patients with homozygous β-thalassemia who presented before BMT without liver fibrosis, no hepatomegaly and a history of regular chelation. The tick marks represent survivors. Reprinted by permission of the New England Journal of Medicine, (329; 4843, 1993)

Even though hepatitis B and C virus seropositivity are not contraindications for BMT, serological tests for these blood borne viral infections must be done. It has been proven that HCV seropositivity does not modify BMT outcome while the effect of BMT on HCV remains to be determined. Chronic hepatitis did not appear as a significant risk factor but it must be considered for post-transplant follow up. The possibility of pre-BMT intensive chelation and anti-HCV therapy is under investigation.

B) SPLEEN

Many thalassemia patients have splenomegaly. Hypersplenism can complicate the posttransplant course by increasing platelet and red blood cell requirements. A pretransplant splenectomy should be considered for patients with severe splenomegaly (We consider pre-BMT splenectomy for patients with a spleen extending to the traversal umbilical line.)

C) CARDIAC

Evaluation must include 12 lead ECG, 24 hour holter monitor, and echocardiogram or MUGA. Patients with severe impairment of left ventricle systolic function (ejection fraction < 30%) are not eligible for BMT.

D) ENDOCRINE

Complete endocrine evaluation including thyroid function tests, growth hormone releasing hormone (GRH) stimulation test for anterior hypothalamic-pituitary function, and β-islet cell function (oral glucose tolerance test—OGTT) should be performed in patients more than 10 years old. Hypogonadism is a frequent complication (50% of patients). In patients with normal semen analyses, sperm bank storage is recommended. Impaired glucose tolerance or insulin dependent diabetes frequently occurs in thalassemic patients. This should be kept in mind when these patients are treated with glucocorticoids.

E) HEMATOLOGY

Patient and donor hemoglobin synthesis must be studied to identify the status of the patient (β^0, β^+, β/s) and of the donor (normal, thalassemia minor). A sibling with thalassemia minor can be a bone marrow donor, but this condition should be known to monitor posttransplant course. The posttransplant course for recipients of marrow from a thalassemia minor donor is similar to engraftment from a normal donor except for mild anemia and microcytosis.

V) BONE MARROW TRANSPLANT FOR THALASSEMIA

In regards to BMT, thalassemia patients are characterized by two points making the clinical managements different from leukemia patients, chronic blood transfusions and the absence of previous chemotherapy.

A) CONDITIONING REGIMEN

Several conditioning regimens have been used. TBI is generally avoided in children because of long-term effects. The results reported in section III-B have been obtained with a busulfan/cyclophosphamide regimen. This regimen includes busulfan given orally at the total dose of 14 mg/kg (instead of 16 mg/kg used in malignancies) given divided every 6 hours on days -9, -8, -7, -6 followed by cyclophosphamide (CY) 200 mg/kg IV divided 50 mg/kg over 4 consecutive days (-5, -4, -3, -2). Decreasing cyclophosphamide in the high risk group to 120-160 mg/kg has decreased BMT related mortality but significantly increased the rejection rate.

6.12

B) INFECTIONS

Despite risk of blood borne infections from prior transfusions, thalassemia patients seem to have the same risk of infection as other patients submitted to BMT. However, because patients with altered glucose metabolism are at higher risk of fungal infection, we use prophylactic anti-fungal therapy (low dose IV amphotericin, 0.3 mg/kg, starting from day +8). A low incidence of CMV mortality has been encountered.

C) CARDIAC

Impaired right heart function secondary to hemochromatosis can lead to fluid accumulation. Specific recommendation is to have patients dry with a negative water balance; low dose continuous infusion dopamine is recommended; digoxin can also be used. A specific problem is cardiac tamponade. This event involves 2% of thalassemia patients undergoing BMT. It is characterized by sudden and unheralded heart decompensation caused by fast fluid accumulation in the pericardial space without concurrent myocardial disease. The fluid is a sterile, colorless transudate. This is an early complication and has been observed immediately after cyclophosphamide infusion until day +62 after BMT. Diagnosis is confirmed by echocardiography. Treatment involves emergency drainage of the pericardial space by percutaneous catheter.

D) LIVER

Despite iron overload and virus contamination, veno-occlusive disease has not been a major problem. On the other hand, transplant can acutely accelerate the course of hemochromatosis-induced liver failure. The differential of hyperbilirubinemia after BMT for thalassemia must include hepatic failure—iron over-load.

E) GRAFT FAILURE

In a variable percentage of cases (10% with standard preparative regimen) the graft can be rejected. In our center, graft failure has never been observed after day +500. Risk factors for graft rejection are cyclophosphamide dose less than 200 mg/kg and number of transfusions received, with patients who received < 100 units of blood before BMT having a higher risk of rejection. A mixed chimera condition (> 25% residual host cells) determined by molecular biology technique (RFLP-VNTR, PCR-VNTR and FISH) in early post-BMT phase (day +50) is highly predictive for graft rejection and autologous reconstitution. Graft rejection can be followed by autologous reconstitution or by persistent aplasia. In the first case, the patient can return to transfusion and chelation therapy while in the second case the only possibility is a second marrow transplant with intensification of the immunosuppressive regimen (e.g. total lymphoid irradiation 750 cGy and cyclophosphamide 200 mg/kg). No large study has been published in this setting. However in our institution success rate in early second transplant does not exceed 30%.

F) GVHD PROPHYLAXIS

Cyclosporine A (CsA) is used in our institution as GVHD prophylaxis with the following schedule: 5 mg/kg/day IV in two divided doses from day -2 to day +5, 3 mg/kg/day IV from day +6 and as soon as possible 12.5 mg/kg/day orally. We schedule CsA up to one year after BMT starting to taper from day +60. Class 1 and 2 patients receive CsA alone while class 3 patients receive a modified short methotrexate (MTX) program derived from Seattle experience in which cyclophosphamide (CY) replaces day +1 MTX (CY 7.5 mg/kg day + 1, MTX 10 mg/m^2 at days +3 , +6 and +11).

6.12

VI) POSTTRANSPLANT FOLLOW UP

The cure of thalassemia in some instances does not cure the patient. This point depends on the reversibility of organ dysfunction due to transfusion associated infections and hemochromatosis. Two different categories of patients must be considered:

A) PATIENTS TRANSPLANTED IN THE EARLY PHASE OF THEIR DISEASE

These patients with mild iron overload are without liver fibrosis, have a normal OGTT, etc. This category of patients probably require no special posttransplant treatment other than careful follow up.

B) PATIENTS TRANSPLANTED WHEN THALASSEMIA-RELATED ORGAN DAMAGE HAS BEEN ALREADY DOCUMENTED

Unfortunately this second category represents the large majority of patients. Iron overload, liver fibrosis, chronic hepatitis, diabetes, cardiac dysfunction acquired during their transfusional therapy are still present. In the first 6 months after BMT, serum ferritin does not reflect iron store because of chemotherapy-induced liver parenchymal damage. We utilize routine liver biopsy (18-24 months after BMT) to assess hemochromatosis. Iron overload can be treated with sequential phlebotomies initiated 18 months after transplantation.

SUGGESTED READING

1. Fosburg MT, Nathan DG. Treatment of Cooley's Anemia. Blood 1990; 76:435-444.
2. Kathryn HE, Giardina PJ, Lesser ML et al. Prolonged survival in patients with beta-thalassemia major treated with deferoxamine. J Perdiatr 1991; 118:540-545.
3. Olivieri NF, Nathan DG, MacMillan JH et al. Survival in medically treated patients with homozygous β-thalassemia. N Engl J Med 1994; 331:574-578.
4. Lucarelli G, Galimberti M, Polchi P et al. Bone marrow transplantation in patients with thalassemia. N Engl J Med 1990; 322:417-421.
5. Lucarelli G, Galimberti M, Polchi P et al. Marrow transplantation in patients with thalassemia responsive to iron chelation therapy. N Engl J Med 1993; 329:840-44.
6. Lucarelli G, Galimberti M, Polchi P, eds. Proceedings of the second international symposium on bone marrow transplantation in thalassemia Pesaro 22-26 September 1992. Bone Marrow Transplant 1993; 12 (Suppl 1).
7. Angelucci E, Mariotti E, Lucarelli G et al. Sudden cardiac tamponade after chemotherapy for marrow transplantation in thalassemia. Lancet 1992; 339:287-289.
8. Lucarelli G, Angelucci E, Giardini C et al. Fate of iron store in thalassaemia after bone marrow transplantation. Lancet 1993; 342:1388-1391.
9. Lucarelli G, Weatherall DJ. For debate: bone marrow transplantation for severe thalassemia. Br J Haematol 1991; 78:300-303.
10. Angelucci E, Baronciani D, Lucarelli G et al. Needle liver biopsy in thalassemia: analyses of diagnostic accuracy and safety in 1184 consecutive biopsies. Br J Haematol 1995; 83:757-761.

6.12

Bone Marrow Transplantation for Sickle Cell Anemia

Christopher L Morris

I) INTRODUCTION

Sickle cell anemia is due to a single amino acid substitution of valine for glutamine at the sixth position of the β globin chain. This substitution destabilizes the hemoglobin molecule causing decreased solubility, and polymerization of deoxygenated sickle hemoglobin. The clinical consequences of the disease include 1) chronic compensated anemia, 2) aplastic crises secondary to parvovirus infection, 3) splenic sequestration and 4) recurrent pain episodes resulting from vaso-occlusion of small blood vessels and tissue infarction. The process ultimately results in organ dysfunction which is the major cause of morbidity and mortality. The most common cause for medical attention is due to pain resulting from infarction of bone and/or bone marrow and muscle. However, microvascular occlusions may occur in virtually any organ and ultimately organ dysfunction involving lung, kidney, brain, liver or spleen may result in death.

II) NONTRANSPLANT THERAPY

A) BLOOD TRANSFUSION

Therapy for patients with severe, symptomatic sickle cell anemia was, until recently, limited to hydration and blood transfusions. Complications of chronic transfusion therapy include infection, allosensitization and hemosiderosis. Chelation therapy is effective at preventing hemosiderosis but it is expensive and requires nightly subcutaneous infusions. RBC transfusions are indicated for symptomatic anemia (e.g. aplastic crises or splenic sequestration), acute chest syndrome (PAO2 < 70 mmHg), acute cerebrovascular accident (CVA), prevention of a recurrent CVA and preoperatively before elective surgeries. The optimal reduction in Hgb S level has not been established.

B) HYDROXYUREA

The Multicenter Study of Hydroxyurea in Sickle Cell Anemia reported a reduction in the incidence of pain crises and hospitalizations for pain crises by 50% and a 33% reduction in use of blood products. Hydroxyurea is an oral medication, which is presumed to work by increasing the fetal hemoglobin content of cells, thereby reducing the tendency to sickle. Since clinical improvement may precede a measurable increase in fetal hemoglobin other mechanisms may be involved. Bone marrow function must be constantly monitored to prevent life-threatening neutropenia. The drug is not a cure, and clinical benefit requires daily therapy. The mutagenic potential and the risk of malignancy from long-term administration of hydroxyurea has not been determined. Thus, hydroxyurea is not currently considered appropriate for pediatric patients or for patients likely to become pregnant.

III) BONE MARROW TRANSPLANTATION

Bone marrow transplantation is the only curative therapy currently available to patients with sickle cell anemia. Current strategies attempt to identify patients with the lowest risk of transplant-related mortality and the highest risk of sickle cell related mortality.

Bone Marrow Transplantation, edited by Richard K. Burt, H. Joachim Deeg, Scott Thomas Lothian, George W. Santos. © 1996 R.G. Landes Company.

A) RISK MODEL FOR BMT OF THE SICKLE CELL PATIENT

Most investigators use risk criteria obtained from the results of BMT for β-thalassemia major. This seems the most appropriate model because, like thalassemia, sickle cell patients have not received prior cytotoxic therapy, and the chemotherapy used to prepare patients with thalassemia is also appropriate for sickle cell patients. The influence of pretransplant transfusion-related complications on BMT mortality has been measured in thalassemia. Since most sickle cell patients considered eligible for transplant have received chronic transfusion therapy, the effect of hemosiderosis on transplant mortality in sickle cell patients should be predicted by the results in thalassemia. In β-thalassemia, patients were defined as low risk for transplant related mortality if they had no hepatomegaly or evidence of portal fibrosis, and high risk if both risk factors were present. Most centers do not transplant sickle patients who are high risk by the above criteria, or who have other significant organ injury such as severe renal, pulmonary or neurological dysfunction.

B) SELECTION OF PATIENTS FOR BMT

There is wide variation between centers in the selection criteria for sickle cell patients most likely to benefit from BMT. Selection criteria fall into two general categories:

1) Organ Dysfunction

Patients with selective organ dysfunction requiring either chronic or repetitive blood transfusions. This category includes patients with neurological complication, sickle cell nephropathy or lung disease and bone necrosis.

2) Pain Crises

Patients who consistently over a several year period suffer multiple severe pain crises. This indicator is based on a natural history study by Platt and co-workers. An analysis of Platt's data shows that among adults ages 20-40 the probability of surviving at age 40 was 80% if the pain rate (number of crises per year) was less than 3, compared to 38% for adults with pain rates that were greater than or equal to 3. No survival predictions could be obtained for pediatric patients due to the low number of deaths in that group. However, when pain rates for individual pediatric patients were compared over 3 years of observation the pain rate increased until patients reached 20 years of age and then remained fairly constant during adulthood. Pediatric sickle cell patients may be selected for transplant based on yearly pain rates, since young patients with high pain rates are at the highest risk for death in early adulthood. Currently we consider patients with 3 or more severe pain episodes per year for 3 or more consecutive years to be candidates for BMT.

C) CURRENT RESULTS OF BMT FOR SICKLE CELL ANEMIA

As of March 1995 there have been 69 BMTs for sickle cell anemia. Fifty-four of these transplants occurred in Europe and 15 in the US. All patients were transplanted using HLA-identical sibling donors. Sixty-four (93%) of patients are currently surviving. Detailed data is only available on 60 of these transplants. Among these 60 patients, 2 were high risk as previously defined, and both of these patients died. Graft rejection occurred in 6 with 2 of these patients successfully re-transplanted. Survival for low risk patients was 96%, event-

6.13

free survival was 86% and disease-free survival 90%, similar to transplant outcomes for low risk patients with β-thalassemia.

IV) PRETRANSPLANT EVALUATION

In our opinion at this time, only patients with matched sibling donors should be offered a transplant. The mortality associated with transplantation using marrow from matched unrelated donors or mismatched family members is on the order of 50%, making this type of transplant unsuitable for conditions like sickle cell anemia. The guidelines given below are broad and represent the minimum level of organ functioning that most centers would accept for transplant candidates.

A) CRITERIA FOR EXCLUSION

1) Pulmonary

The presence of restrictive lung disease measured by pulmonary function testing with a predicted FEV1/FVC less than 50%, or a DLCO less than 50% of predicted after correction for anemia.

2) Liver

Hepatomegaly or elevation of bilirubin or hepatic transaminase unless liver biopsy demonstrates absence of hepatitis or fibrosis.

3) Renal

Glomerular filtration rate less than 60 ml/min/70 kg.

4) Cardiac

Shortening fraction measured by echocardiography of less than 29% or ejection fraction less than 50% by MUGA.

5) Central Nervous System

We do not have specific criteria for considering patients high risk due to prior CNS injury. We require that all patients undergo a neuropsychological evaluation. For patients with a history of stroke, we require cerebral angiography within 2 months prior to BMT in order to evaluate progression of vascular injury following transplant.

V) PREPARATIVE THERAPY

Prior to the start of the preparative chemotherapy we believe all patients should be transfused to reduce the hemoglobin S level to less than 35%. Most centers use a combination of busulfan and cyclophosphamide with or without anti-thymocyte globulin to prepare patients for BMT. There is some variation between centers in the dose and schedule of the preparative therapy. For young patients (less than 8 years) we currently obtain busulfan pharmacokinetics within 1 month of the transplant to choose a dose (generally 0.8-1.5 mg/kg/dose) which would predict an area under the curve of 1000 μmoles-minutes. For older children and adolescents we use a dose of 1 mg/kg/dose. Sixteen doses of busulfan are given on a 6-hour schedule for 4 days, followed by 4 days of cyclophosphamide at a dose of 50 mg/kg/day given as a 2 hour IV infusion. General supportive measures during chemotherapy include IV fluid hydration and Mesna uroprotection during cyclophosphamide therapy. There should be at least 36 hours between the last dose of cyclophosphamide and infusion of marrow.

VI) CARE OF THE PATIENT AFTER TRANSPLANT

Life-threatening complications following BMT include regimen related toxicity, graft-versus-host disease, infection and graft failure. In addition, a recent report suggests that sickle cell patients have a high incidence of CNS injury in the first 30 days following transplant, including seizures and intracranial hemorrhage. Post-

6.13

BMT neurologic complications were most common in patients with a prior history of sickle cell-induced CNS injury. Hypertension and thrombocytopenia were common at the time of the neurological complications. This suggests that sickle patients need special attention to blood pressure control and maintenance of platelet counts following BMT, especially in patients with a history of prior stroke.

A) REGIMEN RELATED TOXICITY

Pre-BMT organ system evaluation is designed to exclude patients with a high likelihood of developing severe regimen related toxicity (RRT). However, even with normal functioning in all critical organ systems a 5-10% incidence of severe RRT should be anticipated.

B) GRAFT-VERSUS-HOST DISEASE

Graft-versus-host prophylaxis regimens most commonly used are combinations of cyclosporine A (CsA) with either steroids or methotrexate (see GVHD chapter).

C) INFECTION

(See chapters on infection prophylaxis, viruses and fungi.)

D) GRAFT FAILURE

Graft failure has occurred in approximately 10% of sickle cell patients. Two recognized patterns of graft failure have been observed after BMT.

1) Primary Graft Failure

Initial engraftment with a rise in neutrophil counts followed within a few days by a sharp fall in counts and complete aplasia. Bone marrow evaluation usually shows aplasia. PCR analysis of extracted DNA shows no donor DNA present. This form of graft failure requires complete repreparation of the patient with immunosuppressive chemotherapy and reinfusion of bone marrow. However, many patients may have autologous recovery of marrow function. In this situation it is prudent to allow the patient to fully recover from organ dysfunction or infectious complications that may be present post-BMT.

2) Delayed Graft Failure

The second pattern of graft failure is characterized by a gradual reduction in bone marrow counts, often many months to a year after BMT. Late graft failure is associated with a mixed chimeric state of variable length in which both donor and host hematopoietic cells are present in the marrow. Over time the host elements may reject the donor marrow. However, long-term stable chimeric states have been documented. Such a chimeric state may not be adequate for a patient with sickle cell anemia, depending upon the level of erythropoiesis that is homozygous for the sickle gene. In rare cases of graft failure when only donor DNA is present, it may be possible to retrieve graft function with a simple reinfusion of donor marrow without immunosuppressive therapy.

SUGGESTED READING

1. Ingram VM. Specific chemical difference between globins of normal and sickle-cell anaemia haemoglobin. Nature 1956; 178:792-794.
2. Harris JW. Studies on the destruction of red blood cells. VIII. Molecular orientation in sickle cell hemoglobin solutions. Proc Soc Exp Biol Med 1950; 75:197-201.
3. Bunn HF, Forget BG. Hemoglobin Molecular, Genetic, and Clinical Aspects. Philadelphia: W. B. Saunders, 1986.

6.13

4. Clinical Alert Drug Treatment for Sickle Cell Anemia Announced. Issued by the National Heart, Lung, and Blood Institute in Reference to the Multicenter Study of Hydroxyurea in Sickle Cell Anemia. Jan. 30, 1995.

5. Lucarelli G, Galimberti M, Polchi P et al. Bone marrow transplantation in patients with thalassemia. N Eng J of Med 1990; 322:417-421.

6. Platt OS, Thorington BD, Brambilla DJ et al. Pain in sickle cell disease, Pates and risk factors. N Eng J Med 1991; 325:11-16.

7. Keith Sullivan. The evolving role of marrow transplantation in sickle cell anemia. Presented at the plenary session of the 20th annual meeting of the national sickle cell disease program, March 1995, Boston Mass.

8. Vermylen C, Cornu G, Ferster A et. al. Bone marrow transplantation for sickle cell disease the Belgian experience. Bone Marrow Transplant 1993; 12:116.

9. Bemaudin F, Souillet G, Vannier JP et al. Bone marrow transplantation (BMT) in 14 children with severe sickle cell disease (SCD) The French Experience. Bone Marrow Transplant 1993; 12:118.

10. Giardini C, Galimberti M, Lucarelli G et al. Bone marrow transplantation in sickle-cell anemia in Pesaro. Bone Marrow Transplant 1993; 12:122.

11. Johnson FL, Mentzer WC, Kalinyak DA et al. Marrow transplantation for sickle cell disease U.S. experience. J of Pediatric Hematology/Oncology 1994; 16:18-21.

12. Walters MC, Sullivan KM, Bernaudin F et al. Neurologic complications after allogeneic marrow transplantation for sickle cell anemia. Blood 1995; 85:897-884.

6.13

7

Approaches to Evaluation of Chimerism and Clonality

Peter A. McSweeney

I) INTRODUCTION

Allogeneic bone marrow transplantation leads to a state of chimerism, i.e. the presence of cells from two different individuals.

1) "Complete chimera" is when hematopoietic cells are entirely donor in origin.

2) "Mixed chimeras" is when both donor and host hematopoietic elements are detected.

3) "Split chimera" has been used in situations where lymphoid and myeloid components are discordant in origin of cells.

II) INDICATIONS FOR EVALUATING CHIMERISM

1) Evaluate origin of lymphoid and hematopoietic cells in patients with inadequate marrow function. In particular, evaluate whether rejection has occurred in patients who are candidates for a second transplant. In cases of poor graft function, determining donor or host origin of hematopoietic elements, and particularly the lymphoid cells, is often critical to determining subsequent treatment options.

2) Determine whether malignant recurrence or EBV lymphoproliferative syndrome is of donor or host origin.

3) Verify genetic identity in twins.

4) Identify whether cells from a transfusion donor are the cause of graft-versus-host disease (GVHD).

5) Identify presence of donor engraftment in relapse patients who are candidates for donor leukocyte infusions.

6) Evaluate whether occult rejection has occurred after long-term follow up in patients at high risk, e.g. aplastic anemia.

7) Study prognostic value of chimerism after transplant with respect to relapse, rejection and GVHD.

7.1

III) EVALUATION OF CHIMERISM

A variety of laboratory techniques have been used to detect chimerism based on the use of genetic markers that distinguish donor and recipient cells (Table 7.1.1). Blood and usually marrow should be simultaneously evaluated when attempting to define the cause of hematopoietic failure or when relapse is a significant possibility. If clonal markers of malignancy are known, then these should be assayed as appropriate. Peripheral blood examination is sufficient when attempting to define chimerism in the setting of graft failure where relapse is not a concern, e.g. aplastic anemia. It may be desirable to isolate PHA-stimulated lymphocytes or fractionated T cells to better evaluate the donor-recipient lymphoid interaction. Assays for cytogenetics and red cell antigen testing are widely available. Fluorescence in-situ hybridization (FISH) analysis using Y-chromosome specific probes allows rapid quantitative chimerism analysis of sex-mismatched donor-recipient pairs. Molecular techniques are becoming more widely available and in centers where they are employed they are the approach of choice in sex-matched donor-recipient pairs.

Bone Marrow Transplantation, edited by Richard K. Burt, H. Joachim Deeg, Scott Thomas Lothian, George W. Santos. © 1996 R.G. Landes Company.

A) CYTOGENETICS

Historically, cytogenetic analysis has been the primary technique for engraftment studies. It differs from other methods in that there is a prerequisite for the presence of dividing cells, a feature which potentially can bias the analysis. In addition, the sensitivity of cytogenetic analysis is limited by the relatively small number of metaphases usually examined. For example, the finding of 20 out of 20 metaphases examined to be donor or host in origin limits the probability of mixed chimerism to only 16.8% with a 95% level of confidence.

Donor engraftment is easily detected in a setting where patient and donor are of differing sex and documented by the appearance of dividing donor cells identified by the presence of the donor sex chromosomes in the marrow 3-4 weeks postgrafting. Lymphoid compartments in the peripheral blood can be evaluated in mitogen-stimulated cultures. Specifically, T cell chimerism can be evaluated by studying metaphases from PHA-stimulated peripheral blood cells. In cases without a donor-recipient sex mismatch, various banding techniques or the presence of constitutional abnormalities have been used to determine the origin of cells. Successful engraftment early posttransplant is indicated by finding 100% donor cells.

The loss of clonal leukemic cytogenetic abnormalities is also an important observation early after transplant. Remission status can be documented by cytogenetic analysis and the same test can be used to monitor relapse. In chronic myelocytic leukemia (CML), Philadelphia chromosome positive (Ph+) metaphases are frequently detected in the months early after transplant and they may disappear without further specific intervention. Rarely has malignancy recurred posttransplant in cells of donor origin as documented in sex-

7.1

Table 7.1.1. Comparisons of markers used to assay chimerism

Assay	Probability of Two-Way Informative Markers*	Technical Difficulty	Assay Sensitivity	Quantitative Accuracy
Cytogenetics	0.50	Moderate	10–20%	Low
FISH	0.50	Moderate	1–2%	High
Red cell antigens	0.75	Low	1–5%†	Moderate
Leukocyte isoenzymes	Unknown	Moderate	≈5%	Moderate
Immunoglobulin allotypes	0.50	High	2–5%†	Moderate
RFLP (five-marker panel)	0.97	High	1–10%	Moderate
VNTR-RFLP (three-marker panel)	0.91	High	1–10%	Moderate
VNTR-PCR (three-marker panel)	0.91	Moderate	1–10%‡	Moderate
VNTR-PCR (five-marker panel)	0.99	Moderate	1-10%‡	Moderate

* with full sibling pairs   FISH = fluorescence in situ hybridization
† transfusion interference   RFLP = restriction fragment length polymorphism
‡ dependent on detection system   VNTR = variable number tandem repeats

mismatched studies. Of note, has been the documentation of an EBV-associated lymphoproliferative syndrome occurring in cells of donor origin, primarily in patients undergoing HLA-mismatched or T cell depleted transplants.

B) FLUORESCENCE IN-SITU HYBRIDIZATION

These probes can be molecularly hybridized to cell preparations and allow specific determination of sex phenotype of individual cells. This allows for rapid analysis by fluorescent microscopy of donor and host populations when a sex mismatch exists between donor and host, i.e. the absence of detectable male cells would confirm 100% chimerism in a male patient with a female donor. The technology is sensitive, quantitative and analyzes nucleated cells irrespective of cell cycle status. When available, this technique can supplant cytogenetic testing as the method of choice for evaluating chimerism in sex-mismatched transplants.

C) DNA POLYMORPHISM ANALYSIS

Two main techniques are used: 1) Southern blotting to detect loci with restriction fragment length polymorphisms (RFLP) and 2) polymerase chain reaction (PCR) to detect variable number of tandem repeats (VNTR).

1) Restriction Fragment Length Polymorphisms

This analysis depends on availability of informative DNA polymorphisms that exist between donor and recipient and which can be resolved by agarose gel electrophoresis and Southern blotting of restriction enzyme digested genomic DNA. RFLP is based on stable traits inherited in a Mendelian co-dominant fashion which can be detected as a result of differences in DNA fragment size (due to differences in the distance between DNA sites cleaved by restriction enzymes) observed after locus-specific probes are hybridized to Southern blots. Fragment size differences found in RFLP analysis result from loss or gain of restriction enzyme cleavage sites or by insertions or deletions of DNA between restriction sites. Insertions or deletions may result from differences in the number of a tandemly repeated sequence. They are usually a repeat of a 2, 3 or 4 base pair sequences e.g. CA or CAG. These repeats, also known as minisatellite regions, are scattered throughout the human genome and are of unknown functional significance. Different size fragments are encompassed by constant flanking restriction enzyme sites that are used for the analysis.

When allogeneic transplants are performed between siblings, identity at a given genetic locus is seen in at least 25% of pairs, and the frequency of identity will be higher if either parent is homozygous or if parents share the same allele. Therefore, examination of multiple polymorphic loci may be needed in order to identify an informative polymorphism that can be used to identify donor and recipient cells. In unrelated donor/recipient pairs informative polymorphisms between donor and recipient occur more frequently. Use of five or more conventional highly polymorphic loci will reveal different RFLP patterns in > 99% of sibling pairs and will reveal a recipient-specific restriction fragment in 98.7% of cases. This method will detect a population of cells with sensitivity to less than 10% and requires at least 10^6 cells for analysis. Disadvantages of this approach include technical difficulty, time and manpower involved, and the requirement for significant amounts of starting tissue.

7.1

2) **Amplification of Hypervariable Regions (variable number of tandem repeats, VNTR) by Polymerase Chain Reaction**

A DNA sequence usually from 8-40 base pairs in length that is tandemly repeated within a genetic locus is known as a region with a variable number of tandem repeats (VNTR); also termed a hypervariable region. These regions have a high degree of polymorphism due to the variable number of tandem repeats (usually 11-60 repeats of the DNA sequence). Hypervariable region DNA amplification has become the method of choice for molecular chimerism analysis in sex-matched transplants. Using oligonucleotide primers specific to conserved DNA sequences that flank the locus, hypervariable tandem repeats can be amplified and allelic differences resolved by gel electrophoresis. This method can be modified to yield greater sensitivity than RFLP analysis for detecting a low percentage of mixed chimerism, but its main advantage is that it can be performed rapidly with a small number of cells. Once again, multiple genetic loci may need to be examined in order to identify informative polymorphisms between siblings. Because PCR allows analysis of small quantities of DNA, this approach can examine small samples and allow cell populations to be fractionated where it otherwise would not be possible. PCR is the technique of choice in situations of graft failure where the number of cells available for analysis is limited.

Markers used for VNTR PCR analysis at the Fred Hutchinson Cancer Research Center are shown in Table 7.1.2. Different genetic loci are used by different laboratories. The likelihood of finding polymorphic markers or heterozygosity at these loci may differ by race and by ethnic group. As can be seen in Table 7.1.1, use of a three-marker panel (DS180, ApoB and 33.6) will yield two-way markers in 91% of sibling pairs. This three-marker panel has proved informative in 100% of unrelated pairs tested to date.

D) **RED BLOOD CELL ANTIGENS**

Red cell phenotypes are highly informative and assays are easy to perform and widely available. Specific markers can usually be found that are polymorphic between donor and host. The most informative loci are MNSs, Duffy, Rh and ABO blood groups. Red cell populations of 1-5% can be detected. These studies are of limited value in transfusion-dependent patients and therefore have little role in the early posttransplant setting.

Table 7.1.2. VNTR markers used for PCR-based chimerism analysis

Locus	Chromosome	Number of Alleles	% Two-way information REL/ UNREL*	% of Heterozygosity
DS180	1	> 29	49/68	87
ApoB	2	> 12	35/82	81
33.6 (D1S111)	1	> 13	35/86	79
pYNZ22 (D17S5)	17	> 11	NE	NE
33.1	NE	?10	NE	NE
SE-33	6	?20	NE	NE

* REL = related, UNREL = unrelated   VNTR = variable number of tandem repeats
NE = not evaluated or unknown   PCR = polymerase chain reaction

E) OTHER LESS COMMONLY USED TECHNIQUES

1) Immunoglobulin Allotypes

Immunoglobulin allotyping reveals donor-specific and recipient-specific markers approximately 50% of the time. This technique reliably evaluates only B cell engraftment. Host antibody production can be prolonged after transplant presumably because of long-lived B lineage cells and may outlast other evidence of mixed chimerism. The reagents for this technique are not generally available and the assay is not routinely performed.

2) HLA Typing

In the absence of other technology, this can be performed to evaluate lymphoid engraftment in HLA-mismatched pairs.

3) Isoenzyme Analysis

Both red cell and leukocyte isoenzyme patterns have been used for evaluating chimerism. Although they have proven informative, they have been replaced by the newer techniques described above.

IV) INTERPRETATION OF CHIMERISM TESTING

Interpretation of results may be heavily influenced by clinical circumstances prompting the test. The detection of certain proportions of cells of donor and host origin does not directly indicate anything about functional characteristics of these cells. Before sampling bone marrow and blood, discussions with laboratory personnel are often desirable, particularly if fractionation of cells may be required. In absence of cell fractionation, peripheral blood analysis is a better indicator of lymphoid engraftment and marrow analysis can better define relapse of marrow-based malignancy. At the time of sampling for chimerism it may be important to perform studies including morphology, flow cytometry, etc., to detect residual host tumor cells if markers for these exist. It may be important to undertake cytogenetic analysis at the same time as other chimerism tests are performed in order to obtain maximum information.

The data from a single test may not be definitive for a pathological process and may only provide a snapshot of an evolving process which becomes clear only by serial chimerism analysis or clinical observation over periods of weeks or months (see Table 7.1.3). Certain situations can pose considerable diagnostic difficulty. Rejection is defined by the absence of donor cells in a patient with pancytopenia. However, occult rejection with host reconstitution can occur in patients receiving nonmyeloablative regimens. CML patients have occasionally reconstituted hematopoiesis after graft rejection despite receiving what would be considered myeloablative regimens. The presence of donor cells can be obscured in florid relapse or in situations where malignant cells have a myelosuppressive effect. This issue can be resolved by testing of peripheral blood mature T cells. Evaluation of chimerism can be difficult in patients with low cell counts. Rejection may occur without detection of host hematopoietic elements. Results of cytogenetics may show poor correlation with other tests because examination of a small number of metaphases may reflect a highly selected cell population.

It is not recommended that repeated chimerism analyses be undertaken merely because mixed chimerism is detected. Testing is expensive and in the absence of tumor recurrence or graft rejection, nothing definite can be said about the implications of mixed chimerism for a given patient, and there is no specific safe therapy to reverse it. Outside of the research setting there is little value in routine chimerism assays in patients who have uncomplicated transplants. Cytogenetic studies, however, can be performed to confirm loss of clonal leukemic markers.

7.1

Table 7.1.3. Interpretation of peripheral blood chimerism testing—likely diagnosis

Cells Tested		Clinical Setting	
PMN	T Cells	Normal Peripheral Blood and Marrow	Pancytopenia and Marrow Aplasia
Donor	Donor	Donor engraftment	Graft failure (nonrejection)
	Mixed	Mix chimerism	Rejection
	Host	Split chimerism (rare)	Rejection
Mixed	Donor	Relapse (CML/MDS)	Relapse (CML/MDS)
	Mixed	Mixed chimerism	Rejection
	Host	Rejection/evolving host reconstitution	Rejection
Host	Donor	Relapse (CML/MDS)	Relapse (CML/MDS)
	Mixed	Relapse (CML/MDS) after mixed chimerism	Rejection
	Host	Rejection/host reconstitution	Rejection

PMN = neutrophils
CML = chronic myelogenous leukemia
MDS = myelodysplastic syndrome

V) EVALUATION OF CLONALITY: X-LINKED CLONAL ANALYSIS

In some malignancies well-defined clonal markers can be identified that allow sensitive and specific disease monitoring. Examples of these include disease-specific clonal cytogenetic abnormalities such as the Philadelphia chromosome, which may be invoked in the etiology of CML, and what may be considered as bystander markers such as antigen receptor rearrangements found in lymphoid malignancies, which reflect the tissue origin of the disease and do not appear to have etiological significance. In diseases lacking a specific clonal marker the technique of X-linked clonal analysis can be used to define the clonal and therefore possibly neoplastic nature of the disease. X-linked clonality can only be used in females because the technique requires the presence of two X-chromosomes. It is based on the lyonization phenomenon occurring early in embryogenesis whereby each female cell randomly inactivates one of the two X-chromosomes. Therefore, approximately half of the cells will inactivate the paternally derived X-chromosome and half the maternally derived X-chromosome. Inactivation of the chromosomes is associated with differing methylation patterns between specific genetic loci found on both X-chromosomes. These differing methylation patterns are exploited for the purpose of DNA-based clonal analysis.

Primarily a research tool, X-linked clonal analysis has been used to confirm that neoplasms are clonal in nature, i.e. have a single cell origin. Further application has involved the use of X-inactivation patterns to determine clonality in cell populations lacking other defined clonal markers. Studies have been performed in a transplant setting to study hematopoietic reconstitution posttransplant.

Original X-linked clonal analysis studies were performed analyzing tissue expression of glucose-6-phosphate dehydrogenase isoenzymes, but these techniques have now been replaced by DNA-based molecular analysis. Application of the technique depends on finding informative polymorphisms between two alleles at one of several defined X-chromosome loci (Table 7.1.4). These genes include phosphoglycerate kinase (PGK), hypoxanthine phosphoribosyltransferase (HPRT), DSX255 (M27β) and human androgen receptor (HUMARA). The primary techniques used are Southern blotting and PCR. The polymorphisms are revealed by restriction enzyme site differences or variable numbers of tandem repeats which lead to different sized DNA fragments that can be resolved by gel electrophoresis. Clonality is determined by cutting DNA with a methylation-sensitive enzyme such as HpaII which cleaves at unmethylated CCGG motifs. At a given genetic locus in a female cell one allele is methylated and therefore protected against HpaII digestion, whereas the other allele is unmethylated and therefore susceptible to enzyme cutting. Therefore, the banding pattern, and in particular the relative intensity of bands from the two alleles, can alter after digestion with HpaII. In a clonal population of cells, each cell has an identical pattern of methylation and the same allele will be susceptible to HpaII digestion, giving rise to a characteristic clonal pattern seen after restriction enzyme digestion and gel electrophoresis.

The ability to apply these techniques as a research tool to the majority of females was initially limited by the relatively low rate of heterozygosity detected with the HPRT and PGK loci. This problem has been overcome to some extent by the use of the DSX255 (M27β) locus and more recently the HUMARA locus, both of which derive polymorphism from differences in the number of tandem repeats between two alleles. The M27β locus, although showing approximately 90% heterozygosity, has been limited by difficulty interpreting banding patterns in a significant proportion of cases. The HUMARA locus relies on a PCR assay and with high resolution gels almost all females can be studied. At this time, the best results are found by using the PGK locus for which both PCR and Southern blotting can be employed. Fairly accurate quantitation appears possible at this locus by both techniques, with the one drawback being the limited rate of heterozygosity.

Studies using X-linked clonal analysis have shown that hematopoietic reconstitution after marrow transplantation almost always originates from a large pool of marrow stem cells. This occurs with both autologous and allogeneic transplantation. Recent evidence has shown a disturbing incidence of myelodysplasia in autograft recipients, suggesting that further clonal analysis studies of hematopoiesis in these patients is indicated. Clonal analysis has been used to study patients with aplastic anemia where long-term follow up has revealed a significant incidence of clonal dysplastic hematopoiesis in ATG-treated patients. This has been associated with the development of myelodysplastic syndrome. Some investigators have found

7.1

Table 7.1.4. X-linked markers used for molecular analysis of clonality

Locus	% Informative	Method	Polymorphism
HPRT	29	Southern blot	RES
PGK	33	Southern blot/PCR	RES
DSX255(M27β)	> 90	Southern blot	VNTR
HUMARA	> 95	PCR	VNTR

RES = restriction enzyme site
VNTR = variable number of tandem repeats
HUMARA = human androgen receptor
PCR = polymerase chain reaction

PGK = phosphoglycerate kinase
HPRT = hypoxanthine phosphoribosyltransferase

clonal recurrance in chemotherapy-treated acute myelocytic leukemia (AML) patients, implying a pre-neoplastic hematopoietic state. However, the incidence of this is less than was initially thought and the prognostic significance of this finding has not been determined.

In interpreting these studies it is important that baseline lyonization status be examined in cases where clonal tissue is found, in order to exclude a pre-existing skewed lyonization pattern that has been found in as many as 20% of normal females. Although a clinical utility for this technique has not been determined, the finding of clonal hematopoiesis in a female patient with clinical evidence of hematopathology may prove diagnostically and clinically informative. As PCR can potentially be employed in many cases, cells can be fractionated by phenotype and examined for clonality. Limitations of this technology include a lack of sensitivity in detecting clonal populations and the requirement for isolation of a phenotypically homogeneous cell population in order to obtain definitive results.

Acknowledgment

I would like to thank Dr. Paul Martin for providing data from chimerism studies performed at the Fred Hutchinson Cancer Research Center.

Suggested Reading

1. Petz LD. Documentation of engraftment and characterization of chimerism following bone marrow transplantation. In: Forman SJ, Blume KG, Thomas ED, eds. Bone Marrow Transplantation. Boston: Blackwell Scientific Publications, 1994:136-148.
2. Busque L, Gilliland DG. Clonal evolution in acute myeloid leukemia. Blood 1993; 82:337-342.
3. Nakamura Y et al. Variable number of tandem repeat (VNTR) markers for human gene mapping. Science 1987; 235:1616-1622.

7.1

Graft Failure

H. Joachim Deeg

I) INTRODUCTION

Hematopoietic stem cells from marrow, peripheral blood, cord blood and fetal liver cells may be obtained from various donors (Table 7.2.1). It is important to distinguish these sources since with autologous and syngeneic cells, no histocompatibility barrier exists between donor and recipient, whereas minor (non-HLA) or major (HLA) barriers exist with allogeneic donors. Xenogeneic donors have recently also been used in clinical marrow transplantation. There may be concurrent reappearance of recipient cells, as is generally the case in immune-mediated graft rejection, or donor cells may persist but fail to "thrive," e.g. in autologous recipients whose stem cells may have been damaged by prior therapy or where defects in the microenvironment, pre-existing or induced by transplant conditioning or by posttransplant events, may occur. Factors contributing to graft failure include the underlying disease, pretransplant management, the transplant (conditioning/cell inoculum) and posttransplant events (Table 7.2.2).

A) AUTOLOGOUS TRANSPLANTS

With autologous and syngeneic transplants, immunological graft rejection is theoretically not possible. Nevertheless, the graft may fail as evidenced by transient, incomplete or absent hematopoietic reconstitution posttransplant. Causes include damaged marrow microenvironment, insufficient or damaged stem cells, infections (e.g. CMV) and posttransplant therapy (e.g. ganciclovir).

B) ALLOGENEIC TRANSPLANTS

Following allogeneic transplantation, the graft may fail through mechanisms similar to those observed with autologous and syngeneic grafts, or the graft may be destroyed by recipient cells recognizing histocompatibility differences on donor cells. Destruction of donor cells by host cells may be due to a memory response following pretransplant allosensitization (pregnancy in female recipients or transfusion of blood products and in some instances, a previous transplant). In addition, donor cells may initiate a GVH reaction which is also directed at the marrow (note that marrow aplasia is a prominent feature of transfusion-induced GVHD).

II) DEFINITION OF GRAFT FAILURE

The definition of graft failure has remained somewhat controversial. With an uncomplicated marrow transplant from an HLA-identical donor, there is generally a rise in peripheral blood granulocyte counts in the third week after transplantation. Recovery of platelets as measured by transfusion independence is more delayed (3-4 weeks or later). Red blood cells (life span normally 100-120 days) are not a sensitive indicator of hematopoietic recovery. Also, patients transplanted from an ABO blood group-incompatible donor (for example, A donor into B recipient) continue to produce host-type isoagglutinins which may result in severe reticulocytopenia and even red cell aplasia in the marrow (single-lineage graft failure) for periods of months or occasionally years posttransplant. With autologous transplants, especially with the use of mobilized peripheral blood stem cells, recovery of granulocyte counts to 0.5×10^9/L by days 10 to 12 and platelet transfusion

7.2

Bone Marrow Transplantation, edited by Richard K. Burt, H. Joachim Deeg, Scott Thomas Lothian, George W. Santos. © 1996 R.G. Landes Company.

independence by days 12 to 15 have been reported. Engraftment is generally delayed in patients who receive in vitro purged marrow (treated by monoclonal antibodies, chemicals or other means) and possibly in patients receiving enriched "stem cell" preparations.

Table 7.2.1. Potential sources of marrow or stem cells

		Histocompatibility Barrier[1]	
		Minor	**Major**
Autologous:	Patient's own marrow	–	–
Syngeneic	Monozygotic (identical) twin	–	–
Allogeneic:	HLA genotypically identical sibling	+	–
	HLA phenotypically identical donor		
	–Sibling	++	–
	–Parent	++	–
	–Other family member	+++	–
	– Unrelated volunteer	++++	–
	HLA-nonidentical donor		
	–Family member	++	+
	–Unrelated volunteer	++++	+
Xenogeneic:	Different species donor	++++	++

[1] By definition no histocompatibility barriers exist with autologous and syngeneic marrow. For allogeneic transplants "-" indicates absence of barrier based on conventional typing; differences may be detectable at the molecular level.

Table 7.2.2. Causes of graft failure

Underlying Disease
- – Immune-mediated marrow suppression (aplastic anemia)
- – Defects of the microenvironment (aplastic anemia,
      myelofibrosis, leukocyte adhesion defects, etc.)
- – Virus-induced failure (hepatitis, EBV, parvovirus)

Pretransplant Management
- – Allosensitization by transfusions
- – Chemoradiotherapy
      –Stem cell damage (autologous)
      –Stromal cell damage (all transplants)

Transplant Procedure
- – Intensity of conditioning regimen
- – Histocompatibility barriers
- – Cell inoculum
      –Inadequate number
      –Damaged stem cells
      –T cell depletion

4. Posttransplant Events
- – Marrow suppressive drugs
      –GVHD prophylaxis
      –Therapeutic agents
- – Infections
      –Viral (CMV)
- – GVHD
      –Microenvironmental damage

7.2

A) HEMATOPOIETIC GRAFT FAILURE

Hematopoietic graft failure is defined as failure of granulocyte counts to reach 0.2×10^9/L by day 21 (or at the latest, day 28). The diagnosis is further substantiated by biopsy findings of an empty marrow or low cellularity without identifiable myeloid, erythroid or megakaryocytic precursors. Hematopoietic graft failure may be secondary to insufficient numbers or damaged hematopoietic precursor cells, damaged microenvironment, infections (e.g. CMV), possibly autoimmune mechanisms and posttransplant therapy (e.g. ganciclovir, methotrexate). In some patients with aplastic anemia who receive cyclosporine as GVHD prophylaxis, late graft failure involving ill-defined mechanisms may occur upon tapering cyclosporine.

B) GRAFT REJECTION

Graft rejection is defined as cellular or serological evidence (or both) of re-emergence of host lymphoid cells, generally T lymphocytes, with or without concurrent re-emergence of host hemopoiesis, involving either normal hematopoietic or malignant precursors. Currently available tests to document cellular origin include cytogenetic analysis of peripheral blood and marrow cells (unmanipulated or mitogen-stimulated), examination for variable number tandem repeats (VNTR), X-linked chromosome analysis, HLA determination if an HLA nonidentity was present, and others (Table 7.2.3). (See chapter on chimerism.) In some instances, host cells, either persisting throughout the procedure or reemerging posttransplant, may be present only transiently. In some patients, a lasting "mixed chimerism" may develop, that is, donor and host cells coexist over years without adverse effects on patient survival. In fact, there is evidence that a transient or lasting mixed chimerism is beneficial as reflected in a reduced incidence of GVHD. In patients transplanted for malignant disorders, however, the risk of relapse also may be increased.

C) PRIMARY VERSUS SECONDARY GRAFT FAILURE

Graft failure may be "primary," i.e. the patient experiences the conditioning regimen-induced decline in blood cells without ever showing a rise in counts.

7.2

Table 7.2.3. Documentation of donor cell engraftment and persistence/re-emergence of host cells

---

Cytogenetic analysis of metaphase spreads
(constitutive or after stimulation)
       –Sex chromosome
       –Autosomal chromosome marker
HLA Typing[1]
       –Serological
       –Restriction fragment length polymorphism (RFLP)
       –Sequence-specific oligonucleotide probes (SSOP)
Typing for variable number tandem repeats (VNTR)
Complement typing
Immunoglobulin allotyping
Erythrocyte typing[2]
       –Antigens
       –Enzymes

---

[1] Especially helpful with HLA-nonidentical transplants; however, polymorphic DNA sequences outside HLA can be recognized by RFLP, SSOP, VNTR or other probes.
[2] Used only infrequently.

Alternatively, graft failure may be "secondary," i.e. a transient rise of blood cell counts (donor-derived) is seen but this is followed by a secondary (delayed) decline. The time interval from transplantation to "graft failure" is important; patients with early graft failure (< 30 days) have a poor prognosis, while patients failing later generally fare better.

III) RISK AND PREVENTION OF GRAFT FAILURE

A) UNDERLYING DISEASE

1) Immune-Mediated Marrow Failure

Currently used regimens include combinations of cyclophosphamide (CY) and anti-thymocyte globulin (ATG) or CY plus total lymphoid irradiation (TLI) or other modified forms of TLI. With aplastic anemia, failing grafts have been observed despite initial good engraftment upon tapering cyclosporine used for GVHD prevention. Reinstituting cyclosporine has rescued hemopoiesis, either of allogeneic or autologous origin, in some patients.

2) Marrow Microenvironment

Defects in the microenvironment currently can be diagnosed only by in vitro techniques (e.g. long-term marrow cultures). Approximately 5% of patients with severe aplastic anemia are found to have a defective environment. In the future, preventive measures against graft failure may include the transplantation of donor-derived microenvironmental (stromal) cells.

B) PRETRANSPLANT MANAGEMENT

1) Blood Transfusions

In patients with severe aplastic anemia, among untransfused patients given HLA-identical marrow grafts, only 5% rejected the graft; whereas among transfused (allosensitized) patients more than 30% rejected the transplant. Thus, it is advisable to avoid transfusions if at all possible. Intensified conditioning regimens, in particular the use of ATG plus CY or CY plus TLI, have reduced the incidence of graft failure in transfused patients to 5-10%.

2) Prior Chemotherapy

In the autologous setting, it is desirable to harvest hematopoietic stem cells (marrow, blood or both), if possible, early on before multiple cycles of potentially stem cell-toxic regimens have been given. The use of peripheral blood stem cells, in particular after mobilization with hematopoietic growth factors, has facilitated this approach.

C) TRANSPLANT PROCEDURE

1) Conditioning Regimen

Intensive conditioning is a prerequisite for immunosuppression to achieve engraftment of donor cells and eradicate the patient's disease. Exceptions are often seen in children with severe combined immune deficiency where the patient's immune deficiency prevents rejection of donor T cells, so that T cell reconstitution is donor-derived, whereas hemopoiesis remains of host type. Higher doses of TBI are more effective, and single dose TBI provides more effective immunosuppression, although with fractionated TBI, toxicity is reduced. Inclusion of ATG or monoclonal antibodies directed at T cells (e.g. CD3, CD5 and others) or at leukocyte adhesion molecules (e.g. CD18/CD11) facilitates engraftment.

7.2

2) **HLA**

With an unmanipulated HLA-compatible graft, the probability of graft failure is 1-2%, compared to 10-15% with an HLA-incompatible graft (Table 7.2.4).

3) **T Cell Depletion**

If the donor marrow is T cell depleted, the probability of not achieving a functional graft may increase to 20% with HLA-compatible and 40-60% with HLA-incompatible transplants.

4) **Number of Cells**

Generally, 2-4 x 10⁸ mononuclear cells/kg recipient weight are infused. If autologous peripheral blood stem cells are utilized, approximately 1-2 x 10^6 CD34+ cells/kg are required to assure hematopoietic reconstitution.

5) **Growth Factors**

Hematopoietic growth factors that have been tested for their potential in accelerating hematopoietic reconstitution include G-CSF, GM-CSF, PIXY321, IL-3, SCF. Pretreatment of the donor prior to harvesting peripheral blood stem cells enhances the number of early precursor cells (CD34+ cells) with a high potential of rapidly reconstituting the patient. A rapidly developing field is one using growth factors for in vitro expansion of hematopoietic precursor cells. Posttransplant in vivo administration of hematopoietic growth factors to the patient has been associated with an accelerated recovery of granulocytes. There is little evidence (except perhaps with PIXY321, a fusion protein of IL-3 and GM-CSF) that platelet recovery is affected. The availability of the recently cloned mpl-1 ligand (thrombopoietin) may change this. The use of erythropoietin in a prophylactic fashion may reduce red blood cell transfusion requirements.

D) **POSTTRANSPLANT EVENTS**

1) **Medications**

Methotrexate given for in vivo GVHD prophylaxis or intrathecally as prophylaxis of CNS leukemia may be myelosuppressive. It appears, however, that the administration of citrovorum factor (12 mg IV every 12 hours starting 12-24 hours after the methotrexate dose and continued

7.2

Table 7.2.4. Probability of graft failure with unmanipulated marrow grafts

Source of Marrow (Indication)	Probability of Failure
Autologous	0%[1]
Syngeneic	0%[2]
HLA genotypically identical sibling	
—Malignancies	1-2%
—Aplastic anemia	5-10%
HLA genotypically nonidentical related donor	5-20%[3]
Unrelated donor	5-10%

[1] Poor graft function has been observed in heavily pretreated patients. Problems appear to be less severe with peripheral blood stem cells.
[2] Among patients with aplastic anemia infused with syngeneic marrow without prior conditioning approximately 50% have failed to show hematopoietic reconstitution.
[3] With a positive crossmatch between donor and patient, the incidence may be as high as 75%.

for 24-72 hours) prevents this undesired side effect while preserving the immunosuppressive (anti-GVHD) effect. This approach is generally taken in patients in whom serum methotrexate levels at 24 hours are $> 0.05 \times 10^{-6}$ molar. Bactrim given for *Pneumocystis carinii* prophylaxis is myelosuppressive and may delay engraftment .

2) Viruses

CMV infections may be marrow suppressive, probably by means of stromal cell infection. Treatment or prophylaxis of CMV infection with ganciclovir is also myelosuppressive. The concurrent use of growth factors (e.g. G-CSF, 5-15 µg/kg/day IV) may be useful.

3) GVHD

The effect of GVHD on hemopoiesis is still incompletely understood.

IV) TREATMENT OF GRAFT FAILURE

If hematopoietic failure without evidence of immunological rejection of the graft occurs, the application of growth factors or a booster infusion of marrow are often useful. In most other instances, reconditioning of the patient for a second stem cell infusion is necessary. The time interval from original transplant to graft failure is important. Patients who show graft failure within 30 days of transplantation have a very poor prognosis, while patients who develop graft failure later have a substantially better outlook. Similarly, results with second transplants depend upon the time at which they are being carried out.

A) "BOOSTER" GRAFTS

If the graft fails, either de novo or following an initial rise in granulocyte count, and host cells re-emerge, the prognosis is poor. This finding indicates that host cells have not been eradicated; a second marrow infusion (from the same or a different donor) is likely to be successful only if additional conditioning is given to the patient. Conditioning regimens for a second transplant usually consist of chemotherapy only if TBI was used for the first transplant, and, vice versa, a TBI-containing regimen if chemotherapy only was used initially. In a recent study, 74 of 77 second transplant patients engrafted and disease-free survival was 14% at 3 years. However, patients whose grafts fail within 6-12 months are often too ill to tolerate an aggressive second conditioning regimen.

One experimental regimen, the usefulness of which is currently being explored, involves an anti-CD3 monoclonal antibody, BC3, which is given peri-transplant (e.g. days -4 to +19) in combination with glucocorticoids. Preliminary data are encouraging. ATG has been used instead of monoclonal antibodies. A combination of ATG plus cyclophosphamide has been quite successful in preparing patients with severe aplastic anemia for a second transplant (see chapter on aplastic anemia).

B) HEMATOPOIETIC GROWTH FACTORS

Growth factors have been given for the treatment of graft failure; however, there is no obvious reason why they should stimulate preferentially donor rather than host cells, and they are successful only with hematopoietic failure, not with rejection. Growth factors are generally instituted by day 21 or day 28 at the latest if 0.5×10^9 granulocytes/L have not been reached. Most experience exists with GM-CSF at a dose of 250 µg/m²/day. G-CSF, given at doses of 5-15 µg/kg/day subcutaneously or IV, has been effective in several studies. Studies using PIXY321 at doses of 500-1000 µg/kg are also encouraging. This

7.2

Table 7.2.5. Prevention of graft failure

Underlying disease
- —Immunosuppressive conditioning
- —Stromal cell transplants (into the marrow cavity)
- —Inclusion of anti-LAF antibodies in the conditioning regimen
- —? Allow for healing of virus-induced damage?

Pretransplant management
- —Avoid transfusions
- —Immunosuppressive conditioning
- —Harvest autologous stem cells as early as possible (before multiple cycles of therapy are given)

Transplant procedure
- —Intensify conditioning regimens
- —Select histocompatible donors
- —Provide sufficient stem cells
- —Harvest stem cells early
- —Selective T cell depletion
- —Hematopoietic growth factors

Posttransplant events
- —Avoid marrow suppressive drugs
- —Adjust doses
- —Treat viral infections
- —Prevent GVHD

approach may be supplemented by a second infusion of hematopoietic stem cells without additional conditioning of the patient. In the autologous setting, the option of a second infusion would exist only if a back up marrow has been cryopreserved.

C) IMMUNOSUPPRESSION

The administration of high-dose glucocorticoids or the reinstitution of cyclosporine in patients in whom graft failure occurred upon cyclosporine tapering may rescue the graft in patients with aplastic anemia. In patients transplanted for a hematological malignancy, graft failure is associated with the concern of recurrence of the underlying malignant disease. Such a course has been particularly frequent after aggressive T cell depletion of the donor marrow.

7.2

SUGGESTED READING

1. Barge AJ, Johnson G, Witherspoon R, Torok-Storb B. Antibody-mediated marrow failure after allogenic bone marrow transplantation. Blood 1989; 74: 1477-1480.
2. Donohue J, Homge M, Kernan NA. Characterization of cells emerging at the time of graft failure after bone marrow transplantation from an unrelated marrow donor. Blood 1993; 82:1023-1029.
3. Fischer A, Friedrich W, Fasth A, Blanche S, Le Deist F, Girault D, Veber F, Vossen J, Lopez M, Griscelli C, Hirn M. Reduction of graft failure by a monoclonal antibody (anti-LFA-1 CD11a) after HLA nonidentical bone marrow transplantation in children with immunodeficiencies, osteopetrosis, and Fanconi's anemia: A European Group for Immunodeficiency/European Group for Bone Marrow Transplantation Report. Blood 1991; 77:249-256.
4. Gluckman E, Horowitz MM, Champlin RE, Hows JM, Bacigalupo A, Biggs JC, Camitta BM, Gale RP, Gordon-Smith EC, Marmont AM, Masaoka T, Ramsay NKC, Rimm AA, Rozman C, Sobocinski KA, Speck B, Bortin MM. Bone marrow transplantation for severe aplastic anemia: influence of conditioning and graft-versus-host disease prophylaxis regimens on outcome. Blood 1992; 79:269-275.
5. Hows JM. Mechanisms of graft failure after human marrow transplantation: a review. Immunology Letters 1991; 29:77-80.

6. Kernan NA. Graft failure following transplantation of T cell-depleted marrow. In: Burakoff SJ, Ferrara J, Deeg HJ, Atkinson K, eds. Graft-vs.-Host Disease: Immunology, Pathophysiology, and Treatment. New York: Marcel Dekker Inc, 1990; 557-568.

7. Martin PJ. Determinants of engraftment after allogeneic marrow transplantation. Blood 1992; 79:1647-1650.

8. Mehta J, Powles RL, Mitchell P, Rege K, De Lord C, Treleaven J. Graft failure after bone marrow transplantation from unrelated donors using busulphan and cyclophosphamide for conditioning. Bone Marrow Transplantation 1994; 13:583-587.

9. Nemunaitis J, Singer JW, Buckner CD, Durnam D, Epstein C, Hill R, Storb R, Thomas ED, Appelbaum FR. Use of recombinant human granulocyte-macrophage colony-stimulating factor in graft failure after bone marrow transplantation. Blood 1990; 76:245-253.

10. Quinones RR. Hematopoietic engraftment and graft failure after bone marrow transplantation. Am J Ped Hematol Oncol 1993; 15:3-17.

11. Simmons P, Kaushansky K, Torok-Storb B. Mechanisms of a cytomegalovirus-mediated myelosuppression: Perturbation of stromal cell function versus direct infection of myeloid cells. Proc Natl Acad Sci USA 1990; 87:1386-1390.

12. Storb R, Deeg HJ. Failure of allogeneic canine marrow grafts after total body irradiation: Allogeneic "resistance" vs transfusion induced sensitization. Transplantation 1986; 42:571-580.

7.2

Minimal Residual Disease

F. Marc Stewart

I) DEFINITION OF MINIMAL RESIDUAL DISEASE (MRD)

Minimal residual disease refers to the presence of residual diseased cells (usually malignant) despite absence of any clinical or histologic (light microscope) evidence of disease. The most common methods employed to assess minimal residual disease include cytogenetics, gene rearrangement, clonogenic culture, polymerase chain reaction, fluorescent-activated cell sorting and fluorescence in situ hybridization. These approaches are summarized in Table 7.3.1.

II) TECHNIQUES TO ASSESS MINIMAL RESIDUAL DISEASE

A) CYTOGENETICS

Karyotype analysis of specific chromosomal abnormalities occasionally may provide for the detection of disease prior to histologically defined relapse. The sensitivity of karyotypic analysis ranges from 10^1 to 10^2. The advantages of cytogenetic approaches include its widespread availability and the fact that a number of neoplastic diseases have defined karyotypic abnormalities that are quite specific.(e.g. t9;22 [Philadelphia chromosome] for chronic myelocytic leukemia; t8;14 for Burkitt's lymphoma). Disadvantages of cytogenetic analysis include its lack of sensitivity, variation in expertise among laboratories, application only to diseases that manifest a stable, specific chromosome abnormality, expense and necessity to analyze cells in metaphase.

B) SOUTHERN BLOT ANALYSIS

Neoplastic lymphoid cells (acute lymphoblastic leukemia or various non-Hodgkin's lymphomas) may exhibit clonogenic rearrangement of immunoglobulin genes or T cell receptor genes. A clone of neoplastic B lymphocytes has a unique rearrangement of immunoglobulin DNA identical in each cell. In contrast, a pool of normal lymphocytes has many different rearrangements of immunoglobulin DNA which constitute the "germ-line" configuration. Restriction enzymes digest DNA into fragments that are separated on agarose gel. Specific radiolabeled DNA probes designed to hybridize or bind to rearranged immunoglobulin DNA fragments accumulate prominently in regions of the gel where DNA fragments are identical and thus define the clonality of large populations of neoplastic B lymphocytes. The sensitivity of this approach is limited and approaches 10^2. The technique is laborious, time-consuming and requires significant expertise. This technique has been replaced with the more sensitive polymerase chain reaction.

C) FLUORESCENCE-ACTIVATED CELL SORTING ANALYSIS (FACS)

Labeling malignant cells with tumor-specific antibodies coupled to a fluorescent dye permits the use of FACS to detect minimal residual disease. Individual cells travel in a single file in a fine stream past a laser beam where the fluorescence of each cell is measured.

Downstream tiny droplets containing either one or no cells are automatically given a positive or negative charge depending on whether they contain fluorescent dye. A strong electrical field serves to deflect them into an appropriate container. In order to minimize the possibility of achieving more than

7.3

Bone Marrow Transplantation, edited by Richard K. Burt, H. Joachim Deeg, Scott Thomas Lothian, George W. Santos. © 1996 R.G. Landes Company.

one cell per droplet, the concentration of the droplets is adjusted so that most contain no cells. The technique, which detects one in 10^3 cells, is limited by the specificity and sensitivity of the antibody preparation since even the most "specific" anti-tumor antibody may have cross-reactivity with a population of normal cells. Malignant cells may also change phenotype over time.

D) CLONOGENIC CULTURE

Samples of mononuclear cells embedded in agar or methylcellulose may give rise to colonies of cells derived from clonogenic precursors. In the clinical setting these colonies may exhibit normal differentiation into functionally normal cells, e.g. granulocytes, monocytes or morphologic/phenotypic char-

Table 7.3.1. Minimal residual disease in marrow or blood

Test	Specimen	Sensitivity	Advantages	Disadvantages
Morphology	Marrow aspirate/biopsy	1:20-1:100	Routinely used	Low sensitivity
Cytogenetic analysis	Viable cells	1:20	Multiple chromosomal abnormalities defined in a single test	Low sensitivity; Time-consuming and laborious
Southern blot analysis	DNA from viable cells	1:20-1:100	Rapid	Relatively low sensitivity; Limited number of probes; Variation in breakpoints
Fluorescence–activated cell sorting analysis (FACS)	Viable cells	10^3-10^5	High sensitivity	Rely on tumor–specific markers; Markers may change over time
Clonogenic culture	Viable cells	10^5	High sensitivity	Time-consuming and laborious; Interpretation sometimes subjective
Polymerase chain reaction (PCR)	DNA from viable cells	10^5-10^6	High sensitivity; Rapid	Complicated technique with potential for contamination; Variation in breakpoints
Reverse transcriptase PCR	mRNA from viable cells	10^5-10^6	High sensitivity; Rapid; Detects fusion genes after translocation	Complicated technique with potential for contamination; Levels of mRNA may vary in cells

7.3

acteristics of malignant cells. Single colonies plucked from culture may be studied for morphologic analysis, antibody or immunochemical staining characteristics, and polymerase chain reaction. Unlike other techniques, the detection of malignant cells by clonogenic techniques may be influenced by the cells' qualitative growth characteristics, i.e. the differential capacity of the malignant cells to grow and multiply over normal cells. These techniques may detect up to 1 in 10^5 malignant cells. However, the techniques are cumbersome, time consuming and interpretations are sometimes subjective.

The results may vary with different culture techniques. These approaches have been used to successfully detect lymphoma cells and carcinoma cells.

E) POLYMERASE CHAIN REACTION (PCR)

PCR allows the DNA from a selected region of a genome to be amplified by more than a million fold. Portions of the nucleotide sequence that surround the region to be amplified must be known. The sequences of these two regions permit the construction of two complementary DNA oligonucleotides (primers), one for each DNA strand. These primers provide sites for the initiation of DNA synthesis which is catalyzed by the enzyme DNA polymerase. The reaction is temperature dependent. Heat separates the two strands of DNA. The primers then anneal to each DNA strand and DNA replication occurs downstream on each strand. After DNA replication is completed (about 5 minutes), heat is applied again, separating the newly synthesized strands from the previous ones, and a second cycle begins. For effective DNA amplification, 20-30 cycles are required. The procedure is automated but requires care to avoid contamination with other oligonucleotides and primers from previous reactions or other samples since these will also be amplified unknowingly and inadvertently. The PCR method is extremely sensitive. It can detect a single, specific DNA molecule in a sample of DNA from up to 1 in 10^5 cells.

The application of PCR to the detection of malignant lymphoid cells differs substantially from the use of Southern blot analysis in these disorders. In lymphocytes, normal physiologic rearrangement of DNA within the immunoglobulin and T cell receptor genes produces unique sequences that are detectable with PCR. In lymphoid malignancies these junctions, once defined, serve as unique markers of neoplastic cells. The amplification of these junctions with sequence-specific primers permits the detection of clonally expanded populations of neoplastic lymphoid cells. Initially, "consensus" primers are used to identify these junctions. "Consensus" primers anneal to homologous or conserved regions in the immunoglobulin heavy chain variable and joining regions or T-cell receptor gamma region. The PCR product produced with these primers is then cloned and sequenced. Once the sequence of the product is known, sequence-specific primers may be synthesized. If prepared from DNA from neoplastic lymphoid cells, these primers in the polymerase chain reaction may detect tumor-specific DNA if present in other cell samples, e.g. bone marrow.

F) REVERSE TRANSCRIPTION PCR (RT-PCR)

Small amounts of specific RNA may be detected in a similar manner. After exposure to reverse transcriptase, RNA is converted to DNA which is then amplified as outlined above (by PCR). A number of neoplastic diseases have tumor-specific DNA or RNA sequences amenable to detection by PCR or RT-PCR. Unique DNA sequences amplifiable by PCR are produced by breakage and

7.3

rejoining of DNA from deletions or translocations. Similarly, distinct RNA fusion sequences may be detected rapidly by RT-PCR.

III) APPLICATION OF TECHNIQUES TO DETECT MINIMAL RESIDUAL DISEASE

Many molecular abnormalities in malignancies and non-neoplastic hematologic diseases have been defined recently. Eventually, clinically useful classifications of neoplastic diseases may be based on molecular variations. The list of defined cell surface markers and gene alterations for various diseases continues to expand. A partial list is provided in Table 7.3.2. After treatment, the persistence or absence of cells detectable with these techniques may confer prognostic significance. Emergence of modified genetic rearrangements during the course of disease and variations in the translocation breakpoint may limit the use of this test in some patients.

IV) SIGNIFICANCE OF MINIMAL RESIDUAL DISEASE

A) MRD FOLLOWING CONVENTIONAL (NON-MYELOABLATIVE) THERAPY

The definitive implications of MRD await the results of carefully implemented long-term studies. In patients with MRD, the efficacy of and/or indications for subsequent treatment are presently unknown. Since patients with MRD may be at high risk to relapse, the development of aggressive treatment approaches to salvage this subset of patients may be appropriate. In applying these tests one must consider that, following conventional (non-ablative) treatment:

1) **Persistence of MRD After Conventional (Non-myeloablative) Therapy**

    Immediately after treatment MRD does not appear to be a predictor of relapse since months are often required before eradication of malignant cells is complete. In acute leukemia after conventional chemotherapy, MRD usually (e.g. AML-M3, [t15;17]), but not always (e.g. AML-M2, [t8;21]) predicts for subsequent relapse

2) **Increasing MRD After Conventional (Non-myeloablative) Therapy**

    Quantitative assays that indicate an increase in MRD usually predict for recurrence.

3) **Long-Term Persistence of MRD After Conventional (Non-myeloablative) Therapy**

    In patients with follicular lymphoma ([t14;18], bcl-2) persistent MRD bears no relation to risk of clinical relapse.

4) **Absence of MRD After Conventional (Non-myeloablative) Therapy**

    This usually, but not always (e.g. ALL-B cell), predicts for disease-free survival in acute leukemia and lymphoma.

B) MRD FOLLOWING MARROW OR STEM CELL TRANSPLANTATION (MYELOABLATIVE) THERAPY

The principles outlined above in section IV may also apply to marrow or stem cell transplantation. The majority of studies evaluating MRD posttransplant have focused on chronic myelogenous leukemia (CML). For example:

1) After allogeneic transplant for CML an initial positive PCR for the Philadelphia chromosome in the first 3 months after BMT does not appear to confer a higher risk of relapse if subsequent specimens remain PCR negative.

7.3

Table 7.3.2. Chromosome aberrations and malignancies

Malignancy	Translocation	Deletion	Gene/Comment
Burkitt's lymphoma	(8;14), (2;8), (8;22)		MYC/Ig enhancer
Large cell lymphoma	3q27		BCL6(LAZ3)
Ki-1 anaplastic large cell lymphoma	t(2;5)		NPM/ALK fusion
Mantle cell lymphoma	t(11;14)		BCL 1– cell cycle cyclin (cyclin-D1)
Follicular lymphoma	(14;18)		BCL 2–an inner mitochondrial membrane protein
B-CLL (chronic lymphocytic lymphoma)	(11;14) (14;19) (8;12)	13q14	Most common chromosome abnormality in CLL BCL 1– cell cycle cyclin (cyclin-D1) BCL 3–cell cycle cyclin MYC/BTG1 locus
AML–M2	(8;21) t(6;9)		AML1-ETO (AML–1 is α subunit of core binding factor) AML with basophilia Can-Dek fusion
AML–M3	(15;17)		APL gene fused with retinoic acid receptor alpha
AML–M4Eo	Inversion 16		Core binding factor β subunit (CBFB)– myosin heavy chain (MYH11)
AML–M5	(9;11) (11q23)	11q-	Proto-oncogene Hu-ETS1 on chromosome 11 MLL–mixed lineage leukemia transcription factor
Myelodysplastic syndrome		20q-,5q-,7q- monosomy 7, monosomy 5	Chromosome 5 contains GM-CSF, IL-3, IL-4, IL-5, CSF-1, proto-oncogene FMS

7.3

7.3

Table 7.3.2. Chromosome aberrations and malignancies (cont'd.)

Malignancy	Translocation	Deletion	Gene/Comment
Pre-B-ALL	(5;14)		IL-3 growth factor/Ig enhancer
	(1;19)		PBX homeobox/E2A DNA binding protein
	(4;11)		HRX-FEL fusion
B-ALL	(8;14)		MYC/Ig enhancer
	(9;22)		BCR/ABL tyrosine kinase
T-ALL	(8;14)		MYC/Ig enhancer
	(7;19)		LYL1-HLH transcription factor
	(1;14)		TCL5 HLH transcription factor
	(11;14)		RBNT1/RBNT2
	(7;9)		TAN1-Notch gene homolog
	(10;14)		HOX11-homeodomain
	(14q11)		
CML	(9;22)		BCR/ABL tyrosine kinase
Colorectal adenocarcinoma		17p,	p53 gene
		18q,	DCC gene
Bladder adenocarcinoma		monosomy 9, 1q	
Breast adenocarcinoma		1p11-13, 3p11-13, 3q11-13	
Ewing's sarcoma	(11;22)		Also observed in PNET and small cell osteosarcoma
Leiomyoma	(12;14)	6p21, 7q 21-31	
Lipoma	(3;12)	13q12-13	

Tumor	Translocation	Location	Function
Liposarcoma	(12;16)		
Lung adenocarcinoma		3p13-23	
Lung small cell		3p13-23	
Melanoma	(1;6), (1;19)	1p11-22, 6q11-27	
Meningioma		Monosomy 22, 22q12-13	
Mesothelioma		1p11-22	
Myxoid chondrosarcoma	(9;22)		
Neuroblastoma		1p32-36	
Ovarian adenocarcinoma	(6;14)	3p13-21, 6q15-23	
Parathyroid adenoma	Inversion 11		
Renal cell carcinoma	(3;8)	3p13-21	
Retinoblastoma		13q14	Nuclear phosphoprotein
Rhabdomyosarcoma (alveolar)	(2;13)		
Synovial sarcoma	(X;18)		
Wilms tumor		11p13	Probable transcription factor

Adapted from Science 1991; 254:1153-1159. Deletion of 3p13-23 found in small cell of lung, adenocarcinoma of lung, ovarian adenocarcinoma of lung, ovarian adenocarcinoma and renal cell carcinoma; deletion of 1p11-22 found in melanoma, breast adenocarcinoma and mesothelioma; deletion of 6q11-27 found in melanoma, ovarian adenocarcinoma.

7.3

2) For patients with CML the risk of relapse after allogeneic BMT correlates with PCR positivity:

PCR Status	Relapse
persistently PCR -	0%
intermittently PCR+	20%
persistently PCR+	80%

3) In CML posttransplant, the interval between first positive PCR and hematologic relapse is approximately 5 months (range 3 months to greater than 24 months).

4) Persistent (stable or increasing) cytogenetic evidence of Ph+ cells posttransplant for CML indicates a high risk for clinical/hematologic relapse (> 90%), particularly in patients who have undergone T cell depletion. No definite criteria have been established for diagnosing "cytogenetic relapse" in the absence of clinical/hematologic relapse.

V) MIXED CHIMERISM

Mixed chimerism is the presence of both host and donor hematopoiesis and may occur in one or more hematopoietic lineages. In malignancies without defined chromosomal abnormalities, continued host hematopoiesis is being evaluated to predict relapse. Persistent host cells of the same lineage (monocyte, myeloid, etc.) as the hematopoietic malignancy, appears to correlate with relapse. The relationship between persistence of host T lymphocytes and leukemia relapse is much more controversial. However, at least for CML and T cell depleted allogeneic transplants, continued host T cell lymphopoiesis increases the risk of relapse.

VI) THERAPEUTIC INTERVENTION FOR MRD AFTER STEM CELL TRANSPLANTATION IN THE TREATMENT OF HEMATOLOGIC DISEASES

Although patients who have MRD after stem cell transplantation may be at greater risk for relapse, it is unclear whether further pharmacologic or immunologic therapy prolongs survival and increases cure rates. In this setting possible (unestablished) treatment approaches include:

A) STOP IMMUNOSUPPRESSION

After allogeneic BMT, discontinuing immunosuppressive (cyclosporines and prednisone) may be used to induce a graft-versus-leukemia effect. However, this approach may be complicated by GVHD.

B) INTERFERON-ALPHA

In patients with MRD after transplant (particularly CML), low dose interferon; initially, 3×10^6 units/m² subcutaneously 3 days per week or 1×10^6 units/m² daily may be administered. Dose escalate if no hematologic or other toxicity.

C) G-CSF (FILGASTRIM)

After allogeneic transplant, G-CSF at a dose of 5 µg/kg/day may preferentially stimulate donor hematopoiesis.

D) DONOR BUFFY COAT INFUSIONS

In patients with MRD who do not have GVHD, add back of previously cryopreserved T cells or peripheral blood T cells may induce a graft-versus-tumor effect. Infusion of $> 1 \times 10^7$ CD3+ cells/kg (usually obtained from 1-3 apheresis collections) has been used at some centers although the optimum T cell dose remains undetermined. T cell doses up to 5 x 10^8/kg have been given. Severe graft-versus-host disease and marrow hypoplasia are potential complications.

E) **SECOND TRANSPLANT**

This is the most aggressive approach and is associated with substantial morbidity and mortality. Its use in the setting of MRD is not established. In patients who have received a TBI containing regimen, usually non-TBI containing conditioning regimens are indicated for second transplants, e.g. cyclophosphamide/busulfan ("little BU/CY" see chapter on conditioning regimens.)

VII) **MRD IN THE MARROW AND PURGING**

Assays for MRD allow comparison between marrow purging techniques; however, in patients who receive purged marrow, the ultimate outcome may relate to the effectiveness of both the purging technique and/or conditioning regimen. In monoclonal antibody purged marrow from bcl-2 positive lymphoma patients, positive PCR for bcl-2 after marrow purging predicted for relapse, while patients with marrows purged to PCR negative sustained prolonged disease-free survival. In marrow purged with 4-HC, increased culture positive AML progenitors (CFU-L) was also associated with relapse.

SUGGESTED READING

1. Hupkes PE, Dorssers LCJ, van't Veer MB. Clinical significance of t(14;18) positive cells in the circulation of patients with stage III or IV follicular Non-Hodgkin's lymphoma during first remission. J Clin Oncol 1994; 12:1541-46.
2. Gribben JG. Attainment of molecular remission: A worthwhile goal? J Clin Oncol 1994; 12:1532-34.
3. Gribben JG, Nadler LM. Monitoring minimal residual disease. Semin Oncol 1993; 20:143-155.

7.3

Relapse After Bone Marrow Transplantation

F. Marc Stewart

I) SPECTRUM OF RELAPSE

A) CYTOGENETIC OR MOLECULAR RELAPSE

In leukemias or lymphomas, cytogenetic, Southern blot, Northern blot or polymerase chain reaction (PCR) molecular abnormalities characteristic with the original disease may be present after BMT. The presence of these abnormalities does not necessarily indicate eventual clinical relapse. However, persistent or progressive increments in disease detected by these techniques often herald clinical relapse.

B) PATHOLOGIC RELAPSE

Microscopic Documentation of Recurrent Disease

C) CLINICAL RELAPSE

Clear Clinical Signs of Recurrent Disease

II) TREATMENT APPROACHES FOR RELAPSE AFTER BONE MARROW TRANSPLANTATION

A) GENERAL

1) Ultimately, in most patients the prognosis for cure remains poor with any additional treatment modality.

2) Prognosis and response to additional treatment generally correlates directly with the duration of remission after the first BMT. In some patients the remission duration with salvage treatment after transplant failure may exceed the remission duration after first transplant. This phenomenon is referred to as "inversion."

3) In some patients additional treatment (chemotherapy or immunotherapy) may induce a complete remission and result in long-term survival short of a cure. Second bone marrow transplants may occasionally be curative.

4) The toxicity of additional treatment may be substantial, particularly with second bone marrow transplantation, where up to 40% may die of early treatment-related complications. Due to toxicity, a second BMT cannot, in general, be tolerated if relapse occurs within 6-12 months after the first BMT.

B) CHEMOTHERAPY

1) Patients who relapse in the first 100 days after first BMT should not undergo aggressive induction chemotherapy.

2) Regimens including cytarabine are preferred for relapsed AML or CML in myeloid accelerated phase. Median survival is approximately 6 months.

Dosage:

Cytarabine 100 mg/m^2 continuous infusion IV daily x 7 days.
Daunorubicin 45 mg/m^2 IV daily x 3 days.

3) Regimens including daunorubicin, vincristine, prednisone for ALL or CML with ALL blast crisis are preferred. Median survival is approximately 10 months.

7.4

Bone Marrow Transplantation, edited by Richard K. Burt, H. Joachim Deeg, Scott Thomas Lothian, George G. Santos. © 1996 R.G. Landes Company.

Dosage:

Daunorubicin 45 mg/m^2 IV daily x 3 days.

Vincristine 1.5 mg/m^2 IV weekly x 4.

Prednisone 60 mg/m^2 PO daily.

4) Extramedullary (testicular or CNS) relapse is a harbinger of systemic relapse and besides local therapy (irradiation, intrathecal chemotherapy) should be treated with systemic chemotherapy.

C) IMMUNOTHERAPY (SEE CHAPTER ON GRAFT-VERSUS-LEUKEMIA)

1) Interferon-Alpha

In patients with CML who relapse into a second chronic phase after the first BMT, 10-25% may achieve complete hematologic and cytogenetic remissions with interferon-alpha. Most patients respond clinically to treatment. Posttransplant patients are extremely sensitive to the hematologic effects of interferon, therefore, patients should be treated with low doses initially. Interferon-alpha: 1 x 10^6 IU/m^2/day, daily or 3 x 10^6 IU/m^2/day three times per week, subcutaneously, to achieve maximum clinical and cytogenetic response.

2) Withdrawal of Immune Suppression

Posttransplant withdrawal of immunosuppression (i.e. cyclosporine) may result in a second remission.

3) IL-2 and All *trans* Retinoic Acid

In acute promyelocytic leukemia these agents have been tried in patients who relapse after transplant with some responses noted.

4) G-CSF

Filgastrim at 5 µg/kg/day subcutaneously, has been used to treat patients in early relapse with some responses.

5) Donor Buffy Coat

Donor leukocyte infusions from the original marrow donor may produce sustained remissions in up to 70-80% of CML patients who relapse into chronic phase, 30% of patients with relapsed AML and < 5% of patients with relapsed ALL (Table 7.4.1). The optimal cell (leukocyte) dose has not been determined. However, 10^7 to 5 x 10^8 T cells have been infused without a clear difference in antileukemic effect. The onset of response is generally delayed until 1-3 months after donor leukocyte infusion. Severe graft-vs-host disease and marrow hypoplasia are potential complications.

D) SECOND BONE MARROW TRANSPLANTATION

1) Timing of Second BMT

Patients who relapse after first BMT may be candidates for second BMT unless the disease has recurred within 6 months to 1 year after the first transplant. Patients should have good performance status and favorable disease-status (acute leukemia in remission, chronic myelocytic leukemia in chronic phase). Carefully selected patients have a disease-free survival rate greater than 20% at 2 years.

2) Conditioning Regimen for 2nd BMT

Second BMT conditioning regimens are limited based on whether or not total body irradiation was used to condition recipients in the first BMT. In patients who have received a TBI-containing regimen, usually non-TBI-containing conditioning regimens are indicated for the second transplant,

7.4

7.4

Table 7.4.1. Clinical efficacy of donor leukocyte infusions for leukemia relapse after allogeneic bone marrow transplant

Disease	Donor T Cell Dose (10^8 cells/Kg)	Other Treatment	Myelosuppression	GVHD	Responses	Reference
3 CML-CP	4.4-7.4	IFNα	NR	2/3	3/3	Kolb, Blood 1990; 76:2462
4 AML	3.3-9.1	Cytarabine, Amsacrine	NR	3/4	1/4	Szer, BMT 1993; 11:109
5 CML-CP 1 CML-Cyto	0.3-3.4	IFNα, Busulfan	2/6	5/6	4/6	Bar, JCO 1993; 11:513
6 CML-AP 2 CML-BC	2.5-5.0	IFNα	4/8	7/8	6/8	Drobyski, Blood 1993; 82:2310
39 CML-CP 12 CML-AP 10 AML 13 ALL	NR	IFNα	NR	NR	36/51 4/10 4/13	Kolb, Blood 1993; 82:214A
8 CML-CP 3 CML-AP	0.9-8.4	IFNα	5/11	9/11	6/11	Poster, NEJM 1994; 330:100
2 CML-AP 5 CML-CP 2 CML-Cyto 2 CML-Mole	0.6-10.1	IFNα Hydrea	2/14	9/14	10/14	van Rhee, Blood 1994; 83:3377
6 CML 13 Acute leukemia	1.1-16.4	IFNα	NR	NR	3/6 3/13	Collins, Blood 1994; 84:333A
84 CML 23 AML 22 ALL	0.1-15.0	IFNα	28 8 5	55 13 8	54/84 5/23 0/22	Kolb, Blood 1995; 86: 2041

CP-chronic phase; AP-accelerated phase; Cyto-cytogenetic relapse; Mole-molecular relapse; NR-not reported, IFNα = interferon α e.g. cyclophosphamide/busulfan ("little BU/CY"—see chapter on conditioning regimens). If TBI was not used in the initial regimen, strong consideration should be given for including it in the second regimen. No ideal drug combination has been determined for second BMT.

3) **Donor Selection for 2nd BMT**

Efforts should be directed toward maximizing a graft-versus-leukemia effect if possible by selecting alternative donors (e.g. a different HLA-matched sibling, if available).

4) **Marrow Processing for 2nd BMT**

Efforts should be directed toward maximizing a graft-versus-leukemia effect if possible by minimizing graft-versus-host prophylaxis (e.g. avoidance of T cell depleted marrow grafts in complete geneotypically matched grafts).

5) **Regimen-Related Toxicity from 2nd BMT**

A high rate of severe hepatic veno-occlusive disease and other regimen-related toxicities have been observed in second BMTs. Depending on selection criteria for patients, up to 40% of patients may die in the first 100 days from treatment-related complications (VOD, interstitial pneumonia, GVHD).

SUGGESTED READING

1. Arcese W, Iori AP, Di Nucci G, Martinez-Rolon J, Pinto RM, Mandelli F. What does one do for the CML patient in relapse after allogeneic bone marrow transplantation? Leukemia and lymphoma 1993; 1:213-9.
2. Kumar L. Leukemia: Management of relapse after allogeneic bone marrow transplantation. J Clin Oncol 1994; 12:1710-17.
3. Darrington DL, Vose JM, Anderson JR et al. Incidence and characterization of secondary myelodysplastic syndrome and acute myelogenous leukemia following high-dose chemoradiotherapy and autologous stem-cell transplantation for lymphoid malignancies. J Clin Oncol 1994; 12:2527-34.

7.4

8.0 General Medical Care

8

8

8

Diagnosis and Management of Complications Related to Long-Term Venous Access Devices

H. Richard Alexander

I) INTRODUCTION

Approximately 500,000 long-term venous access catheters are placed in the United States annually. It is incumbent upon all house officers to maintain expertise on a variety of topics related to the diagnosis and management of complications related to long-term venous access catheters to minimize patient morbidity and maximize the life span of these devices.

II) LONG-TERM VENOUS ACCESS CATHETERS

Long-term venous access catheters may be left in place for months or years without complications. There are two basic types of long-term venous access devices currently in use, external catheters and implantable catheters. External devices are tunneled subcutaneously from an exit site usually 8-10 cm from the actual insertion of the catheter into the vein. A dacron cuff on the catheter is positioned in the subcutaneous tunnel which promotes fibrous ingrowth around the catheter thereby reducing the chances for accidental dislodgement of the catheter and retarding migration of bacteria along the catheter. The devices are made of silicone rubber (Silastic) which is barium-impregnated so they can be easily visualized on x-ray. For patients about to undergo BMT, multiple lumen (not single lumen) devices are necessary and an external catheter not an implantable port should be used. Implantable catheters are similar to an external catheter except, instead of exiting the skin, a port housing is anchored in a subcutaneous pocket and accessed with a special noncoring Huber needle. When these devices are not in use they are contained completely under the skin and have the advantages of requiring very little routine maintenance and interfering minimally with life style. The disadvantages of these devices are that they must be accessed through the skin each time the device is used, and whereas external catheters such as the Hickman or Groshong can be removed easily in an office setting, the implantable ports require a second surgical procedure for removal. Due to concern about infection and subsequent surgical removal in transplant patients who may have neutropenia and refractory thrombocytopenia, implantable devices are not used in the early BMT period.

8.1

A) HICKMAN (FIG. 8.1.1)

The most commonly used external catheter in adult patients is the 10 French double lumen Hickman catheter. Catheter hubs are color coded and the internal diameter of the corresponding lumen is marked on the catheter adjacent to the hub. They are available in a variety of French sizes and with one, two or three lumens.

B) GROSHONG (FIG. 8.1.2)

The Groshong catheter is an external venous access device that has several modifications designed to improve the safety and reduce the maintenance

Bone Marrow Transplantation, edited by Richard K. Burt, H. Joachim Deeg, Scott Thomas Lothian, George W. Santos. © 1996 R.G. Landes Company.

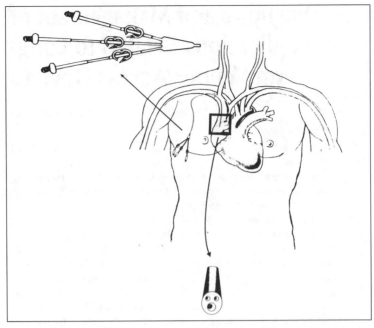

Fig. 8.1.1. Triple lumen 12.5 Fr Hickman catheter inserted into subclavian vein, positioned in the distal superior vena caval–right atrial junction, and tunneled subcutaneously to an entrance site on the precordium. The external hubs are of different lenth (arrow) and the respective internal diameters marked near the hub so that each lumen can be distinguished (inset).

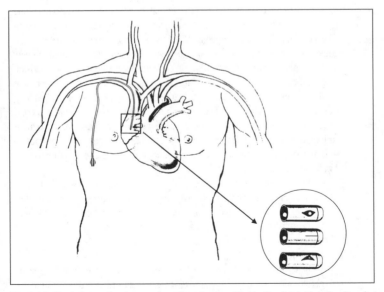

Fig. 8.1.2. Single lumen Groshong catheter in a site described in Fig. 8.1.1. Note the slit-valve tip (inset) designed to prevent passive reflux of blood into the lumen and reduce the incidence of luminal clot formation.

8.1

requirements of the device. The tip of the Groshong catheter is rounded and the lumens open through a slit valve on the sides of the catheter. The slit valve is designed to prevent passive reflux of blood into the catheter and thereby prevent intraluminal clots from developing. In addition, the catheter has a slightly higher durometer (silicone content) which means that a Groshong catheter with internal lumens comparable in size to the standard double lumen Hickman catheter has a slightly smaller external diameter. Groshong catheters are available with one or two lumens.

C) PORT-A-CATH (FIG. 8.1.3)

Probably the most frequently used implantable device is a port-a-cath. Implantable ports are available with one and two lumens; however, the double lumen device has a relatively large port housing which may not be suitable for pediatric or asthenic patients. It is important to appreciate that long-term venous access catheters are also used to access a variety of body cavities other than the central venous system. Therefore, it is important not to assume that an implanted port housing is necessarily accessing the venous system; it is possible that it may be connected to a catheter in the intraperitoneal or intrathecal space.

III) PHERESIS CATHETERS

Pheresis catheters are external double or triple lumen large-bore catheters which, due to their large caliber, are capable of handling the high flow rates of blood (5-10 liters per hour) required for collection of peripheral stem cells. In addition, infusion and draw ports are staggered at different levels to avoid repeated pheresis of the same blood. Due to their large caliber, pheresis catheters have a higher incidence of infections and thrombosis.

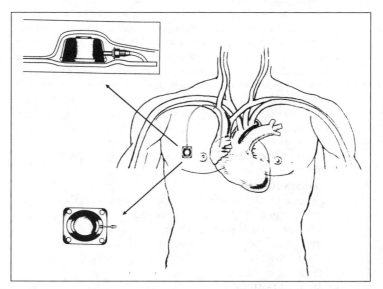

8.1

Fig. 8.1.3. Typical site for an implantable port with the catheter inserted into the subclavian vein. Note the four eyelets in the base used to suture the port to underlying fascia and prevent the port from flipping in the subcutaneous pocket. When not in use, the port is contained completely under the skin (inset) and is accessed with a noncoring Huber needle.

A) **SHORT-TERM PHERESIS CATHETERS**

Temporary pheresis catheters (e.g. Quinton, Shiley, Arrow) may be placed under local anesthesia at bedside into the femoral or subclavian vein and are removed after stem cell collection. These catheters are made of polyurethane and tend to be rigid. If inserted into the femoral vein, the patient may still ambulate but remains in the hospital. Outpatient pheresis may be done if placement is in the subclavian vein.

B) **LONG-TERM (TUNNELED) PHERESIS CATHETER**

Pheresis catheters that are tunneled under the skin and into the internal jugular or subclavian vein (e.g. Vascath, Cook, Davol, Quinton Percath) are made of Silastic, have a dacron cuff and are generally inserted in the operating room. They allow for outpatient collection of stem cells, and, thereafter, these catheters may be used for long-term venous access.

IV) **PICC LINES** (FIG. 8.1.4)

Peripherally inserted central catheters (PICCs) are inserted into an antecubital vein and positioned so that the tip resides in the superior vena cava. PICCs are currently made with less thrombogenic material which has resulted in renewed interest of these devices as an alternative to centrally inserted devices. PICCs may be very suitable in a number of clinical settings including: patients requiring long-term antibiotic therapy; patients with malignancy of the chest wall; patients with radiation injury to the neck or precordium; patients with a tracheostomy; or patients physically unable to tolerate insertion of a central catheter. In general, PICCs have a smaller diameter and are stiffer than long-term central catheters so that they can be advanced through an antecubital vein. PICCs are available in a variety of Fr sizes and lengths, are available with one or two lumens and as external or implantable devices. The implantable PICC housing is a specially designed low profile port. Because the septum of the port housing is small there is less area skin to rotate sites for inserting the access needle. The clinical performance of PICCs has been quite favorable. PICCs can be successfully placed in about 90% of patients and have an overall complication rate of 15%. By far the greatest complication associated with PICCs is catheter-induced mechanical phlebitis in the proximal upper extremity during the first week after insertion. This can be effectively controlled with local treatment (usually warm compresses over the area). The major advantages of PICCs are the fact that they can be inserted at the bedside often by trained nursing personnel, have a high patient acceptance, low rate of complications, low cost and may be left in place for weeks to months.

V) **PERSISTENT WITHDRAWAL OCCLUSION (PWO)—INABILITY TO ASPIRATE**

A) **DIFFERENTIAL DIAGNOSIS**

Occasionally it is possible to infuse fluids into a long-term venous access catheter without being able to aspirate blood from it. This may be secondary to catheter tip malposition, a catheter kink, pinching of the catheter as it traverses the space between the first rib and clavicle, a fibrin sheath at the catheter tip acting as a flap valve or a primary venous thrombosis extending around the tip of the catheter.

B) **APPROACH/TREATMENT**

If one is unable to aspirate blood from the device, one should connect a saline-filled syringe directly to the hub of the device to verify the correct placement of the Huber needle in the implantable port and attempt to irrigate with

8.1

10 ml of normal saline. If there is episodic or no withdrawal, it is sometimes helpful to attempt to withdraw from the catheter with the patient in various positions such as supine, lateral, with arms raised or during a valsalva maneuver. If it is still impossible to draw blood from the catheter a chest x-ray should be obtained to verify normal catheter position. If the catheter has migrated to an abnormal location or has flipped out of the superior vena cava then an interventional radiology procedure should be considered in order to reposition the device. If malposition has been excluded, then attempts to restore the function of the catheter using a variety of thrombolytic agents may be used (Table 8.1.1).

VI) PARTIAL CATHETER OCCLUSION

A) DIFFERENTIAL DIAGNOSIS

If a catheter has partial or complete occlusion when one attempts to infuse the differential diagnosis includes catheter or tip malposition, intraluminal clot,

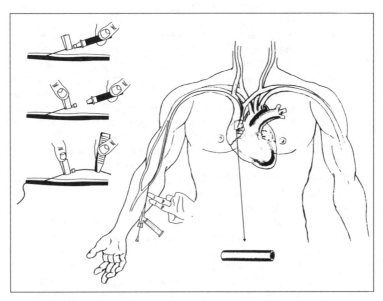

Fig. 8.1.4. A PICC is inserted via an antecubital vein and advanced to a final position as shown. Initial access to the antecubital vein is made by using a break-away needle. Once the catheter has been threaded into the vein, the needle is withdrawn from the vein and the flanges are squeezed to break the needle.

8.1

Table 8.1.1. Thrombolytic therapy of central venous catheter occlusion

Persistent Withdrawal Occlusion
Urokinase, 250,000 U in 150 ml D$_5$W over 90 minutes.
Partial or Complete Catheter Occlusion
Urokinase, 5000 U/ml, inject 0.4-1.0 ml into the device and leave in place for approximately 30-60 minutes. This can be repeated two or three times until catheter patency is restored. The success with restoring catheter function using urokinase in this setting is over 95%.

intraluminal drug precipitate, catheter kink, pinching between the clavicle or first rib, subclavian venous thrombosis or other mechanical etiology.

B) **APPROACH/TREATMENT**

For an implantable catheter, initially one should again verify the correct position of the Huber needle by insuring its contact with the needle stop in the port housing. For the external devices, connect a saline filled syringe directly to the hub and attempt to irrigate the device with 10 ml of normal saline. If the patient complains of pain or discomfort, it is possible that the catheter may be fractured and infusing into an extravascular site. In this instance, a chest x-ray should be obtained to verify normal catheter position followed by catheter contrast study to verify normal infusion through the device. If the catheter flushes without resistance or discomfort to the patient, then one should hang a maintenance IV solution and document that it infuses normally and without variation, depending upon patient position.

VII) **COMPLETE CATHETER OCCLUSION—INABILITY TO INFUSE**

If complete catheter occlusion is encountered, initial steps include verifying correct Huber needle placement in an implantable port housing and checking to insure that the catheter clamp has been released on an external device. If one is unable to irrigate at all, a chest x-ray should be obtained to document normal catheter position. If the catheter is malpositioned but appears to be intravascular in location, then an interventional radiology procedure should be considered to reposition the catheter tip. If the catheter has withdrawn and suspicion exists that it may be extravascular in location, it should be removed. If the chest x-ray shows that the catheter is in its normal intravascular position, then attempts to clear the catheter using thrombolytic therapy are indicated. Numerous thrombolytic regimens have been reported to be highly successful in restoring catheter function (Table 8.1.1).

VIII) **DIAGNOSIS AND MANAGEMENT OF LONG-TERM ACCESS-ASSOCIATED INFECTIOUS COMPLICATIONS**

A) **EXIT SITE INFECTION**

An exit site infection is characterized by localized erythema and tenderness at the exit site of an external catheter. Frequently systemic and laboratory signs of infection are not present. If an exit site infection develops secondary to extrusion of the dacron catheter cuff, then the catheter must be removed. Otherwise, the exit site should be cultured and aggressive local wound care and oral or IV antibiotics against Gram positive organisms is normally successful in treating the infection.

B) **TUNNEL INFECTION**

A tunnel infection represents a more advanced infectious process in which the subcutaneous track of the catheter has developed into a suppurative infection. Surrounding cellulitis is frequently present and the patient will have erythema, induration and tenderness along the subcutaneous catheter track. Typically, purulence can be expressed from the exit site by milking the tract. Systemic signs of infection are frequently present. The catheter must be removed. The catheter track should be cultured and intravenous antibiotics should be instituted. If the catheter is left in situ, it is possible that significant tissue necrosis and worsening cellulitis can develop.

C) **POCKET INFECTION**

A pocket infection is a suppurative infectious process around an implantable port housing and is comparable to a tunnel infection of an external venous

8.1

access device. Often skin changes over the port housing can be very subtle and the diagnosis may be confirmed by aspirating infected fluid from around the port housing. In this setting the port must be removed promptly and the subcutaneous pocket should be packed open. Intravenous antibiotics are usually indicated.

D) LINE SEPSIS

Line sepsis represents a potentially serious complication of long-term venous access devices. The etiology of line-related bacteremia appears to be secondary to the migration of Gram positive organisms that have colonized the catheter hub or have seeded the port housing from skin contaminants when accessed. Bacteria gain access to the intravascular portion of the catheter and can colonize on the surface of the device which is typically coated with a glycocalyx biofilm. The most common organisms which are isolated from catheter tips during line-related bacteremia include *Staphylococcus epidermidis*, *Staphylococcus aureus* and *Streptococcus spp.* The diagnosis of line-related bacteremia is based upon the criteria listed in Table 8.1.2.

In patients with line-related sepsis, successful treatment with antibiotics alone while the catheter remains in situ, even in the presence of neutropenia, is extremely high. Therefore, initial treatment of line-related septicemia should include appropriate empiric antibiotic coverage to include Gram positive organisms and supportive measures for the patient. Deteriorating clinical condition while the patient is on antibiotics is an indication for catheter removal. Recurrent infections and persistently positive blood cultures after completion of treatment are other indications for catheter removal. Infections secondary to *Staphylococcus aureus* appear to carry a worse prognosis and have a lower likelihood of successful therapy while the catheter remains in situ. Other types of pathogens such as *Candida spp.*, *Mycobacterium spp.* and *Bacillus spp.* are also more difficult to eradicate and early catheter removal may be considered in these patients who are seriously ill.

IX) CATHETER-RELATED VENOUS THROMBOSIS

The presence of a long-term venous access device in the subclavian or other vein and the associated intimal injury at the insertion site predisposes patients to the development of venous thromboses. Central venous catheters may account for up to 40% of all deep venous thromboses of the upper extremity. Many subclavian vein thromboses remain asymptomatic and this may be secondary to several factors including the presence of extensive venous collaterals in the upper extremity that minimize the hemodynamic effect of an intramural thrombosis and the hydrostatic forces which favor venous drainage from an upper extremity to a much greater extent than that of the lower extremity. After anticoagulation treatment, the short- and long-term sequelae of upper central venous thromboses appear to be minimal. Catheter-related venous thrombosis has been reported to occur

8.1

Table 8.1.2. Commonly used criteria for the diagnosis of catheter-related bacteremia

1) Clinical signs of infection; no other remote site identified.
2) Temporal relationship between catheter manipulation and development of infection.
3) Positive blood cultures from the catheter show a 5- to 10-fold (or greater) colony count vs peripheral cultures.
4) Greater than 100 CFU/ml from catheter blood cultures.

in 4-40% of patients by clinical assessment or autopsy series. The time course in the development of catheter-related venous thrombosis is highly variable, however, it is most likely that over half of all the thromboses will develop within 2 weeks after catheter placement and two-thirds will develop within 1 month.

A) PREDISPOSITION

Factors which predispose patients with long-term venous access catheters to venous thrombosis include a high platelet count, some solid tumors such as adenocarcinoma (in general, catheter-related thrombosis is more of a problem in BMT patients with breast adenocarcinoma compared to leukemia, lymphoma or aplastic anemia), low flow states and the presence of bilateral versus unilateral venous access lines. Age, sex or technique of placement do not appear to predispose to venous thrombosis.

B) SYMPTOMS AND DIAGNOSIS

Patients who develop symptoms compatible with subclavian vein thrombosis such as arm swelling, pain and evidence of collateral venous flow on the chest should be evaluated with peripheral venography, duplex doppler sonography or computed tomography. The use of contrast instilled through the catheter will demonstrate a thrombosis that is distal to the catheter tip but may not identify a thrombosis that is located in a more peripheral position in the central venous system.

C) THERAPY

In the past catheter removal and systemic anticoagulation with heparin and subsequently coumadin has been considered fairly standard treatment for catheter related venous thrombosis. However, there are several management approaches that have been described while the catheter remains in situ that have been very successful in resolving catheter related venous thrombosis and preserving the life of the catheter. Systemic streptokinase has been used as thrombolytic therapy for catheter-related subclavian vein thrombosis (Table 8.1.3). Hemorrhagic complications associated with the use of systemic streptokinase are not inconsequential and therefore close monitoring of the patient is essential. More recently the use of urokinase administered through a forearm vein on the side ipsilateral to the subclavian vein thrombosis has been reported to be highly successful and safe in resolving catheter-related venous thrombosis (Table 8.1.3). Complete dissolution of the thrombus can be achieved in over 90% of patients if the interval of symptoms is less than one week but in only 56% of patients whose symptoms are present for longer than 1 week. This suggests that once the diagnosis of subclavian vein thrombosis is made, prompt treatment is warranted. Typically if only partial lysis can be achieved with local urokinase therapy then a recurrent thrombosis is most likely. The hemorrhagic complications associated with the local infusion of urokinase are usually minor in nature.

8.1

Table 8.1.3. Thrombolytic therapy of catheter-related central vein thrombosis

Streptokinase: administered systemically 250,000 U intravenously followed by 100,000 U per hour for 72 hours.

Urokinase: administered in the forearm vein on the side ipsilateral to the subclavian vein thrombosis as a loading dose of 500 U/kg intravenously followed by a dose of 500-2000 U/kg per hour for an interval of 1-4 days.

Table 8.1.4. Catheter care guidelines

Catheter	Sterile NaCl Flush Prior to IV Infusion or Blood Draw	Blood Discarded Prior to Blood Draw	Sterile NaCl Flush After IV Infusion or Bood Draw	Heparin Lock After IV Infusion or Blood Draw	Frequency of Heparin Lock when Catheter Not in Use	Catheter Site Dressing Change[1]	Injection Cap Change[2]
Pheresis (e.g. Cook)	No blood draws[3]; 10 ml prior to infusion	No blood draws[3]	No blood draws[3]; 10 ml after IV infusion	1.5 ml of 5000 U/ml = 7500 U	q 72 hours	transparent tegaderm q week; or gauze 3 times a week	q week
Hickman	10 ml	5 ml	10 ml	3.0 ml of 100 U/ml = 300 U	q day	same	q week
Groshong	10 ml	5 ml	10 ml	N/A	N/A (q week with sterile normal saline)	same	q month
Port-a-cath	10 ml	5 ml	10 ml	5 ml of 100 U/ml = 500 U	q month	same; if no access needle is inserted no dressing is needed	N/A

8.1

N/A = not applicable
[1] Dressing change should be done while wearing sterile gloves with the care giver and patient wearing a mask. Prior to covering the catheter with tegaderm or sterile gauze, clean the skin with alcohol first then betadine second then alcohol again. After cleaning the skin, clean the external catheter line with same sequence of alcohol and betadine. Finally cover with tegaderm or gauze.
[2] Frequency of injection cap change varies by institution. Sterile gloves should be worn and the care giver and patient should wear a mask. Prior to cap change, clamp catheter line, scrub cap with provodone iodine, remove new cap with 4x4 sterile gauze, insert new cap with 4x4 sterile gauze, apply new dressing as above.
[3] Because of a pheresis catheters's large diameter and perceived increased risk of infection and the generally short duration (days to weeks) needed for mobilization of peripheral blood stem cells, blood draws from pheresis catheters are generally avoided by most centers.

D) PROPHYLAXIS

Because certain patient populations are at high risk of developing venous thrombosis, it is reasonable to consider prophylactic treatment with low-dose warfarin in select patients. A dose of 1 mg per day for 3 months after insertion of a long-term venous access catheter has been shown to significantly reduce the incidence of catheter-related venous thrombosis. However, it is not clear whether this is reasonable for all patients undergoing placement of a vascular access device.

X) REPAIR OF A CATHETER FRACTURE

External catheters are at risk for leaks or cracks because they are manipulated so frequently. If a break occurs in either a single or multilumen catheter it can be repaired with repair kits available from the manufacturer. Because catheters have different diameters, repair kits are not typically interchangeable, only the repair kits specifically designed for the type and diameter of the catheter that is broken should be used. If a crack or leak in a catheter has been identified, it should be cleaned and secured temporarily and the device should not be used until the catheter has been repaired. Instructions are available with the various repair kits. However, the catheter should ideally be repaired by appropriately trained nurses or physicians.

XI) CATHETER REMOVAL

A) IMPLANTABLE CATHETER

Removal of an implantable port requires a surgical procedure. An incision is made over the housing, the pseudocapsule is incised and the device is carefully excised. This procedure requires familiarity with operative techniques and judgment, and should be performed by an individual with surgical training.

B) EXTERNAL CATHETER

Removal of an external device can be potentially difficult and should only be performed by a physician who has had training in the procedure. The dacron cuff causes considerable fibrosis in the subcutaneous tunnel and this dense tissue must be sharply incised either through the exit site or through an incision directly over the cuff if it is high in the tunnel. If one attempts to simply pull the catheter out there is a good chance that the catheter will break. The manipulations to remove the catheter require planning based on its location, sterile technique, local anesthesia and a set of surgical instruments. If there is significant bleeding from the exit site after catheter removal it is important to appreciate that direct pressure to achieve hemostasis should be applied at the site where the catheter was inserted into the vein (usually the infraclavicular approach to the subclavian vein).

8.1

SUGGESTED READING

1. Alexander HR, Lucas AB, Steinhaus EP, Torosian MH. Vascular Access in the Cancer Patient. H. Richard Alexander, ed. Philadelphia: JB Lippincott Co., 1994.

Nutritional Support in Bone Marrow Transplant Recipients

Sara L. Bergerson

I) INTRODUCTION

Recipients of bone marrow transplantation (BMT) routinely require nutritional support due to increased metabolic needs and decreased oral intake. Increased metabolism occurs from fever, infection, graft-versus-host disease (GVHD) and chemotherapy or radiation-related tissue breakdown. Decreased oral intake arises from gastrointestinal disturbances such as mucositis, enteritis, nausea, vomiting, diarrhea and altered salivation; disturbances which often occur in multiples. Non-relapse survival for BMT patients less than 95% of ideal body weight is significantly worse than for patients who are at or exceed ideal weight. The goal of nutritional support is to correct pre-existing deficiencies, minimize treatment-associated losses and ultimately enable the BMT recipient to resume an oral diet to meet posttransplant needs. Nutritional management should begin with a pretransplant evaluation to assess baseline nutritional status and potential risk factors, determine nutrient requirements and develop a nutrition treatment plan.

II) CALORIE REQUIREMENTS

Calorie needs are dependent on factors such as baseline nutritional status, age, weight and degree of metabolic stress. Calorie needs are usually based on actual body weight, but for obese adults (> 20% ideal body weight) calorie needs should be based on adjusted body weight (Table 8.2.1). In children, growth parameters should be used to develop goal calorie needs to promote normal growth and development. Calorie needs can be calculated using several predictive formulas. For clinical purposes, a quick bedside calculation can be used to determine empiric calorie needs (Table 8.2.2). When using the bedside calculation, initial calorie needs are generally based on the lower end of the range, and are adjusted upward only if ongoing metabolic and nutritional monitoring indicates that needs are higher. The ability to achieve goal calorie needs may be limited by fluid constraints and/or substrate intolerance associated with demands of the transplant process. A more exact measure of calorie needs is rarely necessary, but may be determined by indirect calorimetry. Using a metabolic cart, indirect calorimetry provides a measure of resting energy expenditure and respiratory quotient (substrate utilization).

III) PROTEIN REQUIREMENTS

The protein requirements of BMT recipients are higher than the normal recommended dietary allowance. If organ function is normal, protein should be encouraged after cytoreductive therapy to restore or preserve lean body mass and provide substrate for the posttransplant hypermetabolic state. Protein requirements are based on age and body size (Table 8.2.2). For obese adults, protein needs should be based on adjusted ideal body weight (Table 8.2.1). If kidney, liver or neurological dysfunction is present, protein should be limited to 0.6-1.2 g/kg, depending on the severity of organ dysfunction. The degree of protein restriction should be tailored to a patient's individualized need with the aim of decreasing uremic toxicity and other metabolic derangements while preventing malnutrition.

8.2

Bone Marrow Transplantation, edited by Richard K. Burt, H. Joachim Deeg, Scott Thomas Lothian, George W. Santos. © 1996 R.G. Landes Company.

IV) FLUID REQUIREMENTS

Maintenance fluid needs are derived from a standard formula based on weight (Table 8.2.3), but should be adjusted to take into account any clinical problems or circumstances that increase or decrease fluid needs. In calculating daily fluid needs, all oral and intravenous intake (e.g. medications, blood components) and body fluid losses (e.g. diarrhea, emesis) including insensible fluid losses (e.g. perspiration from fever) should be included. Although the volume of insensible loss can vary significantly, a useful general formula for insensible water loss is 0.5 ml/kg/hr. This calculation must be increased 10% for each degree fever above 99°F. Fluid overload due to cardiac, kidney or liver dysfunction after transplant will necessitate an adjustment to decrease fluid needs. For adequate renal perfusion, fluid

Table 8.2.1. Adjusted ideal body weight

Adjusted weight = IBW + [0.25 x (actual body weight- IBW)]
    IBW = Ideal body weight
    Unit of weight = kg

Table 8.2.2. Estimation of calorie and protein needs

Age (yrs)	Calories (total kcal/kg*)	Protein (g/kg IBW)
Adult > 19	30-45	1.5
15-18	40-50	1.8
11-14	45-60	2.0
7-10	50-75	2.4
4-6	60-90	2.5-3.0
1-3	70-100	2.5-3.0

IBW = ideal body weight, kg = kilograms, kcal = kilocalories
* Includes protein calories
Source: Adapted with permission from Nutritional Assessment and Management During Marrow Transplantation: A Resource Manual. Lenssen P, Aker SN, eds. Seattle: Fred Hutchinson Cancer Research Center;1985.

8.2

Table 8.2.3. Maintenance fluid needs

Weight (kg)	Fluid Needs/24 hours
< 10	100 ml/kg
10-20	1000 ml plus 50 ml/kg for each kg > 10 kg
21-40	1500 ml plus 20 ml/kg for each kg > 20 kg
> 40	1500 ml/m²

kg = kilograms
Source: Adapted with permission from Nutritional Assessment and Management During Marrow Transplantation: A Resource Manual. Lenssen P, Aker SN, eds. Seattle: Fred Hutchinson Cancer Research Center;1985.

needs may be increased in excess of maintenance during use of nephrotoxic agents. Maximum fluid intake tolerated is approximately twice the maintenance needs.

V) **DIET**

Food is a possible source of pathogenic bacteria which may colonize and seed the blood through a gastrointestinal tract damaged from chemotherapy, radiation, GVHD and neutropenia. A diet with reduced bacterial content is generally provided, especially to patients in laminar air flow rooms and to patients with less than 500 granulocytes. Table 8.2.4 describes the three types of modified bacterial diets available. The protective benefit of reducing the bacteria content in diet is unproven for non-laminar air flow, and the trend is away from strictly sterile foods and toward more liberal food service in BMT units. More commonly used are either a low bacteria diet or a modified house diet without fresh vegetables and fruits.

VI) **PARENTERAL NUTRITION (PN)**

A) **INDICATIONS**

PN should be initiated when oral intake decreases to less than 1000 kcal/day or < 60% of estimated needs for 3-5 days or when nutrition monitoring parameters, such as weight and pre-albumin (transthyretin), fall below admission level. Situations where bowel rest is indicated, such as intestinal GVHD or ileus, are also indications for PN. Any patient with debilitation due to underlying disease or active infection should begin PN support immediately upon admission.

B) **CONTRAINDICATIONS**

PN is a supportive therapy and should not automatically be initiated in BMT recipients who are able to maintain acceptable nutritional status. In the presence of a functional gastrointestinal tract when oral intake is > 60% of estimated needs, concurrent use of PN may suppress appetite.

C) **TYPE OF ACCESS**

Central access is suggested over peripheral access since most marrow transplant patients require long-term support, and the desired caloric concentration is not achievable in a reasonable volume via the peripheral route.

D) **DELIVERY METHODS**

PN is usually infused continuously. A cyclic infusion (over 8-20 hr) can be used if the line must be free for drug or blood product administration or during the transition from PN to oral diet. Cyclic PN at night allows the patient greater mobility during the day and may facilitate improved oral intake. Cyclic PN may also put less stress on the liver. To avoid hyper- or hypoglycemia, cyclic PN solution should be gradually tapered up and down over the first and last hours of administration, respectively.

8.2

E) **SOLUTION DESIGN**

There is not a universal method for formulating PN. The hospital pharmacy PN system will dictate the method to be used. For example, the base PN solution can be a total nutrient admixture of amino acids-dextrose-lipid or an admixture of amino acids-dextrose with lipids separated. The design of PN can be based on a "custom formula approach" where the PN solution is tailored to nutrient needs or on a "standard formula approach" where nutrient needs are fitted to a standard solution. Additionally, the PN solution can be expressed on the basis of weight (e.g. grams of substrate) or on the basis of volume (e.g. final concentration of substrate, milliliters of substrate). Most hospitals have established policies, guidelines and protocols related to the administration of

Table 8.2.4. Modified bacteria diet for BMT patients

Diet	Description	Comments
Modified House Diet	1) Regular diet excluding fresh vegetables and most fresh fruits 2) Washed, freshly peeled, thick-skinned fruits are often acceptable	1) Diet contains a large number of pathogens from fresh dairy products, spices and other pathogen-containing foods 2) Use of general kitchen facilities 3) Offers variety and increased palatability 4) More commonly being utilized in transplant centers as there is a lack of data demonstrating decreased pathogen acquisition and infection with additional dietary restrictions
Low Microbial Diet (also called Low Bacteria or Cooked Food)	1) Foods with high gram (–) bacilli (> 500/ml of *Bacillus spp*), yeasts and molds are excluded 2) Allowed foods are well-cooked 3) Food handling and delivery techniques are modified to reduce the possibility of contamination after preparation 4) Commercial canned or packaged foods are used or portions are "first cut", wrapped and delivered soon after preparation	1) Specific foods allowed, and foodhandling practices can vary widely among transplant centers, as this diet is often based on practical knowledge of the distribution of pathogens in food rather than microbiological testing 2) Specialized kitchen or general kitchen facilities can be used 3) Used at centers with laminar air flow isolation
Sterile Diet	1) Food and water with no bacterial or fungal growth on culture 2) Foods are commercially canned or packaged or sterility is achieved by steam auto-claving, irradiation or oven baking 3) Ongoing microbiological testing 4) Food is prepared and assembled aseptically, using a laminar air hood 5) All tray service items and packaging are sterilized	1) Rarely used, even at centers with laminar air flow isolation 2) High meal cost 3) Need for specialized separate kitchen 4) Increased training needed for food service personnel 5) Food is less palatable due to changes in appearance, taste and texture from the sterilization processes

8.2

PN. Planning and daily decisions regarding PN administration should be made in a collaborative manner with the assistance of the pharmacist and dietitian. Table 8.2.5 summarizes the generic process for approaching PN administration.

The first step in designing a PN formulation is to develop goal nutrient needs that are patient-specific (Tables 8.2.2, 8.2.3). Based on the patient-specific needs, a PN formulation is developed to meet goals. In general, for BMT recipients, PN is designed to provide a 5% amino acid solution-20% dextrose solution with 30% of the total calories as lipids in a volume to meet maintenance fluid requirements. Initiating and advancing a PN regimen is different in children than in adults (Table 8.2.6), with a need to give incremental increases in all the macronutrients in children whereas, only dextrose needs to be incrementally increased in adults. Like children, debilitated adults need slower PN advancement to reach the goal formulation, to prevent possible re-feeding syndrome.

Multivitamins, trace metals, electrolytes and minerals should be added to the PN solution on a daily basis. Standard doses of micronutrients are initially added to PN (Table 8.2.7). Factors such as GVHD, antibiotics, metabolic stress, immunosuppressive drugs, diarrhea and vomiting may alter micronutrient requirements upward or downward, therefore, serum levels should be checked daily to make adjustments in PN. Table 8.2.8 gives an overview of guidelines for formulating PN that are specific to BMT recipients.

F) PN MONITORING

Equally as important as PN formula design, is the PN monitoring (Table 8.2.9). Monitoring PN minimizes the risk of metabolic and septic complications while maximizing the goal of attaining acceptable nutritional status.

G) TERMINATION OF PN / TRANSITION TO ORAL INTAKE

Oral intake improves as neutropenia and mucositis resolve. PN should continue until oral calorie intake exceeds 60% of estimated needs (approximately 1000 calorie/day in adults). PN should not be abruptly terminated, but should be reduced and/or cycled while oral intake is encouraged. When oral calorie intake meets 60% of estimated needs, it is safe to discontinue PN without tapering. Table 8.2.10 provides suggestions to encourage oral intake for BMT recipients with complications affecting oral intake.

VII) TUBE FEEDING (TABLE 8.2.11)

A) INDICATIONS

Low-volume tube feeding can provide gut stimulation and nutrition for gut mucosal regeneration. Goal-volume tube feeding can nourish the patient and provide gastric stimulation associated with normal eating. While tube feeding is not routinely used in the early transplant phase, it has been successfully used 30-40 days after marrow transplant to feed patients who fail to thrive on oral intake alone.

8.2

Table 8.2.5. Process for approaching parenteral nutrition administration

1) Check with the pharmacy to orient to the hospital PN system
2) Consult appropriate resource people for assistance—e.g. pharmacist,dietitian or nutrition support team
3) Develop patient-specific nutrient needs
4) Design patient-specific PN formulation
5) Develop PN progression schedule
6) Provide daily monitoring and adjustments to PN formulation

---

8.2

Table 8.2.6. Suggested parenteral nutrition progression schedule

Component	Initiation	Advancement	Maximum
Dextrose			
Older infant/child	10% final dextrose concentration (100 g dextrose/liter soln)	5% per day	35% final dextrose concentration (350 g dextrose/liter soln)
Adolescent	10% final dextrose concentration (100 g dextrose/liter soln)	5-10% per day	35% final dextrose concentration (350 g dextrose/liter soln)
Adults	10-15% final dextrose concentration (100-150 g dextrose/liter soln)	5-10% per day	35% final dextrose concentration (350 g dextrose/liter soln)
Protein			
Older infant/child	0.5-1.0 g/kg/d	0.5-1.0 g/kg/d	3.0 g/kg/d
Adolescent	1.0 g/kg/d	1.0 g/kg/d	2.5 g/kg/d
Adults	No need to progress. Start at goal.		2.5 g/kg/d
Lipids			
Older infant/child	1.0 g/kg/d	0.5-1.0 g/kg/d	3.0 g/kg/d (not to exceed 60% of total calories
Adolescent	1.0 g/kg/d	1.0 g/kg/d	2.5-3.0 g/kg/d (not to exceed 60% of total calories)
Adults	No need to progress. Start at goal.		Usually 1.5 g/kg/d, max 2.5 g/kg/d (not to exceed 60% of total calories)
Fluids			
Older infant/child	Start at goal unless severely malnourished or depleted.		2 x maintenance fluid needs*
Adolescent	(same as above)		2 x maintenance fluid needs*
Adults	(same as above)		2 x maintenance fluid needs*

* Identify clinical problems or circumstances that increase or decrease fluid needs. Take all fluids into account.

Table 8.2.7. Parenteral micronutrient requirement summary

	5-10 kg	10-30 kg	Adolescent	Adult
Sodium (mEq/kg/d)	2.0-4.0	2.0-4.0	2.0-3.0	1.0-2.0
Potassium (mEq/kg/d)	2.0-3.0	2.0-3.0	1.5-3.0	0.7-1.0
Calcium (mEq/kg/d)	0.5-1.0	0.5-1.0	0.5-1.0	0.2-0.4
Magnesium (mEq/kg/d)	0.3-0.6	0.3-0.6	0.2-0.3	0.1-0.3
Phosphorus (mM/kg)	1-2	1-2	1-1.3	0.33-0.67
Trace Elements				
Ped (ml/kg/d)	0.2	0.2	0.2 (10 ml max)	—
Adult (ml/d)	—	—	—	—
Multivitamins				
Adult(ml/d)**	—	—	10	10
Selenium (µg/d)	2.0 (40 µg/max)	2.0 (40 µg/max)	2.0 (40 µg/max)	40

**Does not contain vitamin K

Potassium and sodium are available as phosphorus, chloride or acetate salts. Calcium is available as gluconate. Magnesium is available as sulfate. Sodium or potassium acetate may be substituted for the respective chloride salt for patients with metabolic or mixed acidosis. In vivo, acetate is metabolized to bicarbonate. Sodium bicarbonate may not be added to dextrose-amino acid-electrolyte solutions.

B) CONTRAINDICATIONS

Tube feeding should not be used in BMT recipients with severe mucositis, esophagitis, ileus (due to GVHD and/or use of narcotics for mucositis) or diarrhea. The risk of aspiration in critically ill and severely immuno-compromised patients with refractory nausea and vomiting also precludes its use. Tube feeding should not be attempted in a thrombocytopenic patient who cannot achieve a platelet count of 50,000 µl during, and sometimes after, the tube feeding access procedure. If there is concern about a patient's ability to maintain acceptable bacterial-risk precautions in preparing and delivering tube feeding, especially in the outpatient setting, then the tube feeding route should not be chosen.

C) TYPES OF ACCESS

A naso-enteric tube into the small bowel can be used for short-term support and a gastrostomy tube or jejunostomy tube can be used for long-term support.

D) DELIVERY METHOD

Continuous feeding is often better tolerated than intermittent or bolus feeding, but the latter two can be used successfully. Nocturnal tube feeding is often used in children.

8.2

Table 8.2.8. Guide for formulating parenteral nutrition in BMT patients

Component	Marrow Transplant Considerations

Carbohydrate (dextrose)	• Final dextrose concentration of 20% to 35%
	• Maximum rate of glucose oxidation:
	  infants/children = 12-14 mg/kg/min
	  adolescents/adults = 4-6 mg/kg/min
	• Provides 45-55% of total kilocalories
	• Carbohydrate intolerance is associated with steroid treatment, infection and high flow rate parenteral nutrition and may require insulin support
	• Carbohydrate intolerance may also be associated with liver enzyme abnormalities and pulmonary dysfunction

Protein (amino acids)	• Standard amino acid solutions (8.5-10% initial concentration) are generally used • Branched chain amino acid solutions can be considered in hepatic encephalopathy with veno-occlusive disease, especially if the plasma amino acid is abnormal • Concentrated amino acid solution (15% initial concentration) can be considered to maximize protein intake but minimize fluid volume • Concentrated amino acid solution has a higher level of phenylalanine (BMT recipients have been shown to already have increased phenylalanine levels), therefore this solution may effect the amino acid profile and be contraindicated in liver/kidney failure • Recent research indicates that the addition of glutamine to parenteral nutrition may be beneficial in promoting nitrogen balance, decreasing the incidence of infection, decreasing mucositis and improving fluid balance in BMT recipients • Complications of substrate intolerance include azotemia, hyperammonemia and liver enzyme abnormalities
Lipid (emulsion of soybean or safflower oil)	• High-density calorie source in patients with glucose/fluid intolerance • Most marrow transplant centers recommend no more than 30% of total calories from lipids (range 10-30%). Up to 60% of total calories can be given as lipids in patients who have ventilatory problems due to high carbon dioxide production. The recommendation to favor moderate lipid calories is based on research that suggests that large amounts of linoleic acid from lipid emulsion are immunosuppressive. • Lipid intolerance may be associated with hypertriglyceridemia and liver function abnormalities
Fluid	Consider oral and other intravenous solutions when determining parenteral requirements. Concentrated parenteral nutrition solution may be used to minimize the fluid load.
Micronutrients	Provide standard daily recommendations of multivitamins and trace elements • Vitamin K is routinely added to adult solutions, unless contraindicated, since it is not included in adult parenteral multivitamin preparations • Monitor and correct all electrolyte deficiencies or excesses as needed • Phosphorus, calcium and potassium can become depleted rapidly, because of poor intake, medications or malabsorption • Supplement with zinc (20 mg/d) for stool output greater than 500 ml Supplement with chromium (20-40 µg/d) for glucose levels greater than 200 mg/dl • Provide 40 µg/d selenium • Iron is not supplemented since marrow transplant patients get frequent transfusions and iron may contribute to oxidant injury of organs
Other additives	• Insulin may be added for glucose levels greater than 200 mg/dl • H2-antagonists (e.g. famotidine 40 mg/d or rantidine 150 mg/d) may be added to parenteral solutions for stress ulcer prophylaxis • Heparin is not usually added to the parenteral solution because of the risk of inducing thrombocytopenia • The pharmacist should be consulted for any other additives/medications to assess compatibility

8.2

Table 8.2.9. Suggested monitoring parameters for PN

Parameters	Initial *	Daily	Twice Weekly	Weekly
Weight	X	X		
Fluid Balance (intake and output)	X	X		
Pre-albumin	X		X	
Electrolytes	X	X		
Magnesium	X		X	
Calcium	X			X
Phosphorus	X			X
Glucose	X	X		
Liver Function Tests	X		X	
Triglycerides	X			X
Bun and Creatinine	X	X		
Nitrogen Balance	as indicated			as indicated

* Initial = Initiation of parenteral nutrition

Table 8.2.10. Eating problems and suggested interventions

Problem	Suggested Diet Interventions
Nausea/ Vomiting	Frequent, small meals; cool or room temperature foods; increased fluid intake, especially clear cool liquids; eat and drink slowly; limit activity after meals **Avoid:** Strong odors; overly sweet foods; high fat foods; high fiber foods and roughage
Mucositis/ Esophagitis/ Stomatitis	High protein, high calorie supplements; cool or room temperature foods; bland and soft/puree foods **Avoid:** Rough and raw foods; highly seasoned foods, citrus and tomato products; highly salted foods
Xerostomia (Dry Mouth)	Soft, moist foods; increased fluid intake; added gravy, sauces or broth; sugar-free hard candies; high protein, high calorie supplements; artificial saliva **Avoid:** Dry foods; bread products; oily foods; excessively hot foods; alcohol
Thick, Viscous Mucus Production	Increased fluid intake; beverages such as club soda, hot tea with lemon; sugar-free hard candies **Avoid:** Thick liquids; oily foods; milk or milk products may be difficult to swallow and cause perceived phlegm build-up
Dysgeusia/ Hypogeusia (Taste Alterations/ Mouth Blindness)	Encourage variety; use aroma to stimulate appetite; highly spiced and strongly flavored food, as tolerated; high calorie, high protein supplements **Avoid:** Bland foods; unsalted food; red meat; coffee; tea
Diarrhea/ Steatorrhea	Increased fluid intake; active-culture yogurt to increase bowel flora; low osmolar fluids; light foods; MCT oil when steatorrhea is documented **Avoid:** High fat foods; high fiber foods and roughage, gasforming foods, lactose-containing foods
Anorexia/ Early Satiety	Frequent, small meals; high protein, high calorie supplements; calorically dense foods; planned meal times; emphasis on presentation of foods and pleasant mealtime atmosphere **Avoid:** Noncaloric liquids at meals; strong odors

8.2

8.2

Table 8.2.11. Examples of enteral tube feeding formulas

	Osmolite HN	Jevity	Respalor	TwoCal HN	Sustacal PLUS	Perative	Vital HN	Nepro	Suplena	Nutrihep
	Low residue, high nitrogen	Fiber, high nitrogen	Trauma, pulmonary	High density	PO supplement	Semi-elemental	Semi-elemental	Renal	Renal	Hepatic
cal/cc	1.06	1.06	1.5	2.0	1.5	1.3	1.0	2.0	2.0	1.5
protein (g/L)	44.4	44.4	76	83.7	61	66.6	41.7	70	30	40
carb (g/L)	141	152	148	217	190	177	185	215	255	290
fat (g/L)	37	36	71	90.9	58	37.4	10.8	96	95	21
cal/N_2 ratio	150:1	150:1	126:1	150:1	160:1	122:1	150:1	179:1	418:1	235:1
NPC/N_2 ratio	125:1	125:1	102:1	125:1	134:1	97:1	125:1	154:1	393:1	210:1
sodium (mEq/L)	40	40	55	57	37	45	20	36	34	14
potassium (mEq/L)	40	40	38	63	38	44	34	27	28.6	34
phosphorus mg/L	746	298	710	1052	850	858	667	686	728	1000

	Osmolite HN	Jevity	Respalor	TwoCal HN	Sustacal PLUS	Perative	Vital HN	Nepro	Suplena	Nutrihep
magnesium (mg/L)	298	298	280	421	340	343	267	211	211	400
cc to meet RDI	1321	1321	1440	950	1180	1155	1500	950	950	1000
free water (cc/L)	842	835	770	712	780	789	867	703	712	760
osmolarity (mOsm/kg)	300	300	580	690	600	425	500	635	600	690
Fiber (g/L)	0	14.4	0	0	0	0	0	0	0	0
indications	a standard house formula	maintenance of normal bowel function	elevated calorie/ protein needs, volume restriction	elevated calorie/ protein needs, volume restriction	high calorie/ protein needs, volume restriction, low potassium	critically malab-sorption, maldi-gestion	low fat	renal failure, dialysis, volume restriction, low potassium	nondialyszed renal failure, volume restriction, low potassium	hepatic failure with enceph-alopathy

cc = cubic centimeter
cal = calories
carb = carbohydrates
g = gram
kg = kilogram

L = liter
mEq = milliequivalents
mOsm = milliosmole
N_2 = nitrogen
NPC = nonprotein calories
RDI = recommended daily allowance

8.2

E) **FORMULA SELECTION**

The use of a small peptide/elemental formula or isotonic polymeric formula may facilitate tolerance and minimize gastrointestinal dysfunction. A concentrated (2 cal/ml) formula can be tried if tube feeding tolerance is volume dependent. The tube feeding formula should also ideally be lactose-free and have a high nitrogen content. Consult the dietitian for specific enteral products that meet these criteria at each hospital (Table 8.2.11).

F) **PROGRESSION SCHEDULE**

Begin the tube feeding at an isotonic (300 mOsm) dilution and low rate (10-20 ml/hr). Increase to full strength, if not isotonic, and incrementally increase rate to goal.

G) **MEDICATIONS TO IMPROVE TOLERANCE**

Antiemetic, antidiarrheal and promotility agents should be aggressively used to treat BMT-related gastrointestinal side effects that may interfere with tube feeding.

H) **BACTERIAL-RISK PRECAUTIONS**

Measures to reduce the risk of infection from bacterial contamination of enteral products should be undertaken. Only commercially produced tube feeding products should be used and, if necessary, should be diluted/reconstituted with sterile water using aseptic preparation technique. A closed tube feeding system could also be considered to limit bacterial exposure. Tube feeding should hang for no more than 8 hours.

VIII) **POSTTRANSPLANT COMPLICATIONS AFFECTING NUTRITIONAL STATUS**

A) **HEPATIC VENO-OCCLUSIVE DISEASE (VOD)**

Fluid and sodium intake need to be restricted in response to sudden weight gain or increasing abdominal girth. Sodium intake should be reduced in PN, oral intake and antibiotics. Antibiotics should be administered in minimal dextrose solutions. Fluid intake should be minimized by concentrating PN, minimizing flushes and restricting oral fluid intake. Restriction of protein should be based on the severity of symptoms, with minimal protein restriction for mild VOD to a severe protein restriction when encephalopathy occurs. If PN is used, a branched chain amino acid solution can be considered if the plasma amino acid profile is abnormal or there is no response to reduction in protein. Stopping the PN amino acid infusion has been suggested to manage severe encephalopathy. There is a controversy in the literature as to whether intravenous fat is effectively utilized in patients with liver failure. Patients with VOD who receive fat emulsion should have triglyceride levels monitored closely to check fat clearance.

B) **ACUTE GASTROINTESTINAL GVHD**

Gastrointestinal involvement can result in nausea, vomiting, anorexia, abdominal cramping and pain, secretory diarrhea, bloody stools, altered intestinal motility, malabsorption and ileus. Optimal nutritional support is an important complement to the therapy of acute gastrointestinal GVHD. A 5 stage nutrition plan, as developed by Gauvreau and colleagues, provides empiric guidelines for progression of nutritional support. Advancement through the stages is based on improvement of clinical symptoms and diet tolerances (Table 8.12).

8.2

Table 8.2.12. Gastrointestinal GVHD nutrition plan

Stage	Oral Diet	Parenteral Nutrition
1	Bowel rest-NPO	Calculate solution formulation using stress calorie and protein needs
2	Liquids (isotonic, lactose-free, low residue); allow 60 ml every 2-3 hour	same as stage 1
3	Introduce solid foods, but remain low lactose, low fiber, low fat (20-40 g/d) low acidity, low gastric irritants; multiple, small feedings every 3-4 hours	same as stage 1
4	Solid foods as in stage 3, but fat intake is slowly increased	Taper PN as oral intake increases
5	Advance to regular diet by adding restricted foods, one per day, to assess tolerance	Discontinue PN when oral intake meets needs

Source: Adapted from Gauvreau JM, Lenssen P, Cheney CL, Aker SN, Hutchinson ML, Barale KV. Nutritional Management of patients with intestinal graft-versus-host-disease. JADA 79: 673-675, 1981. (used with permission)

C) CHRONIC GVHD

The nutrition-related problems associated with chronic GVHD can occur in multiples and have long-range effects. These problems include oral sensitivity/stomatitis, xerostomia, anorexia/poor oral intake, reflux symptoms, dysgeusia, weight loss and steroid-induced nitrogen loss, weight gain, diabetes and fluid retention. The specific nutrition intervention will depend on the BMT recipient's nutrition-related problem(s). Oral diet should be modified to manage eating problems (Table 8.2.10). If oral intake is depressed, nutritional supplements should be given. Tube feeding or PN may be indicated in patients who are unable to maintain acceptable nutritional status by the oral route. Management of the side effects associated with high-dose steroid treatment may include increasing protein intake to 1.5-2.0 g/kg IBW, providing calcium/vitamin D intake to 1-1.5 times the recommended dietary allowance, restricting concentrated carbohydrate intake if glucose level is elevated and restricting sodium for fluid retention.

IX) OUT-PATIENT NUTRITIONAL MANAGEMENT

A multi-vitamin supplement should be given for 1 year following BMT. After discharge from the hospital, ongoing nutritional assessment and monitoring by the dietitian should be provided to all BMT recipients on a weekly basis until day 100, and yearly thereafter. Nutritional care should focus on attaining or maintaining acceptable nutritional status. In children, issues related to growth and development should be addressed. Optimizing oral intake is the first line approach, however, tube feeding or PN may be indicated in BMT recipients who, despite intervention, fail to thrive.

8.2

SUGGESTED READING

1. Cunningham BA, Lenssen P, Aker SN et al. Nutritional considerations during marrow transplantation. Nurs Clin North Am 1983; 18:585-596.
2. Gauvreau JM, Lenssen P, Aker SN et al. Nutritional management of patients with intestinal graft-versus-host-disease. JADA 1981; 70:673-675.
3. Herrmann VM, Petruska PJ. Nutritional support in bone marrow transplant recipients. NCP 1993; 1:19-27.
4. Lenssen P, Aker AN eds. Nutritional assessment and management during marrow transplantation: A resource manual. Seattle, 1985.
5. Moe G. Enteral feeding and infection in the immunocompromised patient. NCP 1991; 6(2):55-64.

8.2

Management of Pain in the Bone Marrow Transplant Patient

F. Peter Buckley

I) INTRODUCTION

Patients receiving bone marrow transplants (BMT) may suffer pain of varying types and severities during the course of their BMT. Common causes of pain include:

1) Primary disease, e.g. bone infiltration by tumor
2) Chemoradiotherapy, e.g. mucositis
3) Immunosuppressive regimes, e.g. cyclosporine neuropathy
4) Opportunistic infections, e.g. herpes zoster
5) Diagnostic procedures, e.g. bone marrow aspirations
6) Surgical procedures, e.g. line placement
7) Pharmacologic treatments, e.g. bone pain from growth factors
8) Common pain problems unrelated to the patient's disease or therapy, e.g. headache, low back pain.

Patients presenting for BMT have often been ill for long periods of time; their physical, psychic and familial resources may be depleted; and they face a prolonged, arduous and perilous course of treatment. Hence patients may be in poor shape to cope with the demands and aversive experiences, including pain, experienced during a BMT. Effective pain relief is important to help preserve the patient's ability to perform essential functions, especially those whose loss may lead to complications, e.g. inadequate oral hygiene leading to oral infection.

II) ANATOMY, PHYSIOLOGY AND PHARMACOLOGY OF NOCICEPTION

A brief summary of nociceptive transmission and its modulation is helpful in understanding pain and its treatment.

A) AFFERENT NOCICEPTIVE TRANSMISSION

Nociceptors signal centrally in response to physical or biochemical stimulation and may be sensitized by the algogenic products of tissue damage to signal at sub-nociceptive levels of stimulation. Nociceptive traffic is transmitted via peripheral nerves which synapse in the dorsal horn of the spinal cord, cross the midline and ascend to cerebral and subcerebral structures. This traffic follows two routes. A "fast pathway" via the A delta fibers direct to the thalamus and cerebrum which is responsible for "first pain", e.g. the initial, brief sharp pain with a burn. The second, "slow pathway" via C fibers ascends through a multisynapse pathway and is responsible for the "second pain", e.g. the persistent burning pain after a burn. Because of its multiple synapses, particularly with the reticular formation and limbic system, this pathway initiates the complex autonomic and affective component of nociception.

B) NOCICEPTION MODULATING SYSTEMS

The afferent nociceptive system can be modulated by:

1) **Alteration of the Type, Quantity or Effects of Alogogenic Substances**

This may be produced by modulating peripheral tissue inflammation, e.g. aspirin.

8.3

Bone Marrow Transplantation, edited by Richard K. Burt, H. Joachim Deeg, Scott Thomas Lothian, George W. Santos. © 1996 R.G. Landes Company.

2) Peripheral Inhibition of Nociceptive Transmission

Inhibition occurs in the dorsal horn by peripherally generated stimuli, e.g. rubbing, transcutaneous electrical stimuli carried to the dorsal horn by large diameter fibers.

3) Central Inhibition of Nociceptive Transmission

Inhibition may occur in the dorsal horn by activation of the descending modulating system, which originates in the hindbrain. This multisynapse system has serotonin, epinephrine and endogenous opioids (morphine-like) as its neurotransmitters. This system can be activated by:

1) Learned behavioral or cognitive strategies

2) Pharmacologic agents that stimulate this system, e.g. serotonin enhancement by tricyclic antidepressants; or mimic the activity of the system, e.g. opioids

3) Direct electrical stimulation of the system via implanted electrodes—not a choice for the BMT patient.

C) GATE CONTROL THEORY OF PAIN

The modulation of nociceptive traffic is the physiologic expression of the "Gate Control Theory" of pain. This theory states that there are various gating systems within the CNS which can be opened or closed to nociceptive transmission by behavioral or pharmacologic interventions. The gate theory is a useful concept to bear in mind when attempting to alleviate pain, and to teach patients using both pharmacologic and nonpharmacologic pain relief techniques.

III) MONITORING PAIN AND TREATMENT EFFICACY

The amount of pain BMT patients suffer often changes rapidly, thus pain and treatment efficacy should be evaluated at least daily. Elements of this evaluation should include:

A) HISTORY OF THE PATIENT'S PAIN

Has the patient any long-standing history of pain unrelated to their disease (headache, low back pain)? Has the patient any history of pain associated with the current disease or its therapies? Has the patient had any surgical procedures? How successfully was the postoperative pain managed? What non-pharmacologic strategies, if any, has the patient used to cope with pain, or other aversive phenomena? What pharmacologic agents has the patient taken to relieve pain or other aversive phenomena? How effective were those agents? Did the patient experience any side effects? Does the patient have a history of emotional or affective disorders (anxiety, depression)? Does the patient have a history of CNS active substance use (opioids, anxiolytics, alcohol)? What is the patient's understanding of the current pain, its significance, its likely duration? What is the patient's preferred mode of dealing with the current pain, both pharmacologic and nonpharmacologic?

B) PAIN SCORE

Pain scores may not accurately reflect nociception, but are useful guides to the efficacy, or lack of efficacy, of analgesic regimens. For children the faces scale (Fig. 8.3.1) is more useful than using a scale of 1-10.

C) DEGREE OF FUNCTIONAL IMPAIRMENT

Is this a consequence of the patient's disease process? Is this a consequence of inadequate pain relief? Is it a consequence of analgesic therapy, e.g. excessive sedation with opioids?

D) SIDE EFFECTS

Are there any side effects of analgesic therapy, and how severe are they?

8.3

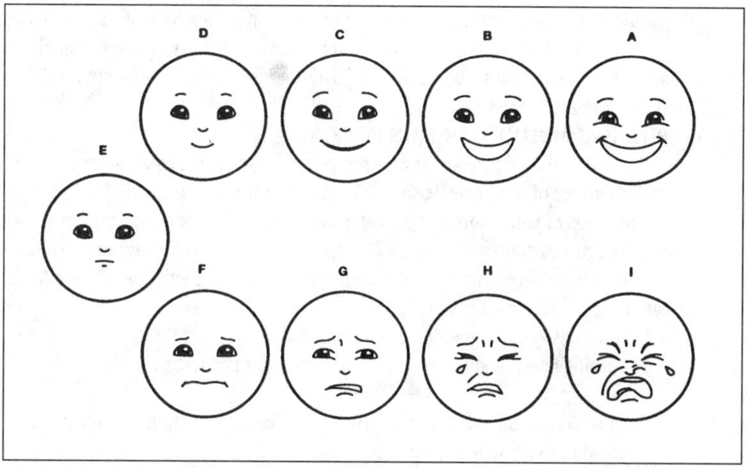

Fig.8.3.1. Faces scale for a child to score severity of pain. Reprinted with permission from McGrath et al. Advances in Pain Research and Therapy. Fields ML et al, eds. 1993; 9:391 © Raven Press.

E) PRESENCE OF DISEASE

For example, is there any renal or hepatic dysfunction, or therapy (e.g. sedation from antiemetic therapy) which may influence analgesic therapy or its side effects?

F) ANALGESIC CONSUMPTION

Is analgesic use rising or falling? Is that consistent with the expected course of the pain (see later)? Does the rise/fall in consumption correlate with improving/worsening pain control/side effects?

G) PATIENT SATISFACTION

Though complete pain relief is a worthy goal in the BMT patient, it is rarely achieved. Patients usually opt for a level of pain that they balance against the side effects of therapy and their psychic comfort at using analgesics. Thus the physician may encounter a patient who gives a pain score of 8/10, a pain level at which one might expect the patient to desire further efforts at pain relief. However, this may not be the case. The patient may be quite satisfied with where they are. To help discern how to continue with the current regime, or use more aggressive therapies, supplementary questions are often helpful: "Tell me how satisfied you are with pain control at present using a 1-10 scale where 1 = completely dissatisfied and 10 = completely satisfied". "In what ways do you think we can do a better job of providing pain control?" It is not unusual to find a patient with a 8/10 pain score give high satisfaction ratings, be quite content to continue with the current analgesic regime and desire no changes in analgesic regime.

IV) COGNITIVE BEHAVIOR TECHNIQUES OF PAIN RELIEF

Nonpharmacologic coping behavior (imagery, muscle relaxation, self hypnosis, active distraction, education) can provide, or enhance analgesia by acting physiologically in various ways, such as impeding the transmission of nociceptive information by activation of the descending modulatory systems or impeding transmission of nociceptive information by competing sensory messages from the limbs or body surface.

8.3

Patients who have moderate anxiety, a moderate need for control and who have not devised a BMT coping strategy benefit the most. Patients who are very anxious and who expect staff to be in control have the most difficulty in learning these skills and benefit least.

V) PHYSICAL METHODS OF PAIN CONTROL

Many patients, particularly those who wish to have some elements of control of their methods of pain relief, wish to use, and get substantial pain relief from physical methods of pain control, e.g. heat, cold, massage. Provided such methods of control are not contraindicated, e.g. massage in the thrombocytopenic patient, they should be encouraged and allowed. Specific pain problems and the methods often used for their relief include:

1) Mucositis—cold saline rinses, ice packs to face, jaw and throat

2) Burning hands and feet—cold, wet towels and a fan, ice packs to the affected areas

3) Musculoskeletal pains—heating pads or ice packs to the affected areas. Patients will usually state their preferences.

VI) NONSTEROIDAL ANTI-INFLAMMATORY DRUGS (NSAIDS) (TABLE 8.3.1)

Nonopioids are suitable for solo use in mild to moderate pain, and when combined with opioids for moderate to severe pain, improve analgesia and produce an opioid sparing effect, i.e. reduce the dose of opioid needed to produce a given analgesic effect.

A) MECHANISM OF ACTION OF NSAIDS

The effect of NSAIDs is to inhibit the enzyme cyclo-oxygenase, part of the arachodonic acid cascade, reducing the release of prostaglandins, one of a number of algogenic substances released by tissue damage, which may stimulate or sensitize nociceptors. Classically, the effect of NSAIDs has been believed to be entirely peripheral. However, evidence from both animal and human studies implies that prostaglandins have a role in nociceptive transmission in the spinal cord and that spinally applied NSAIDs may influence nociceptive transmission.

B) PHARMACOLOGY OF NSAIDS

Nonopioid analgesics have an efficacy ceiling. Increasing blood levels of drug up to a certain point produces increasing analgesia, but increasing blood levels of drug further produces no improvement in analgesia. This is a point to emphasize to patients who have access to over the counter drugs and who may escalate dosage, beyond that recommended, in an effort to obtain further pain relief.

C) DIFFERENCES BETWEEN NSAIDS

There is little hard data on the comparable efficacy of the various drugs. The differences between drugs are in the realms of duration of action, incidence of side effects and the individual patient's tolerance.

D) ROLE OF NSAIDS IN BMT

During BMT the role of nonopioids is limited by the hematologic, renal and gastrointestinal side effects and by the fact that few are available in a parenteral formulation. Ketorolac is a recently introduced NSAID which may be administered both orally and parenterally. It has only been used in short-term trials in postoperative pain but appears to be potent; 30-45 mg IM ketorolac produces analgesia of similar intensity and duration to 8-10 mg of morphine IM. Ketorolac has been given after bone marrow harvest to decrease pain at the harvest site.

8.3

E) SIDE EFFECTS OF NSAIDS
   1) **Interference with Platelet Function**
      Virtually all NSAIDs impair platelet function often for prolonged periods, choline magnesium trisilicate and salsalate allegedly have the least platelet effects. Acetaminophen has virtually no hematologic effects.
   2) **Gastric Irritation and Upper GI Bleeding**
      This is a feature of all NSAIDs, choline magnesium trisilicate and salsalate allegedly being the most benign.
   3) **Renal Effects**
      In high dosage, NSAIDs have direct renal toxic effects, the effects of modest dosage on patients with moderate renal compromise is not clear.

Table 8.3.1. Dosing data for nonopioid analgesics

Drug	Usual Dose for Adults and Children ≥ 50 kg Body Weight	Usual Dose for Adults and Children ≤ 50 kg Body Weight
Acetaminophen and oral over-the-counter NSAIDs		
Acetaminophen	650 mg q 4 h 975 mg q 6 h	10-15 mg/kg q 4 h 15-20 mg/kg q 4 h (rectal)
Aspirin	650 mg q 4 h 975 mg q 6 h	10-15 mg/kg q 4 h 15-20 mg/kg q 4 h (rectal)
Ibuprofen (Motrin, others)	400-600 mg q 4 h	10 mg/kg q 6-8 h
Prescription oral NSAIDs		
Choline magnesium trisilicate (Trilisate)	1000-1500 mg TID	25 mg/kg TID
Choline salicylate (Arthropan)	870 mg q 3-4 h	
Diflunisal (Dolobid)	500 mg q 12 h	
Ketorlac tromethamine (Toradol)	10 mg q 4-6 h to a maximum of 40 mg/day not to exceed 5 days	
Naproxen (Naprosyn)	250-275 mg q 6-8 h	5 mg/kg q 8 h
Prescription parenteral NSAIDs		
Ketorolac tromethamine (Toradol)	60 mg initially, then 30 mg q 6 h; intramuscular dose not to exceed 5 days	

Adapted from: Management of Cancer Pain. Clinical Practice Guideline No. 9. Agency for Health Care Policy and Research, Public Health Services.

8.3

4) Hepatic Effects
In high dosage, acetaminophen is hepatotoxic.

5) Antipyretic Effects
All nonopioids posses this facility to some extent. Suppression of fever, a cardinal sign in the detection of infection, may be unwanted in the BMT patient.

VII) TRICYCLIC ANTIDEPRESSANTS (TCA) FOR PAIN CONTROL
TCAs are clearly effective in relieving some types of neuropathic pain, e.g. diabetic neuropathy, post-herpetic neuralgia. They have a prophylactic role in treatment of common daily headache and migraine. Their analgesic efficacy for chronic low back pain or fibromyalgia is not clear, but they do improve sleep in these conditions.

A) TCA MECHANISM OF ACTION
TCAs are believed to produce their effects by altering levels of serotonin (ST) and norepinephrine (NE) in the descending nociceptive modulating system. Animal studies imply that ST selectivity is the important factor but human studies show that drugs with both ST and NE enhancement are efficacious.

B) TCA PHARMACOLOGY
Tricyclic antidepressants have analgesic effects, at blood levels of drug and dosage lower than those needed to treat depression. The most effective drugs are amitryptiline and desimpramine, with effective dosage ranging from 10-100 mg per day. TCAs have long half lives hence once a day dosage, usually at bedtime to minimize sedative side effects, is satisfactory. Tricyclics have a slow onset of effect, 1-3 days for pain control, 7-14 days for depression.

C) ROLE OF TCA IN BMT
There is no data on the use of TCAs for appropriate pain in BMT patients as there is considerable concern about the marrow suppressant effects of TCAs. If marrow suppression is not a concern they are a reasonable modality to consider for some types of neuropathic pain associated with BMT, e.g. cyclosporine neuropathy.

VIII) OPIOIDS FOR PAIN CONTROL

A) OPIOID MECHANISM OF ACTION
Opioids produce their effects by binding to specific cell membrane receptors (Table 8.3.2). Opioids that bind to receptors and produce effects are termed agonistic, e.g. morphine. Opioids that bind to receptors and produce no effect, or antagonize the effect of previously administered opioid, e.g. naloxone, are termed antagonists. Other opioids, which are agonistic at one receptor but antagonists at another, e.g. nalbuphine—a mu antagonist and kappa agonist, are termed agonist/antagonists . Opioids produce analgesia by inhibiting nociceptive input in the dorsal horn (epidural administration); activating the descending nociceptive inhibitory system (systemically administered opioids); and activating the limbic and reticular systems which alters the emotional and affective response to pain.

B) OPIOID PHARMACOLOGY
When used to provide analgesia, the limiting factor is often the occurrence of unwanted dose-dependent effects (sedation, respiratory depression). The dose/clinical response curve of opioids is depicted in Figure 8.3.2. Note the wide standard deviations of the means in Figure 8.3.2 indicating that individual patients need widely differing plasma levels of drug to produce a given effect.

8.3

Table 8.3.2a. Opiate receptor effects

Receptor			
μ	κ	σ	δ
Morphine β Endorphin	**Bremazocine Dynorphin**	**N-allylcyclazocine**	**Morphine Leuenkephalin**
Analgesia	Analgesia	No analgesia	No analgesia
Apnea (?)	Apnea ±	Tachypnea	Apnea ++
Indifference	Sedation	Delirium	?
Miosis	Miosis	Mydriasis	
Nausea and vomiting			Nausea and vomiting
Constipation			
Urine retention	Diuresis		
Pruritis	No change	No change	Pruritus
δ Cross-tolerance	No cross-tolerance		μ Cross-tolerance

Table 8.3.2b. Opiate receptor interactions

Drug	μ	κ	σ	δ
Morphine	+++	+		++
Fentanyl	++++			±
Meperidine	++			++
Hydromorphone	+++			++
Naloxone	───	─	─	─
Nalbuphine	───	+++	+	
Butorphanol		++	++	

+ = agonist effect; - antagonist effect
Adapted with permission from: Benedetti C. Acute pain: A review of its effects and therapy with systemic opioids. In: Benedetti C, Chapman CR, Giron G eds. Opiod Analgesia: Recent Advances in Systemic Administrration. (Advances in Pain Research and Therapy, vol 14.) New York: Raven Press 190; 367-424.

8.3

When providing analgesia with opioids, a useful concept is the minimum effective analgesic concentration (MEAC). This is the blood level of opioid at which the patient requests further pain relief and may be defined experimentally. In Figure 8.3.3 the MEAC would be at the bottom of the shaded area. In an individual patient with a given degree of nociception, MEAC, determined by repetitive experimental measurements, is very consistent, with a standard deviation of about 5-10%. However, between patients with a given degree of nociception, the MEAC will vary 4-5 fold. This variation in MEAC plus the interpatient variability of opioid pharmacokinetics means that for individual patients there is a wide variation in the dose requirement to maintain blood levels of drug at or above MEAC.

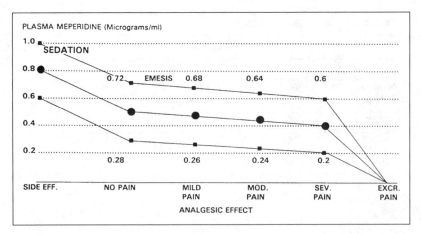

Fig 8.3.2. Mean plasma level of meperidine plus or minus one standard deviation the mean plotted against pain relief. An analgesic effect is not seen until a mean plasma level of meperidine of 0.4 µg/ml. As plasma levels rise, analgesia improves linearly to a mean plasma level of 0.5 µg/ml, when pain disappears. Raising plasma levels of drug beyond this point results in undesirable dose limiting side effects, such as nausea and sedation. Modified from Benedetti C. Acute pain: A review of its effects and therapy with systemic opioids. Reprinted with permission In: Benedetti C, Chapman CR, Giron G. eds. Advances in Pain Research and Therapy. New York: Raven Press 1990; 14:367-424.

C) MODES OF OPIOID ADMINISTRATION (FIG. 8.3.3, TABLE 8.3.3)

The basic precept of opioid administration is to give the patient sufficient drug, at appropriate time intervals, so as to maintain blood levels of drug within the therapeutic window portrayed in Figure 8.3.3.

1) Oral Administration of Opioids

Opioids are well absorbed by the gut but undergo considerable first pass metabolism by the liver to the extent that to produce a given blood level of drug the oral dose may be two to six times the parenteral dose (Table 8.3.3). After oral administration blood levels of drug rise slowly, reach the effective range in about 30-40 minutes and stay within the therapeutic window for about 3-4 hours. Thus repeat dosing will be necessary every 3-4 hours. The period of time during which blood levels of drug stay within the therapeutic window may be prolonged by the use of sustained release preparations. Although oral opioids are appropriate drugs to use in the BMT population, their use is somewhat restricted by the patients' inability to retain the drug (nausea and vomiting) or absorb the drug (gut GVHD).

2) Rectal Administration of Opioids

The time profile and duration of analgesia produced by rectally administered opioids is similar to that produced by oral opioids. This mode of administration is useful when patients are unable to take by mouth. It is seldom used in the BMT patient for fear of producing a bacterial shower when the suppository is inserted.

8.3

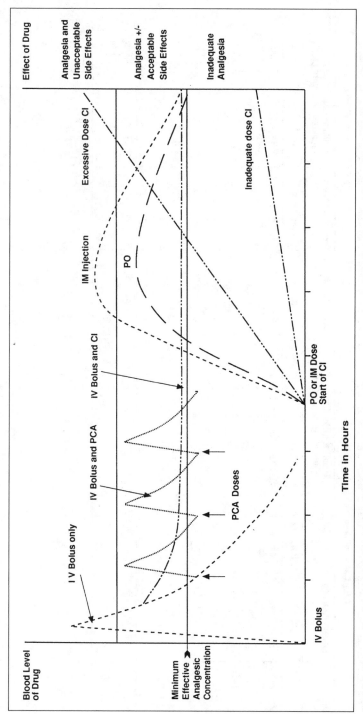

Fig 8.3.3. *Pharmacokinetics of opioid administration. The "therapeutic window" of opioids is the range of blood levels of opioids within which they produce satisfactory pain relief without unacceptable side effects. CE = continuous infusion; IM = intramuscular; IV = intravenous; PCA = patient controlled analgesia; PO = oral.*

8.3

8.3

Table 8.3.3. Dose equivalents for opioid analgesics

Drug	Adults and Children ≥50 kg Body Weight				Adults and Children ≤50 kg Body Weight			
	Approximate equianalgesic dose		Usual starting dose for moderate to severe pain		Approximate equianalgesic dose		Usual starting dose for moderate to severe pain	
	Oral	Parenteral	Oral	Parenteral	Oral	Parenteral	Oral	Parenteral
Morphine	30 mg q 3-4 h (repeat around-the-clock-dosing) 60 mg q 3-4 h (single dose or intermittent dosing)	10 mg q 3-4 h	30 mg q 3-4 h	10 mg q 3-4 h	30 mg q 3-4 h (repeat around-the-clock-dosing) 60 mg q 3-4 h (single dose or intermittent dosing)	10 mg q 3-4 h	0.3 mg/kg q 3-4 h	0.1 mg/kg q 3-4 h
Morphine controlled-release	90-120 mg q 12 h	N/A	90-120 mg q 12 h	N/A	90-120 mg q 12 h	N/A	N/A	N/A
Hydro-morphone (Dilaudid)	7.5 mg q 3-4 h	1.5 mg q 3-4 h	6 mg q 3-4 h	1.5mg q 3-4 h	7.5 mg q 3-4 h	1.5 mg q 3-4 h	0.06 mg/kg q 3-4 h	0.015 mg/kg q 3-4 h

Drug	Adults and Children ≥ 50 kg Body Weight				Adults and Children ≤ 50 kg Body Weight			
	Approximate equianalgesic dose		Usual starting dose for moderate to severe pain		Approximate equianalgesic dose		Usual starting dose for moderate to severe pain	
	Oral	Parenteral	Oral	Parenteral	Oral	Parenteral	Oral	Parenteral
Meperidine (Demerol)	300 mg q2-3 h	100 mg q3 h	N/R	100 mg q3 h	300 mg q2-3 h	100 mg q3 h	N/R	0.75 mg/kg q2-3 h
Methadone	20 mg q6-8 h	10 mg q6-8 h	20 mg q6-8 h	10 mg q6-8 h	20 mg q6-8 h	10 mg q6-8 h	0.2 mg/kg q6-8 h	0.1 mg/kg q6-8 h
Codeine	180-200 mg q3-4 h	130 mg q3-4 h	60 mg q3-4 h	60 mg q2 h (IM/SC)	180-200 mg q3-4 h	130 mg q3-4 h	0.5-1 mg/kg q3-4 h	N/R
Hydro-codone	30 mg q3-4 h	N/A	30 mg q3-4 h	N/A	30 mg q3-4 h	N/A	0.2 mg/kg q3-4 h	N/A
Oxycodone	30 mg q3-4 h	N/A	10 mg q3-4 h	N/A	30 mg q3-4 h	N/A	0.2 mg/kg q3-4 h	N/A

h = hour
N/A = not applicable

mg = milligram
kg = kilogram

8.3

3) **Intramuscular and Subcutaneous Administration of Opioids**

Intramuscular (IM) and subcutaneous (SC) opioid administration results in a rapid rise in blood levels of drug, which stay within the effective range for 2-3 hours and repeat doses will be needed at that time interval. The speed and extent of opioid uptake from the various IM sites varies—deltoid > quadriceps > gluteus. IM and SC opioids are little used in the BMT population as the mixture of thrombocytopenia and IM injections results in IM hematomata and patients would require many IM doses during the long periods of time they need opioid analgesia.

4) **Intravenous Administration of Opioids**

Intravenous (IV) opioids are widely used in the BMT population—the patients all have indwelling central IV lines and the drug reliably and rapidly reaches the bloodstream. A bolus of intravenous opioid results in a very rapid rise in blood level of drug reaching effective range in minutes, but then rapidly falling, with a duration of effect of some 30-40 minutes. To maintain blood levels within the effective range bolus injections would have to be given frequently.

5) **Continuous Infusion of Opioids** (Fig. 8.3.3)

A pure continuous infusion (CI) technique will result in blood levels of drug rising slowly and taking a long time to reach the therapeutic window. This slow onset of effect can be circumvented by the use of an initial bolus of drug to bring blood levels within the therapeutic window and then maintaining them at that level by a CI. However, because of the variability in individual patient's MEACs, pharmacokinetics and changing degrees of nociception, it is difficult to empirically choose an infusion rate which will reliably maintain blood levels of drug within the effective range. Starting infusion rates are 1-2 mg/h (15-30 µg/kg/hr) morphine in normal adults and 20-40 µg/kg/hr in children. For inadequate pain relief CI rates should be increased by 50%.

6) **Patient Controlled Analgesia (PCA) Administration of Opioids**

PCA has proven to be a very safe and effective means of delivering opioid analgesia in the BMT patient. By individualizing dosage it circumvents many of the problems posed by other modes of administration, i.e. variations in MEAC, variations in pharmacokinetics, changes in degree of nociception and convenience of administration. Virtually all adult patients have the cognitive and discriminative capacity to use PCA. Children as young as 5-6 years may also be able to use PCA. Drugs used in PCA should be restricted to the relatively rapid acting and short acting drugs, i.e. morphine, hydrographies, meperidine and fentanyl. A specimen PCA order sheet is given in Figure 8.3.4.

  a) *PCA Parameters*

  PCA devices are computer controlled pumps in which a number of delivery parameters may be selected.

  i) *PCA bolus size (incremental dose)*

  Initial boluses are 0.5-2 mg (10-30 µg/kg) in normal sized adults and 20-30 µg/kg in children. As nociception and pain worsen, bolus size should be increased or if nociception and pain are improving, the bolus size should be decreased so that the patient self administers 25-40 doses in 24 hours. This can be achieved by ob-

8.3

1. PCA PRESCRIPTION:

MODE	☐ PCA Only	☐ PCA + Continuous Infusion		☐ Continuous Infusion Only
DRUG	☐ MORPHINE 1 mg / ml	☐ MEPERIDINE 10 mg / ml	☐ HYDROMORPHONE 0.2 mg / ml	☐ Other: _____ mg - mcg / ml
PCA MODE Incremental dose Lockout 4 Hour Limit	_____ mg ___ minutes 30 mg or _____ mg (Adult) (Pediatric)	_____ mg ___ minutes 300 mg or _____ mg (Adult) (Pediatric)	_____ mg ___ minutes 6mg or _____ mg (Adult) (Pediatric)	_____ mg / mcg ___ minutes _____ mg
LOADING DOSE On initiation of therapy give:	_____ mg q _____ min to a maximum _____ mg	_____ mg q _____ min to a maximum _____ mg	_____ mg q _____ min to a maximum _____ mg	_____ mg/mcg q _____ min to a maximum _____ (mg / mcg)
FOR INADEQUATE RELIEF After 1 hr with initial pump settings:	Increase incremental dose to _____ mg	Increase incremental dose to _____ mg	Increase incremental dose to _____ mg	Increase incremental dose to _____ mg / mcg
FOR INADEQUATE RELIEF After 1 additional hr:	Decrease lockout to _____ min	Decrease lockout to _____ min	Decrease lockout to _____ min	Decrease lockout to _____ min
OPTIONAL INFUSION MODE Recommended starting rate: Instructions for infusion:	_____ mg/hr Adult (0.5 mg/hr) Pediatric (0.01 mg/kg/hr) ☐ Continuous ☐ Infuse from 2200 to 0600 nightly	_____ mg/hr Adult (5 mg/hr) Pediatric (0.1 mg/kg/hr) ☐ Continuous ☐ Infuse from 2200 to 0600 nightly	_____ mg/hr Adult (0.1 mg/hr) Pediatric (1.5 mcg/kg/hr) ☐ Continuous ☐ Infuse from 2200 to 0600 nightly	_____ mg / mcg/hr (Consult PRS) ☐ Continuous ☐ Infuse from 2200 to 0600 nightly
FOR INCIDENT OR BREAKTHROUGH PAIN Give loading dose:	_____ mg q _____ min to a maximum _____ mg per event	_____ mg q _____ min to a maximum _____ mg per event	_____ mg q _____ min to a maximum _____ mg per event	_____ mg/mcg q _____ min to a maximum _____ per event (mg / mcg)

2. **No systemic opioids or other CNS depressants** to be given except on direct one-time order from HO/PRS.

3. **MONITORING:** Respiratory rate, Pain Level (0-10), Sedation Scale (0-3) - q 2 hr for 8 hrs; then q 4 hr while patient is on PCA. Document on "Pain Management Flowsheet".

4. **TREATMENT OF SIDE EFFECTS:**

   Check desired box(es) and fill in weight/age appropriate dose.

   Call HO/PRS for Sedation Scale = 3, RR < 8 per min or < 50% of normal RR.
   ☐ NALOXONE _____ mg IV stat for Sedation Scale = 3 and RR < 8 per min or < 50% of normal RR.
   May repeat x _____ q 5 min. Call HO/PRS.
   ☐ METOCLOPRAMIDE _____ mg IV q 4 hr prn for nausea / vomiting.
   ☐ TRANSDERMAL SCOPOLAMINE PATCH to either mastoid area (1/2 patch for 12 > age > 60 yrs). Change q 72 hr prn.
   Not recommended for children < age 6 yrs.
   ☐ DIPHENHYDRAMINE _____ mg IV or PO q 6 hr for severe itching.
   ☐ For urinary retention, "in and out" catheter, prn.

5. **For sleep** _____

Fig 8.3.4. A sample PCA order form.

taining the patient's 24 hour opioid consumption from nursing records or the PCA memory and dividing that into the requisite number of doses. As a patients' pain decreases, bolus size should be titrated down aiming at 25-40 doses in 24 hours.

ii) Lock out interval

After a bolus of opioid has been delivered the pump will not deliver a further dose within the lock out interval. This is an essential safety feature as it stops the administration of a further dose of drug until such time as the previous dose has had chance to take effect. The usual lock out intervals are from 8-15 minutes, the shortest that can be used on current PCA pumps is 5 minutes.

iii) Four hour dose limit

A further safety feature, a dose limit per time can usually be set at about 30 mg of morphine per 4 hours for postoperative pain control. For the BMT patient, who may well use substantial amounts of drug, this limit may well be inadequate and larger limits set.

8.3

MONITORING	SEDATION SCALE	PAIN SCALE	INCIDENT PAIN

PCA: Q ___ hr x ___ hrs;
then Q ___ hrs
	0 = None
	1 = Mild
	  (occasionally drowsy; easy to arouse)
	2 = Moderate
	  (frequently drowsy; easy to arouse)
	3 = Severe
	  (somnolent; difficult to arouse)
	S = Normal sleep
	  (easy to arouse)
		0 = No pain
		10 = Worst pain imaginable
			painful procedure
			dressing change
			mobilization
			turn, cough, deep breath

Date														
Time														
MONITORING														
Respiratory Rate														
Level Of Sedation														
Level Of Pain														
PATIENT CONTROLLED ANALGESIA														
___ Morphine 1mg/ml ___ Hydromorphone 0.2mg/ml ___ Meperidine 10mg/ml ___ Other- ___ mcg - mg/ml														
ROUTE ___ IV ___ SC														
___ Incremental dose														
Lockout interval ___ min														
4 hour limit														
On initiation of therapy give ___ x1														
For inadequate relief:														
After 1 hour, increase incremental dose to —														
After 1 additional hour may decrease lockout to ___ min														
Optional infusion /hr														
___ Continuous														
___ Noc only (2200 - 0600)														
Loading dose for incident pain														
During therapy may give:														
___ every ___ minutes or														
___ x ___ to max of ___ per event														
Cumulative dose total														
OTHER PAIN MANAGEMENT TX														
Initials USC RN														
SIGNATURE DATE SIGNATURE DATE SIGNATURE DATE														

Fig 8.3.5. Pain management monitoring form for patients receiving opioids, in this case PCA opioids.

8.3

iv) Continuous infusion option (optional infusion mode)

This may be used to deliver a constant, predetermined, CI of drug, in addition to what the patient self administers. The premise behind the use of CI is that it will smooth the peaks and troughs of blood levels of drug associated with PCA and will save the patient from having to self administer many doses. It is unclear if the use of CI in addition to PCA produces extra benefits in terms of analgesia or side effects. If a CI is used it should be no greater than about 50% of the patient's predicted opioid need.

D) OPIOID SIDE EFFECTS AND THEIR MANAGEMENT

1) Gastrointestinal

Opioids slow gastric emptying and intestinal transit cause constipation. For many BMT patients constipation is not a problem. For those in who it is, the routine addition of a stool softener and a cathartic is useful.

2) **Genitourinary**

Opioids increase tone and contractility of the ureters and increase detrusor tone causing retention; rarely a problem in the BMT population.

3) **Sedation**

Opioids cause sedation, particularly at high dosage. However not all sedation in the BMT patient receiving opioids is due to the opioids; other potential causes of sedation include: other CNS active drugs (antiemetics, anxiolytics, antifungal agents, cyclosporine), metabolic derangements (renal dysfunction, hypo- or hypernatremia, hepatic dysfunction), fatigue and sleep deprivation. In the circumstances where opioids are providing satisfactory pain relief but at the cost of unacceptable sedation, this sedation may be relieved by small doses of Ritalin (5-10 mg BID or TID usually given by day and avoided by night).

4) **Euphoria**

Opioids may cause euphoria, but this is rare in the BMT population.

5) **Dysphoria**

Dysphoria usually occurs with initial use and declines with more prolonged exposure. Change to an alternate opioid.

6) **Nausea and Vomiting**

Nausea and vomiting due to opioids tends to occur more frequently in patients who have a history of motion related sickness as opioids sensitize the vestibular apparatus. If nausea and vomiting are caused by opioids, it is usually most apparent on initial opioid use and occurs in a strictly temporal relationship to a dose of opioid. If this is the case, a smaller dose given more frequently, giving doses by slow infusion or switching to an alternative opioid is appropriate. If nausea and vomiting are due to motion, transdermal scopolamine should be used.

7) **Confusion, Hallucinations and Delusions**

CNS symptoms occur in about 10-15% of BMT patients receiving opioids. Opioids are often not the sole culprit; drugs and conditions categorized in the section on sedation are often involved.

8) **Respiratory Depression**

When measured by sensitive indices (e.g. respiratory response to CO_2 challenge) all opioids blunt respiratory drive in a dose dependent fashion. However, when opioids are used in clinical practice, respiratory depression (defined as $paCO_2 > 42$ torr) is a uncommon event. Patients with pre-existing respiratory disease, sleep apnea-like symptoms and those receiving concurrent benzodiazepines are particularly at risk of respiratory depression.

8.3

9) **Tolerance**

Tolerance may occur with prolonged opioid treatment but there is slim evidence that it occurs in the usual BMT scenario, unless drugs of the fentanyl family are used.

10) **Dependence**

Characterized by the occurrence of an abstinence syndrome if the drug is abruptly ceased, dependence, or at least abstinence syndromes, are rarely seen in BMT patients receiving opioids.

11) **Psychological Dependence**

Otherwise called addiction, psychological dependence is characterized by an abnormal pattern of drug use, a craving for drug effects other than analgesia, an all consuming involvement with drug procurement and consumption even when such activity is contrary to the individual's best interests and a tendency to relapse into drug use after weaning. Addiction is a compendium of genetic, psychological, social and cultural factors which may affect the individual, in addition to drug use. Thus, unlike tolerance and dependence, psychological dependence is a patient issue, not a drug issue. In the absence of risk factors (previous substance use or abuse, psychiatric disorders) the risks of iatrogenic psychological dependence, as a consequence of therapeutic use of opioids, is very low—of the order of 1:2-4000.

E) **MONITORING THE PATIENT RECEIVING OPIOIDS**

When patients are receiving opioids concerns about analgesic efficacy and the occurrence of undesirable side effects (sedation, respiratory depression) mandate that patients are monitored, and such monitoring is recorded on regular and frequent basis. An example of a pain management monitoring form is given in Figure 8.3.5.

F) **COMMONLY PRESCRIBED OPIOIDS FOR THE BMT PATIENT**

A list of drugs, equianalgesic doses, half lives and dosage intervals is given in Table 8.3.3. The choice of opioid should be limited to agonistic. The most widely used drugs are:

1) **Morphine**

The prototypic opioid, morphine, may be given by any route, though it has low oral bioavailability secondary to high first pass metabolism by the liver. Oral morphine has a short duration of effect, 3-4 hours, which may be extended to 8-12 hours in the case of the recently available sustained release preparation. Morphine is conjugated in the liver to the renally excreted end products, morphine 3 glucuronide (M3G) and morphine 6 glucuronide (M6G). With hepatic dysfunction, metabolism of morphine is well preserved until the dysfunction is severe. Both M3G and M6G have long half lives, which will be even more prolonged if renal function is compromised, thus morphine should be used cautiously in patients with renal compromise.

M6G has analgesic actions, accounting for an unknown proportion of morphine's analgesic effects and other CNS effects (sedation, disturbances in mentation, respiratory depression) and thus may be responsible for some of morphine's toxicities, especially in patients with renal compromise. When used initially, morphine has a 15-25% incidence of itch, nausea and vomiting which often diminish with sustained use.

2) **Meperidine**

Meperidine can be given by any route but has poor oral availability. Oral meperidine has a short duration of effect. Meperidine is metabolized in the liver to a number of metabolites which are renally excreted. The clinically important metabolite is normeperidine (NMP) which has CNS executory and proconvulsive effects and a half life which is much longer than meperidine. With short-term, low dose use of meperidine, the presence

8.3

of NMP is unimportant as its levels remain low and its CNS effects are negated by the sedative and anticonvulsive effects of meperidine. However, in longer term use (800-1000 mg/24 hrs for 4-5 days), or in a patients with renal compromise, NMP levels may reach hazardous levels even in the presence of meperidine. Convulsions can also result from NMP after cessation of meperidine administration when short life meperidine levels fall quickly, with long half life NMP levels falling slowly, leading to a convulsion. Thus in the BMT patient, who will often need high dose analgesics for several days, meperidine should be used cautiously and reserved for low dose treatment, e.g. of chills and for premedication before blood products are given.

3) Hydromorphone

Hydromorphone may be given by any route, and has reasonable oral availability. When given by mouth it has a similar duration of effect (3-4 hours) as morphine. Sustained release preparations which permit 8-12 hourly dosing are in the process of being made available. It is metabolized by the liver and metabolites are excreted renally. Metabolites are believed to be pharmacologically inert and thus hydromorphone is a good choice in patients with renal dysfunction.

4) Fentanyl

Fentanyl is a highly potent, rapid onset and short duration of action opioid, hepatically metabolized with metabolites excreted renally. Metabolites are believed to be pharmacologically benign. It is primarily used intravenously for brief intensely painful procedures such as dressing changes. It may be used in PCA but patients become rapidly tolerant and very rapidly escalate their drug use. Recent new formulations include:

a) Transdermal Patches

These are formulated to deliver 25, 50, 75 or 100 μg of fentanyl per hour for a period of about 72 hours. When a patch is applied it takes some 10-14 hours for the drug to penetrate the dermis, build a drug depot in the dermis and achieve stable blood levels of drug. When a patch is removed a depot of the drug is left in the dermis which is absorbed over some 10-14 hours. Fentanyl patches are a relatively inflexible form of opioid therapy and are most useful when dealing with a stable pain problem which requires stable blood levels of opioid. Such a stable circumstance is infrequently encountered in the BMT population, where pain and analgesic needs often change rapidly.

b) Oral Transmucosal Fentanyl

The so called "fentanyl popsicle" is used for opioid administration to facilitate brief painful procedures, or to provide premedication when an intravenous site is unavailable, e.g. in children.

IX) LOCAL ANESTHETICS FOR PAIN CONTROL

Local anesthetics (LA) have a small, but important role, in several aspects of pain relief in the BMT population.

A) TOPICAL ORAL LOCAL ANESTHETICS

Viscous 2% lidocaine or Dyclone are useful remedies for minor degrees of oral pain associated with mucositis. Dyclone is usually rapidly effective. If viscous

8.3

lidocaine is used it should be kept in contact with the affected area for 3-5 minutes to allow the drug to penetrate the mucosa. Each application will have a duration of effect of about 45-60 minutes.

B) TOPICAL CUTANEOUS LOCAL ANESTHETICS

There is only one—eutectic mixture of local anesthetics (EMLA)—a recently introduced mixture of 2.5% lidocaine and 2.5% prilocaine. When applied to skin, under an occlusive dressing (Op site, Tagaderm, shrink wrap) for 1-2 hours, EMLA will produce sufficient skin anesthesia to permit relatively painless skin puncture, skin biopsies or small skin grafts.

C) INJECTABLE LOCAL ANESTHETICS

Suitable drugs include 1% lidocaine and 0.25% bupivacaine. These can be used to provide infiltration anesthesia of skin (use fine needles 27-30G), and deeper structures (use larger needles 22-25G). The LA solutions are acidic and sting when injected. This stinging can be reduced by alkalinizing them with molar sodium bicarbonate (3 ml per 10 ml 1% lidocaine, 1 ml per 10 ml 0.25% bupivacaine). LA does not work instantaneously—allow 5-10 minutes after injection for maximal efficacy. Maximal doses for infiltration use are 5 mg/kg lidocaine (35 ml 1% lidocaine in a 70 kg patient) and 1 mg/kg bupivacaine (26 ml 0.25% bupivacaine in a 70 kg patient). IV lidocaine, in doses that produce cardiac therapeutic blood levels (1-2 μg/ml), is occasionally helpful for the neuropathic burning pain that occurs with immunosuppressive drugs such as cyclosporine and FK 506. If used for these purposes, appropriate clinical monitoring of the patient and laboratory monitoring of blood levels of drug are necessary.

X) INVASIVE PAIN CONTROL TECHNIQUES

These techniques are the province of the pain relief expert, usually an anesthesiologist, and include: peripheral nerve blocks with LA, sympathetic nerve blocks with LA, intra spinal (epidural or subarachnoid) techniques using LA +/- opioids. While these techniques are theoretically feasible in the BMT patient they all involve the insertion of needles and/or catheters into various bodily structures. In the BMT patient, who is likely to be thrombocytopenic and immunosuppressed, the use of such techniques carries a higher than normal risk of producing deep hematomata and of infection and hence should only be used after expert consultation.

XI) COMMONLY OCCURRING PAIN PROBLEMS IN THE BMT PATIENT

A) PAIN PROBLEMS OCCURRING PRETRANSPLANT

1) Pain Related to the Disease

Pain due to leukemic bony deposits occurs infrequently in acute leukemias but may be severe during a blast crisis. Patients with solid tumors (breast, lymphoma) or multiple myeloma are more likely to have pain from tumor deposits which usually subsides with chemoradiotherapeutic regimes.

Common treatment regimens for pain due to cancer are nonopioid analgesics +/- opioids by oral, or other routes. Patients who present for BMT with pain due to their disease and who are on an effective analgesic regime should have that regime continued, assuming that components of the regime are not contraindicated, e.g. tricyclic antidepressants and NSAIDs. Patients who have been taking opioids prior to the commence-

8.3

ment of their BMT will need higher than usual doses of opioids to achieve pain control during and after BMT, presumably as a consequence of tolerance.

B) PAIN PROBLEMS OCCURRING PERITRANSPLANT

1) Mucositis

Painful oropharyngeal mucositis (OPM) is a common phenomenon following BMT chemoradiotherapeutic conditioning regimes. Risk factors for OPM severity are age, poor oral hygiene, smoking, previous chemo-radiotherapy, previous oral infections, radiotherapy and triple alkylating drugs. At Fred Hutchinson Cancer Research Center (FHCRC) about 90% of BMT patients suffer OPM which is sufficiently painful, with pain scores of 6-8 on a 1-10 scale, to require opioid analgesics. OPM is severe enough to preclude swallowing in 80% of patients and requires endotracheal intubation, for airway maintenance and protection, in 1-2%. OPM usually begins at about the time of transplant; early onset implies greater severity and duration, later onset implies less severity and duration. OPM and its accompanying pain peak about 10-14 days post BMT and resolve when the donor marrow shows evidence of engraftment with pain and analgesic use declining over a few days. Systemic methotrexate, given as an immuno-suppressant in allogeneic BMTs, often results in a 1-2 day worsening flare of OPM and pain. On average, BMT patients with OPM need opioid analgesics for about 10-12 days, but this period may be 20-25 days in the case of delayed engraftment.

Irrespective of the degree of OPM and pain, frequent oropharyngeal rinsing with physiologic saline is critical to pain relief and maintenance of oral hygiene. Many patients find the use of chilled rinsing solutions helpful. In the initial stages of OPM, topical analgesics such as Dyclone rinses and topical viscous lidocaine should be used. Dyclone is most useful for pan oral pain, viscous lidocaine is more useful for discrete painful areas. When topical regimes are insufficient begin opioids. Because of dysphagia in OPM, oral opioids are not very useful, begin with nurse administered boluses. Once the need for boluses becomes frequent (4-5 per shift) switch to PCA opioids in adults, children over 10 years, or CI opioids in younger children. Begin PCA opioids at 1-2 mg morphine boluses or equivalent.

On a daily basis, track pain and satisfaction scores, opioid use in the past 24 hours and side effects profile. Adjust PCA dose so that the patient will have to give 25-40 doses to match the previous day's opioid use. Be aware that opioid need varies considerably, and that to maintain acceptable analgesia, substantial doses of drug may be needed. At FHCRC the mean peak morphine use in adults with OPM is about 150-200 mg/24 hrs and many patients will use up to 4-5 times this dose. Opioid use parallels OPM pain rising up to about the time of engraftment and rapidly decreasing thereafter. As opioid use declines, adjust PCA bolus size down so that the patient delivers the previous 24 hour total drug in 25-40 doses. By 4-5 days after engraftment patients usually do not need opioids. The value of the addition of an opioid CI is unclear. If a CI is used it should be of the order of 50% of the patient's 24 hour opioid need, PCA bolus size being set so that the remaining 50% of opioid need is delivered in 25-40 boluses.

8.3

2) Conditioning Regimen Cutaneous Burns

This is usually associated with radiotherapy and occurs predominantly in the hands, feet, groins and axillae but, may be localized to the scrotum and penis in the case of testicular irradiation. The severity ranges from mild cutaneous erythema to frank second degree burns with blister formation, and usually will have healed by 6-10 days post engraftment. For mild burns topical cold is often helpful. Beyond this treat with systemic opioids as for OPM. The pain from burns and OPM often overlap and treating one will often treat the other. Topical treatment with silver containing creams or biologic dressings, e.g. pig skin, may be necessary for second degree burns.

3) Peripheral Burning Pain Associated with Immunosuppressive Drugs

Classically this occurs with cyclosporine but has also been observed with experimental immunosuppressives such as FK 506. The severity is loosely associated with high peak serum levels of cyclosporine. Changing the infusion duration of cyclosporine to 24 hours (i.e. continuous infusion) will decrease peak levels and may diminish symptoms. To clinical observation, symptoms also appear worse when cyclosporine is given IV, as opposed to by mouth. While this pain may start in the peritransplant period it can persist into the post engraftment period. In the initial stages cooling with water, fans and cold packs with ice are effective. If this is not effective a trial of opioids, used as for OPM, is worthwhile. Often opioids are ineffective; it may be necessary to conduct a clinical trial, pushing the opioid dosage to the limits imposed by side effects. For severe and treatment resistant cases, IV lidocaine in doses which produce cardiac therapeutic blood levels of lidocaine (1-2 µg per ml) may be useful.

4) Hepatic Capsule Distention

This may occur with veno-occlusive disease of the liver. IV or PCA opioids, as for OPM are appropriate. Patients with VOD will tolerate opioids until the VOD becomes severe with bilirubin in the 25-30 mg/dl range, or liver failure occurs.

5) Pain Associated with Growth Factors

Growth factors may produce severe pain, of rapid onset, usually some 2-4 hours after their administration which lasts for 3-4 hours. Usual pain sites are abdominal, distal femur and proximal tibia, with patients being essentially pain-free between these episodes. Patients should have access to substantial doses of opioids over short time periods, either by RN administered IV boluses, or by PCA. An example would be morphine 4-8 mg at the start of the episode and 2-4 mg RN administered boluses every 10-15 minutes or equivalent PCA settings.

C) PAIN PROBLEMS OCCURRING POSTENGRAFTMENT

1) Graft-Versus-Host Disease

In patients receiving an allogeneic transplant, graft-versus-host disease (GVHD) is common despite prophylaxis with immunosuppressive drugs and steroids. Commonly occurring sites are:

a) Oral

Pain from oral GVHD often overlaps mucositis. Persisting oropharyngeal pain after engraftment should arouse suspicions of GVHD

8.3

or infection. Pain is commonly symmetrical on the lateral border of the tongue opposite the second premolar and the molar teeth. Minor degrees of pain may respond to 3-5 ml of topical viscous lidocaine. More severe pain may require oral or systemic opioids, titrated to effect.

b) *Skin*

Treatment is oral or systemic opioids titrated to effect.

c) *Gut*

Gut GVHD usually presents as vague, midabdominal background pain which has brief, episodic, crampy, severe exacerbations and may be relieved by a bowel movement.

Gut GVHD pain is difficult to treat. A trial of systemic opioids by CI or PCA is worthwhile. However while opioids are somewhat helpful for background pain, they provide poor relief from the episodic cramping pains. Indeed opioids may make cramping pain worse and contribute to an ileus. Pain from gut GVHD may be a circumstance where the use of invasive analgesic techniques, which can provide profound analgesia with modest doses of drug, may be appropriate.

2) Pain Associated with Steroid Tapering

This usually occurs at the time that steroids are being slowly withdrawn. Typically this pain is of rapid onset, lasting a few hours, occurring at night in the distal femur and tibia, most often in the young and those with previous knee pathology. This pain responds well to opioids.

3) Pain Due to Opportunistic Infections

a) *Herpes Zoster*

Varicella zoster can occur in any dermatome, but the frequency is thoracic > head and neck > cervical > lumbosacral. It may present as radicular pain with rash following, or as rash and pain simultaneously. The acute phase is painful in about 75% of cases. Long standing pain, due to peripheral nerve damage, may result and increases in incidence with increasing age.

In the acute phase nonopioid and opioid analgesics, in addition to anti-infective agents and steroids is appropriate. After 6-8 weeks from the initial infection, the pain process is more likely to be a neuropathic process than a nociceptive process and opioids are unlikely to be very helpful. At this stage three major types of pain occur.

i) *Severe Skin Sensitivity*

Pain occurs with the lightest touch. Try topical EMLA (eutectic mixture of local anesthetics, a recently introduced mixture of 2.5% lidocaine and 2.5% prilocaine) applied every 2-4 hours. If EMLA fails consider dilantin.

ii) *Episodic Shooting Pain*

Perhaps best thought of as an epileptiform discharge from damaged dorsal horn neurones. Use dilantin to achieve therapeutic blood levels.

iii) *Background Burning Pain*

Occasionally responds to opioids. Best response to TCAs, provided these are acceptable and do not cause marrow suppression.

8.3

b) Other Infections

Common infections include cytomegalovirus and herpes simplex. Common sites for pain are the oropharynx and the gut. Use nonopioids and opioids as for oral GVHD.

4) Pain Due to Ischemic Necrosis of Bone

Usually occurs in proximal or distal femur and is often secondary to steroid-induced asceptic necrosis. Titrate up dose of opioids and nonopioids to effect.

XII) PAIN FROM DIAGNOSTIC OR THERAPEUTIC PROCEDURES

A) POSTSURGICAL PAIN

BMT patients may need surgical procedures ranging in magnitude from central line placement to a thoracotomy. The severity and time course of pain following surgery in BMT patients is similar to that seen in the normal population. The mainstay of treatment is PCA opioids. Postsurgical pain is most intense immediately after the surgery, at which time blood levels of opioid are low. Thus, immediately postsurgery it is essential to "load the patient up" to rapidly achieve blood levels of drugs which will be effective. Patients who are currently receiving opioids, or who have recently been receiving opioids will have a much greater dose requirement than opioid naive patients.

B) POSTLUMBAR PUNCTURE HEADACHE (PLPHA)

This is believed to be due to a slow, persistent leak of CSF through the hole in the dura created by the LP. This leak leads to either a low pressure headache, or traction on the base of brain structures. Classically PLPHA is a symmetrical, occipital headache which radiates to the frontal region and is constant, but with a throbbing component. A cardinal feature is that the headache gets rapidly worse (within seconds) when the patient sits or stands and fairly rapidly improves (within a few minutes) when the patients lies down. Nausea and vomiting and photophobia are common. PLPHA is more likely to occur in patients who have a previous history of headache, is unusual in children and most commonly occurs in the 15-40 age group. With supportive therapy alone most PLPHAs will resolve within 5-8 days but may persists for weeks or months.

1) Prophylaxis for PLPHA

Prophylactic maneuvers which minimize the chances of a persistent CSF leak and thus reduce PLPHA incidence are: adequate hydration; minimize the size of the dural hole (use 22G or 25G needles); minimize the number of dural punctures; have the patient lie flat for 1-2 hours immediately after the LP; spend as much time as possible lying down in the next 24 hours; avoid maneuvers which produce raised CSF pressure—e.g. coughing, straining at stool.

2) Treatment of PLPHA

Once a PLPHA has occurred, initial treatment includes: maintaining recumbency as much as possible for 24-48 hours, adequate hydration and oral analgesics (nonopioid and opioid). If at the end of 24-48 hours the PLPHA has not resolved, caffeine may be infused (500 mg in 250 ml of IV fluid) over 3-4 hours and repeated not more frequently than every 12 hours. If the PLPHA still has not resolved consider an epidural blood patch (EBP), normally done by anesthesiologists. EBP should not be considered lightly in the BMT patient who may be thrombocytopenic (risk of an epidural hematoma) and immunosuppressed (risk of epidural infection). EBP in-

8.3

volves an epidural puncture and injecting 12-15 ml of autologous blood at about the site of the dural hole. The blood will initially reverse the transdural pressure gradient, reduce CSF leakage and subsequently act as "physiologic glue" sealing the dural hole. Blood patches successfully resolve PLPHA in 80-90% of cases.

SUGGESTED READING

1. Patt RB. Cancer Pain. Philadelphia: JB Lippincott, 1993.
2. Ibid. Clinical Practice Guideline, Management of Cancer Pain, 1993.
3. Agency for health care policy and research, Public Health Service, US, Department of Health and Human services. Clinical Practice guideline. Acute Pain Management: Operative or medical procedures and trauma, 1992.

8.3

Cardiovascular Support

Robert W. Taylor, Steven J. Trottier

I) INTRODUCTION

The bone marrow transplant patient is at significant risk of cardiovascular instability during the course of his or her illness. This may occur because of the primary oncologic/hematologic disease process, underlying cardiovascular disease, infection/sepsis and/or conditioning regimen-related toxicity or other therapies.

II) PROGNOSIS

BMT patients who require critical care unit (CCU) monitoring generally have a poor prognosis. Survival for those who do not require mechanical ventilation is 10-20%, while for those requiring mechanical ventilation it is roughly 3%. The APACHE II (acute physiology and chronic health evaluation) prognostic scoring system or the reason for intensive care monitoring does not, in general, correlate with outcome for BMT patients. The most significant good prognostic factors for survival of CCU monitoring are younger patient age (< age 40), absence of mechanical ventilation and late CCU admission (> 90 days after BMT).

III) FLUIDS AND WEIGHT

First line therapy for cardiovascular instability in non-BMT patients has traditionally been fluid resuscitation. However, during the early bone marrow transplant period (first 30 days), leakage occurs into the extravascular space across endothelial cells damaged by high-dose chemotherapy and cytokines which are elevated from sepsis and/or and graft-versus-host disease. In addition, intravascular oncotic pressure is often decreased due to hypoalbuminemia while at the same time large volumes of fluid are being administered in order to deliver medications and parenteral nutrition and to prevent hemorrhagic cystitis. Consequently, all patients undergoing BMT have a tendency to gain weight from third spacing of fluids and must be maintained at pretransplant dry weight with diuresis. If these patients become hypotensive, aggressive fluid resuscitation will result in ascites, peripheral edema and pulmonary edema often with little effect on blood pressure. Therefore, emphasis must be on early invasive hemodynamic monitoring. Even then due to endothelial cell damage, pulmonary edema may occur despite a normal pulmonary artery occlusion (wedge) pressure. Therefore, blood pressure should be maintained with pharmacologic pressure support and by attempting to maximize intravascular oncotic pressure with packed red blood cell transfusions keeping the hemoglobin between 12-15 g/dl. In addition, if the patient develops pulmonary edema, our bias (although unproven) is to initiate early dialysis with ultrafiltration, in order to minimize volume overload.

IV) CLASSIFICATION OF SHOCK (TABLE 8.4.1)

A) CARDIOGENIC

Myocardial ischemia and/or infarction can lead to inadequate left and right ventricular performance. Myocardial depression is a frequent occurrence in the septic patient. In the BMT patient, high-dose cyclophosphamide or ifosfamide may cause myocardial depression. If the patient survives, cyclophosphamide-induced left ventricular dysfunction is reversible over a period of weeks to months. Cardiogenic shock is typically associated with elevated pulmonary artery (PA) and pulmonary artery occlusion pressures (PAOP), a

8.4

Bone Marrow Transplantation, edited by Richard K. Burt, H. Joachim Deeg, Scott Thomas Lothian, George W. Santos. © 1996 R.G. Landes Company.

Table 8.4.1. Hemodynamic variables in various shock states

	RAP	PAP	PAOP	CO	SVR
Normals	0-5 mm Hg	25/10 mm Hg	8-12 mm Hg	4.0-5.0 L/min	800-1200 dynes/sec/cm^{-5}
Shocks Classification					
Cardiogenic	increased	increased	increased	decreased	increased
Hypovolemic	decreased	decreased	decreased	decreased	increased
Obstructive					
cardiac tamponade	increased	increased	increased	decreased	increased
massive pulmonary embolism	increased	increased	variable	decreased	increased
Distributive	decreased	decreased	decreased	increased	decreased

RAP = right atrial pressure   PAP = pulmonary artery pressure
PAOP = pulmonary artery occlusion pressure
CO = cardiac output   SVR = systemic vascular resistance

depressed cardiac output and often an elevated systemic vascular resistance index (SVRI). Acute therapy is usually directed toward relieving cardiac ischemia if it exists, enhancing myocardial contractility and lowering SVR.

B) OBSTRUCTIVE SHOCK

Cardiac tamponade and pulmonary embolism are prime examples of this type of shock. The patient with cardiac tamponade commonly presents with tachycardia, hypotension, a narrowed pulse pressure, jugular venous distention and muffled heart tones. The ECG may demonstrate low voltage. If a pulmonary artery catheter is placed, a rapid X descent and a blunted Y descent are seen in the right atrial pressure tracing. An "equalization" of diastolic pressures is often seen, meaning that right atrial, pulmonary artery diastolic and pulmonary artery occlusion pressures are roughly equal. The diagnosis is further confirmed by echocardiography. Treatment involves drainage of the pericardial space either percutaneously by catheter or operatively by the creation of a pericardial window.

The patient with massive pulmonary embolism presents with dyspnea, chest discomfort, cyanosis, tachycardia and hypotension. The chest radiograph is most commonly unchanged from baseline. The ECG commonly demonstrates tachycardia and a right ventricular strain pattern may be seen. Supraventricular tachycardias such as atrial fibrillation are sometimes seen. If a pulmonary artery catheter is placed, a widened pulmonary artery diastolic to PAOP gradient may be seen. Normally the PAOP is 3-5 mm Hg lower than the pulmonary artery diastolic pressure. This gradient may widen substantially in the patient with massive pulmonary embolism. This finding is sufficiently rare, however, such that the clinician should not depend upon it to suspect pulmonary embolism. Ventilation/perfusion lung scanning and/or pulmonary angiography confirm the diagnosis. Standard therapy involves systemic heparin followed by warfarin. Heparin is given as 5,000-10,000 units IV push loading dose followed by a continuous infusion of 700-2,000 units/hr (1,300 U/hr in an average sized adult). Check activated partial thromboplastin time (aPTT) at

8.4

6 hours and adjust to an aPTT ratio of 1.5-2.5 times the control value. Heparin and warfarin therapy overlap for 5-7 days.

C) HYPOVOLEMIC SHOCK

Hypovolemic shock implies insufficient intravascular volume leading to inadequate preload. Patients may have intravascular volume deficits for a variety of reasons including blood loss due to hemorrhage, diuretic use, inadequate fluid intake and/or loss of intravascular volume to the extravascular space. The hemodynamic "fingerprint" of hypovolemic shock is low right atrial, pulmonary artery and pulmonary artery occlusion pressures, and a low cardiac output. Therapy in the acutely ill non-BMT patient involves rapid volume challenges with crystalloid or colloid solutions. However as discussed under fluids and weight (above) due to excessive capillary leakage, administration of fluids to the BMT patient must be done with caution. This volume loading is facilitated by placement of a pulmonary artery catheter.

D) DISTRIBUTIVE SHOCK

Distributive shock implies dysregulation of the distribution of blood flow to the tissues. Some tissues receive inadequate blood flow while other tissues may receive blood flow far in excess of need (luxury perfusion). Prime examples of distributive shock include septic shock, anaphylactic shock and spinal shock. The septic patient often has normal to elevated right atrial, pulmonary artery and pulmonary artery occlusion pressures. Cardiac output is characteristically elevated despite the fact that myocardial performance is depressed. A reversible cardiac dilation occurs that helps to maintain stroke volume. Tachycardia is almost universally present and accounts for the elevated cardiac output. Hemodynamic management of septic shock involves expansion of intravascular volume, maintenance of adequate perfusion pressures (mean pressure > 65-70 mm Hg) and inotropic augmentation of ventricular performance. The hemodynamic profile and management strategies of anaphylactic shock and spinal shock are similar to septic shock.

E) OVERLAPPING TYPES OF SHOCK

Many patients exhibit conditions that overlap several of the above described categories. For example, the patient with septic shock has a primary problem with distribution of blood flow to be sure (distributive shock), but also has venous capacitance vessel dilatation and leaking capillaries leading to reduced effective intravascular volume (hypovolemic shock). Myocardial depression also occurs secondary to circulating myocardial depressant factor(s) (cardiogenic shock).

V) DETERMINANTS OF CARDIAC FUNCTION

A) PRELOAD

Preload may be simply defined as end diastolic volume. Unfortunately at the present time no practical method exists to measure end diastolic volume on a continuous basis in the intensive care unit. End diastolic pressure is substituted as an indicator of volume as the two are related by the compliance of the ventricle, since compliance is defined as V/P. The pulmonary artery catheter is useful in estimating right and left ventricular end diastolic pressure (preload). Pulmonary artery occlusion pressure (PAOP) is measured from the distal port of a balloon tipped catheter "wedged" in a pulmonary artery. Under these circumstances the catheter tip "looks" downstream as right ventricular pressure events are blocked by the balloon. If a continuous column of blood exists

8.4

between the catheter tip and the left ventricle, for an instant at end diastole with the mitral valve still open, the PAOP and left ventricular end diastolic pressure (LVEDP) move toward equilibrium. The pressures are not exactly the same (PAOP is slightly higher than LVEDP) but close enough to use PAOP as a reflection of LVEDP. The clinician must recognize that a variety of clinical circumstances violate the premise that a continuous column of blood exists between the balloon tip and the left ventricular cavity. Examples include severe mitral stenosis, atrial myxoma, abnormal pulmonary vasculature (chronic lung disease, adult respiratory distress syndrome) and institution of positive-pressure mechanical ventilation with positive end expiratory pressure. Under these circumstances the relationship of PAOP and LVEDP is tenuous. Right ventricular preload is estimated in a similar fashion by monitoring right atrial pressure.

B) AFTERLOAD

Afterload is the tension that develops in the ventricular wall during systole. Afterload is influenced by aortic pressure, aortic compliance, ventricular wall thickness, ventricular volume, peripheral vascular resistance, blood mass and viscosity. Afterload is increased by hypertension, ventricular hypertrophy and increased blood viscosity (polycythemia). It is decreased by lowering of blood pressure, vasodilatation (sepsis, medication), peripheral shunting of blood (A-V fistulas) and reduced blood viscosity (severe anemia). Clinically we estimate afterload by calculating systemic vascular resistance (SVR) (see below).

C) CONTRACTILITY

Contractility may be thought of as myocardial strength. It refers to a change in the velocity of muscle shortening at any given tension level. An increase in contractility (starting dobutamine for example) will be reflected in an increase in the volume of blood ejected from the ventricle with each heart beat (stroke volume). Clinically we can estimate contractility by calculating the amount of work performed by the ventricle with each heart beat (left and right ventricular stroke work see below).

D) CARDIAC OUTPUT

Preload, afterload and contractility are primary determinants of stroke volume. Cardiac output (liters of blood pumped by the heart to the body per minute) is the product of stroke volume and heart rate. Slow heart rates may be associated with reduced cardiac output. A great deal of individual variability exists however. A trained athlete may have a resting heart rate of 40 beats per minute yet have a perfectly adequate cardiac output because of an increased stroke volume. This is not the case, however, for most patients in the ICU. Under these circumstances we are concerned by heart rates below 60 beats per minute if the patient is hemodynamically compromised. As heart rate increases above 120-140 for older people and above 170-180 for normal young people, diastolic filling time is significantly shortened, stroke volume may decline and cardiac output decreases.

VI) PULMONARY ARTERY CATHETER

The pulmonary artery balloon floatation catheter has channels opening in the right atrium (proximal channel), right ventricle (optional) and pulmonary artery (distal channel). A separate channel exists for inflation of the 1.5 cc balloon at the distal tip of the catheter. A final channel connects the thermistor (distal catheter tip) to the cardiac output computer. Catheters are most commonly inserted via a central vein (femoral, internal jugular or subclavian) utilizing a modified Seldinger

8.4

technique (guidewire technique). Peripheral IV access is obtained if possible. Oxygen is applied as necessary if not previously done. The pulse oximeter, ECG and vital signs are monitored throughout the procedure. Sedation is given as the clinical situation dictates (many patients will not require this). If necessary, small titrated doses of a short acting benzodiazepine such as midazolam (1-2 mg IV push, titrated to effect) are effective. Insertion via the femoral vein typically requires fluoroscopy and generally is confined to the cardiac catheterization laboratory. Internal jugular and subclavian vein insertion is often done at the bed side and will be described below.

A) INTERNAL JUGULAR VEIN APPROACH

The patient is placed in a 15 degree head down (Trendelenberg) position to insure filling of the internal jugular vein. Some patients do not tolerate this position. Under these circumstances a flat (supine) or slight head up position is utilized. Stand at the head of the patient's bed. Turn the patient's head away from the side to be cannulated. Both right and left internal jugular veins may be cannulated; however, the right side has several distinct advantages over the left. The right side is a more direct route to the superior vena cava and right atrium. The sharp angle at the junction of the left internal jugular vein and the left subclavian vein is often difficult to negotiate. The apex of the left lung is higher than the right making the possibility of pneumothorax greater on the left. The thoracic duct is on the left, increasing the possibility of thoracic duct injury. The medial (sternal) and lateral (clavicular) bellies of the sternocleidomastoid muscle form a triangle with the clavicle at the base. The internal jugular vein lies within the carotid sheath just beneath the apex of the triangle formed by the bellies of the sternocleidomastoid. The carotid artery also lies within the carotid sheath just medial and deep to the internal jugular vein. Puncture the skin at the apex of the triangle. Direct the needletip caudally at a 45-60 degree angle to the frontal plane and laterally toward the ipsilateral nipple. Advance the needletip to a depth of 3-5 cm depending upon patient size. If the vein is not entered, redirect the needletip slightly more medially and repeat. Do not direct the needletip across the midline as the carotid artery may be punctured.

B) SUBCLAVIAN VEIN APPROACH

Positioning of the patient for the infraclavicular subclavian vein approach is identical to patient positioning for the internal jugular vein approach described previously. A towel roll is placed between the scapula with the long axis parallel to the spine. This facilitates posterior positioning of the shoulders and improves exposure to the subclavian vein. Stand at the side of the bed. Turn the patient's head away form the side to be cannulated. Both right and left subclavian veins may be cannulated. Puncture the skin 1 cm caudal to the junction of the medial and middle thirds of the clavicle. Advance the needle beneath the clavicle parallel to the frontal (horizontal) plane directed toward the sternal notch. Take care never to allow the needletip to dip beneath the frontal plane as the risk of pneumothorax increases significantly. Advance the 18 gauge needle at the specific angle and direction to a predetermined depth while applying suction to the syringe. Entry into the vein will be signaled by a rapid flush of venous blood into the barrel of the syringe. If a rapid flush of blood does not occur as the needle is advanced to the predetermined depth, continue to apply suction to the syringe and continue to withdraw the needle slowly along the

8.4

same pathway as entry. Often a flush of venous blood will occur during needle withdrawal implying that the needle collapsed the vein and perforated both anterior and posterior vein walls during advancement. If the vein is not encountered do not change needle direction midcourse, rather retract the needletip to the skin and redirect the tip.

C) CATHETER INSERTION

Once in the internal jugular or subclavian vein, immobilize the needle with the free hand. Remove the syringe from the needle and cover the hub of the needle with the thumb (some needle/syringe units are made to insert the guidewire directly through the syringe and needle without disconnecting the needle and syringe). Advance the guidewire through the needle, minimal resistance should be met. Many guidewires are long enough to reach the right ventricle and cause ectopy. Monitor the ECG carefully during passage of the guidewire. With the guidewire in place, remove the needle. One should keep a firm grip on the guidewire at all times. Use the scalpel and dilator to open the skin and dilate the subcutaneous tissue. Using a rotating motion, advance the introducer over the guidewire and dilator. Remove the guidewire and the dilator leaving the introducer in place. Connect the introducer side port and connect IV fluids. Secure the introducer to the skin with suture.

Fill the pulmonary artery catheter lumens with saline, check the integrity of the balloon by inflating it with 1.5 cc of air. Pass the sheath over the catheter tip to the most proximal portion of the catheter. Connect the proximal and distal lumens to the pressure transducer. Insure that the monitoring equipment is properly connected prior to proceeding. This might be accomplished by gently shaking the catheter tip at mid chest position. The oscilloscope should oscillate with the catheter tip. When this is accomplished advance the pulmonary catheter (balloon deflated) to approximately 15 cm. Begin to record a printed pressure tracing (this will be scrutinized later) and monitor the oscilloscope pressure throughout the procedure. Inflate the balloon to 1.5 cc. Advance the catheter until a right atrial pressure tracing is seen. This should occur at approximately 20 cm insertion depth (variations occur with site of entry and patient size). Advance the catheter until a right ventricular pressure tracing is seen. This should occur at 30-35 cm insertion depth. Advance the catheter through the right ventricle and pulmonic valve until a pulmonary artery pressure tracing is obtained. This should occur at 40-45 cm insertion depth. Gently advance the catheter until the PAOP is seen. This should occur at 45-50 cm insertion depth. Deflate the balloon. The pulmonary artery pressure tracing should reappear on the oscilloscope. Pass the catheter sheath over the external portion of the catheter and affix to the introducer. Obtain a chest radiograph to confirm the correct placement of the catheter. Read right atrial, right ventricular, pulmonary artery and PAOP from the printed pressure tracing (see below).

Rarely does insertion of a pulmonary artery catheter proceed as smoothly as outlined above. A few words of caution here might prevent complications. If the appropriate pressure tracing is not observed at the insertion depths given above, do not continue to advance the catheter! Deflate the balloon, withdraw the catheter, insure that the pressure monitoring equipment is correctly connected, then begin again. Always deflate the balloon before withdrawing the catheter. When properly positioned, the catheter tip should rest in the main

8.4

right or left pulmonary artery. Catheters advanced beyond this point place the patient at risk of pulmonary infarction and pulmonary artery rupture.

VII) PULMONARY ARTERY CATHETER WAVEFORMS

Notice in Figure 8.4.1 the waveform change from the right atrium to the right ventricle to the pulmonary artery and to the occlusion position. Pressure measurements are made during end expiration and are made most accurately from a printed tracing of the pressure waveform. Attention to this detail is important as intrapleural pressures often vary greatly in the critically ill patient and cause large variations of vascular pressures during the respiratory cycle.

VIII) PULMONARY ARTERY CATHETER MEASUREMENTS

A) CARDIAC OUTPUT

The thermodilution technique is used to measure cardiac output. A thermistor is imbedded in the tip of the pulmonary artery catheter and measures pulmonary artery blood temperature continuously. A known volume of fluid (usually 10 ml of D_5W) at a known temperature is injected into the right atrial catheter port. The fluid is colder than the patient's blood and drops blood temperature as it mixes in the right ventricle. As the "bolus of cold" passes the thermistor in the pulmonary artery the blood temperature is seen to drop. Because of generous blood flow, the patient with a high cardiac output will dilute the effect of the cold bolus and the pulmonary artery blood temperature will fall only slightly. However, the patient with a low cardiac output has less blood flow relative to the cold bolus and therefore the pulmonary artery blood temperature changes more dramatically. Following injection of the cold bolus a computer integrates the area under the pulmonary artery blood temperature change curve. The patient's cardiac output is inversely proportional to the area. The measurement is given in liters per minute. It is convenient to index the cardiac output to the patient's body size.

B) CARDIAC INDEX

Cardiac index equals cardiac output divided by the patient's body surface area. Body surface area (BSA) is estimated from the patient's height and weight using the Dubois body surface area normogram (Appendix V) or formula: BSA (m^2) = [Ht (cm) + Wt (kg)-60]/100.

C) MEAN PRESSURE

Mean pressures may be measured electronically at the bedside or may be calculated. Because we are most interested in end expiratory pressures it is often necessary to calculate mean pressure from a systolic and diastolic pressure obtained from a printed pressure tracing. Vascular pressures are usually expressed in mm Hg.

$$MP = DP + \frac{SP - DP}{3}$$

MP is mean pressure
DP is diastolic pressure
SP is systolic pressure

8.4

Fig. 8.4.1. (Facing page) Pressure waveforms obtained during passage of a pulmonary catheter are demonstrated. Note the waveform change from right atrium to right ventricle to pulmonary artery and the occlusion position.

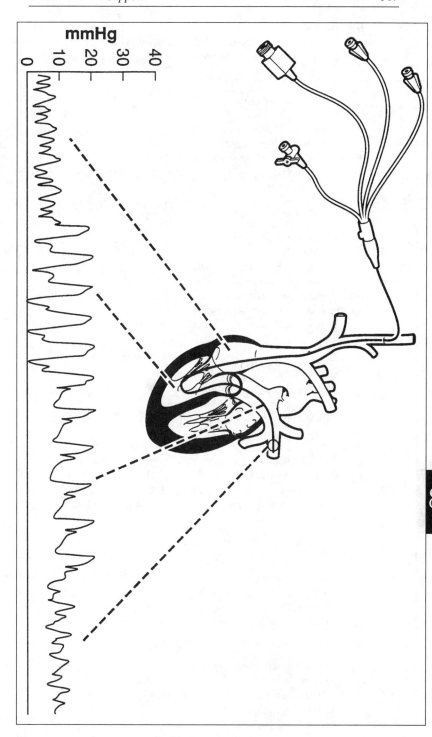

8.4

D) STROKE VOLUME

Stroke volume is the volume of blood in ml ejected from the left or right ventricle per heart beat. Stroke volume may be indexed by dividing by BSA.

SV (ml/heart beat) = CO/HR

SV is stroke volume
CO is cardiac output
HR is heart rate in beats per minute
normal: 30-60 ml/beat/m^2

E) SYSTEMIC VASCULAR RESISTANCE

Left ventricular afterload is clinically measured as systemic vascular resistance.

$$\text{SVR (dyne / sec / cm}^{-5}) = \frac{(MAP - MRAP) \times 79.9}{CO}$$

SVR is systemic vascular resistance
MAP is mean arterial pressure
MRAP is mean right atrial pressure = (CVP = central venous pressure)
CO is cardiac output
normal: 800-1200 dyne /sec /cm^{-5}

F) PULMONARY VASCULAR RESISTANCE

Right ventricular afterload is clinically measured as pulmonary vascular resistance.

$$\text{PVR (dyne / sec / cm}^{-5}) = \frac{(MPAP - PAOP) \times 79.9}{CO}$$

PVR is pulmonary vascular resistance
MPAP is mean pulmonary artery pressure
PAOP is pulmonary artery occlusion pressure = (PCWP) pulmonary capillary wedge pressure
CO is cardiac output
normal: 150-250 dyne/sec/cm^{-5}

G) VENTRICULAR STROKE WORK

Stroke work is an estimate of contractility. Left ventricular stroke work may be indexed by dividing by BSA. Both LVSW and RVSW may be indexed by dividing by BSA.

LVSW (g/m/m^2) = SV x (MAP-PAOP) (0.0136)

LVSW is left ventricular stroke work
SV is stroke volume
MAP is mean arterial pressure
PAOP is pulmonary artery occlusion pressure
normal: 43-61 g/m/m^2

RVSW (g/m/m^2) = SV x (MPAP-RAP) (0.0136)

RVSW is right ventricular stroke work
SV is stroke volume
MPAP is mean pulmonary artery pressure
RAP is right atrial pressure
normal: 7-12 g/m/m^2

8.4

H) OXYGEN DELIVERY (DO$_2$)

Oxygen delivery to peripheral tissues equals the product of cardiac output (CO) and arterial content of oxygen (CaO$_2$).

$$DO_2 = CO \times CaO_2 \times 10$$

CO = cardiac output
CaO$_2$ = 1.36 x Hgb x SaO$_2$ + 0.003 PaO$_2$ (ml O$_2$/dl)
Hgb=hemoglobin
SaO$_2$= arterial oxygen saturation (normal 93-98%)
PaO$_2$= arterial oxygen tension (normal 70-100 mm Hg)
normal: 650-1000 ml O$_2$/min at rest

IX) PULMONARY ARTERY CATHETER COMPLICATIONS

Reported complications with the pulmonary artery catheter are numerous. They vary from trivial to life-threatening. Every effort must be made during pulmonary artery catheterization to avoid complications by paying strict attention to detail. Most ICUs have hemodynamic monitoring protocols that should be followed carefully. An important and underrecognized complication involves acting upon incorrect information obtained from the catheter. A common example is recording of a PAOP during inspiration delivered by a positive-pressure ventilator rather than end expiration. The PAOP is elevated by the positive-pressure breath. Administration of a diuretic or other preload reducing agent is a common next step. Disastrous consequences may occur if the patent's intravascular volume is inadequate.

Local complications may occur during catheter insertion such as venous and/or arterial hemorrhage, venous thrombosis, nerve damage and pneumothorax. Catheter induced cardiac arrhythmias and heart block may occur. The possibility of catheter related infections including sepsis and endocarditis encourage placement of the catheter under strict sterile conditions, careful maintenance of the insertion site and removing or changing the catheter at regularly scheduled intervals (usually 3-4 days). Valvular heart damage and myocardial rupture have been reported. Catheters positioned too distal in the pulmonary artery circulation place the patient at risk for development of pulmonary infarction and pulmonary artery rupture. Catheters may become knotted within cardiac structures if strict attention is not paid to the depth of catheter insertion.

X) ARTERIAL PRESSURE MONITORING

Indications for insertion of an arterial catheter include need for on-line, beat to beat assessment of systolic, diastolic and mean blood pressure and the need for frequent blood sampling. Catheters may be inserted via the radial, femoral, axillary, brachial and dorsalis pedis arteries. Radial and femoral insertion sites are by far the most common and are discussed here.

8.4

A) RADIAL ARTERY INSERTION

The patient's hand is dorsiflexed and fixed to an arm board with tape. A roll of gauze is useful to keep the hand and wrist in the correct position. Create a sterile field. Palpate the radial artery at the head of the radius. Infiltrate local anesthetic. Insert needle over the radial artery approximately 1 cm distal to the radial head. Advance the needle at approximately a 20-45 degree angle. Entry into the artery will be signaled by appearance of pulsating arterial blood. Once in the artery immobilize the needle with the free hand. Advance the catheter with a rotating motion to the hub then remove the needle. Connect the catheter to arterial pressure tubing and the pressure transducer. Secure the catheter with suture and apply a sterile dressing.

B) FEMORAL ARTERY INSERTION

   The femoral artery is located midposition between the anterior superior iliac spine and the pubic symphysis. Using sterile technique, infiltrate local anesthetic 1-2 cm below the inguinal ligament. Enter the skin with the needle at this point at approximately a 45 degree angle. Entry into the artery will be signaled by the appearance of pulsating arterial blood. Once in the artery immobilize the needle with the free hand. Advance the guidewire through the needle. Remove the needle leaving the guidewire in place. Pass the arterial catheter over the guidewire. Remove the guidewire. Connect the catheter to arterial pressure tubing and the pressure transducer. Secure the catheter with suture and apply a sterile dressing.

XI) ARTERIAL CATHETER WAVEFORM ANALYSIS

   Additional useful information can be obtained by a careful assessment of the pressure tracing (Fig. 8.4.2). A large variation in systolic arterial pressure during positive-pressure mechanical ventilation signals hypovolemia.

XII) ARTERIAL CATHETER COMPLICATIONS

   An under recognized complication involves acting upon incorrect information. An example is institution of vasoactive medication to increase blood pressure when a technical problem such a kinked catheter or air bubble in the transducer has lead to a gross under representation of the blood pressure.

   Complications such as hemorrhage, arterial or venous damage, arterial or venous thrombosis, nerve damage, extremity ischemia, and infection should be anticipated and avoided by careful insertion technique and adherence to monitoring protocols. Arterial catheters should be changed every 4-5 days.

XIII) PHARMACOLOGIC SUPPORT OF THE CARDIOVASCULAR SYSTEM (TABLE 8.4.2)

   Depressed myocardial contractility might be enhanced by infusion of inotropic agents (beta-1 agonists) such as dobutamine, dopamine, norepinephrine and/or epinephrine (see below).

8.4

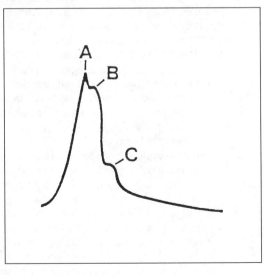

Fig. 8.4.2. The arterial pressure waveform. The upstroke reflects the inotropic state of the ventricle and is related to the rate of acceleration of blood from the aorta. Point A is the inotropic spike. Point B reflects volume displacement within the aorta and represents the balance of stroke volume and the peripheral runoff. Point C is the dicrotic notch. This represents a time-delayed reflection of the volume displacement wave. Reprinted with permission from Veremakis C. Hemodynamic monitoring. In: Kirby RR, Taylor RW, eds. Respiratory Failure. Chicago Year Book 1986:563-582.

Table 8.4.2. Guideline for use of continuous infusion medication

Drug	Concentration	Usual Dose	Patient's Weight 40 kg µg/kg/min	40 kg ml/hr	50 kg µg/kg/min	50 kg ml/hr	60 kg µg/kg/min	60 kg ml/hr	70 kg µg/kg/min	70 kg ml/hr	80 kg µg/kg/min	80 kg ml/hr	90 kg µg/kg/min	90 kg ml/hr	100 kg µg/kg/min	100 kg ml/hr
Dobutamine	250 mg in 250 ml D$_5$W (1000 µg/ml)	2.5-10 µg/kg/min	2	5	2	6	2	7	2	8	2	10	2	11	2	12
			4	10	4	12	4	14	4	17	4	19	4	22	4	24
			6	14	6	18	6	22	6	25	6	29	6	32	6	36
			8	19	8	24	8	29	8	32	8	38	8	43	8	48
			10	24	10	30	10	36	10	42	10	48	10	54	10	60
			40	96	40	120	40	144	40	168	40	192	40	216	40	240
Dopamine	400 mg in 250 ml D$_5$W (1600 µg/ml)	low dose : 1-3 µg/kg/min mod dose 5-15 µg/kg/min high dose > 29 µg/kg/min	2	3	2	4	2	5	2	5	2	6	2	7	2	8
			5	8	5	9	5	11	5	13	5	15	5	17	5	19
			10	15	10	19	10	23	10	26	10	30	10	34	10	38
			20	30	20	38	20	45	20	52	20	60	20	68	20	75
			30	45	30	56	30	68	30	79	30	90	30	101	30	113
			40	60	40	75	40	90	40	105	40	120	40	135	40	150
			50	75	50	94	50	113	50	131	50	150	50	169	50	188
Sodium Nitroprusside	50 mg in 250 ml D$_5$W (200 µg/ml)	0.5-10 µg/kg/min	0.5	6	0.5	8	0.5	9	0.5	11	0.5	12	0.5	14	0.5	15
			1	12	1	15	1	18	1	21	1	24	1	27	1	30
			3	36	3	45	3	54	3	63	3	72	3	81	3	90
			7	84	7	105	7	126	7	147	7	168	7	189	7	210
			10	120	10	150	10	180	10	210	10	240	10	270	10	300

8.4

Table 8.4.2. Guideline for use of continuous infusion medication (cont'd.)

Drug	Concentration	Usual Dose	µg/min	ml/hr
Norepinephrine	2 mg in 250 ml D$_5$W (8 µg/ml)	2-20 µg/min	2	15
			4	30
			6	45
			8	60
			10	75
			20	150
Epinephrine	2 mg in 250 ml D$_5$W (8 µg/ml)	1-8 µg/min	2	15
			4	30
			6	45
			8	60
Phenylephrine	5 mg in 250 ml D$_5$W (20 µg/ml) concentrate sol at higher doses	20-200 µg/min	20	60
			40	120
			60	180
			100	300
			200	600

8.4

A) DIGITALIS

Digitalis preparations have clear therapeutic benefit in certain patients with chronic heart failure. Their role in acute cardiovascular instability due to depressed myocardial contractility is less well defined and in this situation, digitalis is generally relegated to second line therapy. Digoxin is usually given as a 1.0 mg loading dose (0.5 mg IV push, followed by 0.25 mg IV push 4 hours later, followed by 0.25 mg IV push 8 hours after the initial dose).

B) DOBUTAMINE

Dobutamine is a synthetic catecholamine commonly used to improve myocardial contractility. The drug also has mild afterload reducing properties. As such it is highly useful in enhancing cardiac performance in the patient with an acceptable arterial blood pressure (mean arterial pressure 60-70 mm Hg). The drug is administered intravenously and titrated for effect in the range of 1-10 μg/kg/min.

C) DOPAMINE

Dopamine is a naturally occurring catecholamine with dose-dependent pharmacologic effects. The drug is administered by continuous infusion through a central venous catheter. At low-dose (1-2 μg/kg/min) dopaminergic receptors in the renal and mesenteric vessels are stimulated and promote vasodilation. At mid range-dose (2-10 μg/kg/min) myocardial contractility is enhanced. At high-dose (10-20 μg/kg/min) myocardial contractility is enhanced and peripheral vasoconstriction occurs. The drug is useful in enhancing cardiac performance in the patient with a low arterial blood pressure.

D) NOREPINEPHRINE

Norepinephrine is a naturally occurring catecholamine. Like dobutamine and dopamine it enhances myocardial contractility. Compared to dopamine it has much more of a vasoconstrictive effect. Norepinephrine has no effect on dopaminergic receptors. It is given by continuous infusion through a central venous catheter at doses of 2-20 μg/min.

E) EPINEPHRINE

Epinephrine is a naturally occurring catecholamine. It markedly enhances myocardial contractility by stimulation of beta-1 receptors. Epinephrine also has a vasoconstrictive effect by stimulation of alpha receptors. It is given during cardiopulmonary resuscitation as a 1 mg intravenous bolus (repeated every 3-5 min) in the setting of ventricular fibrillation (VF), pulseless ventricular tachycardia (VT), pulseless electrical activity and asystole. In treatment of shock it is often given by continuous infusion through a central venous catheter at doses of 1-8 μg/min.

8.4

F) PHENYLEPHRINE

Phenylephrine is a synthetic catecholamine that has almost pure alpha-adrenergic stimulating effects. It is commonly used intraoperatively to counteract the vasodilatory effects of anesthetic agents and is also used to elevate mean arterial pressure in various forms of shock. It is given by continuous infusion through a central venous catheter at doses of 20-200 μg/min.

G) NITROPRUSSIDE

Nitroprusside is a potent arterial and venous dilator. Its effects on the arterial circulation are more prominent than on the venous circulation. It is used to emergently reduce arterial pressure (afterload) in the patient with dangerous arterial blood pressure elevation. The drug is titrated for effect very carefully at doses of 0.5-10 µg/kg/min. Several cautions need to be mentioned with this drug. Hypotension may occur if titration is not carefully performed. Patients with chronic hypertension may suffer cerebral and other organ ischemia if blood pressure falls too drastically. The drug is very short acting. Should hypotension occur the nitroprusside dose may be adjusted down or discontinued usually with blood pressure returning to acceptable levels within several minutes. Cyanide and thiocyanate toxicity are further concerns particularly if renal function is compromised and the drug is used at high-dose for more than several days.

H) ENALAPRILAT

Enalaprilat is an intravenous angiotension-converting enzyme inhibitor used in the management of hypertension and as an afterload reducing agent in patients with left ventricular failure. The drug is given intravenously in a dose of 1.25 mg over 5 minutes every 6 hr.

I) DILTIAZEM

Diltiazem is a calcium channel blocking agent available for oral and intravenous use. Intravenous diltiazem is highly effective in controlling the ventricular response rate in patients with atrial fibrillation. It is given in a dose of 0.25 mg/kg over 2 minutes, then after 15 minutes 0.35 mg/kg over 2 minutes if needed. The intravenous maintenance dose is 10-15 mg/hour for 24 hours.

J) NICARDIPINE

Nicardipine is a calcium channel blocking agent used in management of hypertension. It is given as a continuous intravenous infusion at an initial rate of 5 mg/hr. This may be increased by 2.5 mg/hr every 5 minutes to achieve desired effect. A maximum dose of 15 mg/hr should not be exceeded.

K) ADENOSINE

Adenosine is a purine nucleoside that slows conduction through the AV node. It is effective in terminating paroxysmal supraventricular tachycardias that involve re-entrant pathways including the AV node. Adenosine does not terminate atrial fibrillation or atrial flutter, but because AV conduction is slowed it is useful diagnostically in patients with these rhythms. The drug is given intravenously as a 6 mg rapid (1-3 seconds) bolus. The dose should be followed by a rapid saline flush (approximately 20 ml). If the expected response is not observed within several minutes a 12 mg dose is given in the same manner.

L) LIDOCAINE

Lidocaine is recommended for patients with pulseless VT and VF refractory to electrical countershock and epinephrine. Lidocaine is appropriate for patients at risk for malignant ventricular arrhythmia (following termination of VT and VF, myocardial ischemia, left ventricular dysfunction). Lidocaine is given in an initial dose of 1.0-1.5 mg/kg. Following restoration of a perfusing rhythm lidocaine is given as a continuous infusion of 2-4 mg/min.

M) PROCAINAMIDE

Procainamide may be effective in suppressing ventricular arrhythmias when lidocaine has failed. Procainamide also slows intraventricular conduction and may be useful in converting supraventricular arrhythmias. The intravenous dose of procainamide is 20-30 mg/min until the arrhythmia is suppressed or to a total loading dose of 17 mg/kg. The loading dose should be slowed or terminated if hypotension occurs or if the QRS complex is widened more than 50% of its baseline. The maintenance infusion of procainamide is given at a dose of 1-4 mg/min.

8.4

N) BRETYLIUM TOSYLATE

Betrylium is a quaternary ammonium compound with antiarrhythmic activity. It is indicated in treatment of VT and VF unresponsive to other therapy such as countershock, epinephrine, lidocaine and procainamide. It is given by rapid intravenous injection at a dose of 5 mg/kg. Countershock is attempted after 30-60 seconds. If the arrhythmia persists, the dose is increased to 10 mg/kg given in the same fashion. Again, countershock is attempted. This dose (10 mg/kg) may be repeated at 5-30 minute intervals to a total loading dose of 35 mg/kg. Following termination of the arrhythmia, a continuous infusion may be started at 2 mg/min.

SUGGESTED READING

1. Tilkian AG, Daily EK, eds. Cardiovascular Procedures: Diagnostic Techniques and Therapeutic Procedures. St. Louis: The C. V. Mosby Company, 1986:111.
2. Varon AJ. Hemodynamic monitoring: arterial and pulmonary artery catheters. In: Civetta JM, Taylor RW, Kirby RR, eds. Critical Care. 2nd ed. Philadelphia:J.B. Lippincott, 1992:255-269.
3. Kett DH, Schein RMH. Techniques for pulmonary artery catheter insertion. In: Sprung CL, ed. The Pulmonary Artery Catheter: Methodology and Clinical Application. 2nd ed. Closter: Critical Care Research Associates, 1992:43-75.
4. Chernow B, ed. The Pharmacologic Approach to the Critically Ill Patient. 3rd ed. Baltimore: Williams and Wilkins, 1994.

8.4

Respiratory Support

Steven J. Trottier, Robert W. Taylor

I) DEFINITION OF RESPIRATORY FAILURE

The pulmonary system is involved in two primary metabolic roles, elimination of carbon dioxide and oxygenation of blood. There are two basic types of respiratory failure, type I, hypoxemic; and type II, hypercarbic, with or without hypoxemia. Respiratory failure may be acute, chronic or an acute exacerbation of chronic respiratory failure. A generally accepted clinical definition of acute respiratory failure is a $PaO_2 < 55$ mm Hg and/or a $PaCO_2 > 50$ mm Hg with a pH of < 7.35 while the patient is breathing room air.

II) RISK OF RESPIRATORY FAILURE IN BMT

The BMT patient's antecedent pulmonary status should have been within normal limits. Nevertheless, pulmonary morbidity may be increased by prior exposure to chemotherapy agents such as bleomycin, or BCNU; or radiation, as may occur in patients with relapsed Hodgkin's disease. The risk of respiratory failure is also greater for allogeneic than autologous grafts; HLA-mismatched compared to matched transplants; and patients with GVHD. Older age and the more aggressive conditioning regimens used for active malignancy have also been associated with an increased risk of respiratory failure.

III) CAUSES OF RESPIRATORY FAILURE AFTER BMT

A litany of etiologies causing hypoxemic (type I) and/or hypercarbic (type II) respiratory failure exist (Table 8.5.1). Several are particularly germane to the bone marrow transplant patient: interstitial pneumonitis, infectious (bacterial, fungal viral, parasitic), adult respiratory distress syndrome (ARDS), alveolar hemorrhage, bronchiolitis obliterans, radiation-induced lung injury, drug-related lung injury, cardiogenic pulmonary edema and volume overload.

The etiology of respiratory failure for the majority of patients undergoing BMT is unknown but probably involves a combination of conditioning regimen-related free radical lung injury, cytokine and conditioning regimen-related endothelial cell damage and donor lymphocyte mediated immune injury. Endothelial cell damage compromises alveolar/capillary integrity and predisposes patients to pulmonary edema especially if volume status is not carefully monitored. Approximately half of patients with respiratory failure have interstitial pneumonitis without an identified infectious etiology which has been termed idiopathic pneumonia syndrome (IPS).

IV) PROGNOSIS OF RESPIRATORY FAILURE IN PATIENTS UNDERGOING MARROW TRANSPLANT

Approximately 15-30% of patients undergoing allogeneic BMT develop respiratory failure requiring intubation. Prognosis for these patients is poor, approximately 80% die on the ventilator. Only 3-5% are extubated and survive more than 6 months. Factors predictive of survival once intubated are younger age (those less than 40 years old do better) and timing of ventilator support (those intubated within the first 90 days do worse).

V) OXYGEN THERAPY

Oxygen therapy is the mainstay treatment for hypoxemic respiratory failure in a spontaneously breathing patient. A myriad of oxygen delivery systems exist. Nasal cannulae deliver approximately a 2-3% increase in the fraction of inspired oxygen

Bone Marrow Transplantation, edited by Richard K. Burt, H. Joachim Deeg, Scott Thomas Lothian, George W. Santos. © 1996 R.G. Landes Company.

8.5

Table 8.5.1. Causes of respiratory failure

I. Disorders Associated with Defective Oxygen Onloading (Type I)
    A. Lower Airway and Pulmonary Parenchyma
        1. Interstitial pneumonitis
            a. idiopathic pneumonia syndrome
            b. other
        2. Infections
            a. bacterial
            b. fungal
            c. viral
            d. parasitic
            e. other
        3. Trauma
            a. pulmonary contusion
            b. pulmonary laceration
        4. Other
            a. drug and/or radiation-induced lung injury
            b. bronchospasm (asthma, COPD)
            c. heart failure (congestive, restrictive, etc.)
            d. ARDS
            e. pulmonary emboli
            f. atelectasis
            g. alveolar hemorrhage
            h. volume overload
            i. neoplasm
II. Disorders Associated with Defective Carbon Dioxide Offloading (Type II)
    A. Central Nervous System
        1. Drugs
            a. opioids
            b. benzodiazepines
            c. propofol
            d. barbiturates
            e. general anesthetics
            f. poisons
            g. others
        2. Metabolic
            a. hyponatremia
            b. hypocalcemia
            c. hypercarbia (excess carbohydrate)
            d. alkalosis
            e. hyperglycemia
            f. myxedema
        3. Neoplasm
        4. Infection
            a. meningitis
            b. encephalitis
            c. abscess
        5. Increased intracranial pressure
        6. Central alveolar hypoventilation
        7. Other

8.5

Table 8.5.1. Causes of respiratory failure (cont'd)

B. Nerves and Muscles
   1. Trauma
      a. spinal cord injury
      b. diaphragm injury
   2. Drugs
      a. neuromuscular blocking agents
      b. aminoglycoside antibiotics
   3. Metabolic
      a. hypokalemia
      b. hypomagnesemia
      c. hypophosphatemia
   4. Neoplasm
   5. Other
      a. motor neuron disease
      b. myasthenia gravis
      c. multiple sclerosis
      d. muscular dystrophy
      e. Gullian-Barrè syndrome
C. Upper Airway
   1. Tissue enlargement
      a. tonsil and adenoid hyperplasia
      b. neoplasm
      c. polyps
      d. goiter
   2. Infections
      a. epiglotitis
      b. laryngotracheitis
   3. Trauma
   4. Other
      a. obstructive sleep apnea
      b. bilateral vocal cord paralysis
      c. laryngeal edema
      d. tracheomalacia
      e. cricoarytenoid arthritis
D. Chest Bellows
   1. Trauma
      a. rib fractures
      b. flail chest
      c. burn eschar
   2. Other contributing factors
      a. kyphoscoliosis
      b. scleroderma
      c. spondylitis
      d. pneumothorax
      e. pleural effusion
      f. fibrothorax
      g. supine position
      h. obesity
      i. pain
      j. ascites

8.5

(FIO$_2$) per each liter/minute. The air-entrainment ("Venturi") mask delivers an FIO$_2$ from 0.25 to 0.60. The partial rebreathing or nonrebreathing reservoir mask delivers an FIO$_2$ from 0.6 to 0.8. Titration of oxygen therapy is facilitated by the use of pulse oximetry. Arterial saturation (SaO$_2$) should be maintained > 90%. Patients failing to maintain an SaO$_2$ > 90% or those who develop hypercarbia likely require intubation and mechanical ventilation. Subsequently the FIO$_2$ should be decreased toward 50% if possible, to avoid oxygen toxicity.

VI) AIRWAY MANAGEMENT

Establishing and maintaining a patent upper airway is the first and most fundamental step in the care of a patient with respiratory failure.

A) OPENING THE AIRWAY

Soft tissues of the upper airway usually collapse and obstruct the airway in patients with a depressed level of consciousness. The chin-lift and jaw-thrust maneuvers are methods utilized for opening the airway. (Note: the head-tilt maneuver may be combined with the above maneuvers provided there is no documented or suspected neck injury). Both of these maneuvers displace the mandible anteriorly but the jaw-thrust is the safer approach if neck injury is suspected because it does not require neck extension. In BMT patients the most common reason for upper airway obstruction requiring emergent intubation is severe oral pharyngeal mucositis.

B) BAG-VALVE-MASK

With a patent airway, a bag-valve-mask apparatus may be applied to a patient temporarily to provide ventilation and oxygenation. In the unconscious patient an oral airway should be inserted carefully. (Note: insertion of an oral airway in a conscious patient stimulates the gag-reflex and may precipitate vomiting.) The bag serves as a reservoir for 100% oxygen and the one-way valve prevents rebreathing. The apparatus is awkward to use and the technique is best performed by two individuals, one maintains a patent airway and mask seal while the other compresses the self-inflating bag.

C) TRANSLARYNGEAL INTUBATION

Prior to translaryngeal intubation, efforts to pre-oxygenate the patient should ensue. Intubation attempts should be limited to 30 seconds to avoid hypoxemia.

1) Nasal Versus Oral Intubation

Coagulopathy is a contraindication to the nasal route. Nasotracheal intubation is difficult or impossible in the apneic patient. The nasal route may be attempted in breathing patients and is facilitated by use of a topical anesthetic. The oral route, under direct laryngoscopic visualization, is mandatory for apneic patients.

2) Equipment

Orotracheal intubation equipment include: oral airway; Yankauer and tracheal suction catheters; suction device; supplemental oxygen source; bag-valve-mask; endotracheal tubes and stylet; water soluble lubricant; 10 ml syringe to inflate cuff; Magill forceps; laryngoscope handle and blade(s); carbon dioxide detector; topical anesthetic spray; intravenous medication for sedation; tape; towel roll; gloves; mask; eye protection; pulse oximeter; ECG monitor; intravenous catheter, tubing and fluid; resuscitation cart.

8.5

3) Intubation Technique

Orotracheal intubation begins with assembling all equipment to ensure proper working order. If the patient is conscious the procedure should be explained. Obtain IV access if not previously done. Apply pulse oximeter. Don gloves, mask and eye protection. Monitor pulse oximeter and ECG throughout procedure. Most adults can accept an 8 Fr endotracheal tube. Several tube sizes should be available however. Check the endotracheal tube cuff integrity with a syringe. Insert the stylet into the tube and bend to a "hockey stick" configuration. Apply water soluble lubricant to the cuff end of the tube. Connect the laryngoscope blade to the handle. Remember that the straight blade (Miller) is used to elevate the epiglottis anteriorly. The curved blade (McIntosh) is inserted into the vallecula. The minimum blade length is the distance between the corner of the mouth and the angle of the mandible.

Position yourself at the head of the patient's bed. After opening the airway as previously described, a tight seal should be formed between the bag-valve-mask and the patient's face with the operator's left hand. The operator's right hand compresses the bag. Insure adequate preoxygenation and ventilation prior to proceeding. Inset a towel beneath the neck and position the head aligning the oral, pharyngeal and tracheal axes. With the laryngoscope in the left hand inset the tip of the blade into the right side of the patient's mouth. Sweep the tongue to the left. Insert a curved blade into the vallecula. If a straight blade is used, insert it beneath the epiglottis. Traction should be applied only along the long axis of the laryngoscope handle. A rocking or rotating motion of the handle and blade may damage teeth, gingiva and/or lips. Visualize the vocal cords and have an assistant apply cricoid pressure. Insert the endotracheal tube through the vocal cords. Carefully remove laryngoscope and stylet. Inset an oral airway. Secure the endotracheal tube and oral airway with tape. Once inserted, confirmation of the endotracheal tube position is required by careful auscultation of the chest and chest radiography. Use of a carbon dioxide detector to insure tracheal tube position is recommended.

VII) MECHANICAL VENTILATION

8.5

Mechanical ventilation is a purely supportive modality used to stabilize the patient until the underlying pathophysiology can be addressed and treated. The goals of mechanical ventilation are to provide adequate oxygenation and ventilation while minimizing complications of this modality.

A) CLASSIFICATIONS OF MECHANICAL VENTILATORS

The advent of new technology has complicated the terminology and classification schema of mechanical ventilators. For practical purposes, this discussion is limited to positive pressure ventilators. Further classification of positive pressure ventilators is based on the mechanism employed to terminate (cycle off) mechanical inspiration. Most mechanical ventilators manufactured in the past few years are capable of performing each of the following cycling methods.

1) Volume Cycling

Inspiration terminates when a preselected tidal volume has been delivered. Volume cycled mechanical ventilation delivers a preselected tidal volume to the patient with each mechanical breath. Based on the tidal volume and respiratory rate a known minute ventilation is delivered.

2) Pressure Cycling

Inspiration terminates when a preselected pressure has been reached. Pressure cycled mechanical ventilation delivers a preset pressure to the patient with each mechanical breath. The tidal volume varies depending upon the resistance and compliance of the system and patient. Thus, minute ventilation can be variable with pressure cycled mechanical ventilation.

3) Flow Cycling

Inspiration terminates when a specific gas flow rate has been achieved.

4) Time Cycling

Inspiration terminates when a preselected inspiratory time has been reached.

B) FULL OR PARTIAL RESPIRATORY SUPPORT

Initially, patients requiring mechanical ventilation should receive full respiratory support. Subsequent decisions regarding full or partial support should be based on clinical judgment. For example, a patient with severe ARDS may require a high minute ventilation and an elevated airway pressure to adequately oxygenate and ventilate. Heavy sedation is often required to achieve these goals. The respiratory muscles in these patients may consume up to 40-45% of the patient's total oxygen utilization. As the patient improves, the requirement for mechanical ventilation falls and subsequently, partial rather than full respiratory support may be used.

C) MODES OF MECHANICAL VENTILATION (TABLE 8.5.2)

1) Controlled Mechanical Ventilation

This mode delivers breaths at preselected intervals, irrespective of the patient's efforts. The tidal volume, respiratory rate and thus the minute ventilation are preprogrammed. Patient's are unable to breath independently. Use of this mode requires heavy sedation and/or neuromuscular blockade. It is presented here for historical reasons only as controlled mechanical ventilation is seldom, if ever used at this time.

2) Assist-Controlled Mechanical Ventilation (AC)

This is a combination mode in which a preprogrammed tidal volume is delivered in response to the patient's inspiratory effort. If the patient becomes apneic, the ventilator provides a preset back up respiratory rate and tidal volume. Thus, this mode of mechanical ventilation allows the patient to set the respiratory rate and provides preset back-up minute ventilation. (Caution: elevated spontaneous respiratory rates may cause hyperinflation, increased intrathoracic pressure, decreased venous return and cardiac output, and barotrauma.)

3) Synchronized Intermittent Mechanical Ventilation (SIMV)

This mode is a means of ventilatory support which provides a preset tidal volume and respiratory rate but also allows the patient to breath spontaneously in an unassisted yet unrestricted fashion between mandatory breaths. If an inspiratory effort occurs within a specified time frame, a mechanical breath is synchronized with the patient's effort. This mode provides a back up minute ventilation and allows the patient to autonomously alter his or her own breathing pattern. SIMV is versatile in that full, partial or no ventilatory support may be provided and it may be combined with pressure support ventilation.

8.5

Table 8.5.2. Mechanical ventilation

Mode	Indication	Contraindication	Advantages	Disadvantages	Complications
AC	RF	High respiratory drive	Synchronous respirations, patient determines RR and MV	Hyperventilation	Barotrauma, ↓ venous return, auto-PEEP
SIMV	RF	Patient requiring high MV with low set MV	Versatile, synchronous respirations	Potential ↑ work of breathing	Barotrauma, ↓ venous return, auto-PEEP
PSV	RF, weaning, supplement SIMV	Apnea	Patient comfort, patient determines RR, TV	Variable MV, requires intact respiratory drive	Barotrauma
PCIRV	RF, ARDS	Conscious patient, COPD, bronchospasm, asthma	↓ PAP, ↑ mean airway pressure, improved oxygenation, reduced shunt	Requires sedation and occasional paralysis, auto-peep	Barotrauma, ↓ venous return, auto-PEEP
HFJV	RF, RDS, broncho-pleural fistula	Conscious patient	↓ PAP, improved V/Q matching, improved cardiac function	Requires sedation	Barotrauma, ↓ venous return, airway desiccation, auto-PEEP, intraventricular hemorrhage

AC=assist-controlled; ARDS= adult respiratory distress syndrome; COPD= chronic obstruction pulmonary disease; HFJV= high frequency jet ventilation; MV=minute ventilation; PAP= peak airway pressure; PCIRV= pressure controlled inverse ratio ventilation; PEEP= positive end expiratory pressure; PSV= pressure support ventilation; RF=respiratory failure; RR=respiratory rate; SIMV= synchronous intermittent mechanical ventilation; TV=tidal volume; V/Q= ventilation perfusion

8.5

4) Pressure Support Ventilation (PSV)

This mode requires that the patient has an intact respiratory drive unlike the previously described modes of mechanical ventilation. With each inspiratory effort, the patient receives a preset pressure. The respiratory rate, flow and inspiratory time are controlled by the patient. Inspiration ceases when the flow decreases to 25% of its maximum value (flow cycled). Tidal volume is variable and depends upon the compliance of the patient's lung and chest wall, resistance of the patient's airways and ventilator circuit and duration of inspiration. PSV does not provide a preset minute ventilation. PSV may be used with SIMV or continuous positive airway pressure (see below). It is frequently used in the weaning process.

5) Pressure-Controlled Inverse Ratio Ventilation (PCIRV)

This is a pressure-limited time cycled mode of ventilation in which there is a preset inspiratory pressure, respiratory rate and a prolongation of inspiratory time. The tidal volume, hence the minute ventilation, varies according to the resistance and compliance of the system and patient. Prolongation of inspiratory time decreases expiratory time. This may lead to "air trapping" which causes an elevation in mean airway pressure and the development of intrinsic positive end expiratory pressure (auto-PEEP) which improve oxygenation. PCIRV has been used primarily in patients with ARDS and usually requires heavy sedation and use of neuromuscular blockade. PCIRV increases intrathoracic pressures and may compromise venous return and cardiac output. Thus, hemodynamic monitoring with a pulmonary artery catheter is recommended when initiating or titrating PCIRV to assure the adequacy of oxygen delivery.

6) High Frequency Jet Ventilation (HFJV)

HFJV is defined as mechanical ventilatory support using higher than normal breathing frequencies (100-600 breaths per minute). This is accomplished by the use of high frequency jets (HFJ) or high frequency oscillators (HFO). Rather than normal "bulk" flow, HFJ or HFO are used to apply pressure on the airway to deliver small tidal volumes. The exact mechanism of gas exchange remains controversial but at least five theories have been described: bulk convection; cardiogenic mixing; Taylor dispersion; molecular diffusion and pendelluft. High frequency ventilation has been studied in a variety of clinical settings and may be beneficial in decreasing neonatal barotrauma and in minimizing air leaks in adults with bronchopleural fistulas.

7) Positive End Expiratory Pressure/Continuous Positive Airway Pressure (PEEP/CPAP)

CPAP is the application of positive airway pressure throughout the respiratory cycle. This may be accomplished by the use of a mask or more commonly through an endotracheal tube. CPAP does not provide ventilation. PEEP is the airway pressure measured prior to the next mechanically delivered breath. PEEP and CPAP serve the same purpose, both elevate baseline breathing circuit pressures. An elevation in baseline circuit pressure is beneficial for patients with diffuse lung pathology such as ARDS. Typically these patients have decreased lung volumes, atelectasis, intrapulmonary shunting and ventilation/perfusion mismatching with resultant hypoxemia. Use of PEEP or CPAP increases lung volumes, recruits collapsed alveoli,

8.5

decreases shunt, improves ventilation and perfusion matching and thus improves oxygenation. The amount of PEEP or CPAP applied to the airways (usual range 0-20 cm H_2O) varies depending upon the clinical circumstances and is a highly controversial subject.

D) Initial Ventilator Settings (Table 8.5.3)

Initial ventilator settings should provide full respiratory support for the patient which includes the following: tidal volume 7-10 ml/kg, respiratory rate 14 (depends upon estimated minute ventilation), $FIO_2 = 1.0$, positive end- expiratory pressure of 5 cm water pressure (provided the patient is not hypotensive), mode (usually AC or SIMV).

E) Weaning

Strict guidelines or uniform protocols facilitating the weaning process are nonexistent. Rather, patients require an individual assessment and plan which must be flexible and guided by clinical judgment. Parameters listed in Table 8.5.4 provide a rough guideline to facilitate the initiation of weaning. Several methods of weaning from mechanical ventilation have been advocated: SIMV, T-piece, PSV, SIMV coupled with PSV. Each method involves a gradual decrease in mechanical support accompanied by an increase in the patient's work of breathing to build endurance. Respiratory muscle fatigue should be avoided. Successful weaning requires a multidisciplinary approach: a focused plan, active communication, adequate patient nutrition, patient participation and physical therapy. Several aspects of patient care should be reassessed when weaning difficulties occur: work of breathing, mode of weaning, secretions, infections, nutrition status, electrolyte balance, cardiac function and neuromuscular status.

VIII) PHARMACOLOGIC SUPPORT OF THE RESPIRATORY SYSTEM

Currently available pharmacologic agents administered to support the respiratory system center around bronchodilation, anti-inflammatory, respiratory stimulation and mucolytic action. In general, mucolytics and respiratory stimulants have a limited role in the care of mechanically ventilated patients. A clear indication should exist prior to the administration of these agents to the bone marrow transplantation patient.

A) β2 ADRENERGIC AGENTS

The main action of the β2 adrenergic agent is relaxation of bronchial smooth muscle, hence, the main indication is bronchospasm. These agents may be administered orally, intravenously, subcutaneously or inhaled which is the preferred route. Inhalation is accomplished by meter dose inhaler (MDI) or nebulized solution. Bronchodilation achieved by these agents must be balanced with the potential systemic toxicity (tachyarrhythmia, tremor, hypokalemia). Suggested dosing for mechanically ventilated patients: albuterol 5 mg nebulized every 4-6 hours.

B) ANTICHOLINERGIC AGENTS

In general, anticholinergic agents are administered via the inhaled route resulting in bronchodilation but to a lesser extent compared to β2 agonists. Currently, ipratropium bromide is the preferred agent which may be administered by MDI or nebulized solution. Due to limited absorption through the pulmonary tract, ipratropium bromide rarely causes systemic toxicity. Suggested dosing for mechanically ventilated patients: ipratropium bromide 0.5 mg nebulized every 4-6 hours.

Table 8.5.3. Mechanical ventilation initial settings

Mode	RR (breaths per minute)	Tidal Volume	PEEP (cm water pressure)	Inspiratory Pressure	Inspiratory to Expiratory (I: E), Ratio	Flow (liters/ minute)
AC	12	7-12 ml/kg	5	Not set	Not set	60-100
SIMV	12	7-12 ml/kg	5	Not set	Not set	60-100
PSV	Not set	Measured 7-10 ml/kg	5	5-60 cm water pressure (usual range 10-25)	Not set	Not set
PCIRV	12	Measured 7-10 ml/kg	Set at 50% prior level of PEEP	Pressure set to achieve desired tidal volume (20-60 cm water pressure)	1:1, 2:1, 3:1, 4:1 change ratio every 2-4 hours if required	Not set
HFJV	100-600	Not set	5	30 PSI	Inspiratory time 0.33	not set

AC=assist-controlled; cm= centimeter; HFJV= high frequency jet ventilation; kg= kilogram; PCIRV= pressure controlled inverse ratio ventilation; PEEP= positive end expiratory pressure; PSI= pounds per square inch; PSV= pressure support ventilation; RR= respiratory rate; SIMV= synchronous intermittent mechanical ventilation

8.5

Table 8.5.4. Weaning parameters

Tidal volume	≥ 5 ml/kg
Respiratory rate	< 25 minute
Minute ventilation	$\leq 10\text{-}15$ L/min
Vital capacity	$\geq 10\text{-}15$ ml/kg
Negative inspiratory force	> 25 cm H_2O
Oxygenation	$PaO_2 > 60$ mmHg on $FIO_2 \leq 0.4$
PEEP	≤ 5 cm H_2O
Central nervous system	alert/conscious

C) METHYLXANTHINES

Methylxanthines are administered orally (theophylline) or intravenously (aminophylline) for the relief of bronchospasm. The mode of action of the methylxanthines is unknown but may be due to phosphodiesterase inhibition and adenosine receptor antagonism. The actions of methylxanthines are reported to be many: bronchodilation; increased diaphragmatic contraction; increased respiratory drive; anti-inflammatory properties, just to mention a few. The use of methylxanthines in mechanically ventilated patients remains controversial and needs to be balanced with its toxicity (cardiovascular, central nervous system). For suggested dosing see Table 8.5.5. (Note: because of great individual variation, monitoring of theophylline blood levels during aminophylline or theophylline use is mandatory.)

D) CORTICOSTEROIDS

Corticosteroids may be indicated in pulmonary disorders such as acute asthma, bronchiolitis obliterans or interstitial pneumonitis/idiopathic pneumonia syndrome. Dosing and duration of therapy remain controversial. In general, corticosteroids may be administered systemically or via the inhaled route MDI. Immune suppression, hyperglycemia and adrenal suppression are just a few of the risks of corticosteroids. For suggested dosing see Table 8.5.5. Corticosteroids decrease cyclosporine clearance; therefore, plasma levels of cyclosporine may be increased by corticosteroids.

8.5

Table 8.5.5. Respiratory pharmacology agents

Corticosteroids

	Triamcinolone (inhaled)	4 puffs every 12 hr (spontaneous resp.) 8-10 puffs every 12 hrs (ventilator)
	Prednisone (PO)	1 mg/kg/day single or divided dose
	Methylprednisolone (IV)	60-125 mg every 6-8 hrs
Methylxanthines		
	Theophylline (enterally)	100 mg PO q4-6 hours
	Theodur (PO)	300 mg q12 hours
	Aminophylline (IV) infused over 30 minutes	6 mg/kg loading dose, followed by 0.5 mg/kg/hr infusion

SUGGESTED READING

1. Stauffer JL. Medical management of the airway. Clin Chest Med 1991; 12:449-482.
2. Tobin MJ, Yang K. Weaning from mechanical ventilation. Crit Care Clin 1990; 6:725-747.
3. Perel A, Stock MC eds. Handbook of Mechanical Ventilatory Support. Baltimore: Williams and Wilkins, Inc., 1992.
4. Tobin MJ, ed. Principles and Practice of Mechanical Ventilation. New York: McGraw-Hill, Inc. ,1994.
5. Clark JG, Hanson JA, Hertz MI et al. Idiopathic pneumonia after bone marrow transplantation. Am Rev Respir Dis 1993; 147:1601-1606.
6. Faber-Langendoen K, Caplan AL, McGlave PB. Survival of adult bone marrow transplant patients receiving mechanical ventilation: a case for restricted use. Bone Marrow Transplantation 1993; 12 (5):501-7.

8.5

Disorders of Renal Function and Electrolytes

Bruce Kaplan, Salim Mujais

I) ACUTE RENAL FAILURE

Acute renal failure can be defined as a sudden decline in the kidneys' ability to clear metabolic waste products and to maintain volume, electrolyte and acid-base homeostasis. Classically, acute renal failure is classified into pre-, intra-(parenchymal) and postrenal causes. In fact, these entities often overlap and one etiology does not preclude the existence of a separate cause (e.g. volume depletion on top of drug toxicity). Acute renal failure is classified further into oliguric and nonoliguric (> 400 ml/day). This separation usually implies a lesser severity of insult in nonoliguric and has both prognostic and practical relevance.

A) PRERENAL AZOTEMIA

Prerenal azotemia represents a reversible functional decrease in renal blood flow and glomerular filtration rate either due to an effective decrease in intravascular volume, to extreme low cardiac output states or intense renal vasoconstriction. This may be due to a variety of factors:

1) Hemorrhage
2) Gastrointestinal losses
3) Poor PO intake with concomitant use of diuretics
4) Third space losses e.g. sepsis, burns, pancreatitis, extreme hypoalbuminemia, capillary leak with high-dose chemotherapy or certain drugs e.g. interleukin 2
5) Congestive heart failure, cardiomyopathy
6) Hepatic veno-occlusive disease

In prerenal azotemia, avid sodium reabsorption is usually present, characterized by a low fractional excretion of sodium. Fractional excretion of sodium is (FeNa+) calculated as:

$$FeNa^+ = \frac{\dfrac{urine(Na^+)}{plasma\ (Na^+)}}{\dfrac{urine\ (Cr)}{plasma(Cr)}} \times 100$$

The FeNa+ in prerenal azotemia is usually less than 1%. A low FeNa+ is indicative of a prerenal state with the notable exception of patients with acute glomerulonephritis, hepatorenal syndrome, early dye-induced toxicity, veno-occlusive disease or in patients on cyclosporine therapy. Prerenal azotemia is usually reversible with volume replacement, correction of the underlying capillary leak or by improvement in cardiac output, depending on the respective etiology.

In patients who are felt to have prerenal azotemia secondary to volume loss or third spacing and who have reasonably maintained cardiac function, a fluid challenge of anywhere from 200-1000 mlof NS over 30-60 minutes should be

Bone Marrow Transplantation, edited by Richard K. Burt, H. Joachim Deeg, Scott Thomas Lothian, George W. Santos. © 1996 R.G. Landes Company.

8.6

attempted. If no response is elicited a decision must be made whether effective blood volume has been restored or not. If it is felt on clinical grounds that volume status has now been restored, a suspicion of intrinsic renal dysfunction must then be raised. In a few cases, addition of a loop diuretic after intravascular volume has been replenished may be of help. Dopamine in renal doses (1-3 μg/kg/min) may be of benefit if renal vasoconstriction is suspected as a cause of unsatisfactory response to volume challenge or if volume challenge is unsafe (CHF, capillary leak syndrome).

B) **INTRINSIC RENAL FAILURE**

Intrinsic renal failure occurs secondary to direct damage to the kidney from a variety of insults. This damage can be characterized as primarily tubular (acute tubular necrosis), glomerular (acute glomerulonephritis), interstitial (acute interstitial nephritis) or vascular (vasculitis, atheroembolic).

1) **Acute Tubular Necrosis (ATN)**

ATN in most cases is a misnomer as frank necrosis of tubules is rarely seen. It has come to be synonymous with acute renal failure secondary to either ischemic or direct toxicity to the renal tubules. The hallmark of ATN is a deterioration of renal function after a toxic/ischemic insult. The $FeNa^+$ is usually high and urine sediment often contains muddy brown and hyaline casts. Most cases of ATN are nonoliguric, however, severe injury may produce an oliguric state which is a harbinger of poorer renal prognosis. Some causes of ATN relevant to BMT are listed below.

Nephrotoxins:

    aminoglycosides

    amphotericin

    NSAID (particularly in the setting of poor renal blood flow)

    cisplatinum

    radio contrast dye

    interleukin 2

    foscarnet

    pentamidine

Ischemia:

    shock (septic, hypovolemic)

Pigment induced:

    rhabdomyolysis with myoglobinuria

8.6

2) **Glomerulonephritis**

Acute glomerulonephritis in the setting of BMT is almost always associated with infection and has been described with almost every type of bacteremia, however, it is most often associated with either streptococcal or staphylococcal infections. In these cases low complement levels are frequently seen. The $FeNa^+$ may be low and the UA may reveal red blood cell casts as well as proteinuria. This type of glomerulonephritis usually resolves when the underlying infectious process is controlled.

3) **Interstitial**

Interstitial damage can be mediated either through an allergic process as in acute allergic interstitial nephritis (AIN), or in the case of both cyclosporine and FK 506 through unknown mechanisms. In AIN there is often concomitant tubular dysfunction leading to various electrolyte disorders out of proportion to the degree of renal dysfunction (e.g. hyperkalemia, sodium wasting, etc.) The FeNa$^+$ is usually high and the UA may reveal white blood cells and WBC casts. The appearance of urine eosinophils either by Wright's stain or preferably by Hansel's stain is highly suggestive of AIN. Treatment of AIN is removal of the offending agent the use of steroids in this condition is controversial. A short course of steroids (moderate dose 30-40 mg) may hasten the recovery of some types of AIN. Some causes of AIN follow.

1) β lactams, cephalosporines
2) Rifampin
3) H2 blockers
4) Fluoroquinolones
5) NSAIDs (often with significant proteinuria)

4) Vascular

Acute renal failure may occur secondary to either damage to the renal vasculature or secondary to agents that cause severe renal vasoconstriction. These include:

1) Vasculitis
2) Cyclosporine/FK 506
3) Endothelial damage, e.g. with the hemolytic uremic syndrome
4) Hepatorenal syndrome
5) NSAIDs
6) Atheroembolic

If vasoconstriction is severe enough or prolonged, renal ischemia with ATN may ensue. Treatment for each disorder is specific (e.g. plasma exchange for hemolytic uremic syndrome, liver transplantation for hepatorenal syndrome).

C) Post Renal Failure

Post renal failure is due to the obstruction of urine flow anywhere from the tubule to the bladder outlet. Since obstruction is a very readily reversible cause of acute renal failure (except in the case of intra-tubular obstruction), an investigation is prudent in all individuals in whom even the smallest suspicion of this etiology exists. For obstruction to cause significant renal dysfunction it must be bilateral or present in a single functioning kidney. Anuria in the adult patient is of obstructive etiology until proven otherwise. Anyone in whom the suspicion of obstruction is high (e.g. elderly males, patients with pelvic malignancy) should at the very least have a foley catheter placed and have a renal ultrasonogram performed. Aside from obstruction at the ureteral, bladder or prostate level, certain drugs or clinical situations may cause intra-tubular obstruction as follows:

1) Acyclovir
2) Ciprofloxocin
3) Triamterene
4) Acute urate nephropathy
5) Ureter or urethral clotting from hemorrhagic cystitis

Prevention of acyclovir toxicity includes adequate volume and appropriate adjustment of dosage. Urate nephropathy frequently occurs after cytotoxic therapy and its incidence can be reduced with allopurinol and a forced alkaline diuresis. Alkaline diuresis to keep urine pH > 7.0 can be accomplished by acetazolamide, NaHCO$_3$ IV replacement or both. Alkali

8.6

therapy, however, is contraindicated if hyperphosphatemia is present (tumor lysis syndrome) as it may precipitate intrarenal calcifications. A saline diuresis or dialysis is required in the latter setting. (In case of tumor lysis syndrome, alkaline diuresis should be avoided secondary to the possibility of calcium phosphate formation and hemodialysis should be instituted as soon as possible).

II) SODIUM

A) HYPONATREMIA

Hyponatremia is defined as a serum sodium concentration less than 135 mEq/L. Symptoms of hyponatremia are related both to the degree of hyponatremia and the rate of decline in serum sodium. These symptoms include anorexia, nausea, lethargy, muscle cramps, depressed deep tendon reflexes, somnolence and seizures.

1) Causes of Hyponatremia

Hyponatremia can be divided into three forms: hypovolemic, euvolemic and hypervolemic (Table 8.6.1).

a) Hypovolemic Hyponatremia

Hypovolemic hyponatremia may occur through severe gastrointestinal losses or related to diuretic use. In GI losses urinary sodium is usually low (< 10 mEq/L), while in diuretic use the urine sodium may be considerably higher. Thiazide diuretics interfere with the final dilution of urine and are particularly prone to cause hyponatremia. Salt wasting states, e.g. Addison's disease or salt-losing nephropathy can also produce a hypovolemic form of hyponatremia. These forms of hyponatremia can be treated with normal saline.

b) Euvolemic Hyponatremia

Euvolemic hyponatremia can occur secondary to hypothyroidism or excessive antidiuretic hormone (ADH) secretion relative to serum osmolality. ADH can be stimulated by nausea, pain, hypotension or various drugs (e.g. chlorpropamide, tolbutamide,

Table 8.6.1. Hyponatremia Etiology

Volume Status	Etiology of Hyponatremia	Therapy
Hypovolemic	GI losses (low urine Na$^+$) Diuresis Salt losing nephritis, ($\uparrow$ urine Na$^+$) Osmotic diuresis	Saline
Euvolemic	Hypothyroidism Pain, drugs, etc… (physiologic stimulus for ADH) SIADH	Water restriction Loop diuretics demeclocycline
Hypervolemic	Nephrotic syndrome Cirrhosis CHF Renal failure	Fluid restriction, Diuretics

ADH = antidiuretic hormone; CHF = congestive heart failure; GI = gastrointestinal; Na$^+$ = sodium; SIADH = syndrome of inappropriate antidiuretic hormone

8.6

cyclophosphamide, vincristine, opiates, etc.). If the urine osmolality is > 300 mOsm/L it is safe to assume ADH is present. If the patient is hyponatremic and no other cause of ADH release can be found then the diagnosis of syndrome of inappropriate antidiuretic hormone (SIADH) can be made. Serum uric acid is usually low in SIADH and urinary sodium is variable depending on dietary sodium intake. Treatment of SIADH includes fluid restriction. If fluid restriction is not sufficient, judicious use of a loop diuretic can be tried. Finally, demeclocycline at a dose of 600 mg/d can be given to patients resistant to fluid restriction and loop diuretics. However, the drug itself is nephrotoxic with chronic use, and it should be reserved for patients in whom symptomatic hyponatremia is resistant to other maneuvers.

c) Hypervolemic hyponatremia

Hypervolemic hyponatremia occurs in states where there is both an excess of total body sodium and total body water (total body water > total body sodium). In these states the patient is usually edematous. Conditions associated with hypervolemic hyponatremia include nephrotic syndrome, congestive heart failure and renal failure. The treatment in these patients should include both fluid restriction and loop diuretics. Patients with hepatic cirrhosis often are hyponatremic with excess total body sodium. The management of these patients often is difficult and should be aimed at correcting their volume and tonicity problems without precipitating acute renal failure.

2) Correction of Hyponatremia

Correction of hyponatremia should not exceed 0.5 mEq/L/hr and not more than 12 mEq in any 24 hour period. Because of cellular adaptation to extracellular osmolality changes, faster correction can lead to neurologic problems and rarely central pontine myelinosis. Correction of hyponatremia can usually be accomplished with the above mentioned therapies. Hypertonic saline should be reserved for symptomatic patients with serum sodiums less than 120 mEq/L. In cases where isotonic or hypertonic (3%) saline is used a loop diuretic should also be used. In calculating the negative water balance the formula:

$$\frac{\text{Actual plasma } (Na^+)}{\text{Desired plasma } (Na^+)} \times TBW$$

can be used where TBW (total body water) is equal to 0.6 of total body weight. For example, in a 70 kg man the TBW = 42 liters (70 kg X 0.6); if the sodium is 115 and you want to raise it to 125 in 24 hours:

$$\frac{115}{125} \times 42 \text{ liters } = 38.6 \text{ liters}$$

The excess free water is: 42 - 38.6 = 3.4 liters. Therefore, lasix diuresis of 3.4 liters in 24 hours should be replaced with isotonic (0.9%) saline at 141 ml/hr (3.4 liters/24 hours) or 47 ml/hour of 3% hypertonic saline.

However, formulas can never take the place of serial plasma sodium monitoring as the estimate of TBW has a wide margin of error. Initial correction should aim at moving Na^+ out of the critical level. Once Na^+ reaches the low 120 range (122-124) a slower correction can be followed.

8.6

B) HYPERNATREMIA

Hypernatremia is defined as a serum sodium concentration greater than 150 mEq/L. Since the thirst mechanism is exquisitely sensitive to changes in serum osmolality, hypernatremia only occurs in the setting of a defect in the thirst mechanism or in the severely ill patient who cannot access water on their own. Most patients with hypernatremia are either euvolemic or hypovolemic. Therapy of hypernatremia is almost always with the use of hypotonic fluid either H_2O, D_5W or $\frac{1}{2}$ NS. To avoid acute brain edema the serum Na^+ should not be corrected faster than 0.5-1.0 mEq/hour. Free water deficit can be calculated by:

$$\frac{\text{Plasma (Na}^+) \text{ actual}}{\text{Plasma (Na}^+) \text{ desired}} \times \text{Total body water}$$

For instance if plasma Na^+ mEq/L is 170 and in a 24 hour period you wish to decrease plasma Na^+ to 155 mEq/L, free water replacement for a 70 kg man would equal:

$$\frac{170}{155} \times 42 \text{ liters} = 46 \text{ liters}$$

Where for a 70 kg man, total body water = 70 kg x 0.6 = 42 liters. Therefore, the deficit is 46-42 liters or 4 liters to achieve the desired plasma Na^+.

III) POTASSIUM

Potassium is the major intracellular cation, therefore it influences several important cellular processes including membrane polarization and many enzymatic processes. Intracellular potassium concentration is far higher than extracellular potassium concentration. Intracellular potassium concentration is approximately equal to 120 mEq/L while extracellular potassium concentration is about 4 mEq/L. In a 70 kg man, 42 liters will be total body water and approximately 30 liters will be intracellular water and 12 liters extracellular water. From this estimate 3600 mEq will be found in the intracellular space while only 60 mEq will be found in the extracellular space. This small amount of K^+ in the extracellular fluid (ECF) highlights the precarious state of ECF and the ease with which wide fluctuations can occur unless the regulatory mechanisms are intact. Because of this tremendous difference in intra- and extracellular concentrations of potassium, several mechanisms are in place to insure the adequate distribution of potassium in the various bodily fluids is maintained. Acid base status, insulin, and beta adrenergic tone are all important in the maintenance of the potassium gradient between the intra- and extracellular space.

8.6

A) HYPERKALEMIA

1) Etiology

Hyperkalemia can be traced either to a defect in the shifting of the potassium or to a defect in the cortical collecting tubule.

a) Cellular Shift

Mineral acidosis causes shifting of potassium from inside the cell to outside the cell. The magnitude of this change is not great (on the order of 0.4 mEq/liter for every 0.1 unit change in pH below a pH of 7.4). It seems that organic acidosis does not cause a shifting of potassium from the intra- to extracellular space as previously believed. Insulin is also important in potassium entry into cells. It works by activating the sodium potassium ATPase pump shifting potassium into the cell in exchange for sodium. In addition, beta 2 adrenergic agents will cause a shifting of potassium from outside the cell to inside the cell.

b) Renal Tubular Defect in Secretion

Almost all disorders involving hyperkalemia relate to a defect in the cortical collecting tubule. Potassium is freely filtered at the glomerulus, however, by the late distal tubule almost no potassium remains. The kidney secretes potassium in the cortical collecting tubule. The cortical collecting tubule secretes potassium under the influence of aldosterone and sodium and chloride delivery. In the cortical collecting tubule, sodium is selectively reabsorbed through discrete amiloride-sensitive sodium channels, creating an electronegative potential inside the urine lumen facilitating potassium secretion into the tubule. For this process to work optimally, you must have normal sodium channels, normal sodium chloride delivery and normal aldosterone. A defect in cortical collecting tubule potassium secretion may occur through lack of aldosterone, a lack of aldosterone effect or to a defect in either the sodium channel or sodium chloride delivery to this section of the tubule. Causes for a defect in potassium secretion are listed in Table 8.6.2.

2) **Evaluation**

After a careful search for causes that may cause shifting of potassium (mineral acidosis, diabetic ketoacidosis) and the exclusion of severe renal failure, it is reasonable to assume that there is a tubular defect in the secretion of potassium. Urine electrolytes can be obtained at this time. Potassium, urine sodium, urine creatinine, urine osmolality should be obtained. A urine potassium less than 20 mEq/liter, a fractional potassium excretion less than 30% or a transtubular potassium gradient (TTKG) of less than 8 are all indicative of a defect in potassium secretion (Table 8.6.3). Once a defect in potassium secretion is documented the search for the cause is indicated. Drugs known to cause hyperkalemia such as cyclosporine should be excluded. In the absence of these drugs, a simultaneous plasma renin activity and plasma aldosterone level can be obtained. A low plasma aldosterone with a low plasma renin activity would be indicative of a hyporeninemic hypoaldosterone state while a high plasma renin with a low plasma aldosterone would be indicative of a primary hypoaldosteronism state.

8.6

Table 8.6.2. Causes of hyperkalemia

Low Aldosterone
    1° adrenal failure
    hyporenin hypoaldosteronism
    ACE inhibitors
    NSAIDs
    heparin
Lack of Aldosterone Effect
    spironalactone
    tubulointerstitial disease
Na^+ channel Blockade
    amiloride
    pentamidine
    triamterine
    trimethoprin

---

3) Treatment

Treatment of hyperkalemia include: removing iatrogenic sources (e.g. K^+ containing intravenous fluids or TPN), increasing urine flow, increasing sodium delivery by the use of a loop diuretic and a low K^+ diet. Potassium exchange resins such as Kayexalate are also useful. In some cases of hypoaldosteronism the use of mineralcorticoid supplements are helpful. Treatment of severe hyperkalemia with EKG changes is outlined in Table 8.6.4.

B) HYPOKALEMIA

1) Etiology

a) Urinary Loss of Potassium

Urinary losses of potassium may be due to an excess aldosterone state. Certain drugs, such as aminoglycoside, amphotericin and cisplatin, can all cause significant potassium wastage as well as magnesium wastage. Renal tubular acidosis, particularly proximal renal

---

Table 8.6.3. Hyperkalemia/urine tests indicative of a tubular defect in potassium secretion

Urine Test	Result Indicating Defect in Tubular Secretion of K^+
urine K^+	< 20 mEq/liter
$FeK^+ = \dfrac{\dfrac{urine}{plasma}(K^+)}{\dfrac{urine}{plasma}(CR)} \times 100$ (must be corrected for ↓GFR, i.e. FeK+↑ with ↓ GFR)	< 30%
transtubular potassium gradient (TTKG) $(TTKG) = \dfrac{\dfrac{urine}{plasma}(K^+)}{\dfrac{urine}{plasma}(Osmolality)} \times 100$ Only applicable for a concentrated urine (urine osm > 300)	< 6-8%

8.6

Table 8.6.4. Treatment of hyperkalemia with EKG changes

Calcium gluconate 10 ml 10% solution IV push over 5-10 minutes

Regular insulin 10 units + 50 ml D50 IV push

Cation exchange resin: sodium polystyrene sulfonate (Kayexelate 30 g, can repeat). **Avoid enemas in BMT patients with mucositis or neutropenia.**

In patients with severe renal impairment, hemodialysis.

tubular acidosis, (RTA) or classic distal RTA, which may occur from chemotherapeutic drugs such as ifosfamide, are associated with severe hypokalemia. A frequent cause of hypokalemia in BMT is the massive diuresis these patients undergo after periods of large fluid overload, a common occurrence. Urinary electrolytes in hypokalemia secondary to urinary loss should show a high urine potassium, a high transtubular potassium gradient (TTKG) and a high fractional potassium excretion in the hyperaldosterone states. A high urinary chloride in the setting of metabolic alkalosis and particularly in the presence of hypertension, should point in the direction of a primary hyperaldosterone state. High-dose glucocorticoids may also produce a situation somewhat similar to that found in a hyperaldosterone state or diuretic use. Plasma renin levels and aldosterone levels may be obtained in certain cases of hypokalemia. High aldosterone levels in the presence of a low plasma renin activity is indicative of a primary hyperaldosterone state, while high plasma renin activity with high aldosterone is more indicative of secondary aldosteronism.

b) *GI Losses of Potassium*

GI losses, particularly through vomiting, may cause hypokalemia. A low urinary chloride in the setting of a hypokalemic metabolic alkalosis should point towards vomiting. Diarrhea will tend to produce a metabolic acidosis with hypokalemia and will tend to have a low urinary sodium.

c) *Shifting of Potassium from the Extracellular to Intracellular Space*

Intracellular shifting of potassium as a cause of hypokalemia most often occurs in the setting of either metabolic alkalosis or by the institution of insulin therapy.

2) **Treatment of Hypokalemia**

Treatment is by vigorous potassium repletion. The use of potassium-sparing diuretics may be needed. In such cases, amiloride or triamterene are often helpful. Amiloride (high dose 20 mg/day) is useful even in cases of drug toxicity related hypokalemia such as with amphotericin.

8.6

IV) **CALCIUM**

Calcium exists in three forms in the plasma: 1) protein bound (40%), 2) free (50%), 3) bound to anions, e.g. phosphate (10%). Alkalosis decreases ionized calcium while acidosis raises ionized calcium. Total plasma concentration of calcium fluctuates with albumin such that an increase or decrease of 1.0 g/dl of serum albumin will change total serum calcium 0.8 mg/dl in the corresponding direction.

A) HYPERCALCEMIA

Hypercalcemia is present when the total corrected calcium is greater than 10.1 mg/dl.

1) **Etiology**

Causes of hypercalcemia are: malignancy, immobilization, granulomatous disease, primary hyperparathyroidism, drugs (thiazides, lithium) and Addison's disease. The etiology of hypercalcemia of malignancy varies according to the type of malignancy. In squamous cell carcinomas and renal cell carcinoma the tumor produces an abnormal parathyroid hormone (PTH) related peptide that acts on the PTH receptor to promote both the bone and kidney to reabsorb calcium. In these patients urinary cyclic AMP is high but PTH immunoassay levels are suppressed, since the PTH-related peptide is not detected by immunoassays. Hematologic diseases such as myeloma and marrow infiltrate diseases such as breast cancer cause hypercalcemia by release of local cytokines that promote osteoclast activation.

2) Symptoms

Hypercalcemia causes drowsiness, lethargy, anorexia, nausea, constipation, muscle weakness, delusions, hallucinations and impaired renal concentration.

3) Treatment

For symptomatic hypercalcemia or a corrected calcium > 13 mg/dl several maneuvers may be attempted.

a) Normal Saline

First line therapy is hydration at a rate of at least 200 ml/hr provided cardiovascular status is not compromised. If fluid overload is a concern, loop diuretics (e.g. furosamide 10-40 mg) may be used but careful attention to avoid volume depletion is necessary.

b) Calcitonin

Dose is usually 200 units (4-8 IU/kg) every 6-12 hours IM or SQ. Onset of action is several hours.

c) Pamidronate

Normal dose is 60-90 mg in one liter NS or D_5W over 4 hours. Onset of effect is generally several days but lasts for weeks. Contraindicated in renal failure.

d) Mithramycin (plicamycin)

Mithramycin is a chemotherapeutic drug that prevents bone reabsorption. The usual dose is 25 μg/kg in 500 ml D_5W, infused over 4-6 hours. Onset of action is over 2-4 days and lasts 1-2 weeks. Contraindicated in renal or hepatic failure, coagulopathy or myelosuppression.

e) Glucocorticoids

The dose is usually 20-50 mg of prednisone PO BID. Onset of action may be as long as 1-2 weeks.

f) Gallium Nitrate

Dose is usually 200 mg/m^2 in one liter NS or D_5W IV over 24 hours for 5 days. Contraindicated in renal insufficiency.

g) Phosphates

Dose is usually 0.5-1.0 g orally TID. Contraindicated in renal insufficiency, serum phosphorous > 3 mg/dl or the product of serum calcium and phosphorous > 60.

B) HYPOCALCEMIA

Hypocalcemia is present when total corrected serum calcium is less than 8.5 mg/dl. Hypocalcemia frequently occurs in the setting of severe illness. Ionized calcium is affected by acid base disorders (acidosis increases and alkalosis decreases ionized calcium), sepsis, heparin and citrate.

1) Etiology

Hypocalcemia may be due to hypoparathyroidism, hypomagnesemia, malabsorption, drugs such as phenobarbital or phenytoin that increase

8.6

metabolism of 25(OH) vitamin D, nephrotic syndrome, acute pancreatitis, renal failure, rhabdomyolysis, osteoblastic metastasis, hyperphosphatemia and tumor lysis syndrome.

2) Symptoms

Hypocalcemia may cause neuromuscular hyperexcitability, tetany, seizures, prolonged QT and S-T intervals and parathesias.

3) Treatment

Urgent treatment of hypocalcemia is called for when the corrected serum calcium is less than 7.0 mg/dl or any patient showing signs of neuromuscular hyperexcitability and a corrected serum calcium less than 8.0 mg/dl. Therapy is calcium gluconate 10 ml of 10% solution intravenously over 20 minutes. Slow infusion is especially important when the patient is hypokalemic, hypomagnesemic or on digoxin. Patients on digoxin should be monitored during calcium infusion.

V) ACID BASE BALANCE

Acid base balance is frequently deranged in the bone marrow transplant patient. Acidemia is said to be present when the blood pH is less than 7.35. Alkalemia is present when the blood pH is above 7.45. Acid base balance has a critical dependence on the carbonic acid-bicarbonate buffer system. The relationship of blood pH, serum bicarbonate and serum PCO_2 is related by the Henderson-Hasselbach equation. This equation states that:

$$pH = 6.1 + \text{the log of } \frac{(HCO_3^-)}{(0.03 \times PCO_2)}$$

In normal individuals the blood bicarb is 24 mEq/L while the PCO_2 is 40 mm Hg.

VI) METABOLIC ACIDOSIS

Normal compensation for metabolic acidosis involves an increase in ventilation and a subsequent decrease in the PCO_2. This compensation is on the order of a drop of 1.25 mm Hg of PCO_2 for every drop of 1 mEq/L of bicarbarbonate (Table 8.6.5). Metabolic acidosis can be divided into two types. Normal anion gap and increased anion gap acidosis.

A) Normal Anion Gap Acidosis

Hyperchloremic acidosis (normal anion gap acidosis) can occur either through loss of bicarbonate or through a defect in renal tubular function. The causes of normal anion gap acidosis can be diarrhea, vomiting, ileal conduit, the administration of carbonic anhydrase inhibitors, renal tubular acidosis or tubular interstitial renal disease (Table 8.6.6). In BMT, recurrent vomiting, sustained gastric suction and chronic diarrhea from antibiotic use or GVHD are common causes of metabolic acidosis.

8.6

Table 8.6.5. Compensation of acid base disorders

Acid Base Disorder	Δ HCO_3^-	Δ PCO_2
Metabolic acidosis	1 mEq/liter	$\downarrow$ 1.25 mm Hg
Metabolic alkalosis	1 mEq/liter	$\uparrow$ 0.75 mm Hg
Respiratory acidosis	Acute $\uparrow$ 1 mEq/liter	10 mm Hg
	Chronic $\uparrow$ 4 mEq/liter	10 mm Hg
Respiratory alkalosis	Acute $\downarrow$ 2 mEq/liter	10 mm Hg
	Chronic $\downarrow$ 4 mEq/liter	10 mm Hg

Table 8.6.6. Causes of metabolic acidosis and alkalosis

Metabolic Acidosis		Metabolic Alkalosis
Normal Anion Gap	Increased Anion Gap	
Diarrhea	Diabetic ketoacidosis	Vomiting
Renal tubular acidosis	Lactic acidosis	NG tube drainage
Hyperparathyroidism	Alcoholic ketoacidosis	Diuretics
Hypoaldosteronism	Salicylate overdose	Hyperaldosteronism
Obstructed ileal conduit	Methanol intoxication	Villous adenoma of colon
Dilutional acidosis	Ethylene glycol ingestion	Bartter's syndrome
Recovery from diabetic ketoacidosis	Renal failure	Exogenous bicarbonate Administration
Renal failure		Excess glucocorticoids (exogenous and endogenous)

B) ANION GAP ACIDOSIS

Acidosis with an increased anion gap occurs when there is a large amount of unmeasured organic anions. This leads to a suppression of chloride and a subsequent calculation of a large anion gap. The causes of anion gap acidosis include lactic acidosis, diabetic ketoacidosis, starvation, alcohol ketoacidosis and severe renal insufficiency (Table 8.6.6). Lactic acidosis and acidosis of renal failure are common causes of anion gap acidosis in BMT patients.

VII) METABOLIC ALKALOSIS

Metabolic alkalosis can be conveniently divided into two forms, sodium chloride responsive type and sodium chloride resistant type. Chloride responsive metabolic alkalosis is typically caused by vomiting, gastric suction, use of diuretics and after acute compensation for chronic hypercapnia. Chloride resistant metabolic alkalosis is typically due to hyperaldosteronism (Bartter's syndrome, Cushing's syndrome and occasionally severe potassium depletion). The normal compensation for metabolic alkalosis is listed in Table 8.6.5.

VIII) RESPIRATORY ALKALOSIS

Respiratory alkalosis occurs as a primary decrease in the PCO_2 content of the blood. The process that stimulates ventilation can cause respiratory alkalosis. This would include anxiety, central nervous system lesions, trauma, sepsis and pulmonary disease. In addition, fever, salicylate intoxication and hepatic insufficiency are frequent causes of a primary respiratory alkalosis. Compensation for respiratory alkalosis is listed in Table 8.6.5.

IX) RESPIRATORY ACIDOSIS

Respiratory acidosis occurs as a primary decrease in alveolar ventilation. This occurs either secondary to central nervous system lesions or to severe pulmonary disease.

X) RENAL REPLACEMENT MODALITIES

Indications for renal replacement therapy in the setting of acute renal failure include severe hyperkalemia, severe metabolic acidosis and severe volume overload which cannot be controlled by other measures. As renal failure becomes more prolonged, signs and symptoms of uremia may become more prominent and indicate the need for renal replacement therapy. These include neurologic toxicity,

8.6

malaise and uremic pericarditis. In general three forms of renal replacement therapy are available, hemodialysis, continuous hemofiltration and peritoneal dialysis.

A) HEMODIALYSIS

Hemodialysis works by diffusion of small molecular weight substances across a semi-permeable membrane. Removal of substances occurs via a concentration gradient. Removal of fluid occurs via ultrafiltration. For hemodialysis an access must be placed somewhere in the vasculature, usually either in a central vein or through an arterial venous graft or fistula. Advantages of hemodialysis are the ability to rapidly remove solutes, toxins and fluid. This rapid removal of fluid and electrolytes may not be tolerated in certain patients. Hemodialysis is the treatment of choice in patients who are hemodynamically stable and who need a rapid removal of electrolytes or fluid.

B) HEMOFILTRATION

Continuous hemofiltration, either continuous venousfiltration (CVVH) or continuous arterial venous filtration (CAVH) is an alternative to hemodialysis and peritoneal dialysis. As opposed to dialysis, hemofiltration removes solutes via the removal of large volumes of fluid with replacement of relatively electrolyte-free and solute-free solutions. Approximately 1-2 liters of plasma is removed per hour with lower volume replacement solution to equal net fluid loss. The advantages of continuous hemofiltration include the ability to remove large amounts of fluid slowly over time and the ability to adjust fluid losses hour by hour. In addition, it allows for slow solute and toxin removal and thus causes smaller osmotic shifts. Continuous hemofiltration is an excellent modality in the hemodynamically unstable patient. Disadvantages of continuous hemofiltration include the need for anticoagulation and the time intensive nursing care necessary to carry this procedure out. Ultimately, the choice of renal replacement therapy will depend on the clinical condition of the patient and the resources available in the particular hospital in which it is being done.

C) PERITONEAL DIALYSIS (PD)

PD uses the peritoneum as a dialysis membrane. PD may be performed either acutely or on a chronic basis. Removal of toxins in PD, occurs via a concentration gradient from the body into the peritoneal space where a dialysis solution is indwelling. Fluid removal occurs through ultrafiltration via an osmotic gradient, usually with a high concentration of dextrose. Solute removal in PD is dependent on the rapidity and frequency of exchanges. The removal of fluid is dependent on the concentration of dextrose in the dialysate fluid. Advantages include its ability to remove fluid slowly and thus be tolerated better in hemodynamically unstable patients. In addition, it allows for the avoidance of central venous access in patients who may already have several indwelling central catheters. The disadvantage of peritoneal dialysis include its rather poor clearance, significant amount of protein loss into the PD fluid and the need to violate the peritoneal space.

XI) CYCLOSPORINE (CsA)

Cyclosporine is an 11 amino acid cyclic peptide. It is commonly used to prevent graft-vs-host disease in BMT patients. Its mechanism of action is to decrease calcium-dependent IL-2 production of lymphocytes. The major limiting toxicity of CsA is renal. CsA renal toxicity is manifest as a decrease in both renal blood flow (RBF) and glomerular filtration rate (GFR). The decrease in RBF and GFR is mediated via afferent arteriolar vasoconstriction. In addition to its hemodynamic toxic-

8.6

ity, CsA also produces histologic renal damage manifested by interstitial fibrosis, tubular atrophy and vasculopathy. In addition to the above, CsA also produces tubular abnormalities manifest as hyperkalemia, hyperuricemia and hypomagnesemia. CsA levels can be measured by various methodologies (Table 8.6.7). Knowledge of the center's methodology is critical in interpreting CsA levels. Some drugs known to affect the metabolism of CsA are listed in Table 8.6.8.

XII) DRUG DOSING IN RENAL IMPAIRMENT

Renal impairment can both affect the clearance and action of many drugs. In general, the loading dose of a drug is dependent on the volume of distribution of that drug and is not affected by renal dysfunction. Certain drugs do have an altered volume of distribution in renal failure e.g. digoxin and fluoroquinalones, and these may need adjustment in loading dose. Obviously drugs that have predominant elimination via the kidney will be affected most by changes in renal function. In general, if a drug has greater than 40% elimination by the kidney, a dosage adjustment should be made in renal insufficiency. The normalized Cockcroff-Gault equation can be used to calculate estimated creatinine clearance (CrCl)per 70 kg (for use with dosing nomograms).

$$CrCl\ (ml\ /\ min\ /\ 70kg)\ =\ \frac{140-age}{Cr}\ (x\ 0.85\ for\ women)$$

This equation is only valid when the serum creatinine is not changing. Common drugs used in BMT and their dosage adjustment in renal impairment are listed in Table 8.6.9.

Table 8.6.7. Methods of monitoring CsA levels

Method	Therapeutic Level (12 hour trough)	Comment
Monoclonal radio-immunoassay	200-400 ng/ml	30% cross-reactivity with metabolites
Monoclonal fluores-cence polarization	200-400 ng/ml	20-50% cross-reactivity with metabolites
Immunoassay (TDX) polyclonal assays	400-900 ng/ml	large cross-reactivity with metabolites
HPLC	150-250 ng/ml	specific for CsA (no metabolites)

8.6

Table 8.6.8. Drugs altering CsA levels

↑ Levels	↓ Levels
Macrolide antibiotics	Phenobarbital
Azole antifungals	Phenytoin
Verapamil	Rifampin
Diltiazem	Carbamezipine
± Nicardipine	Isoniazide
Cimetidine	

Table 8.6.9. Common drugs in BMT that need dosage adjustment

Drug	CrCl (ml/min/70 kg)				
	> 80	50-80	20-50	< 20	Dialyzed
Acyclovir	5 mg/kg q8°	same	q12-24°	2.5 mg/kg q24°	Yes
Allopurinol	300 mg/d	75%	50%	25%	Yes
Aminoglycosides#	100% (q8-12°)	60-90% (q12°)	30-60%(q12-24°)	10-30% (q24-48°)	Yes
Amphotericin	0.5-1.0 mg/kg/d	same	same	q36-48°	No
Ampicillin	0.5-2 g q4-6°	same	75%	50%	Yes
Azathioprine	0.5-3 mg/kg/d	same	75%	50%	Yes
Aztreonam	1-2 g q8°	same	0.5-1g q12°	0.5g q12°	Yes
Cefazolin	1-2 g q8°	q8°	0.5-1g q12°	0.5g q24°	Yes
Cefaperazone	1-2 g q12°	same	same	same	No
Cefotaxime	1 g q6°	q6°	q8-12°	q24°	Yes
Cefoxitin	1-2 g q6-8°	q8°	q12°	q24°	Yes
Ceftazidime	1-2 g q8°	q8-12°	1g q12-24°	0.5g q24°	Yes
Ceftizoxime	1-2 g q6-8°	q8-12°	1g q12-24°	0.5g q24°	Yes
Ceftriaxone	0.5-1 g q12-24°	same	same	same-q24°	No
Clindamycin	600-900 mg q8°	same	same	same	No
Cimetidine	400 mg q12°	same	50%	25%	Yes
Ciprofloxocin	250-750 mg q12°	same	50-75%	50%	No
Compazine	5-10 mg q6°	same	same	same	NA
Cyclosporine*	3-10 mg/kg/day	same	same	same	No
Clarithromycin	500 mg q12°	same	75%	50%	Yes
Erythromycin	250-750 mg q6°	same	same	50-75%	No
Ethambutol	15 mg/kg/d	same	same	q48°	Yes
Famotidine	20-40 mg q24°	same	50%	25%	No
Fluconazole	200-400 mg/d	same	50-100 mg/day	50 mg/day	Yes
FK 506*	0.1-0.5 mg/kg/day	same	same	same	No
Ganciclovir	5 mg/kg q12°	same	25-50%	12%	Yes
Imipenim/Cilastin	0.25-1 g q6°	same	50%	< 25%	Yes
Isoniazide	300 mg/d	same	same	50%	Yes
Itraconazole	100-200 mg q12°	same	same	same	No
Methotrexate	up to 12 g/m²	same	50%	?	No
Metoclopramide	10-40 mg QID	same	75%	50%	No
Metronidazole	250-750 mg q8°	same	same	50%	Yes
Norfloxacin	400 mg q12°	q12-24°	q24°	avoid	No
Odansetron	0.15-0.45 mg/kg	same	same	same	No
Omeprazole	20 mg/d	none	none	none	No
Pentamidine	4 mg/kg/d	same	q36°	q48°	No
Piperacillin	3-4 g q4-6°	same	75%	50%	Yes
Ranitidine	300 mg/d PO	75%	50%	25%	Yes
Ticarcillin	3-4 g q4-6°	same	50%	25%	Yes
Trimethoprim-sulfamethoxazole	5-20 mg/kg/d of TMP	same	50-75%	25%	Yes
Ursodeoxy-cholic acid	300-600 mg PO BID-TID	same	same	same	No
Vancomycin#	15 mg/kg q12°	15 mg/kg q24-36°	15 mg/kg q36-48°	15 mg/kg q3-7 days	No

8.6

*Some centers adjust FK 506 and cyclosporine dose on renal function—refer to GVHD chapter
Refer to drug-specific dosing information   ° = hour

9

Growth Factors

Lori A. S. Hollis, Carole B. Miller

I) INTRODUCTION

More than 20 hematopoietic cytokines and growth factors are approved, currently under investigation or near approval. Figure 9.1 summarizes the regulation of hematopoiesis by growth factors. Colony stimulating factors, in general, act by binding to specific cell surface receptors and stimulating proliferation, differentiation, commitment and selected end-cell functions. Cytokines have multiple complex effects other than hematopoiesis giving rise to the term cytokine network. Some of the various effects of cytokines are listed in Table 9.1.

II) FILGRASTIM (HUMAN GRANULOCYTE-COLONY STIMULATING FACTOR, G-CSF)

A) PHARMACOLOGY OF G-CSF

G-CSF is a nonglycosylated recombinant human protein produced by *E. coli* using recombinant DNA technology. The activity of G-CSF is limited to the neutrophil pathway of stem cell development. It promotes neutrophil proliferation, differentiation and enhanced phagocytic ability. It has no effect on other cell lines including red cells or platelets.

There appears to be a dose-related effect of G-CSF on the recovery of neutrophil counts over a range of 1 - 70 μg/kg/day. The dose-related effects are demonstrated by both a decrease in the duration of neutropenia after cytotoxic therapy and in the peak granulocyte count achieved after recovery. Increasing the dose will increase the response to growth factor only up to a certain point, then higher doses begin to have either a decreasing effect or no additional benefit over lower doses. For example, after autologous BMT for breast cancer, patients who received G-CSF doses of 32 and 64 μg/kg/day had lower white cell counts on day 15 after transplant compared to patients who received 16 μg/kg/day.

B) CLINICAL APPLICATIONS OF G-CSF IN BONE MARROW TRANSPLANTATION

1) Peripheral Blood Progenitor Cell (PBPC) Mobilization

Two agents, chemotherapy and/or growth factors, have been used to increase the number of progenitor cells collected when harvesting PBPCs. After chemotherapy, apheresis begins when the peripheral white blood cell count reaches 1000/μl, and is generally performed daily for 2-4 days. After growth factors, apheresis is generally done for 2-4 consecutive days, beginning day 4 after G-CSF. Some small studies have indicated that the use of CSF-mobilized PBPCs may hasten hematologic recovery and improve clinically measurable transplant-related toxicities such as transfusion requirements, mucositis, antibiotic use and duration of hospitalization over transplants performed without CSF- mobilized cells.

2) Neutrophil Recovery—Use of G-CSF After Marrow or PBPC Infusion

Although no controlled randomized studies using G-CSF in BMT have been completed at this time, its use does appear to have some benefit compared to historical controls. The time to neutrophil recovery after autologous BMT has decreased to approximately 15 days with the addition

Bone Marrow Transplantation, edited by Richard K. Burt, H. Joachim Deeg, Scott Thomas Lothian, George W. Santos. © 1996 R.G. Landes Company.

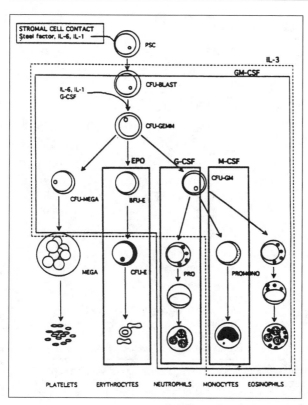

Fig. 9.1. Schema of hematopoiesis with growth factor roles. Abbreviations: PSC-primitive stem cell; CFU-GEMM-colony-forming unit, granulocyte erythroid-monocyte-megakaryocyte; CFU-MEGA-colony-forming unit, megakaryocyte; BFU-E-burst-forming unit, erythroid; CFU-GM-colony-forming unit, granulocyte-macrophage; CFU-E-colony-forming unit, erythroid; mega-megakaryocyte; pro-promyelocyte; promono-promonocyte; EPO-erythropoietin.

of G-CSF after bone marrow infusion. While the addition of G-CSF does not decrease the duration of absolute neutropenia (absolute neutrophil count (ANC) < 100/μl), it has shortened the overall period of neutropenia (ANC < 500/μl) in patients receiving both autologous and allogeneic transplants. It does so by rapidly increasing counts once neutrophil recovery in the bone marrow has begun.

A small number of studies have indicated that G-CSF after autologous BMT not only shortens the duration of neutropenia, but also decreases the number of febrile days. G-CSF administration after allogeneic transplantation has decreased the duration of fever and antibiotic use in one trial but, as of yet, no trials have demonstrated any significant decreases in hospitalization duration.

Initial studies of growth factors in BMT avoided the allogeneic setting based on concern over immune system activation having a detrimental effect on graft-versus-host disease, graft rejection or relapse. However, studies of G-CSF given after allogeneic BMT have demonstrated no increase in complications of GVHD, graft rejection or relapse.

3) Graft Failure

Currently, there is no evidence to suggest that G-CSF is useful in graft failure or delayed engraftment.

9

Table 9.1. Cellular effects of cytokines

Effect	Cytokine
T cell proliferation	IFN-α, IL-1, IL-2, IL-4, IL-6, IL-7, IL-9, IL-10, IL-12, GM-CSF, TGF-β
T cell activation	IL-1, IL-2, IL-4, IL-6, IL-7, IL-8, TGF-β
T cell differentiation	IFN-γ, IL-2, IL-4, IL-6, IL-7, TGF-β
B cell proliferation	IFN-α, INF-β, IFN-γ, TNF, LT, IL-1, IL-2, IL-3, IL-4, IL-5, IL-7, TGF-β
B cell activation	IFN-γ, TNF, IL-4, IL-5
B cell differentiation	IFN-α, IFN-γ, TNF, LT, IL-1, IL-2, IL-3, IL-4, IL-5, IL-6, IL-7, IL-10, TGF-β
Macrophage activation	IFN-α, INF-β, IFN-γ, TNF, IL-2, IL-4, IL-7, IL-10, M-CSF, GM-CSF
NK cell activation	IFN-α, INF-β, IL-1, IL-2, IL-3, IL-4, IL-6, IL-7, IL-12, TGF-β
Endothelial cell activation	IFN-γ, TNF, LT, IL-1, IL-4, TGF-β
Hematopoietic cell proliferation	TNF, LT, IL-1, IL-3, IL-4, IL-6, IL-11, G-CSF, M-CSF, GM-CSF, LIF, C-kit ligand, mpl ligand
Other effects:	
upregulate MHC class I	LT, IL-5
upregulate MHC class II	IFN-α, INF-β, IFN-γ, IL-3, IL-4, IL-10, TGF-β
upregulate adhesion molecules	IFN-γ, TNF, IL-1, IL-2, IL-3, IL-4, IL-5, IL-6, IL-7, GM-CSF, TGF-β
angiogenesis	TNF, LT, GM-CSF, TGF-α, TGF-β
antitumor effect	IFN-α, INF-β, IFN-γ, TNF, IL-1, IL-2, IL-4, IL-6, M-CSF

Abbreviations: c-kit ligand = stem cell factor, G-CSF = granulocyte-colony stimulating factor, GM-CSF = granulocyte macrophage-colony stimulating factor, INF = interferon, IL = interleukin, LT = lymphotoxin, mpl ligand = thrombopoietin, M-CSF = macrophage-colony stimulating factor, TGF = tumor growth factor, TNF = tumor necrosis factor

C) DOSE OF G-CSF

Studies of G-CSF for PBPC mobilization have appeared to show no incremental release of PBPCs above doses of 3 μg/kg/day given subcutaneously. Subsequent trials have failed to show any increased yield at higher doses. Many studies, however, have used higher doses given intravenously in an attempt to improve response for patient convenience. The dose of G-CSF is often rounded to the nearest vial size, either 300 μg or 480 μg.

The current FDA approved dosing guidelines for G-CSF after BMT recommend starting at a dose of 10 μg/kg/day and tapering the dose to 5 μg/kg/day once the ANC reaches 1,000/μl for at least 3 days. G-CSF should be discontinued once the ANC is greater than 1,000/μl for 3 more days. In clinical practice, however, different institutions choose different doses and routes to administer growth factors. Most commonly in the bone marrow transplant setting, intravenous infusions of 5-15 μg/kg/day are infused over 4 hours. Higher doses of G-CSF used in the BMT setting have indicated similar activity at doses of greater than 10, 20, and 30 μg/kg/day. Therefore, it is believed that higher doses of growth factor are not necessary.

G-CSF may be given by daily subcutaneous injections, intravenous infusion or as a continuous infusion, but the latter route has increased toxicity. With the exception of shorter infusions (i.e. over less than 1 hour) having diminished effects, all other routes appear to have similar efficacy.

In general, administration of G-CSF is started on the day that bone marrow or PBPCs are reinfused and continue to up to 21 days or until neutrophil counts recover. There is some preliminary data that suggest delaying the start of G-CSF up to 8 days after bone marrow infusion may retain similar efficacy while reducing the cost of growth factor administration.

D) TOXICITIES/RISKS OF G-CSF

The most common side effect seen with G-CSF administration is medullary bone pain. This pain appears to occur more frequently at higher doses and when the drug is given by prolonged intravenous infusions. The pain can sometimes be severe requiring analgesics. There appears to be less bone pain when G-CSF is given subcutaneously. Other rare side effects of G-CSF include exacerbations of pre-existing inflammatory conditions including eczema, psoriasis or vasculitis. Some laboratory abnormalities that may be related to the use of G-CSF include elevations in LDH, uric acid and alkaline phosphatase.

III) SARGRAMOSTIM (HUMAN GRANULOCYTE-MACROPHAGE COLONY STIMULATING FACTOR, GM-CSF)

A) PHARMACOLOGY

There are several DNA recombinant forms of GM-CSF. Molgramostim rhGM-CSF, derived from *E. coli*, is similar to that derived from yeast except the toxicities may be more serious. First dose reactions including flushing, tachycardia, hypotension, dyspnea and rigors are more common especially when given intravenously. Molgramostim is available in Europe. Regramostim rhGM-CSF is another GM-CSF which is derived from Chinese hamster ovaries. This discussion will consider only sargramostim rhGM-CSF which was derived from yeast through recombinant DNA technology.

GM-CSF stimulates the proliferation and differentiation of several hematopoietic progenitor cells including granulocyte, monocyte, erythroid and multilineage hematopoietic colonies. It, therefore, has broader activity in the hematopoietic system than G-CSF. GM-CSF is also a potent activator of mature macrophages and granulocytes, thereby inducing adhesion molecules and enhancing superoxide production, degranulation, phagocytosis and intracellular killing. Theoretically, these properties may make it a more useful agent in severely immunocompromised hosts. Similar to G-CSF, GM-CSF increases neutrophil, eosinophil and monocyte counts in a dose-dependent manner.

B) CLINICAL APPLICATIONS IN BONE MARROW TRANSPLANT

1) Peripheral Blood Progenitor Cell (PBPC) Mobilization

To date it is unclear which colony stimulating factor (G-CSF or GM-CSF) will prove most effective as priming for PBPC harvest. In general, leukapheresis is usually done daily for 3-4 consecutive days between 5 and 9 days after starting GM-CSF.

2) Neutrophil Recovery—Use of GM-CSF After Marrow or PBPC Infusion

In general, GM-CSF has been shown to decrease the time to engraftment, shorten duration of hospitalization and decrease antibiotic use when given after autologous BMT. In some institutions, this has translated to cost savings. The use of GM-CSF after autologous bone marrow transplant for lymphoid malignancies decreases the time to reach neutrophil counts greater

9

than 500/µl although the duration of severe neutropenia (ANC less than 100/µl) is not affected. Essentially, GM-CSF has no effect on platelet recovery.

In the allogeneic transplant setting, GM-CSF has the theoretical potential to increase the severity of graft-versus-host disease through activation of monocytes and upregulation of macrophage-produced inflammatory mediators such as interleukin-1 and tumor necrosis factor. However, effects such as increasing the incidence or severity of GVHD have not been observed in clinical trials.

GM-CSF has also been shown to be beneficial after allogeneic BMT but the evidence is not as strong as in the autologous setting. In a trial of patients receiving cyclosporine and prednisone for GVHD prophylaxis after transplants from HLA- matched donors, GM-CSF significantly decreased the time to neutrophil engraftment compared to historical controls. There was no increase in GVHD and the infection rate was lower.

In a phase I trial of patients receiving transplants from unrelated donors, the addition of GM-CSF after transplant did not affect the time to neutrophil recovery compared to historical controls; however, there were fewer episodes of sepsis in the patients who received growth factor. Also, while the overall incidence of GVHD was no different than historical controls, there was a significantly lower incidence of severe forms of GVHD in patients who received GM-CSF.

3) Graft Failure

Overall, approximately 50% of patients with graft failure responded to one or two 2 week courses of GM-CSF with increases in their ANC to greater than 500/µl. The survival of these patients is better than the historical rate of 20-30%, but these have not been controlled studies. Some practitioners feel that, given the dismal prognosis of graft failure and the lack of serious

Table 9.2. Growth factor dose

Growth Factor	Clinical Situation	Dose
G-CSF	Peripheral blood progenitor cell (PBPC) mobilization	3-10 µg/kg/day subcutaneously or intravenously over 4 hours
	Hematopoietic recovery after BMT	5-15 µg/kg/day intravenously infused over 4 hours or subcutaneously
GM-CSF	Peripheral blood progenitor cell (PBPC) mobilization	250 µg/m²/day subcutaneously, continuous infusion or 2-4 hour intravenous infusion
	Hematopoietic recovery after BMT	250 µg/m²/day subcutaneously, continuous infusion or 2-4 hour intravenous infusion
	Graft failure	250–500 µg/m²/day subcutaneously, continuous infusion, or 2-4 hour intravenous infusion
Epoietin	Hematopoietic recovery after BMT	150–300 units/kg subcutaneously 3-7 times per week

9

adverse effects, a trial of GM-CSF is warranted in any patient who presents with this condition.

C) **DOSE OF GM-CSF**

Dose escalation is usually limited by toxicity to less than 500 μg/m²/day. The current FDA approved dosing guidelines for GM-CSF after BMT are limited to the autologous setting. The FDA recommended dose is 250 μg/m²/day given as a 2 hour intravenous infusion. Initiation of therapy should begin 2-4 hours after autologous bone marrow infusion and not less than 24 hours from the last dose of chemotherapy and 12 hours from the last dose of radiotherapy. Therapy should continue for 21 days and should be discontinued or dose reduced by half if the ANC increases to greater than 20,000/μl. While a small study demonstrated that doses of GM-CSF as low as 60 μg/m² have appeared to promote earlier recovery of functional levels of neutrophils, most studies have used doses of 250 mcg/m²/day. There are no data to suggest that increasing the dose beyond this increases effectiveness; it may simply increase toxicity.

GM-CSF may be given by continuous infusion, intravenous infusions over 2-4 hours or as a subcutaneous injection. All routes of administration appear to have similar efficacy. Daily subcutaneous injections are often avoided in transplant patients due to the risk of bleeding or purpura when platelet counts are low. Some authors have suggested that continuous infusions of GM-CSF may be more effective than 2 hour infusions, but this remains to be proven in a clinical trial.

The FDA approved dosing guidelines for GM-CSF in graft failure are 250 mμg/m²/day for 14 days as a 2 hour intravenous infusion. The treatment can be repeated after 7 days off therapy if engraftment has not occurred. If engraftment still has not occurred, a third course of 500 μg/m²/day for 14 days may be tried after another 7 days off therapy. Prolonged administration of these high doses may cause increased toxicity and the dose should be reduced or temporarily discontinued until adverse effects resolve.

D) **TOXICITY/RISKS OF GM-CSF**

The toxicities of GM-CSF are somewhat dose, route and duration related. Continuous infusions induce more frequent and more severe toxicities than shorter infusion rates or subcutaneous doses. In high doses (greater than 250 μg/m²/day), many patients may experience myalgias or bone pain. Sometimes the bone pain may be severe enough to require narcotic analgesics. Few patients will experience reversible pleural or pericardial effusions after intravenous administration. These side effects appear to be more common when GM-CSF is given by continuous infusion as opposed to short infusions of 2-4 hours.

Many studies have reported cases of fever, nausea, fatigue and injection site reactions. Less commonly reported adverse effects include diarrhea, anorexia, arthralgias, skin rashes, flushing and hypotension. Also, capillary leak has been reported but may be associated more with the *E. coli*-derived form of GM-CSF that is not currently approved for use in the United States. The studies that reported these effects had a high rate of background toxicities due to other therapy given during the trials, so the true relationship of these effects to administration of GM-CSF is unclear.

9

Some laboratory abnormalities that may be related to the use of GM-CSF include increases in LDH, uric acid, and alkaline phosphatase, mild thrombocytopenia and eosinophilia. Decreases in serum cholesterol and albumin may also be seen.

IV) EPOIETIN (RECOMBINANT HUMAN ERYTHROPOIETIN, rHuEPO)

A) PHARMACOLOGY

rHuEPO is a 165 amino acid glycoprotein produced by recombinant technology. Because the post-translational glycosalation is required for clinical efficacy, rHuEPO is manufactured in mammalian cells. rHuEPO contains the identical amino acid sequence as natural human erythropoietin. Erythropoietin is a single chain glycoprotein hormone produced mainly by the peritubular cells of the kidneys in response to hypoxia. A small amount of erythropoietin is produced in the adult liver. Erythropoietin has a very restricted range of target cell activity. It stimulates the growth and differentiation of both primitive erythroid progenitors (burst forming units–erythroid, BFU-E) and committed erythroid progenitors (colony forming units—erythroid, CFU-E). In contrast to other circulating mature blood cells which require interaction with their growth factors for function, mature red blood cells lack receptors for erythropoietin and the hormone cannot effect the mature progeny of stimulated erythroid progenitors.

As there are no preformed stores of erythropoietin, erythropoietin production is an early response to tissue hypoxia. When hypoxia is detected in the oxygen sensing cells of the kidney, more erythropoietin-producing cells are recruited to produce erythropoietin. Production of erythropoietin is tightly regulated by hypoxia at the level of the gene and there is no feedback inhibition based on serum levels of the hormone.

B) CLINICAL APPLICATIONS OF EPOIETIN IN BONE MARROW TRANSPLANTATION

rHuEPO has been approved for use in chemotherapy-associated anemia based on two large randomized trials that showed an improvement in hematocrit and a decrease in transfusion related to rHuEPO. The role of rHuEPO after bone marrow transplantation is still being defined. Anemia is universal after both autologous and allogeneic bone marrow transplantation and red blood cell transfusion is common in the posttransplant period. The etiology of the anemia after bone marrow transplantation is multifactorial with both increased utilization and inadequate production of red blood cells playing roles.

While many factors influence erythroid recovery after bone marrow transplantation, the prolonged reticulocytopenia and resultant large transfusion requirement after bone marrow transplantation is in part related to a relative erythropoietin deficiency at a time of marrow recovery. This depressed erythropoietin response to anemia persisted up to 1 year after bone marrow transplant. Sixty percent of the patients seen at the 6 month follow up visit remained anemic (hemoglobin < 11 grams per deciliter). It is important to understand that a "normal erythropoietin level" as described when lab test results are returned, means a level which is normal for someone who is not anemic. In order to determine the adequacy of the erythropoietin response to anemia, a nomogram of the normal erythropoietin response is needed (Fig. 9.2).

Four randomized trials of rHuEPO after allogeneic bone marrow transplant have been completed. All of the allogeneic studies showed accelerated eryth-

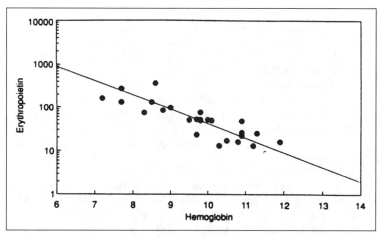

Fig. 9.2. Erythropoietin-hemoglobin relationship in 27 iron deficient patients. This can be used as a nomogram for determining the adequacy of erythropoietin response to anemia in BMT.

roid engraftment but only two showed a significant decrease in transfusion requirement. Whether rHuEPO is cost effective early postallogeneic transplant is still unknown. None of the autologous BMT studies showed a benefit to early rHuEPO use.

We have completed a pilot trial of rHuEPO in delayed erythropoiesis (hemoglobin < 9 g/dl after day 21) after bone marrow transplant. Patients were treated with 150 units per kilogram three times a week with escalation to 300 units per kilogram. Twenty patients were entered (13 allogeneic, 7 autologous) at a median day 44 after bone marrow transplant. Fifteen of eighteen evaluable patients (83%) responded with an increase in hemoglobin of 1 g/dl or decrease in transfusion requirement. Nine patients had a complete response (hemoglobin increase greater than 2 g/dl or 100% decrease in transfusion requirement). Response correlated with baseline erythropoietin level with a complete response being achieved in 7 of 8 patients with a baseline erythropoietin level of less than 50 milliunits/ml but only 2/9 with baseline level of greater than 50 milliunits/ml. These data suggest that rHuEPO may be helpful in ameliorating anemia and decreasing transfusion requirement in patients with delayed erythropoiesis after bone marrow transplant.

C) NORMAL DOSING OF EPOIETIN

A standard dose has not been used in all studies. The recommended dose to start is approximately 150 units per kilogram 3 - 7 times a week. Dose can be escalated in nonresponders to 300 units per kilogram. rHuEPO is generally given subcutaneously because data from dialysis patients showed an improved response when the drug is given by that route. Several studies showed this route to be safe even in thrombocytopenic patients. If desired, rHuEPO can be given as an I.V. push.

D) TOXICITY/RISKS OF EPOIETIN

rHuEPO has been well tolerated in all nondialysis patients. In dialysis patients, a worsening of hypertension was seen in responding patients. This is

9

thought to be due to changes in plasma volume as rHuEPO has no intrinsic pressor effect. In BMT patients, no significant side effects were associated with rHuEPO use, with the exception of some burning at injection site.

V) OTHER CYTOKINES UNDER DEVELOPMENT OR IN TRIALS FOR BONE MARROW TRANSPLANTATION

A) C-KIT LIGAND

C-kit ligand is also known as Steel factor, stem cell factor or mast cell growth factor. It is thought to stimulate very immature hematopoietic progenitor cells. It is synergistic with other growth factors in promoting the expansion and maturation of blood cells. The toxicities of this growth factor are as yet unknown.

B) INTERLEUKIN-1

Interleukin-1 stimulates hematopoiesis and acts as a radioprotectant. It also may be synergistic with both GM and G-CSF. The maximum tolerated dose after autologous BMT is 0.3 - 0.5 µg/kg/day given as a 30 minute intravenous infusion. Dose limiting toxicity is hypotension. Other toxicities are fever, chills, arrhythmias, hypertension and bone pain. The role of rhIL-1 in BMT is currently undefined.

C) INTERLEUKIN-2

Interleukin-2 is a T cell growth factor. Human bone marrow has been activated in long term cultures with IL-2 as a method of in vitro purging. Animal models suggest that infusion of IL-2 for a short period after allogeneic marrow infusion inhibits GVHD, although the mechanism is unclear. IL-2 is also being used in various schedules and doses as an attempt to activate host lymphocytes, induce a graft-versus-tumor effect and eradicate minimal residual disease after autologous BMT for lymphomas and leukemias. Toxicity is dose related and includes hypotension, thrombocytopenia, fever, nausea, rash, fluid retention, arthalgias, confusion and a capillary leak syndrome.

D) INTERLEUKIN-3

Interleukin-3 has been shown in vitro to stimulate growth of granulocytes, megakaryocytes, eosinophils and erythroid cells. It is thought to be potentially synergistic with both GM and G-CSF and may stimulate platelet recovery. There is, currently, no evidence that rhIL-3 is superior to G-CSF or GM-CSF in terms of PBPC mobilization or hematopoietic recovery after BMT. Toxicity of rhIL-3 includes fever, confusion, headache, diarrhea, mucositis, rash, rigors, dyspnea and edema. After autologous transplant, the maximum tolerated dose is 2 µg/kg/day when given by 2 hour intravenous infusion.

E) INTERFERON-α

Interferon-α is a lymphokine that inhibits cell multiplication, activates macrophages and NK cells and has antiviral, antineoplastic and other immune modulating effects. It has been approved for use to treat hairy cell leukemia, AIDS-related Kaposi's sarcoma, chronic myelogenous leukemia (CML), multiple myeloma (MM) and low grade non-Hodgkin's lymphomas. After autologous BMT for CML or MM, interferon-α is being used as maintenance therapy in attempt to prolong remission or prevent relapse. It is generally given at a dose between 1-10 x 10^6 units/m² subcutaneously 3 times a week for 2 years. Toxicity includes flu-like symptoms, anorexia, nausea, vomiting, confusion, parathesias, dizziness, rash and dose-related hepatotoxicity and hematopoietic suppression, predominately thrombocytopenia.

9

Table 9.3. Cytokine application in bone marrow transplantation

Agent	Examples of Dosing	Potential (but unproven) BMT Applications	Side Effects
Interferon-α2A Leukemia Lymphoma 1992; 8(6):421.	1-10 x 10⁶ units/m² SQ 3 times a week	Immune modulation, e.g. post allogeneic or autologous BMT as maintenance therapy for CML, myeloma, indolent lymphomas	Frequent - fever, chills, headache, fatigue, anorexia, weight loss, alopecia Less common—nausea, vomiting, diarrhea, lymphopenia, neutropenia, thrombocytopenia, elevated transaminases, proteinuria, depression, confusion, hypotension
Interleukin-1 Blood 1994; 83(12):3473.	MTD after autologous BMT given over 30 minutes IV daily was 0.05 μg/kg	Immune modulation to decrease relapse or hematopoietic growth factor to accelerate engraftment	Frequent - fever, chills, headache, myalgias, nausea, vomiting Less common— somnolence, phlebitis, abdominal pain, dyspnea, hypotension
Interleukin-2 J Hematotherapy 1995; 4(2):113.	18 million IU/m²/day IV continuous infusion	Immune modulation used on marrow ex vivo or posttransplant (autologous or allogeneic) in vivo to decrease relapse or modulate GVHD, e.g. IL-2 increases Th1 T cell helper phenotype	Flu-like illness, fever, chills, rigors, myalgias, arthalgias, headache, fatigue, nausea, vomiting, diarrhea, tachycardia, hypotension, arrhythmias, edema, weight gain, vascular leak syndrome, dyspnea, pulmonary edema, congestive heart failure, renal insufficiency, elevated bilirubin and transaminases, rash, pruritis, confusion, depression, change in mental status, leukopenia, thrombocytopenia
Interleukin-3 Blood 1993; 82(11):3273.	30-500 μg/m² SQ daily; after autologous BMT MTD was 2 μg/kg/d IV over 2 hours	Mobilization of peripheral blood stem cells or hematopoietic growth factor to accelerate engraftment	Fever, headache, chills, bone pain, diarrhea, rash
PIXY321—fusion protein of IL-3 and GM-CSF Stem Cells 1994; 12(3):253.	SQ once or twice daily or IV over 2 hours at doses up to 1000 μg/m²/day	Mobilization of peripheral blood stem cells or hematopoietic growth factor to accelerate engraftment	Fatigue, nausea, vomiting, diarrhea, headache, fever, rigors, myalgias, dyspnea, edema, pericarditis

9

Table 9.3. Cytokine application in bone marrow transplantation (cont'd.)

Agent	Examples of Dosing	Potential (but unproven) BMT Applications	Side Effects
Interleukin-4		Modulation of immune function to decrease relapse or alter GVHD, e.g. IL-4 increases Th2 T cell suppressor phenotype	Clinical BMT studies pending
Interleukin-10		Modulation of immune function to decrease relapse or alter GVHD, e.g. IL-10 increases Th2 T cell suppressor phenotype	Clinical BMT studies pending
Interleukin-12		Immune modulation used on marrow ex vivo or posttransplant (autologous or allogeneic) in vivo to decrease relapse or modulate GVHD, e.g. IL-12 increases Th1 T cell helper phenotype	Clinical BMT studies pending
M-CSF monocyte-macrophage growth factor Blood 1993; 82(5):1422.	100-2000 µg/m²/day	Treatment of fungal infections after BMT	Dose-limiting toxicity is thrombocytopenia
C-kit ligand		Mobilization of peripheral blood stem cells or hematopoietic growth factor to accelerate engraftment	Clinical BMT studies pending
Thrombopoietin		Treatment of thrombocytopenia	Clinical BMT studies pending
TNF-tumor necrosis factor		Modulation of immune function to decrease relapse or alter GVHD	Clinical BMT studies pending

MTD = maximum tolerated dose

9

F) MONOCYTE-MACROPHAGE GROWTH FACTOR (M-CSF)

Since M-CSF increases the functional activities of monocytes and macrophages, it may prove useful in treating fungal infections. A nonrandomized trial of rhM-CSF at 100 $\mu g/m^2$/day to 2000 $\mu g/m^2$/day administered by 2 hour intravenous infusion for 28 consecutive days in BMT patients with fungal infections suggested benefit for patients with invasive candidiasis. The dose-limiting toxicity occurring at 2000 $\mu g/m^2$ is thrombocytopenia.

G) PIXY321

PIXY321 is a recombinant fusion protein of IL-3 and GM-CSF. In vitro, PIXY321 has a 10-fold greater proliferative effect on hematopoietic progenitors than the same concentration of GM-CSF and IL-3 combined. PIXY321 may be given subcutaneously or intravenously and is well tolerated at doses up to 1000 $\mu g/m^2$/day. Adverse reactions are fatigue, nausea, vomiting, diarrhea, headache, fever, rigors, myalgias, dyspnea, edema and pericarditis. Trials of PIXY321 in PBPC mobilization and hematopoietic reconstitution after BMT are in progress.

H) THROMBOPOIETIN

Thrombopoietin also known as mpl ligand is a growth factor for platelets which causes in vitro megakaryocytopoietic differentiation of early progenitor cells. Thrombopoietin has homology to erythropoietin perhaps explaining why anemic patients with high erythropoeitin levels may develop thrombocytosis. Clinical studies are pending.

SUGGESTED READING

1. American Society of Clinical Oncology. American Society of Clinical Oncology recommendations for the use of hematopoietic colony-stimulating factor: evidence-based, clinical practice guidelines. J Clin Onc 1994; 12(11):2471-2508.
2. Appelbaum FR. The use of colony stimulating factors in marrow transplantation. Cancer 1993; 72:3387-3392.
3. Miller CB, Mills SR. Erythropoietin after bone marrow transplantation. Hem/Onc Clin N Amer 1994;8 :975-991.
4. Nemunaitis J, Singer JW. The use of recombinant human granulocyte macrophage colony stimulating factor in autologous and allogeneic bone marrow transplantation. Cancer Invest 1993; 11(2):224-228.
5. Peters WP, Rosner G, Ross M et al. Comparative effects of granulocyte-macrophage colony-stimulating factor (GM-CSF) and granulocyte-colony stimulating factor (G-CSF) on priming peripheral blood progenitor cells for use with autologous bone marrow after high-dose chemotherapy. Blood 1993;8 1(7):1709-1719.
6. Petros WP, Peters WP. Hematopoietic colony stimulating factors and dose intensity. Sem Oncol 1993;2 0(1):94-99.

9

10.0 Infections

10

10

Infection Prophylaxis in Bone Marrow Transplant Recipients— Myths, Legends and Microbes

Richard K. Burt, Thomas Walsh

I) RISK FOR INFECTION

BMT patients are susceptible to infections from neutropenia, depressed T and B cell function, disruption of anatomical barriers (mucositis, indwelling intravenous and urinary catheters) and immunosuppressive medications (steroids, cyclosporine, methotrexate, anti-thymocyte globulin). Susceptibility to infection also depends on the type of BMT.

A) NEUTROPENIA

Risk of infection in neutropenic patients increases with depth of absolute neutrophil count (ANC) nadir (especially if $< 100/\mu l$), rapid rate of ANC decline and protracted duration of granulocytopenia (> 10 days). All of which are anticipated complications of BMT. The expected duration of neutropenia after hematopoietic stem cell transplantation (HSCT) depends on prior treatment, use of growth factors and source of hematopoietic stem cells (i.e. peripheral blood or bone marrow). The mean duration of an ANC less than $500/\mu l$ after bone marrow HSCT (autologous, conventional allogeneic and T cell depleted allogeneic BMT) is 2 to 3 weeks. The mean duration of an ANC less than $500/\mu l$ is shorter after peripheral blood HSCT (10-14 days). However, in general, neutropenia is protracted for autologous HSCT regardless of source (peripheral blood or marrow) if the patient is heavily pretreated with prior chemotherapy or the hematopoietic cells are manipulated ex vivo (e.g. chemical purging). Allograft failure is likely if the ANC remains less than $100/\mu l$ after day 21.

B) T CELL FUNCTION

Defects in T lymphocyte reconstitution, even in the absence of GVHD, are present for months to years after allogeneic BMT (Table 10.1.1). Natural killer (NK) cells which appear as large granular lymphocytes are increased in the peripheral blood following allogeneic BMT. The reason for increased NK cells is unclear but may be part of the normal sequence in recapitulation of lymphocyte ontogeny. CD4+ helper cells are decreased for 4-8 months. CD8+ suppressor cells although decreased for the first month are increased for the next 6-12 months. In healthy allogeneic grafts (i.e. no chronic GVHD), an inverted CD4/CD8 ratio is present for 6-12 months. In patients with chronic GVHD, a decreased CD4/CD8 ratio generally persists as long as chronic GVHD is present. An inverted CD4/CD8 ratio is normal after BMT and does not necessarily indicate evaluation for HIV infection. An exception to the inverted CD4/CD8 ratio after BMT occurs in recipients of allografts selectively depleted of CD8+ T cells. These patients generally have a normal CD4/CD8 ratio after BMT.

T lymphocyte proliferative response to phytohemagglutinin (PHA) mitogen is diminished for 6 months after conventional allogeneic BMT and for 18 months

Bone Marrow Transplantation, edited by Richard K. Burt, H. Joachim Deeg, Scott Thomas Lothian, George W. Santos. © 1996 R.G. Landes Company.

10.1

after T cell depleted (TCD) allogeneic BMT. Abnormalities in T cell number and function predispose patients to viral illnesses, especially herpes simplex, cytomegalovirus (CMV), varicella zoster and Epstein-Barr virus. Increased risk of viral infection in T cell depleted BMT may be explained by persistent abnormalities in T cell reconstitution. Less has been reported about T lymphocyte recovery after autologous BMT. However, similar to allogeneic BMT, NK cells predominate during early recovery. CD4+ helper T cells are decreased and CD8+ supressor cells are increased for several months. An inverted CD4/CD8 ratio may be present for 4-12 months following autologous BMT.

C) B CELL FUNCTION

B lymphocyte number and function are impaired for months to years after allogeneic BMT (Table 10.1.2). Mature B lymphocytes (CD20 phenotype) and their response to *Staphylococcus aureus* Cowen strain A (SAC) mitogen are diminished for 2-6 months after allogeneic BMT.

Immunoglobulin levels following BMT are also abnormal. IgE peaks at 3-4 weeks posttransplant and remains elevated in patients with GVHD. The explanation for increased IgE is unclear but may be due to elevated cytokines such as IL-4, which is involved in IgE isotype switching. In healthy allografts, IgG is diminished for up to 9 months after conventional BMT and 18 months after TCD BMT. Patients with chronic GVHD remain IgG deficient for as long as GVHD persists. Since T lymphocyte help is necessary for B lymphocyte isotype (IgM to IgG) switching, prolonged IgG immunoglobulin deficiency after TCD BMT may be due to T lymphocyte defects. Persistent hypogammaglobulinemia in patients with chronic GVHD predisposes to infections with encapsulated bacteria. Prolonged IgA deficiency increases susceptibility to infections at mucosal barriers (e.g. sinusitis).

Less is known about B cell reconstitution after autologous BMT. However, similar to allogeneic BMT, B cell number and response to SAC stimulation are decreased for 2-3 months after autologous BMT. IgE is also increased and peaks 3-4 weeks after autologous BMT. Defects in B cell production of immunoglobulins are common in multiple myeloma, both before and after BMT.

D) MUCOSITIS

Breakdown of mucosal barriers due to radiation or chemotherapy (especially anthracyclines, paclitaxel, etoposide, cytarabine, melphalan, thiotepa, busulfan) allows a portal of entry for bacteria. Medical management is largely

Table 10.1.1. T Cell reconstitution after allogeneic BMT

Phenotype	Conventional Allogeneic BMT	T Cell Depleted Allogeneic BMT
Natural Killer—NK	↑ for 6 months	↑ for 24 months
T Cell—CD3	↓ for 3 months	↓ for 3 months
Helper T Cell—CD4	↓ for 4-8 months	↓ for 4-8 months
Suppressor T Cell—CD8	normal or ↑ for 1 year	normal or ↑ for 1 year
CD4/CD8	↓ for 1 year	↓ for 1 year
T CELL Response to Phytohemagglutinin	↓ for 4-6 months	↓ for 16-18 months

10.1

supportive, consisting of bowel rest, total parental nutrition, good oral hygiene and analgesia (Table 10.1.3). All BMT candidates should have a pre-BMT dental consultation. Prophylactic acyclovir (200 mg TID-QID PO) is generally used in patients with positive herpes type I serology. The best therapy for mucositis is neutrophil recovery. Other drugs such as methotrexate, used as GVHD prophylaxis, may exacerbate chemotherapy induced mucositis.

E) TYPE OF BMT

GVHD impairs immune reconstitution and is treated with immunosuppressive drugs, both of which increase the risk of bacterial and fungal infection after allogeneic BMT. Therefore, once neutropenia resolves, the risk of infection correlates with the type of BMT: unrelated and mismatched allogeneic > HLA-matched sibling > autologous. The risk of viral disease is increased with T cell impairment. Therefore, recipients of T cell depleted allografts or patients with a greater chance for severe GVHD (i.e. recipients of unrelated or mismatched allografts) are at increased risk for viral diseases.

II) SOURCE OF ORGANISMS

The source for infection may arise from either normal endogenous body flora (e.g. the gastrointestinal tract) or exogenous acquisition of organisms from the environment. Entry is facilitated by neutropenia, immunosuppressive drugs, and anatomical barrier breakdown from mucositis, GVHD, conditioning regimen and catheters.

A) GRAM NEGATIVE RODS

Infections due to aerobic Gram negative rods (*Escherichia coli, Klebsiella pneumonia, Pseudomonas aeruginosa*, etc.) arise primarily from the gastrointestinal tract.

B) GRAM POSITIVE ORGANISMS

The portal of entry for Gram positive cocci (*Staphylococcus epidermidus, Staphylococcus aureus, Viridans streptococci* such as *Streptococcus mitus* and *Enterococcus* species) is usually the skin, oropharynx or through indwelling catheters.

Table 10.1.2. B cell reconstitution after allogeneic BMT

Phenotype	Conventional Allogeneic BMT	T Cell Depleted Allogeneic BMT
CD20	↓ for 3 months	↓ for 3 months
SAC[1] response	↓ for 2 months	↓ for 2 months
IGM	↓ for 4-6 months	↓ for 4-6 months
IGG	↓ for 6-9 months	↓ for 16-18 months
IGA	↓ for 6-36 months	↓ for > 36 months
IGE	↑ for 1-2 months	↑ for 1-2 months
SAC[1] stimulated IGM production	↓ for 4-6 months	↓ for 4-6 months
SAC[1] stimulated IGG production	↓ for 6-9 months	↓ for 16-18 months

[1] SAC = *Staphylococcus Aureus* Cowen strain, a B cell mitogen

Table 10.1.3. Mouth care

Mouth Care	Dose / Interval / Comment
Oral hygiene	
soft toothbrush	removes debris, use with platelets < 50,000/µl
toothettes	removes debris, use with platelets < 50,000/µl or mucositis
biotene mouthwash	rinse and spit QID
hydrogen peroxide (Peridex)	rinse and spit QID
saline mouthwash	one teaspoon salt with 1 quart warm water; rinse and spit QID
sodium bicarbonate	one teaspoon baking soda with 1 pint warm water; rinse and spit QID
carafate slurry	10-15 ml swish and swallow QID; aqueous solution of carafate 2 g/15 ml, agoral 2.5 ml/15 ml diphenhydramine 12.5 mg/15 ml
Antimicrobial	
nystatin (Mycostatin)	500,000 units as oral suspension, swish and swallow QID
clotrimoxazole (Mycelex)	10 mg oral troche dissolved over 15-30 minutes in mouth, 5 times daily
acyclovir	200 mg PO TID-QID or 60 mg/m^2 IV q4 hours
Analgesic	
xylocaine viscous 2% dyclonine (Dyclone)	5 ml swish and swallow QID 3 mg lozenges (sucrets) dissolve slowly in mouth as needed; topical solution (Dyclone) swish and spit QID
lidobenalox	5 ml swish and swallow; suspension is an equal mixture of xylocaine viscous 2%, diphenhydramine 12.5 mg/ 5 ml, and maalox
benzocaine (Cepacol, Hurricaine,Cetacaine, Orabase, Orajel	10 mg Cepacol lozenge dissolved over 15-30 minutes in mouth, repeat as necessary; (Hurricane and Cetacaine are aerosol preparations,Orajel and Orabase are gels)
Moisturizer/lubricant	
petroleum jelly, K-Y jelly	topically to lips as necessary
Moi-Stir swabsticks	prn
Salix lozenges	prn
Salivert spray	spray into mouth for 1-2 seconds, prn
Xero-Lube spray	spray into mouth prn
pilocarpine tablet	5 mg PO TID, cardiovascular side effects of hypotension,bradycardia

10.1

C) FUNGI

The most common BMT fungal pathogens are *Candida spp.* and *Aspergillus spp.*. Other fungi (*Trichosporon beigelii, Fusarium, Alternaria, Cryptococcus neoformans,* etc.) are less frequently involved.

1) *Candida spp.*

Colonization of cutaneous and mucosal surfaces (oropharynx, GI tract, vagina) which act as a portal of entry occurs from use of broad spectrum antibiotics. Vascular catheters may also be a portal of entry, as well as a target of hematologic seeding for the GI tract.

2) *Aspergillus spp.*

Airborne spread from air conditioning, ventilation systems, fireproofing materials in ceilings or walls, and construction activity causes nosocomial infection with *Aspergillus*. The portal of entry are usually the sinuses or upper airway.

D) VIRUSES

Viral infections post-BMT generally arise from reactivation of latent infections. The most common viral infections complicating BMT are herpes viruses (simplex, zoster, CMV), adenovirus and BK virus. Less common are enterovirus (echovirus, cocksackie), rotavirus, JC virus and parvovirus B19.

1) Herpes

The most common viral offenders post-BMT are herpes simplex, varicella zoster and CMV. Infection generally arises from reactivation. In T cell depleted allogeneic BMT, a potentially fatal lymphoproliferative syndrome (LPS) may arise from reactivation or seroconversion (primary infection) with Epstein-Barr virus (EBV). LPS is manifest by fever, hepatosplenomegaly and lymphadenopathy. Paraproteinemia and atypical or plasmacytoid peripheral blood lymphocytes may be present. Without treatment, LPS is fatal within weeks. Treatment for LPS is withdrawal of immunosuppressive drugs and/or infusion of donor lymphocytes.

2) Adenovirus

Adenovirus infection post, BMT usually causes hemorrhagic cystitis, pneumonitis or hepatitis. It may arise from reactivation or be transmitted via body fluids.

3) BK Virus

The polyoma BK virus may be dormant in the kidney and reactivation after BMT may cause hemorrhagic cystitis.

III) TEMPORAL SEQUENCE OF INFECTIONS

Infectious complications after BMT may be schematically viewed to occur in three phases: early (during neutropenia, day 0-30), intermediate (during acute GVHD, day 30-100) or late (during chronic GVHD, day > 100) (Table 10.1.4). If prior to BMT, the patient has a serious infectious process, the preparation for transplant should be delayed until fully eliminated. Otherwise, it is likely that the infectious process may become fatal, most frequently in the middle of marrow aplasia.

A) EARLY INFECTIONS (DAY 0-30)

In general, bacterial infections occur during the interval of neutropenia (day 0-30) and are due to Gram negative organisms (e.g. *Pseudomonas aeruginosa, Klebsiella, Escherichia coli, Enterobacter cloacae*) and/or Gram positive organisms (e.g. *Staphylococcus aureus, Staphylococcus epidermidis, Viridans streptococci, Corynebacterium, Fusobacteria, Proprionibacteria*). Bacterial

10.1

Table 10.1.4. Relationship between time of BMT and organisms causing common medically important infections

Period of Neutropenia (Day 0-30)	Period of Acute GVHD (Day 30-100)	Period of Chronic GVHD (Day > 100)
Gram negative bacteria	Gram negative bacteria	Encapsulated bacteria
Gram positive bacteria	Gram positive bacteria	Varicella zoster
Herpes simplex	CMV-Cytomegalovirus	*Pneumocystis carinii*
Candida spp.	BK virus	*Aspergillus spp.*
Aspergillus spp.	Adenovirus	
	Varicella zoster	
	Candida spp.	
	Aspergillus spp.	
	Pneumocystis carinii	
	Toxoplasma gondii	

infections cause sepsis, pneumonia, oropharyngitis, perianal-perirectal infections, sinusitis and cutaneous infections. For the first 30 days, viral and fungal infections are generally due to herpes simplex virus, and *Candida spp.* (Table 10.1.5).For patients with pre-existing immunodeficiency (e.g. aplastic anemia, severe combined immune deficiency), *Aspergillus* infection may also occur in the first 30 days.

B) **INTERMEDIATE INFECTIONS (DAY 30-100)**

After neutrophil recovery and during the interval of acute GVHD (day 30-100), the spectrum of infections shifts to cytomegalovirus pneumonia, adenovirus or BK virus hemorrhagic cystitis, pneumocystis carinii pneumonia, disseminated or hepatosplenic candidiasis and aspergillosis usually involving the lungs or sinuses.

C) **LATE INFECTIONS (> DAY 100)**

Late infections occurring after day 100 and associated with chronic GVHD are generally secondary to encapsulated bacteria (*Streptococcus pneumonia* and *Hemophilus influenza*), *Aspergillus* and varicella zoster virus.

IV) **PREVENTION OF INFECTION**

Prevention of infection varies between institutions, but generally involves some combination of decontamination and prevention of acquisition (Table 10.1.5). The guidelines of each institution's BMT unit should be respected. Our recommendation for preventing infections during hospitalization for autologous or allogeneic BMT is conscientious hand washing, meticulous oral hygiene, low bacterial diet (cooked food, no fresh fruits, vegetables, diary products or spices), oral antibiotics for selective aerobic anti-microbial gut decontamination and high efficiency particulate air (HEPA) filtration. Some centers question the value of oral gut decontamination due to concern for developing drug resistant organisms. However, in a review of over 500 BMT patients at Johns Hopkins Hospital receiving prophylactic oral norfloxacin, no patients developed resistant Gram negative organisms. In addition, bacterial infections are capable of breaking T cell tolerance and theoretically causing worse GVHD. Since most infections arise from endogenous gut flora, prophylactic gut decontamination may not only decrease infections but may also decrease the severity of GVHD. Nevertheless, some centers do not employ prophylactic gut decontamination, and the value of other measures including LAF

10.1

rooms, masks, gowns, hair and shoe covers, sterile food and skin decontamination is debatable.

A) DECONTAMINATION

1) Cutaneous

Daily baths with povidone/iodine or chlohexidine to prevent infections from Gram positive organisms colonizing the skin are of no proven value.

2) Gastrointestinal Tract

Anaerobic flora in the GI tract are thought to provide colonization resistance to superinfection from pathogenic aerobic gram negative organisms. Therefore, oral gut decontamination regimens are designed to preserve anaerobic colonization. Oral nonabsorbable antibiotics (Table 10.1.6) are generally given as multidrug decontamination regimens, such as gentamicin, vancomycin, and nystatin (GVN) or neomycin, colistin and nystatin (NEOCON). Oral amphotericin B may also be used instead of nystatin. Problems with these regimens are GI irritation and nonpalatable taste resulting in noncompliance especially in patients with mucositis.

More recent selective antimicrobial gut decontamination regimens have used absorbable antibiotics (Table 10.1.6). Absorbable antimicrobial antibiotics are generally given as a combination of a quinolone (ciprofloxacin) and an antifungal (fluconazole, itraconazole, ketoconazole). Quinolones should not be used in children or adolescents due to potential epiphysial growth plate damage. Trimethoprim-sulphamethoxazole (TMP/SMX) generally lacks activity against *Pseudomonas aeruginosa*, may cause superinfection with Gram positive cocci and may be myelosuppressive. Rifampin has been added to TMP/SMX to provide better Gram positive coverage but may cause elevated liver transaminases and cholestasis. Antibiotics for selective antimicrobial gut decontamination are generally started when the patient enters the BMT ward and terminated when neutrophil count recovers (ANC > 500).

B) PREVENTION OF ACQUISITION

1) Diet

Fresh salads and fruits are sources of aerobic Gram negative rods (*Escherichia coli, Klebsiella pneumonia, Pseudomonas aeruginosa,* etc.) and are generally not allowed on the menu of a BMT patient. Any food not cooked serves as a source of pathogens. Therefore, most institutions do not allow uncooked or nonmicrowaved foods to be given during the period of neutropenia. More expensive and elaborate sterile food preparation (e.g. autoclaving) is used by some institutions but is of unproven benefit and the food is tasteless. Pepper may be contaminated by *Aspergillus* and should also be avoided.

2) High Efficiency Particulate Air (HEPA)

Less than 3 per 10,000 atmospheric particles 0.3 micron in size are able to pass through HEPA filters. This is sufficient to remove bacteria and fungal spores from the air. HEPA filtration is the most effective method to prevent hospital-acquired *Aspergillus*. If the entire BMT unit is HEPA filtered by cross flow ventilation, patients may ambulate in the halls. All BMT rooms/units should have high efficiency particulate air.

10.1

Table 10.1.5. Allogeneic BMT antimicrobial prophylaxis

Center	Isolation	Daily Prophylaxis-(usually started upon admission)	First Fever
MD Anderson	Wash hands, gown, Gloves, cap, shoe covers	norfloxacin PO bactrim 1 DS PO until day 2 fluconazole PO or IV acyclovir PO or IV	ceftazidine and vancomycin
City of Hope	wash hands, gown, gloves, cap, shoe covers	vancomycin PO neomycin PO amphotericin 5 cc PO acyclovir PO or IV amphotericin 0.1 mg/kg IV—started when neutropenic	ceftazidine— fever persists add vancomycin—fever persists add ciprofloxacin—fever persists increase amphotericin
Stanford	wash hands, gown, mask	neomycin PO vancomycin PO acyclovir PO or IV amphotericin 0.15 mg/kg IV—started when neutropenic	ceftazidine and vancomycin
Johns Hopkins	wash hands	norfloxacin PO fluconazole PO or IV acyclovir PO or IV	ceftazidine and vancomycin—fever persists change ceftazidine to ticarcillin and gentamicin
Milwaukee	wash hands, mask	acyclovir PO or IV amphotericin 0.2 mg/kg IV—started when neutropenic	ceftazidine and vancomycin and ciprofloxacin (if culture neg for Gram + stop vancomycin)
Fred Hutchinson	wash hands— if not LAF room	intravenous ceftazidine started prophylactically when ANC <500 acyclovir PO or IV fluconazole PO or IV	add gentamicin +/– amphotericin

All transplants performed in BMT unit separate from other patients and hospital traffic.

10.1

Table 10.1.6. Examples of BMT gastrointestinal decontamination regimens

Nonabsorbable Regimens

GVN	gentamicin liquid formula (200 mg PO q4 hours), vancomycin capsule (500 mg PO q8 hours), nystatin suspension (1 million units PO q8 hours)
NEOCON	neomycin (1 g PO BID), colistin (5 mg/kg PO q8 hours), nystatin suspension (1 million units PO q8 hours)
norfloxacin	norfloxacin (400 mg PO q 12 hours),

Absorbable Regimens

fluconazole	fluconazole (400 mg PO q day)
ciprofloxacin/fluconazole	ciprofloxacin (250 mg PO q12 hours), fluconazole (400 mg PO q day)
trimethoprim-sulphamethoxazole (TMP/SMX) / rifampin	TMP/SMX (1 DS tablet PO q12 hours) rifampin (300 mg PO q12 hours)

3) Isolation Rooms

a) Life Island

Life islands are complete enclosure of the patient and his bed in a tent or bubble universe. They are generally obsolete.

b) Laminar Air Flow (LAF) Rooms

LAF rooms combine the principles of gut and skin decontamination, sterile or low microbial diet, laminar air flow, reverse isolation and HEPA filtration (Fig. 10.1.1). LAF rooms utilize the same principles as operating theaters. Visitors and medical personal are isolated from the patient (reverse isolation) by sterile gowns, gloves, masks, and hair and shoe covers before entering the room. Any inaminate object is sterilized before entering the room. Air flows in laminar, non-turbulent, horizontal layers from the head of the bed. Visitors are "down wind" from the patient. Any dust is removed from the room on horizontal currents of air that remain near the floor.

For psychological reasons, approximately 10-15% of patients insist on early termination of LAF isolation. LAF rooms have been reported to decrease infections, GVHD and mortality from BMT in patients with aplastic anemia. The outcome of BMT in patients with severe combined immunodeficiency has also been reported to be improved by a protected LAF environment. LAF rooms decrease infections but with some exceptions have not been shown to alter mortality in patients undergoing BMT for leukemias or solid tumors. Due to expense, inconvenience to the patient and staff, and lack of effect on survival for most diseases, LAF rooms are declining in usage.

V) SURVEILLANCE CULTURES

Except for CMV, the role of surveillance cultures after allogeneic BMT is questionable. Again, except for a positive CMV surveillance culture, whether therapy should be changed or instituted for a positive surveillance culture is unclear.

A) BACTERIAL SURVEILLANCE CULTURES

Bacterial surveillance cultures are done by some institutions from the oropharynx, urine, stool and blood during the period of neutropenia. The interval

10.1

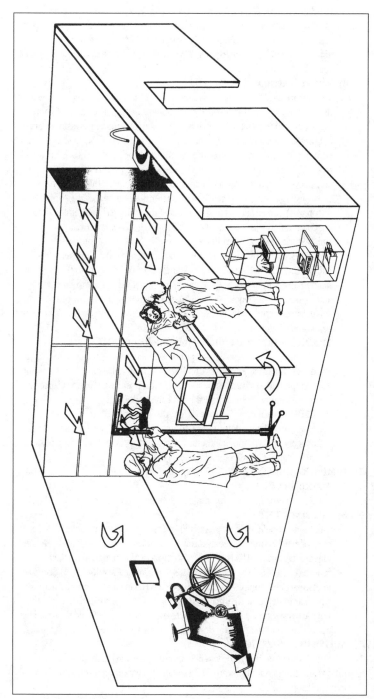

Fig. 10.1.1. Laminar air flow (LAF) room. High efficiency particulate air (HEPA) filtration unit is in the wall at head of patient's bed.

10.1

between cultures is variable and may be daily, twice a week or weekly. However, bacterial surveillance cultures have a low positive predictive value (i.e. are of little value in predicting infection), generally do not result in a change of antibiotic therapy and are expensive. We do not recommend routine bacterial surveillance cultures.

B) FUNGAL SURVEILLANCE CULTURES

Oropharyngeal fungal cultures detect colonization with *Candida spp.*, while nasal (anterior nares) cultures detect *Aspergillus spp.* The value of routine oropharynx fungal cultures remains questionable. Routine nasal surveillance cultures are of limited utility, although perhaps there is a role in selected patient populations. For example, if a nasal surveillance culture in a febrile neutropenic patient is positive for *Aspergillus spp.*, a CT scan of the sinuses and chest is warranted to evaluate pulmonary nodules, sinus thickening or bone erosion suggestive of disease. For patients receiving oral prophylaxic fluconazole, fungal surveillance cultures of the stool may be useful in detecting colonization with resistant *Candida spp.* (e.g. *Candida kruzei*).

C) VIRAL SURVEILLANCE CULTURES

Viral surveillance cultures are usually limited to CMV. A positive surveillance culture (urine, blood or throat) is defined as CMV infection. CMV disease is defined as organ invasion. CMV infection occurs with equal frequency after autologous and allogeneic BMT (25-50%). However, in the setting of a positive surveillance culture, CMV disease rarely occurs after autologous (2-8%) in comparison to allogeneic (25-50%) BMT.

1) CMV Surveillance Culture After Autologous BMT

A positive CMV surveillance culture is generally not predictive for subsequent disease after autologous BMT. CMV surveillance cultures are not recommended after autologous BMT, and a positive CMV surveillance culture is generally not an indication for treatment.

2) CMV Surveillance Culture After Allogeneic BMT

CMV surveillance cultures are done weekly from day 30 to day 100 after allogeneic BMT. Antiviral therapy (gangciclovir) is indicated for a positive surveillance culture after allogeneic BMT.

VI) IMMUNOGLOBULINS (TABLE 10.1.7)

A) AUTOLOGOUS BMT

There is no proven role for immunoglobulins after autologous BMT.

B) ALLOGENEIC BMT

Immunoglobulin therapy is generally recommended for allogeneic BMT at a dose of 400-500 mg/kg weekly intravenous infusion from day 7 to day 100 or longer if chronic GVHD is present. Prophylactic immunoglobulins have been reported to: 1) decrease bacterial sepsis, 2) decrease CMV disease, 3) decrease interstitial pneumonitis, 4) decrease acute GVHD in patients over 20 years old who had undergone HLA-matched sibling BMT, 5) decrease bacterial infections in hypogammaglobulinemic patients with chronic GVHD and in one study, 6) decrease mortality in a randomized trial of HLA-matched sibling allografts.

VII) ACTIVE IMMUNIZATION

Immunizations are deferred until 6-24 months after BMT. Schedule, dose, indications and contraindications are discussed in the chapter on outpatient management and appendix.

10.1

VIII) FEVER/NEUTROPENIA

An extensive discussion of treatment for the febrile neutropenic patient is beyond the scope of this chapter. Briefly, initial therapy for a patient with an ANC < 500/µl and fever (> 38.0°C) or clinical suspicion of infection without an obvious source is broad spectrum empiric antibiotic coverage. Different antibiotic regimens may be chosen (Table 10.1.8): 1) an aminoglycoside and antipseudomonal β-lactam, 2) an aminoglycoside and antipseudomonal β-lactam and vancomycin or 3) single drug monotherapy- ceftazidine or imipenim. The approach used depends on institutional preference and history of prior BMT unit bacterial infections. For patients allergic to penicillin, the combination of vancomycin and aztreonam may be used.

Persistent fever after 4 days without source is indication for empirical amphotericin therapy. Antibiotics may be discontinued when the ANC is > 500/µl. Alternatively, if the patient is afebrile and stable for 5-7 days antibiotics may be stopped

---

Table 10.1.7. Intravenous immunoglobulin (IVIG)

Preparation

IVIG is obtained from pooled plasma of over 1000 donors. Preparation inactivates hepatitis B and HIV, and removes most IgA and vasoregulatory peptides (kinins, kalikrein, etc.). All IgG subclasses and complement activity remain. Several commercial brands are available: sandoglobulin, gammagard, gammar-IV, venoglobulin-I, gamimune-N, Iveegam.

Indications

1) Prophylaxis–No proven role in autologous BMT. For allogeneic BMT, prophylactic IVIG decreases bacterial sepsis, CMV disease, interstitial pneumonitis, acute GVHD in recipients of HLA-matched sibling BMT over age 20, and may also decrease platelet transfusion requirements. In hypogammaglobulinemic patients with chronic GVHD, IVIG decreases bacterial infections.

2) CMV pneumonitis–Decreases mortality when given in combination with ganciclovir.

Contraindication

Contraindicated in persons with a history of hypersensitivity or anaphylaxis or individuals with selective IgA deficiency.

Adverse reactions

Chills in less than 10%; fever, nausea, or headache in less than 2-4%; hypotension or hypertension in less than 1%. Side effects may be ameliorated by slowing rate of infusion or premedication (benadryl, acetaminophen, corticosteroids). Urticaria, angiodema, nephrotic syndrome and anaphylaxis are rare but may be due to low concentrations of vasoactive agents (kinins, kalikren, plasmin, etc.) or IgA infused into a patient with selective deficiency of IgA. Hypersensitivity syndrome (flushing, chest tightness, chills, hypotension, diaphoresis, urticaria, etc.) may result when a hypogammaglobulinemic patient has sufficient quantities of circulating antigen to induce immune complex formation and activation of complement.

Note of caution—IVIG preparations contain antibody titers to viruses (hepatitis A, B, C) and red blood cell antigens (Rh antigen). This may cause false positive viral antibody screens and difficulties in crossmatching blood.

Dose

1) Prophylaxis–Dose for allogeneic BMT recipients is 400-500 mg/kg IV weekly from day 7 to 100 or as long as chronic GVHD persists.

2) CMV pneumonitis–Ganciclovir combined with IVIG 500 mg/kg IV every other day x 7 doses then weekly. No proof exists that CMV hyperimmune globulin is more effective than IVIG.

Administration

Usually reconstituted as 5% solution. Final volume is often a liter or more. Initial infusion rate is 1.0 ml/kg/hour and may be increased by 1.0 ml/kg/hour increments at 15 minute intervals up to 5.0 ml/kg/hour if no adverse reactions occur.

10.1

despite persistent neutropenia. Outpatient management of fever and neutropenia with once daily antibiotic regimens has been shown by some centers to be safe, effective and cost efficient (Table 10.1.8).

IX) PNEUMOCYSTIS/TUBERCULOSIS PROPHYLAXIS

A) *PNEUMOCYSTIS CARINII*

Pneumocystis carinii infection is decreased by trimethoprim-sulfa-methoxazole (TMX-SMP) prophylaxis (e.g. bactrim DS one tablet BID every Saturday and Sunday) beginning when the ANC is >1000/μl and continuing until day 120 or for as long as chronic GVHD is present. Bactrim even with leucovorin rescue may delay engraftment, and since PCP is rare for the first 30 days, we do not start bactrim until day 30 or the ANC is > 1000/μl. If patients cannot tolerate TMX-SMP due to rash or marrow suppression, aerosolized pentamidine 300 mg inhaled every month may be substituted.

Table 10.1.8. Examples of empiric regimens for treatment of fever and neutropenia

Regimen	IV Adult Dose (normal renal and liver function)
Aminoglycoside and β-lactam amikacin or gentamicin or tobramycin plus piperacillin or ticarcillin or mezlocillin or ceftazidine or cefperazone or imipenim	
amikacin	15 mg/kg daily divided q 8-12 hrs
gentamicin	3 mg/kg daily divided q 8 hrs
tobramycin	3 mg/kg daily divided q 8 hrs
mezlocillin	200-300 mg/kg daily divided q 4-6 hrs
piperacillin	200-300 mg/kg daily divided q 4-6 hrs
ticarcillin	200-300 mg/kg daily divided q 3-6 hrs
ceftazidine	1-2 g q 8-12 hrs
cefperazone	2-4 g daily divided q 12 hrs
imipenim	0.5-1.0 g q 6 hrs
Aminoglycoside and β-lactam and Vancomycin addition of vancomycin to above aminoglycoside and β-lactam regimens	
vancomycin	500 mg q 6 hrs or 1 g q 12 hrs
Monotherapy	
ceftazidine	2 g q 8 hrs
imipenim	1 g q 6 hrs
Penicillin Allergy	
vancomycin	1 g q 12 hrs,
aztreonam	2 g q 6-8 hrs
Once Daily Regimens	
ceftriaxone	30 mg/kg IV q 24 hrs (adult) 80 mg/kg IV q 24 hrs (children < 12 years)
amikacin	20 mg/kg IV q 24 hrs

10.1

B) TUBERCULOSIS

If a patient has granulomas on chest radiograph or a positive PPD skin test, isoniazid 300 mg PO daily should be started with the conditioning regimen and continued, if tolerated, until all immunosuppressive drugs are stopped.

SUGGESTED READING

1. Hiemenz JW, Greene JN. Infectious complications of the immunocompromised host: special considerations for the patient undergoing allogeneic or autologous bone marrow transplantation. Hematology Clinics of North America 1993; 7:961-1002.
2. Hughes WT, Armstrong D, Bodey GP et al. Guidelines for the use of antimicrobial agents in neutropenic patients with unexplained fever. The Journal of Infectious Disease 1990; 161:381-396.
3. Karp JE, Merz WG, Dick JD, Saral R. Strategies to prevent or control infections after bone marrow transplantation. Bone Marrow Transplantation 1991; 8:1-6.
4. Walsh TJ, Hathorn J, Pizzo PA. Prevention of bacterial infections in neutropenic patients. Bailliere's Clinical Infectious Disease 1994; 1(3):1-30.
5. Walsh TJ, Lee JW. Prevention of invasive fungal infections in patients with neoplastic disease. Clinical Infectious Diseases 1993; 17(suppl 2):S468-480.
6. Wingard JR. Advances in the management of infectious complications after marrow transplantation. Bone Marrow Transplantation 1990; 6:371-383.

10.1

Approaches to Management of Invasive Fungal Infections in Bone Marrow Transplant Recipients

Thomas J. Walsh, Philip A. Pizzo

I) INTRODUCTION

Patients undergoing bone marrow transplantation (BMT) are predisposed to develop invasive fungal infections as the result of host defense impairments due to intensive cytotoxic chemotherapy, ablative radiation therapy, corticosteroids, cyclosporine and the underlying neoplastic process. *Candida spp.* and *Aspergillus spp.* constitute the most frequent causes of invasive fungal infections in patients undergoing bone marrow transplantation. Common and emerging fungal pathogens in BMT recipients are listed in Table 10.2.1.

II) DEFICITS IN HOST DEFENSE PREDISPOSING BMT RECIPIENTS TO INVASIVE FUNGAL INFECTIONS

A) PHAGOCYTES (NEUTROPHILS, MONOCYTES AND MACROPHAGES)

Intensive cytotoxic chemotherapy or ablative radiation therapy results in neutropenia, the duration of which is directly related to the frequency of invasive fungal infections.

B) T LYMPHOCYTES

Concomitant corticosteroid therapy as part of the antineoplastic regimen or for treatment of graft-versus-host disease has a broad effect upon impairing the function of circulating neutrophils, monocytes and macrophages, as well as profoundly compromising T lymphocyte cell-mediated immunity. Cyclosporine also impairs cell-mediated immunity, resulting in further increased risk for invasive mycoses.

C) MECHANICAL BARRIERS

Both cytotoxic chemotherapy and radiation therapy also may disrupt the integrity of mucosal barriers. For example, disruption of the epithelial surfaces of the alimentary tract facilitates invasion of *Candida spp.* into the submucosal blood vessels and entry into the portal venous system. Insertion of venous catheters also compromises the host, resulting in another portal of entry into the blood stream for yeast-like fungi. Vascular catheters also may be seeded hematogenously from a distal site and become an intravascular source of perpetuating fungemia.

III) CANDIDIASIS

Candidiasis is a spectrum of infections that may be classified as cutaneous, mucosal and deeply invasive. Deeply invasive infection may be further classified as fungemia, tissue proven disseminated candidiasis and single organ candidiasis. Disseminated candidiasis may be further classified as acute and chronic disseminated candidiasis, which constitute two ends of a clinical and pathological spectrum.

Amongst the conditions of mucosal candidiasis commonly encountered in BMT recipients are oropharyngeal and esophageal candidiasis.

A) OROPHARYNGEAL CANDIDIASIS

Although the diagnosis of oropharyngeal candidiasis is often made presumptively by visual inspection of the mucosal surfaces, this practice may not be reliable. Herpes simplex virus infections, bacterial infections and nonbacterial mucosal disruption may easily simulate *Candida*-like lesions. Classic beige plaques and ulcers may or may not be due to *Candida* species. The most assured and practical way of establishing a diagnosis is the examination of a wet mount or Gram stain for pseudohyphae and blastoconidia and culture for identification of the *Candida* species. A culture for herpes simplex virus is also important in excluding concomitant infection. Therapeutic approach to oropharyngeal candidiasis traditionally includes administration of nystatin, clotrimazole or, more recently, fluconazole (Table 10.2.2). These agents may be given therapeutically for proven infection or prophylactically for prevention of infection.

B) ESOPHAGEAL CANDIDIASIS

Due to the broad spectrum of potential infections potentially developing in BMT recipients, endoscopy performed by an experienced gastroenterologist is the optimal technique by which to guide definitive therapy. Herpes simplex virus (HSV) esophagitis, cytomegalovirus (CMV) esophagitis and mixed bacterial infections may each develop either before or concomitantly with esophageal candidiasis. While empirical therapy with fluconazole or amphotericin B for presumptive esophageal candidiasis may be employed (Table 10.2.2), the possible presence of concomitant HSV esophagitis, CMV esophagitis and mixed bacterial infections warrant the need for endoscopy, particularly in patients unresponsive to empirical therapy.

C) CANDIDEMIA

Candida spp. are the most common causes of fungemia in BMT recipients. The detection of candidemia has been greatly improved by blood culture detection techniques, most notably the lysis centrifugation system (Isolator)

Table 10.2.1. Common and emerging fungal pathogens in bone marrow transplant recipients

Opportunistic Yeasts
Candida spp.
Trichosporon beigelii
Cryptococcus neoformans
Hyaline Moulds
Aspergillus spp.
Fusarium spp.
Zygomycetes
Dematiaceous Molds
Pseudallescheria boydii
Scedosporium inflatum
Bipolaris spicifera
Endemic Dimorphic Fungi
Histoplasma capsulatum
Coccidioides immitis
Blastomyces dermatitidis
Penicillium marneffei

10.2

and the BacTAlert system. These systems have been able to detect candidemia earlier and more frequently than conventional broth and biphasic systems. The nonradiometric infrared resin broth system (BacTec) may be similar to lysis centrifugation in the detection of fungemia. Nonculture techniques such as polymerase chain reaction, antigen and antibody detection, and measurement of metabolites for *Candida spp.* are investigational.

Amphotericin B (0.5-1.0 mg/kg/d) is the treatment of choice of fungemia in neutropenic BMT recipients, particularly in patients with hemodynamic insta-

Table 10.2.2. Approaches to treatment of mucosal and invasive candidiasis in bone marrow transplant recipients

Patterns of candidiasis	Treatment	Comments
Mucosal		
Oropharyngeal Candidiasis	• Nystatin suspension • Clotrimazole troches • Fluconazole, PO/IV • Itraconazole, PO • Amphotericin B	Selection depends upon severity of infection, host status, and *Candida spp.* involved
Esophageal candidiasis	• Fluconazole, PO/IV • Amphotericin B, IV	Selection depends upon severity of infection, host status, *Candida spp.* involved, and probability of concomitant dissemination
Vaginal candidiasis	• Nystatin suspension • Clotrimazole troches or cream • Miconazole cream • Fluconazole, PO	Selection depends upon severity of infection, host status, and *Candida spp.* involved
Candida cystitis	• Fluconazole, PO/IV • Amphotericin B, IV	Selection depends upon severity of infection, host status, *Candida spp.* involved
Non-mucosal Deep Tissue		
Fungemia	• Amphotericin B, IV ±flucytosine • Fluconazole, PO/IV	Removal of vascular catheter; selection of compound depends upon severity of infection, host status, *Candida spp.* involved
Acute disseminated candidiasis	• Amphotericin B, IV ±flucytosine	Removal of vascular catheter
Chronic disseminated candidiasis	• Amphotericin B, IV ±flucytosine • Fluconazole, PO/IV	Protracted therapy usually required
Single organ candidiasis	• Amphotericin B, IV ±flucytosine • Fluconazole, PO/IV	Removal of vascular catheter; possible surgical intervention, depending upon site
Candida peritonitis	• Amphotericin B, IV ±flucytosine • Fluconazole, PO/IV	Removal of peritoneal catheter

10.2

bility, patients receiving prophylactic fluconazole and those at risk for azole-resistant *Candida spp.* (Table 10.2.2). As allogeneic BMT recipients are often receiving fluconazole for antifungal prophylaxis, amphotericin B is the logical choice for treatment of fungemia in this setting. Higher doses of amphotericin B (1.0-1.5 mg/kg/d) may be appropriate for infections due to non *albicans Candida spp.* (e.g., *Candida tropicalis*, *Candida parapsilosis*, and *Candida lusitaniae*), which may be more resistant than those due to *Candida albicans*. There is no evidence that a stepwise escalation of the dose to the targeted daily dose decreases toxic effects. Indeed, fungemia may persist during such inadequate dosing, resulting in more deep-seeded infection. Although the issue of removal of catheters is controversial, several series support removal of central venous catheters, where possible, and concomitant administration of amphotericin B. Removal of the catheter alone for patients with fungemia is not tenable, as fungemia in neutropenic patients is generally a marker of deeply invasive candidiasis. Patients who have infection due to *Candida tropicalis* or persistent fungemia may benefit from the combination of amphotericin B (1 mg/kg/d) plus 5-fluorocytosine (100 mg/kg/d in four divided doses). However, 5-fluorocytosine may delay engraftment in the neutropenic BMT patient. Emerging experimental and clinical data indicate that fluconazole may be as effective as amphotericin B in treatment of fungemia due to *Candida albicans* in stable neutropenic patients who have not previously been treated with fluconazole.

D) DISSEMINATED CANDIDIASIS

Disseminated candidiasis in BMT recipients constitutes a spectrum of infections ranging from acute disseminated candidiasis to chronic disseminated candidiasis.

1) Acute Disseminated Candidiasis (ADC)

ADC is characterized by persistent fungemia, hypotension, multi-organ failure, skeletal muscle involvement and cutaneous lesions. Amphotericin B (1.0-1.5 mg/kg/day +/- 5-fluorocytosine 100 mg/kg/day) is administered for acute disseminated candidiasis (Table 10.2.2). Hemodynamic support and monitoring is often required in these critically ill patients.

2) Chronic Disseminated Candidiasis

The most common manifestation of chronic disseminated candidiasis is hepatosplenic candidiasis. Chronic disseminated candidiasis is established during the course of fungemia while the patient is neutropenic and becomes clinically overt as chronic refractory progressive lesions upon recovery from neutropenia. Hepatosplenic and other tissue lesions are radiologically occult during neutropenia but become evident by diagnostic imaging methods upon recovery from neutropenia. Unlike acute disseminated candidiasis, chronic disseminated candidiasis is seldom associated with hypotension, fungemia or multi-organ failure. Earlier studies and more recent reports emphasize the chronicity and refractory nature of this infection. There are various approaches to the management of chronic disseminated candidiasis. A reasonable approach for treatment of chronic disseminated candidiasis consists of amphotericin with or without 5-fluorocytosine followed by fluconazole, once fever is resolved and lesions are stable to resolving.

The administration of lipid formulations of amphotericin B offers an alternative strategy for treatment of this infection. These formulations are

10.2

discussed below in greater detail. For example, a multilamellar liposomal formulation of amphotericin B was effective in achieving a complete response of clearing *Candida* lesions in approximately 70% of patients with hepatosplenic candidiasis. The improved therapeutic index of several of the lipid formulations of amphotericin B permit the administration of higher dosages of amphotericin B with less nephrotoxicity.

IV) ASPERGILLOSIS

Invasive pulmonary aspergillosis continues to be the major opportunistic fungal respiratory pathogen in BMT recipients. *Aspergillus fumigatus* and *Aspergillus flavus* constitute the most frequently isolated pathogenic *Aspergillus* species. Persistent and profound granulocytopenia as well as corticosteroids constitute the major risk factors for development of invasive aspergillosis. Invasive aspergillosis occurs in a biphasic distribution in allogeneic BMT recipients: the first 30 days of transplantation (during neutropenia) and the postengraftment period characterized by graft-versus-host disease (GVHD), corticosteroid therapy and cyclosporine. Pulmonary involvement by *Aspergillus spp.* may develop as an invasive disease, as a saprophytic process or as allergic bronchopulmonary aspergillosis. Invasive disease is the pattern of aspergillosis which develops most frequently in patients undergoing BMT.

Aspergillosis is established in immunocompromised hosts by inhalation of *Aspergillus* conidia into the respiratory tract. These 3-5 μ-diameter particles may be inhaled directly into the alveolar spaces, where germination of hyphal forms occurs. Pulmonary alveolar macrophages form the first line of defense against *Aspergillus* conidia. Once germination occurs, polymorphonuclear leukocytes form the next line of host defense against *Aspergillus* hyphae. While most episodes of nosocomial invasive aspergillosis occur as sporadic events in BMT, a review of the literature clearly demonstrates that outbreaks or clusters of invasive pulmonary aspergillosis continue to emerge in hospitals, where the most common environmental sources of *Aspergillus* conidia are contaminated ventilation systems and construction sites. The critical elements of successful management of invasive pulmonary aspergillosis complicating BMT are: 1) early diagnosis, 2) initiation of aggressive doses of amphotericin B at 1.0-1.5 mg/kg/d and 3) reversal of immunosuppression, including recovery from granulocytopenia and discontinuation of corticosteroid therapy (Table 10.2.3).

A) EARLY DIAGNOSIS

Isolation of *A. fumigatus* or *A. flavus* from respiratory secretions or respiratory passages in febrile granulocytopenic patients is strongly associated with invasive pulmonary aspergillosis. *Aspergillus fumigatus* or *A. flavus* are seldom contaminants when recovered from respiratory secretions in BMT recipients. Computerized tomographic scan may reveal radiologic evidence of invasive pulmonary aspergillosis, characterized by bronchopneumonia, sub-pleural nodules, or cavitation. Cavitation tends to occur more readily in the postengraftment phase. Ultrafast CT scanning may be an effective approach for rapid screening of high-risk patients with equivocal radiological infiltrates. Radiological evidence of sinus opacifications in a persistently neutropenic patient should prompt further otolaryngological evaluation for possible invasive aspergillosis of the paranasal sinuses. Bronchoalveolar lavage (BAL) is an important adjunct in detection of microbiological evidence of *Aspergillus spp.* However, BAL may vary in sensitivity of detection of invasive aspergillosis. Serum markers of invasive aspergillosis, such as galactomannan, may prove to be useful

10.2

Table 10.2.3. Summary of approaches to treatment of fungal infections of the respiratory tract in bone marrow transplant recipients

Disease	Therapy
Invasive aspergillosis • bronchopneumonia • hemorrhagic infarction • cavitation • necrotizing tracheobronchitis • invasive sinusitis • local extension to intrathoracic structures • disseminated aspergillosis • chronic necrotizing aspergillosis	• amphotericin B (1.0-1.5 mg/kg/day) • itraconazole (8-10 mg/kg/day) (alternative in stable patients) • combination antifungal therapy? • lipid formulation of amphotericin B • reversal of immunosuppression • surgery recommended for: (1) hemoptysis from a single cavitary lesion (2) progression of a cavitary lesion despite antifungal therapy (3) infiltration into pericardium, great vessels, bone, or thoracic soft tissue while receiving antifungal therapy (4) progressive sinusitis
Zygomycosis • rhinocerebral • pulmonary • disseminated infection	• amphotericin B (1.0-1.5 mg/kg/day) • lipid formulation of amphotericin B • surgical debridement of rhinocerebral infection to viable tissue • surgery for pulmonary infection: refer to invasive aspergillosis • reversal of immunosuppression (1) correction of metabolic acidosis (2) removal of desferrioxamine and corticosteroids
Pseudallescheria boydii • pulmonary infection • sinusitis • disseminated infection	• antifungal azole [miconazole (20-40 mg/kg/day) or itraconazole (8-10 mg/kg/day)] plus amphotericin B • surgery for pulmonary infection: refer to invasive aspergillosis • reversal of immunosuppression
***Fusarium* infection** • pulmonary infection • sinusitis • disseminated infection	• amphotericin B (1.0-1.5 mg/kg/day) plus 5-fluorocytosine (50-100 mg/kg/day) • investigational triazole? (fluconazole and itraconazole have little or no activity) • lipid formulation of amphotericin B • reversal of immunosuppression

10.2

Table 10.2.3. Summary of approaches to treatment of fungal infections of the respiratory tract in bone marrow transplant recipients (cont'd.)

Disease	Therapy
Bipolaris sinusitis • pulmonary infection • sinusitis • fungemia • disseminated infection	• itraconazole (8–10 mg/kg/day) or amphotericin B (0.5–1.0 mg/kg/day) • surgical resection • reversal of immunosuppression
Histoplasmosis • pulmonary infection • fungemia • disseminated infection	• itraconazole (4–10 mg/kg/day) • amphotericin B (0.5–1.0 mg/kg/day) • reversal of immunosuppression
Coccidioidomycosis • pulmonary infection • disseminated infection	• itraconazole (4–10 mg/kg/day) • fluconazole (5–10 mg/kg/day) • amphotericin B (0.5–1.0 mg/kg/day) • reversal of immunosuppression
Blastomycosis • pulmonary infection • disseminated infection	• itraconazole (4–10 mg/kg/day) • amphotericin B (0.5–1.0 mg/kg/day) for life-threatening infections
Penicilliosis • disseminated infection	• itraconazole (4–10 mg/kg/day) • amphotericin B (0.5–1.0 mg/kg/day)
Trichosporon infection • pulmonary infection • disseminated infection	• amphotericin B (1.0–1.5 mg/kg/day) with 5-FC (50–100 mg/kg/day) plus fluconazole (8–10 mg/kg/day) • reversal of immunosuppression

10.2

but currently remain investigational. Open lung biopsy may yield a diagnosis of invasive aspergillosis when BAL is nondiagnostic. Open lung biopsy, when performed after all other diagnostic studies, still has important therapeutic implications for invasive aspergillosis by providing a definitive basis for use of increased dosages of amphotericin B, the implementation of lipid formulations of amphotericin B or the use of itraconazole, as well as to detect concomitant infections.

B) AMPHOTERICIN B

Amphotericin with or without flucytosine remains the mainstay of treatment of invasive aspergillosis. Invasive aspergillosis may develop during the course of empirical amphotericin B. Such breakthrough cases warrant increasing the dosage of amphotericin B to 1.0-1.5 mg/kg/d. A single lesion may benefit from surgical resection. The addition of flucytosine or rifampin to amphotericin B for invasive aspergillosis is controversial and has not been studied rigorously in controlled clinical trials. As amphotericin B is the most active agent against *Aspergillus spp.* among these three agents, a guiding principle for management of pulmonary aspergillosis is to first optimize the dosage of amphotericin B to maximally tolerated dosages targeted at 1.0-1.5 mg/kg/d.

C) LIPID AMPHOTERICIN

The lipid formulations of amphotericin B are undergoing clinical trials and appear to have important activity against this mycosis. Phase I studies and controlled clinical trails are underway or soon to be initiated in North America and Western Europe.

D) ITRACONAZOLE

Itraconazole, a new antifungal triazole, also demonstrates activity against *Aspergillus spp.* Previous work and current ongoing studies suggest that this agent is active in the treatment of selected immunocompromised patients with invasive pulmonary aspergillosis. Treatment of invasive aspergillosis in neutropenic patients, however, is limited. Impaired bioavailability, particularly in the setting of gastric achlorhydria, mucositis and total body radiation, limit its utility in critically ill BMT recipients. However, the availability of a cyclodextrin formulation may improve the bioavailability of itraconazole. The lack of availability for parental formulation of itraconazole further complicates the management of critically ill patients with pulmonary aspergillosis. Itraconazole ultimately may be more appropriately used for prevention of infection and for treatment of aspergillosis in the more stable non-neutropenic BMT recipient, such as those with chronic GVHD. There are currently no randomized trials demonstrating the successful prevention of invasive aspergillosis in high-risk patients. Episodes of aspergillosis may inexplicably wax and wane in a given center such that the use of these episodes as historical controls may not be reliable. Consequently, nonrandomized trials using historical controls to study preventive or therapeutic strategies against invasive aspergillosis may be difficult to interpret.

E) REVERSAL OF IMMUNOSUPPRESSION

Reversal of immunosuppression, the third arm of management of invasive aspergillosis, may be achieved by earlier recovery from neutropenia and discontinuation of corticosteroids, where applicable. Early recovery from granulocytopenia may be possible with the use of recombinant human cytokines, such as granulocyte-colony stimulating factor (G-CSF) and granulocyte macrophage-colony stimulating factor (GM-CSF).

10.2

F) PATIENTS WITH PRIOR INVASIVE ASPERGILLOSIS

Patients who recover from an episode of invasive aspergillosis have a risk of approximately 50% of relapsing aspergillosis if the patient receives another course of cytotoxic chemotherapy. Such patients may be managed with early empirical administration of high doses of amphotericin B during the next cycle of chemotherapy or with itraconazole as secondary prophylaxis.

V) EMERGING OPPORTUNISTIC FUNGI IN IMMUNOCOMPROMISED BMT RECIPIENTS

There has been increasing recognition of the less common but potentially devastating opportunistic mycoses in children and adults undergoing bone marrow transplantation. Among the organisms recognized more recently for these infections are *Fusarium* species, *Trichosporon* species, and dematiaceous fungi. *Cryptococcus neoformans*, *Zygomycetes* and resistant species of *Candida* are well-recognized pathogens being observed in new hosts and settings. Antifungal strategies are evolving for these infections (Table 10.2.3).

A) *FUSARIUM* INFECTIONS

Fusarium infections, which typically occur in patients with profound persistent granulocytopenia, produce a pattern similar to that of invasive aspergillosis. *Fusarium* infections in neutropenic BMT recipients are characterized by pulmonary infiltrates, cutaneous lesions, positive blood cultures, and sinusitis. Biopsy of the cutaneous lesions often reveals fine, dichotomously branching, acutely angular, septate hyphae. Unlike *Aspergillus spp.*, *Fusarium* species are frequently detected by advanced blood culture detection systems, such as lysis centrifugation. This emerging fungal pathogen often does not respond to conventional doses of amphotericin B and may require substantially higher doses for successful outcome. Some cases of invasive *Fusarium* infection may be completely refractory to amphotericin B, therefore requiring investigational antifungal compounds.

B) *TRICHOSPORON* INFECTIONS

Trichosporon beigelii is the most common of the *Trichosporon* species causing invasive infection. Although invasive *Trichosporon* infections are uncommon in immunocompromised patients, they often produce a fatal disseminated mycosis in patients with profound granulocytopenia or those receiving corticosteroids. Clinical manifestations are characterized by refractory fungemia, funguria, renal dysfunction, cutaneous lesions, pulmonary infiltrates and chorioretinitis. Despite the administration of amphotericin B, fungemia may persist. Recent in vitro and in vivo studies indicate that the organism is inhibited but not killed by safely achievable serum concentrations of amphotericin B. Newer antifungal triazoles, however, such as fluconazole, have been found to be active in vivo against this organism. The combination of amphotericin B plus fluconazole may also be effective without antagonism.

C) ZYGOMYCOSIS

Zygomycosis, most commonly due to *Rhizopus spp.*, is increasingly recognized as being caused by other species including Cunninghamella, and other members of the class of Zygomycetes. These fungi characteristically invade blood vessels resulting in extensive tissue infarction in granulocytopenic- or corticosteroid-treated hosts. Recent studies suggest that patients receiving deferoxamine constitute a newly recognized group of patients at risk for severe pulmonary and disseminated zygomycotic infections. Amphotericin B is

10.2

the treatment of choice for the sino-pulmonary infections caused by these organisms. Surgical resection of lesions, where possible, may be the most critical therapeutic intervention.

D) **CRYPTOCOCCOSIS**

Cryptococcal infections in BMT recipients typically occur in patients with impaired cell-mediated immunity. Characteristically patients with human immunodeficiency virus are at high risk for cryptococcal infection. Granulocytopenia per se is not commonly associated with infections due to *Cryptococcus neoformans*. Instead, patients with HIV infection, corticosteroid therapy or other impairments of cell-mediated immunity are more likely to develop pulmonary, disseminated or meningeal cryptococcosis.

The effective management of cryptococcosis depends upon the pattern of disease and the level of immunosuppression. Amphotericin plus fluorocytosine remains the treatment of choice for treatment of meningeal and disseminated cryptococcosis in immunocompromised patients. Those patients who will continue to receive ongoing immunosuppression after clearing their cryptococcal disease should be considered candidates for fluconazole suppression during the course of their immunosuppression.

E) *PSEUDALLESCHERIA BOYDII*

Pseudallescheria boydii is an uncommon but highly aggressive organism in granulocytopenic patients which produces a pattern of infection similar to that of *Aspergillus* species with invasion of blood vessels and infection of the respiratory tract. This organism may be completely resistant to amphotericin B. Treatment with miconazole may be successful; however, in neutropenic patients this agent also may not be effective. Recent in vitro studies and limited clinical experience suggest the combination of amphotericin B plus an antifungal azole (e.g. miconazole).

F) *CANDIDA KRUSEI*

As fluconazole is used increasingly in the oncology setting, emergence of fungi resistant to this triazole becomes probable. *Candida krusei* was recently reported to emerge as a resistant pathogen in bone marrow transplant recipients. Amphotericin B is active against *C. krusei* and is the appropriate treatment for infections due to *C. krusei*.

G) **PHAEOHYPHOMYCOSIS**

Phaeohyphomycosis (infections due to dematiaceous or pigmented fungi), are caused by such organisms as *Bipolaris spicifera*, *Cladosporium (Xylohypha) bantianum (bantiana)*, *Wangiella dermatitidis* and *Dactylaria constricta* var. *gallopava*. These dematiaceous fungi are uncommon but frequently fatal causes of invasive mycoses, particularly involving the central nervous system in immunocompromised hosts. These organisms may be initially treated with amphotericin B but are also amenable to therapy by itraconazole.

H) *MALASSEZIA FURFUR*

Malassezia furfur may occur in the setting of parenteral administration of lipids. Fungemia due to *M. furfur* may be manifest as persistent fever, pulmonary infiltrates and thrombocytopenia. Laboratory diagnosis is facilitated by addition of olive oil or other long chain carbon nutritional supplement. Management of this infection includes discontinuation of the lipid, removal of the vascular catheter where possible and administration of an antifungal azole.

10.2

I) ENDEMIC MYCOSES

Bone marrow transplantation recipients who live in endemic areas of the Southwestern United States and the Ohio-Mississippi river valley basin areas, are at risk, respectively for infection due to *Coccidioides immitis* and *Histoplasma capsulatum*. Disseminated histoplasmosis and coccidioidomycosis often develop in high risk patients. Amphotericin B should be considered the first line therapy for patients with these disseminated mycoses.

VI) RECOMBINANT HUMAN CYTOKINES IN THE MANAGEMENT OF FUNGAL INFECTIONS IN BMT RECIPIENTS

Reversal or amelioration of immunosuppression, which is essential for optimal management of invasive fungal infections in BMT recipients, can be accomplished through several strategies (Table 10.2.4). Cytokines such as G-CSF and GM-CSF appear to have ameliorated one of the important risk factors, neutropenia, for development of invasive fungal infections. However, patients with profound persistent granulocytopenia such as those undergoing allogeneic bone marrow transplantation still carry a high risk for invasive fungal infections. Patients undergoing repeated cycles of intensive cytotoxic therapy may become colonized with *Candida* species during the course of repeated cycles, resulting in the potential for invasive candidiasis despite an abbreviated course of granulocytopenia. Whether cytokines such as GM-CSF or M-CSF (macrophage colony stimulating factor) are effective in treatment of proven fungal infections in BMT recipients warrants further investigation.

VII) SYSTEMICALLY ADMINISTERED ANTIFUNGAL COMPOUNDS IN BONE MARROW TRANSPLANTATION

A) AMPHOTERICIN B

The cornerstone of therapy in most critically ill patients with deeply invasive fungal infections is amphotericin B. It is amphoteric, forming soluble salts in both basic and acidic environments, and is virtually insoluble in water. The principal mechanism of action of amphotericin B, as well as other polyenes, is due to binding to ergosterol, the principal sterol present in the cell membrane of sensitive fungi. This binding alters the membrane permeability, causing leakage of sodium, potassium and hydrogen ions, eventually leading to cell death. Amphotericin B also binds to a lesser extent to other sterols, such as cholesterol, which accounts for much of the toxicity associated with its usage.

Toxicity of amphotericin B may be classified as acute or chronic. Acute or infusion-related toxicity is characterized by fever, chills, rigor, nausea, vomiting

Table 10.2.4. Reversal of immunosuppression: immunologic adjuncts to prevention and treatment of invasive fungal infections in bone marrow transplant recipients

10.2

- Recombinant cytokines
  granulocyte-colony stimulating factor (G-CSF)
  granulocyte macrophage-colony stimulating factor (GM-CSF)
  interferon-gamma
  macrophage-colony stimulating factor (M-CSF)
- Stem cell reconstitution
- Immune reconstitution
- Granulocyte transfusions
- Adoptive immunotherapy
- Discontinuation of corticosteroids

and headache. Fever, chills and rigors may be mediated by tumor necrosis factor and interleukin-1, cytokines that are released from human peripheral monocytes in response to the drug. These acute reactions may be ameliorated by corticosteroids, acetaminophen, aspirin, other nonsteroidal anti-inflammatory drugs or meperidine. Corticosteroids should be utilized only in low dosages, such as 0.5-1.0 mg/kg of hydrocortisone. Meperidine in low doses (0.2-0.5 mg/kg) interdicts development of rigors. Acetaminophen may decrease fever but appears to have little effect on rigors. Aspirin should be avoided in thrombocytopenic patients.

Nephrotoxicity is the most significant chronic adverse effect of amphotericin B. Nephrotoxicity may be classified as glomerular or tubular. Glomerular toxicity includes a decrease in glomerular filtration rate and renal blood flow, while tubular toxicity is manifest as the presence of urinary casts, hypokalemia, hypomagnesemia, renal tubular acidosis and nephrocalcinosis.

Administration of sodium in the form of a liter of normal saline is often effective in preventing or attenuating the development of azotemia, possibly through the inhibition of tubuloglomerular feedback. However, sodium loading requires close monitoring of patients to avoid hypernatremia, hyperchloremia, metabolic acidosis and pulmonary edema. Furthermore, sodium loading will not ameliorate, and may exacerbate hypokalemia.

Tubular toxicity is most commonly evident as hypokalemia and hypomagnesemia. Hypokalemia, which occurs in the majority of patients receiving amphotericin B, may require the parenteral administration of 5-15 mEq of supplemental potassium per hour. Amphotericin B-induced hypokalemia appears to be a result of increased renal tubular cell membrane permeability to potassium due to direct toxic effects, or it may be caused by enhanced excretion via activation of sodium/potassium exchange. Cautious use of amiloride, the potassium-sparing diuretic, may attenuate the severity of hypokalemia. Magnesium wasting may also occur in association with amphotericin B therapy particularly in patients with a history of divalent cation-losing nephropathy associated with the cisplatin.

Anemia is another common side effect of amphotericin B therapy. It is characterized as a normochromic and normocytic process that is probably mediated by suppression of erythrocyte and erythropoietin synthesis. The anemia may be exacerbated by deterioration of renal function due to a decrease in red blood cell production. Treatment with erythropoietin may reverse amphotericin B-induced anemia.

Among the important drug interactions with amphotericin B, particularly in patients undergoing allogeneic BMT, is the renal toxicity associated with combined use of aminoglycosides, cyclosporine A and amphotericin B. Acute pulmonary reactions (hypoxemia, acute dyspnea and radiographic evidence of pulmonary infiltrates) have been associated with simultaneous transfusion of granulocytes and infusion of amphotericin B.

B) LIPID FORMULATIONS OF AMPHOTERICIN B

The recent introduction of lipid formulations has been an important therapeutic advance in improving the therapeutic index of amphotericin B. As toxicity is the major dose-limiting factor of amphotericin B, lipid formulations of amphotericin B have been developed to reduce toxicity and permit larger doses to be administered. While classically considered as "liposomal" formulations

10.2

of amphotericin B, the investigational and clinically approved formulations of amphotericin B have a wider diversity of lipid structure. The lipid formulation may provide a selective diffusion gradient toward the fungal cell membrane and away from the mammalian cell membrane. The lipid composition, molar ratio of lipid and liposomal size all play a role in toxicity.

Several carefully engineered lipid formulations of polyenes are investigational in North America or are approved in Western Europe: a small unilamellar vesicle formulation (AmBisome), amphotericin B lipid complex (ABLC), amphotericin B colloidal dispersion (ABCD; Amphocil), and liposomal nystatin (LN). AmBisome was the first lipid formulation of amphotericin B approved for use in Western Europe. These polyene lipid formulations are currently being investigated in clinical trials in North America. The use of amphotericin B in parenteral nutrition lipid supplements, in our opinion, should be considered an investigational compound subject to appropriate protocol and regulatory guidelines.

C) FLUCYTOSINE

Flucytosine (5-fluorocytosine, 5-FC), a fluorine analog of cytosine, was first synthesized in the 1950s as a potential antineoplastic agent. While not effective against tumors, it was found to have in vitro and in vivo antifungal activity. Flucytosine is used as an adjunct to amphotericin B therapy. This combination was originally proposed because of the observation that amphotericin B potentiated the uptake of flucytosine by increasing fungal cell membrane permeability. Two mechanisms of action have been reported for flucytosine: disruption of protein synthesis by inhibition of DNA synthesis and alteration of the amino acid pool by inhibition of RNA synthesis.

As a low molecular weight, water soluble compound, absorption of orally administered flucytosine from the gastrointestinal tract is rapid and nearly complete, providing excellent bioavailability. There is negligible protein binding in serum and the drug has excellent penetration with a volume of distribution that approximates that of total body water. Administration of 150 mg/kg/day results in peak serum concentrations of 50-80 mg/L within 1-2 hours in adults with normal renal function. CSF concentrations are approximately 74% of corresponding serum concentrations, accounting for its usefulness in central nervous system mycoses. However, the compound accumulates in patients with impaired renal function, resulting in potentially toxic serum levels unless the dosage is reduced. The plasma $t_{1/2}$ of flucytosine in adults with normal renal function is 3-5 hours. Dosage adjustments are required in patients with renal insufficiency and those on dialysis. As approximately 90% of a given dose is excreted unchanged in the urine by glomerular filtration, dosage adjustment of flucytosine is inversely related to creatinine clearance.

Gastrointestinal side effects, such as diarrhea, nausea and vomiting, are the most common symptomatic side effects associated with flucytosine therapy, occurring in approximately 6% of patients. Abnormally elevated hepatic transaminases have also been reported in approximately 5% of patients receiving the drug. Dose-dependent bone marrow suppression is the most serious toxicity associated with flucytosine administration. Conversion of flucytosine to 5-fluorouracil by gastrointestinal flora may account for the majority of these toxicities. These adverse effects may be controlled by close monitoring of the serum concentrations and adjustment of the dose to maintain peak serum

10.2

concentrations between 40 and 60 mg/L. Since flucytosine is used in combination with amphotericin B, the conventional dosage of 150 mg/kg/day is not recommended in most patients. Instead, we use 100 mg/kg/day as a starting dose in patients with normal renal function. As the glomerular filtration rate decreases due to amphotericin B, flucytosine dosage is reduced to < 100 mg/kg/day in 3-4 divided doses. We have found that these properties are applicable in adults and children.

The combination of flucytosine with amphotericin B is recommended for the treatment of acute disseminated candidiasis, meningeal cryptococcosis, *Candida* endophthalmitis, *Candida* thrombophlebitis of the great veins, renal candidiasis and hepatosplenic (chronic disseminated) candidiasis.

D) ANTIFUNGAL AZOLES

The antifungal azoles are synthetic compounds composed of imidazoles (clotrimazole, miconazole and ketoconazole) and triazoles (itraconazole and fluconazole). The antifungal azoles demonstrate less toxicity than amphotericin B, have flexibility for oral administration and have comparable efficacy under many circumstances. The antifungal azole agents function principally by inhibition of the fungal cytochrome P450 enzyme lanosterol 14α-demethylase, which is involved in the synthesis of ergosterol. This section will focus upon fluconazole and itraconazole.

Substitution of the triazole ring for the imidazole ring confers many structure-function advantages including: 1) greater polarity—improving solubility and reduce protein binding for some compounds, 2) reduced nucleophilicity of the triazole ring—improving resistance to metabolic degradation, 3) increased specificity for fungal enzyme systems, 4) broader antifungal spectrum and 5) increased potency. Itraconazole and fluconazole are the only antifungal triazoles licensed worldwide.

1) Itraconazole

Itraconazole has a broad spectrum of antifungal activity, minimal toxicity, a relatively long plasma half-life, and the capacity to penetrate into brain tissue. The spectrum of itraconazole includes *Candida spp.*, *Cryptococcus neoformans*, *Trichosporon spp.*, *Aspergillus spp.*, dematiaceous molds, and the thermally dimorphic fungi, including *Histoplasma capsulatum*, *Blastomyces dermatitidis*, *Coccidioides immitis*, *Paracoccidioides braziliensis* and *Sporothrix schenckii*. Itraconazole is only soluble at low pH, such as in the normal gastric milieu. There is wide inter-subject variation in the plasma concentration curves of itraconazole in healthy volunteers. Oral bioavailability is compromised and becomes more erratic in patients receiving intensive cytotoxic chemotherapy causing disruption of gastrointestinal mucosal epithelium. Absorption of itraconazole may be markedly diminished in patients receiving antacid therapy, such as oral antacids or H2-receptor blocking agents.

Mean peak serum concentrations of 0.02 mg/L are attained in adults when a single 100 mg dose is administered during fasting, while peak concentrations of 0.18 mg/L are attained when the drug is administered after feeding, suggesting enhanced absorption with feeding. Bioavailability may be further enhanced by administration of itraconazole with acidulin or a carbonated beverage. Initial findings indicate that the bioavailability and inter-patient variation in absorption of itraconazole is improved by

10.2

incorporation of the molecule into cyclodextrin. Studies are currently underway to investigate the safety and plasma pharmacokinetics of this novel formulation of itraconazole. These properties should expand the utility of itraconazole to a wider range of patients undergoing intensive cytotoxic chemotherapy.

Itraconazole follows nonlinear plasma pharmacokinetics. Dosage increases between 100, 200 and 400 mg/day produce nonlinear increases in the area under the plasma concentration-time curve suggesting the possibility of saturable metabolic processes. The drug has a $t_{1/2}$ of 15-20 hours following a single dose, and 30-35 hours following multiple dosing. Further reflecting its nonlinear pharmacokinetic properties, twice daily dosing of itraconazole leads to improved total area under the curve (AUC) in comparison to that of once daily itraconazole. Whether AUC or peak plasma concentrations, however, correlate with antifungal response is not known.

Attainment of adequate plasma concentrations is critical for optimal antifungal effect of itraconazole. In a study of itraconazole in patients with prolonged neutropenia, there was a direct relationship between plasma concentrations of drug and antifungal activity. As itraconazole is highly protein bound (> 99%), with only 0.2% available as free drug, concentrations in body fluids equivalent to body water, such as saliva and CSF, are negligible. However, tissue concentrations are 2-5 times higher than those in plasma, and they persist for longer, explaining the efficacy of the drug despite low plasma concentrations. Itraconazole is extensively metabolized by the liver to hydroxy-itraconazole, which also possesses intrinsic antifungal activity. Less than 1% of the active drug, and approximately 35% of the inactive metabolites are excreted in the urine. As the primary route of excretion is the biliary tract, no adjustment of dosage is necessary in patients with renal impairment.

Several interactions between itraconazole and other drugs bear note, particularly for BMT recipients. Cyclosporine levels may become elevated with the concomitant administration of itraconazole. Cyclosporine concentrations should be monitored closely when these drugs are co-administered. Itraconazole plasma concentrations are diminished by concurrent administration of rifampin and phenytoin. Caution also is warranted in its co-administration with antihistamines, coumadin, cisparide and oral hypoglycemic agents, as competitive inhibition of metabolism may lead to elevated levels of these compounds.

Itraconazole is well tolerated with long term use. Most of the adverse reactions reported are transient, and include gastrointestinal disturbances, dizziness, headache and rarely leukopenia. In comparison, to ketoconazole, itraconazole has a lower incidence of hepatic toxicity, no apparent dose-dependent nausea and vomiting, and no adverse effect on testicular steroidogenesis. A syndrome of hypertension and hypokalemia has been observed in some patients receiving high doses of itraconazole, particularly at > 10 mg/kg/day.

2) Fluconazole

Fluconazole is a water soluble meta-difluorophenyl bis-triazole compound that has been shown to be effective against infections due to *Candida spp.*, *Cryptococcus neoformans*, and other fungi in patients with neoplastic dis-

10.2

eases, HIV infection and other immunocompromised states. In comparison to itraconazole and ketoconazole, which are relatively large lipophilic molecules with erratic bioavailability, fluconazole is a relatively small water soluble molecule with rapid absorption and high bioavailability. Also unlike itraconazole, fluconazole is only weakly bound to serum proteins (12%), and thus most fluconazole circulates as free drug. Itraconazole follows nonlinear plasma kinetics and is extensively metabolized, whereas fluconazole exhibits linear plasma kinetics and is only slightly metabolized. The plasma half life of fluconazole in children is substantially reduced in comparison to that of adults; for example, a mean plasma half life of 17 hours was found in children versus 27 and 37 hours previously reported in adults. In the setting of renal impairment, itraconazole requires no dosage adjustment whereas the dosage of fluconazole is adjusted to reflect glomerular filtration. A 50% reduction of dosage is recommended in those with a creatinine clearance of 21-50 ml/min/70 kg, and a 75% dose reduction with a creatinine clearance < 21 ml/min/70 kg. Unlike itraconazole, oral absorption of fluconazole does not depend upon a low intragastric pH, feeding, fasting or gastrointestinal disease. The pharmacokinetics of fluconazole are independent of both the route of administration and formulation, such that the concentration-time curves of orally and parenterally administered fluconazole are very similar.

Fluconazole penetrates well into CSF. This distribution property results in CSF to serum concentration ratios of between 0.5 and 0.9, increasing to between 0.8 and 0.9 in the setting of meningeal disease. Such CSF penetration may contribute substantially to the important role of fluconazole in the management of cryptococcal meningitis.

Fluconazole has been well tolerated with very few dose-limiting side effects in BMT populations. Nausea, other gastrointestinal symptoms, and elevated hepatic transaminases occur infrequently and are usually reversible. Exfoliative skin reactions (Stevens-Johnson syndrome) have been reported in patients with AIDS, although the exact role of fluconazole in these reactions is unclear.

The drug interactions of fluconazole in principle are potentially similar to those of other azoles. For example, fluconazole has been reported to precipitate phenytoin toxicity due to inhibition of metabolism, thus warranting monitoring of phenytoin concentrations during co-administration of fluconazole. Concentrations of cyclosporine may be increased and the effects of warfarin may be potentiated.

Among allogeneic BMT recipients, fluconazole has achieved an important role as an agent for prevention of invasive candidiasis. Two randomized, double-blind studies of fluconazole (400 mg/day PO or IV initiated on day 1 of marrow-ablative chemotherapy) for prevention of deeply invasive candidiasis in adult bone marrow transplant recipients significantly prevented the development of invasive candidiasis. While the findings for prevention of invasive candidiasis by fluconazole in bone marrow transplant recipients are encouraging, there are several limitations in antifungal activity. For example, fluconazole at the current dosages of 200-400 mg/day has little or no activity against *Candida krusei*,

10.2

Torulopsis glabrata, *Aspergillus spp.*, Zygomycetes, and some hyalo-hyphomycetes, such as *Fusarium spp.*

SUGGESTED READING

1. Hiemenz JW, Greene JN. Special considerations for the patient undergoing allogeneic or autologous bone marrow transplantation. Hematol Oncol Clin North Am 1993; 7:961-1002.
2. Lyman C, Walsh TJ. Systemically administered antifungal agents: a review of clinical pharmacology and therapeutic applications. Drugs 1992; 44:9-35.
3. Vartivarian SE, Anaissie EJ, Bodey GP. Emerging fungal pathogens in immunocompromised patients: classification, diagnosis, and management. Clin Infect Dis. 1993; 17 Suppl 2:S487-91.
4. Walsh TJ, Hiemenz J, Pizzo PA. Evolving risk factors for invasive fungal infections—all neutropenic patients are not the same. Clin Infect Dis 1994; 18: 793-8.
5. Wingard JR. Infections in allogeneic bone marrow transplant recipients. Semin Oncol 1993; 20 (5 Suppl 6):80-7.

10.2

Virus Infections Complicating Bone Marrow Translation

William H. Burns

I) INTRODUCTION

Several viruses, but especially herpesviruses, often complicate BMT (Fig. 10.3.1). Until recently, cytomegalovirus (CMV) was the leading infectious cause of death following allogeneic BMT (fungal infections have supplanted CMV in this regard). Other herpesviruses recognized as significant pathogens in the BMT setting include herpes simplex virus (HSV) and varicella zoster virus (VZV). Epstein-Barr virus (EBV) is associated with post-BMT lymphomas, especially following certain immunosuppressive regimens, and human herpesvirus 6 (HHV6) has been associated with cases of interstitial pneumonitis (IP) and marrow suppression. Other families of viruses pathogenic in this setting include enteroviruses including adenoviruses, rotaviruses and coxsackie viruses, and respiratory viruses such as the influenza and parainfluenza viruses and respiratory syncytial virus (RSV).

II) HERPES SIMPLEX VIRUS (HSV)

Seventy to eighty percent of BMT patients are seropositive for HSV. Before the introduction of acyclovir (ACV), about 70% of these seropositive patients developed HSV lesions, primarily represented as mucositis and frequently with severe and extensive ulcerative lesions on the palate, tongue, lips and nose. HSV type 2 lesions in the genital and perineal regions were less often seen. The herpetic origin of these lesions was especially appreciated in centers that did not employ methotrexate in their GVHD prophylaxis, the latter producing severe mucositis that can obscure the herpetic lesions. Fatal pneumonias due to HSV were also seen.

A) PROPHYLAXIS

In the early 1980s a number of studies were reported in which ACV was shown to be effective in prophylaxing against HSV infections when given intravenously or orally and by different schedules. Today, most centers prophylax against HSV for all seropositive patients undergoing BMT. The dose schedule and route vary among institutions, based on economic factors involved in pharmacy preparations and delivery systems. Examples of effective schedules are given in Table 10.3.1a. Most centers prophylax from just before the transplant through 1 month post-BMT, or until the neutrophil count is recovered and the patient is not receiving intense immunosuppressive therapy for GVHD.

B) TREATMENT AFTER PROPHYLACTIC PERIOD

Treatment of herpetic infections after the prophylactic period is discouraged for several reasons: 1) late infections usually resolve without treatment; 2) treatment in this context frequently (up to 18%) results in emergence of resistant virus; 3) prolonged treatment delays the development of immune responses to HSV. Treatment is justified if the herpetic lesions are painful, interfere with nutrition, produce fever that complicates interpretations concerning other possible infections or are thought to be responsible for clinically

10.3

Bone Marrow Transplantation, edited by Richard K. Burt, H. Joachim Deeg, Scott Thomas Lothian, George W. Santos. © 1996 R.G. Landes Company.

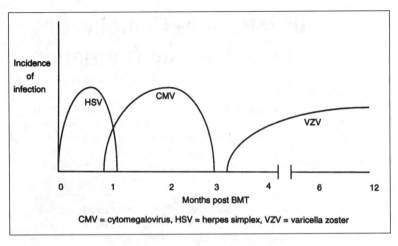

Fig. 10.3.1. Time course of herpes infections after allogeneic BMT.

significant neutropenia or thrombocytopenia. Patients being treated with ganciclovir (GCV) or foscarnet for CMV need not be prophylaxed for HSV with ACV as these drugs, although not formally studied in this context, should be adequate for HSV prophylaxis.

C) **TREATMENT OF ACV-RESISTANT VIRUS**

Virus resistant to ACV rarely occurs during ACV prophylaxis (0.4%) but frequently emerges during treatment of established infection in the post-prophylaxis period (up to 18%).

1) **Acyclovir**

Should treatment of resistant virus infection be deemed clinically desirable, one can initially increase the dose of ACV to 500 mg/m^2 every 8 hours intravenously if renal function is normal.

2) **Foscarnet**

One should switch to foscarnet immediately if there is severe, life-threatening infection with resistant virus. If renal function is adequate, treatment should be with foscarnet at 40-60 mg/kg every 8 hours intravenously (see Table 10.3.3).

3) **Vidarabine**

Vidarabine is not useful in treating ACV-resistant HSV infection in immunocompromised patients.

III) **CYTOMEGALOVIRUS (CMV)**

CMV infections are clinically prominent as IP or enteritis and less often in the BMT setting as retinitis or encephalitis. CMV-related IP is often fatal. Typically, it occurs during the second month (day 50-60) after BMT, presenting as a nonproductive cough, hypoxia, dyspnea and fever, with associated radiological findings of an interstitial process that usually becomes diffuse. It can be of rapid onset with death occurring within several days to several weeks. Factors that are associated with this syndrome include the type of BMT (with allogeneic patients much more prone to developing it than autologous patients and almost never occurring in

10.3

Table 10.3.1a. Acyclovir dose for herpes simplex

1) Acyclovir prophylaxis for herpes simplex virus
Initiate from time of conditioning regimen and continue until absolute neutrophil count greater than 500 / μl:
  250 mg/m^2 q 12 hours IV
  5 mg/kg q 12 hours IV
  200 mg QID PO
  800 mg q 12 hours PO

2) Acyclovir treatment of mucocutaneous herpes simplex infection
  250 mg/m^2 q 8 hours IV for 7-10 days
  5-10 mg/kg q 8 hours IV for 7-10 days
  200-400 mg PO 5 times per day for 7-10 days

3) Dose modification of acyclovir for renal impairment in patients with mucocutaneous herpes simplex

Creatinine Clearance (ml/min/70 kg)	Dose (mg/m^2 IV)	Interval (hours)
> 50	250	8
25-50	250	12
10-25	250	24
hemodialysis	125	24 (post dialysis)

Table 10.3.1b. Acyclovir dose for varicella zoster

1) Acyclovir treatment of varicella zoster infection
  500 mg/m^2 q 8 hours for 7-10 days

2) Dose modification of induction acyclovir for renal impairment in patients with varicella zoster infection

Creatinine Clearance (ml/min/70 kg)	Dose (mg/kg IV)	Interval (hours)
> 50	12.4	8
25-50	12.4	12
10-25	12.4	24
hemodialysis	6.2	24 (post dialysis)

syngeneic patients) the presence of acute GVHD, age, seropositivity and the presence of CMV viremia or CMV in bronchoalveolar lavage (BAL) fluid.

A) CMV DETECTION ASSAY

Effectiveness of the pre-emptive GCV strategy relies on an adequate CMV detection assay predictive for CMV disease. These assays are not standardized and may vary in availability and usefulness in different centers.

1) Shell Vial

Currently, most centers use the shell vial culture technique in which clinical samples are centrifuged onto susceptible cell monolayers and assayed 24-72 hours later for the presence of immediate early or early CMV antigen. Positive cultures from blood cells or BAL fluid, but not throat/saliva or urine, are predictive for developing CMV disease (particularly IP).

10.3

2) **Antigenemia Assay**

   The antigenemia assay (detection of the CMV protein pp65) correlates well with and is positive earlier than viremia and may supplant the shell vial assay in the future as the trigger for beginning GCV administration.

3) **Polymerase Chain Reaction Assay (PCR)**

   The PCR assay can be performed on plasma and is highly sensitive, but prone to false positive results and may not be as useful clinically as a predictor of CMV disease. It may play a more useful role in the future as an indicator of successful clearance of the viral infection and an indicator of whether therapy should be continued.

B) **PREVENTION OF CMV DISEASE**

   About 70% of the adult population is seropositive for CMV. As with all herpesviruses, the virus assumes a latent or persistent state of infection following acute infection. The site(s) of latency of CMV in humans is unknown but almost certainly includes the monocyte, the salivary glands and possibly widespread cells like endothelial cells. Infection post-BMT can be from reactivated endogenous virus or newly acquired virus, the latter by reactivation of the virus from transfused leukocytes.

1) **CMV Negative Blood Products**

   a) *Allogeneic BMT—CMV Seronegative Recipients with Seronegative Donors*

   CMV negative blood products are recommended as a successful strategy to prevent exposure and infection.

   b) *Allogeneic BMT—CMV Seronegative Recipients with Seropositive Donors*

   The utility of providing seronegative blood products to seronegative recipients who have seropositive marrow donors is controversial, but should be done if feasible.

   c) *Allogeneic BMT—CMV Seropositive Recipient*

   It is not necessary to restrict patient to CMV negative blood products.

   d) *Autologous BMT—CMV Seronegative Patient*

   Ideally, the provision of CMV negative products to seronegative autologous patients is desirable if the transplant center can provide such products.

2) **Use of Filters**

   The use of filters to remove leukocytes from blood products is understudy as a supplement of blood products to seronegative recipients when seronegative donors are not sufficient in numbers to provide the required products. Several centers have already replaced CMV seronegative blood products with pall filtered blood products. Should these studies conclude that the available filters are effective, economic factors will determine whether hemopharesis donors should continue to be screened for CMV or whether all blood products given to seronegative recipients should be filtered.

3) **Prophylaxis**

   a) *Prophylaxis with Ganciclovir*

   Studies have examined the question of whether GCV should be started when the WBC reached a certain level post-BMT. CMV dis-

0.3

ease was prevented; however, there was no survival advantage because GCV-induced neutropenia resulted in an increased mortality from fungal or bacterial infections.

b) *Prophylaxis with High-Dose Acyclovir*

ACV at high doses (500 mg/m² every 8 hours, intravenously) has been used to prophylax against CMV disease. This regimen can delay the timing of virus shedding but its usefulness in preventing CMV disease is unconvincing. Because of economic considerations and the introduction of pre-emptive GCV, the use of high-dose ACV for CMV prophylaxis is not recommended.

c) *Prophylaxis with Immunoglobulin*

The administration of immunoglobulin to prophylax against CMV is also not recommended because of its cost, the lack of consistent, convincing data of its efficacy and because of the availability of a proven strategy (pre-emptive GCV).

4) **Pre-emptive Therapy**

a) *Allogeneic BMT*

Two studies clearly demonstrated that administration of GCV at the time of detection of viremia or presence of CMV in BAL fluid will result in the prevention of CMV disease. The standard approach today is to perform surveillance cultures of seropositive allogeneic patients for CMV viremia weekly (or positive BAL fluids performed on day 35 and day 49 in some centers) from day 30-120 post-BMT and to institute GCV pre-emptive therapy when virus is detected. GCV is then continued in doses given in Table 10.3.2a until 100 days after BMT. The drug is discontinued if the neutrophil count drops below 750 or 1000 for 2 consecutive days. This occurs in about 30% of patients receiving GCV and the neutropenia usually lasts about 2 weeks but may last several weeks. GCV is restarted when the neutrophil count rises above 1000 for 2 consecutive days. The administration of G-CSF or GM-CSF should be considered if neutropenia occurs or is prolonged. Foscarnet may be used in place of GCV if there is concern about discontinuing GCV; in this case, the renal function must be adequate and closely monitored.

b) *Autologous BMT*

There is no proven role for CMV surveillance cultures. The use of pre-emptive GCV is not recommended for detection of viremia in asymptomatic patients.

C) **TREATMENT OF CMV DISEASE**

Despite monitoring for viremia and/or BALs, CMV IP sometimes occurs. It might also occur at a later time after discontinuing GCV. The latter may be secondary to a delayed or insufficient immune response as a less viral antigen is presented during GCV administration. Several studies have demonstrated a 30-50% success by treatment with GCV and immunoglobulin. There is insufficient data available to recommend immunoglobulin preparations enriched or selected for antibody to CMV; no specific brand of immunoglobulins has proved superior. Interestingly, it is the combination of GCV and immunoglobulin that is effective whereas singly each is ineffective. In a controlled study, the treatment of CMV enteritis with GCV alone did not result in clinical improvement

10.3

Table 10.3.2a. Ganciclovir dose for pre-emptive therapy

1) **Pre-emptive therapy for positive CMV surveillance culture of blood or bronchioalveolar lavage**
   5 mg/kg IV every 12 hours for 10-14 days then 5 mg/kg 3-5 times per week until day 100 after BMT
2) **Dose modification of pre-emptive ganciclovir therapy in patients with renal impairment**

Creatinine Clearance (ml/min/70 kg)	Dose (mg/kg IV)	Interval (hours)
> 80	5.0	12
50-79	2.5	12
25-49	2.5	24
< 25	1.25	24
hemodialysis	1.25	24 (post dialysis)

Hold ganciclovir for absolute neutrophil count < 500/ul or platelet count < 25,000/µl unless cytopenias are considered secondary to CMV.

Table 10.3.2b. Ganciclovir dose for treatment of CMV pneumonitis or enteritis

1) **Treatment of CMV pneumonitis or enteritis**
   5 mg/kg IV every 12 hours for 21 days then 5 mg/kg 5 days per week until off immunosuppression and
   IVIG 500 mg/kg IV every other day for 7-10 doses then weekly until off immunosuppression
2) **Dose modification of induction ganciclovir for patients with CMV disease and renal impairment**

Creatinine Clearance (ml/min/70 kg)	Dose (mg/kg IV)	Interval (hours)
> 60	5.0	12
31-60	2.5	12
16-30	2.5	24
5-15	1.25	24
hemodialysis	1.25	24 (post dialysis)

Hold ganciclovir for absolute neutrophil count < 500/µl or platelet count < 25,000/µl unless cytopenias are considered secondary to CMV.

despite an antiviral effect. In analogy to CMV IP treatment, the same combination of GCV and immunoglobulin is therefore used to treat CMV enteritis although this has not been formally studied (Table 10.3.2b).

IV) VARICELA ZOSTER VIRUS (VZV)

Zoster occurs in 20-30% of autologous and 20-50% of allogeneic patients during the first year after BMT with a peak incidence around 4 months post-BMT. Cutaneous dissemination occurs in about 25% of cases and visceral dissemination occurs clinically in about 10-15% of cases. Visceral involvement can include the lung, liver or central nervous system, and thrombocytopenia or intravascular coagulopathy may occur. It can present, sometimes without cutaneous involvement, as an acute abdomen with severe abdominal pain and involves the pancreas, intestines or adrenal glands. Death from VZV infection post-BMT is rare since the advent of ACV. Although ACV prophylaxis is effective for VZV, it is generally not done because of

10.3

Table 10.3.3. Foscarnet dose

Foscarnet dose for ganciclovir-resistant herpes simplex or varicella zoster

40-60 mg/kg IV q 8 hours or 60-90 mg/kg IV q 12 hours until clinical resolution

For dose adjustment in renal insufficiency, refer to schedule and dose reduction based on creatinine clearance as provided by the manufacturer's label.

the long period at risk and the cost of the prophylaxis. If zoster occurs, the immediate institution of intravenous acyclovir therapy at 500 mg/m^2 every 8 hours for 7-10 days is recommended. It is usual to see the formation of new lesions for 1-3 days after beginning ACV therapy. Although in placebo-controlled studies the time to defervescence and cutaneous healing was not significantly shortened, varicella pneumonia and mortality from VZV infection are decreased. The introduction of better absorbed antivirals (expected in the near future) may change this recommendation to their use orally. Strict isolation of patients infected with VZV should be practiced, including respiratory isolation procedures for patients with visceral dissemination (see Table 10.3.1b).

V) ENTEROVIRUSES

A) ADENOVIRUS

Infection occurs in about 5% of patients, usually during the second or third month after BMT but occasionally late, especially in the setting of chronic GVHD. Disseminated disease occurs in about 20% of infected patients and can affect multiple organs including the liver, lung and urinary tract. Adenovirus type 11 commonly infects the urinary tract and can result in hemorrhagic cystitis. There is no proven therapy.

B) COXSACKIE VIRUSES AND ROTAVIRUSES

Both can cause severe diarrhea and can contribute to patient mortality. Coxsackie viruses can be transmitted by the respiratory route as well as the fecal/oral route and can cause a rash and respiratory symptoms as well as diarrhea. Diagnosis is by virus culture or more often by antigen detection or by PCR. Coxsackie virus A1 was the cause of an outbreak that resulted in the deaths of 6 of 7 infected BMT patients. As with the adenoviruses, there is no proven therapy. Strict isolation procedures should be instituted for patients with enterovirus infections.

VI) RESPIRATORY VIRUSES

The common respiratory viruses (RSV, the parainfluenza viruses, rhinoviruses, and influenza A and B) may be isolated in the BMT population, and generally the frequency and timing of their isolation reflects the pattern found concomitantly in the community and may vary from year to year.

A) RESPIRATORY SYNCYTIAL VIRUS

RSV infection is most common between December and February. In the Seattle experience during an outbreak in the winter of 1989-90, 18 of 31 infected patients developed pneumonia with a mortality of 78%. The use of aerosolized ribavirin did not appear useful in patients with pneumonia. In the MD Anderson experience, aerosolized ribavirin (20 mg/ml for 18 hours a day by face mask or endotracheal tube) and IVIG (500 mg/kg IV every other day) resulted in a 22% mortality if therapy was started before onset of severe

10.3

respiratory symptoms and 100% mortality if started after intubation. The duration of therapy was individualized according to clinical status. The use of intravenous ribavirin or RSV-specific immunoglobulin, or the combination of the immunoglobulin with ribavirin given by either route, may be useful in the BMT setting but no study results are available.

B) PARAINFLUENZA AND INFLUENZA VIRUS

Infections by parainfluenza and influenza viruses usually cause severe and often fatal pneumonias in BMT patients. There is little data concerning treatment with ribavirin in this setting. Diagnosis of infection by these respiratory viruses depends on culture of nasopharyngeal or throat washings, swabs or immunofluorescent staining of epithelial cells from such specimens. Culture and staining of material from BAL fluid is also usually performed. There is controversy concerning the better source of material. The incubation period of these viruses is short (a few days) and since there is often evidence that the respiratory infection is present prior to BMT, patients with symptoms or positive diagnoses should have their transplants delayed if possible until the symptoms/infections resolve as these infections are currently untreatable and often prove fatal during the BMT period.

VII) EPSTEIN-BARR VIRUS

Infection of B lymphocytes by Epstein-Barr virus (EBV) results in B cell proliferation. In the nonimmunocompromised individual, cytotoxic EBV-specific T lymphocytes prevent uncontrolled B cell proliferation. In immunocompromised allogeneic transplant patients, failure of immune surveillance by EBV-specific T lymphocytes results in a polyclonal or less often monoclonal B cell proliferation. The affected lymphocytes may be of donor or host origin. EBV lymphoproliferative syndrome (EBV-LPS) occurs in approximately 0.5% of allogeneic bone marrow transplant recipients. Risk factors for EBV-LPS are a T cell depleted marrow graft, use of ATG or anti-CD3 antibodies for GVHD, and an HLA disparate transplant complicated by GVHD. The posttransplant interval for development of EBV-LPS ranges from day 45-500 with a median onset between day 70-80. Presentation before day 70 is usually associated with fever and aggressive extranodal disease. Onset after day 70 generally has a more indolent course manifest by fever and adenopathy. Antiviral therapy for EBV-LPS is generally ineffective. Intravenous infusion of anti-B cell antibodies has been effective for poly- or oligoclonal proliferations but not monoclonal EBV-LPS. Administration of unirradiated donor lymphocytes which contain EBV-specific cytotoxic donor lymphocytes has resulted in remissions of both oligoclonal and monoclonal EBV-LPS within 14-30 days of infusion.

10.3

11

Graft-Versus-Host Disease

Nelson J. Chao

I) ACUTE GRAFT-VERSUS-HOST DISEASE (aGVHD)

A) ETIOLOGY

aGVHD is a consequence of donor T cells recognizing host antigens as foreign. It results from an afferent phase of T lymphocyte stimulation (antigen presentation, T cell activation and T cell proliferation) and efferent phase of T cell and secondary effector cell response (cytokine secretion, cytotoxic T cells, natural killer cells). The essential factors necessary for aGVHD are: 1) immunologically competent donor T cells, 2) histoincompatibility between donor and host, 3) inability of host to reject donor lymphocytes.

B) INCIDENCE

The incidence of aGVHD varies (Table 11.1) with degree of genetic disparity at the major histocompatibility complex (MHC) or minor histocompatibility loci, recipient age, number of donor T cells infused, method of prophylaxis, and to a lesser degree with gender, parity, TBI containing conditioning regimens, use of a gnotobiotic (germ free) environment and co-existent herpes infections.

1) Genetic Disparity

T cells via the T cell receptor (TCR) are capable of differentiating self from nonself. The TCR recognizes a peptide/MHC complex. In general, class I molecules (HLA-A, B, C) present endogenous cellular proteins while class II molecules (HLA-DR, DP, DQ) present exogenous (extracellular) proteins in the form of small amino acid peptides. The peptides presented by the MHC molecule are termed non-MHC or minor histocompatibility antigens.

a) MHC Antigens

The incidence and severity of aGVHD directly correlates with MHC disparity. The MHC is a complex genetic loci found in the human leucocyte antigen (HLA) system. Even with intensive immunosuppressive prophylaxis, 10-50% of HLA-matched sibling bone marrow allografts are complicated by clinically significant aGVHD which directly results in death for 5-15% of recipients. The incidence of clinically significant aGVHD in HLA-mismatched related allografts is higher and increases with the number of mismatches. The transplant group in Seattle has reported that single antigen mismatched related bone marrow transplant recipients had a higher incidence of acute GVHD compared to matched related recipients but both groups had the same overall outcome. This was related to a lower incidence of relapse in the mismatched related recipients.

b) Non-MHC (HLA) Antigens

Non-MHC antigens are the peptides presented by the MHC molecules. These peptides are also termed the minor histocompatibility antigens. This term belies their significance since aGVHD in an HLA-

Bone Marrow Transplantation, edited by Richard K. Burt, H. Joachim Deeg, Scott Thomas Lothian, George W. Santos. © 1996 R.G. Landes Company.

Table 11.1. Factors associated with lower acute GVHD

HLA matching:	matched sibling donors
Age:	children
Donor:	same gender and nulliparous women
Type of disease:	aplastic anemia
Stage of disease:	early (i.e. chronic phase of chronic myelogenous leukemia or first remission of acute leukemias)
Dose of radiation:	lower dose (≤ 12 Gy)
Dose of methotrexate or cyclosporine delivered:	ability to deliver the intended dose
Enviroment:	laminar flow rooms (only for patients with aplastic anemia)

matched recipient is due to minor histocompatibility antigens. Little is known about the number or identity of clinically relevant or immunodominant minor antigens. However, work predominately done in mice has identified approximately 40-50 minor antigens. An example of these include male specific minor antigen (HY-Ag), maternally transmitted antigen (Mta), epidermal alloantigen (Epa-1) and viral antigens. The viral antigens presented by the host function as minor antigens in aGVHD and may explain the increased risk of GVHD in BMT patients with CMV or herpes infections.

2) **Recipient Age**

The clearest example is a comparison between the incidence of acute GVHD in the pediatric population compared to those found in adults. The higher incidence in older patients may be related to immunological factors associated with aging, such as an increase in antigen sensitization or thymic involution.

3) **Number of Donor T Cells**

Donor T cells are important for the occurence of GVHD. The issue of the actual number of T cells is likely to be important. With extensive T cell depletion of the marrow graft, there is a marked decrease in the incidence and severity of acute and chronic GVHD but a significantly higher incidence of graft failure and relapse of the underlying disease. Unfortunately, the actual numbers of T cells to add back in an engineered graft has not been established. In the adult patient it appears that $1\text{-}5 \times 10^5$ T cells/kg recipient weight may be safe if cyclosporine is also used.

4) **Method of Prophylaxis**

The choice of prophylaxis for acute GVHD is also associated with the incidence of GVHD. Most drug regimens result in approximately a 10-40% incidence of acute GVHD in matched related bone marrow transplantation. T cell depletion is usually significantly better, resulting in a 5-20% incidence, although some form of immunosuppression such as cyclosporine is frequently but not always administered.

5) **Other**

Other factors which have been associated with a lower incidence of GVHD include transplantation in a gnotobiotic (germ-free) environment, specifically for patients with aplastic anemia, transplantation from a nuliparous donor and transplatation from a herpes simplex virus serologically negative donor.

11

C) CLINICAL PRESENTATION

The usual triad of aGVHD is dermatitis (rash), hepatitis (jaundice) and gastroenteritis (diarrhea, abdominal pain). However, these symptoms may occur alone or together and the hematolymphoid system and mucosal surfaces may also be involved.

1) Skin/Mucosa (Fig. 11.1)

In general, the first and most common clinical manifestation of aGVHD is a maculopapular rash, usually occurring near the time of white blood cell engraftment. The early stages may be pruritic and confined to the nape of the neck, ears, shoulders, palms of the hands or soles of the feet. As the disease progresses, the rash may become confluent and involve the entire integument. In severe cases, bullous lesions (epidermal necrolysis) similar to third degree burns occur and are as life-threatening. Mucosal surfaces are also involved in aGVHD. One of the earliest symptoms of aGVHD (often not appreciated) is a sense of nasal stuffiness and/or sniffles by the patient. Conjunctivitis attributed to the conditioning regimen and/or aGVHD is also common.

a) Differential Diagnosis

Following BMT, the etiology of a skin rash may be difficult to clinically or histologically differentiate from conditioning regimen or drug toxicity, especially from antibiotics.

2) Liver

The second most common organ involved in aGVHD is the liver. Rarely, patients may have moderate or severe hepatic aGVHD without clinical cutaneous disease. The earliest and most common liver abnormality is a rise in conjugated bilirubin and alkaline phosphatase, although transaminases may also be elevated.

a) Differential Diagnosis

Other causes of hyperbilirubinemia include a side effect of hyperalimentation, veno-occlusive disease of the liver, nodular

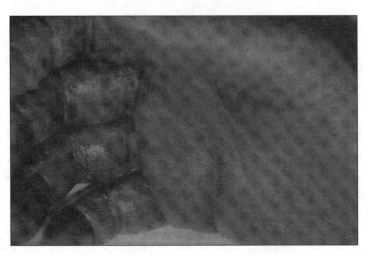

Fig. 11.1. Clinical acute GVHD.

11

regenerative hyperplasia, infections (predominantly cytomegalovirus [CMV], herpes simplex, hepatitis B, etc.) and drug toxicity (especially cyclosporine or methotrexate). Occasionally, a biopsy may be required to establish the diagnosis. However, coexistent thrombocytopenia, coagulopathies or ascites make a percutaneous biopsy prohibitively risky. One possible manner to obtain a liver biopsy is by the transjugular route. This procedure requires a well-experienced operator and may result in insufficient tissue for diagnosis. Abnormal liver function tests and biopsy documented GVHD in another organ system (such as skin) usually suggest the diagnosis of hepatic GVHD.

3) Gastrointestinal

The third important organ system affected by aGVHD is the gut, and is characterized by diarrhea and abdominal cramping. The diarrhea may be voluminous and bloody causing life-threatening fluid and electrolyte losses and a portal of entry for infections. A severe ileus may also occur. Although not as common, upper GI aGVHD presents as anorexia, dyspepsia, food intolerance, nausea and vomiting.

a) Differential Diagnosis

Diarrhea with crampy abdominal pain is suggestive of aGVHD, but medications (e.g. antibiotics), infections (e.g. clostridium difficile, CMV) and conditioning regimen toxicities may also cause diarrhea. Stool cultures are routine and a rectal biopsy may be helpful.

4) Hematolymphoid

GVHD results in lymph node and thymic involution, inversion of the CD4/CD8 ratio, hypogammaglobulinemia, anergy and absence of immune response to vaccination. Thus, independent of immunosuppressive therapy used to treat it, aGVHD increases the risk of infection. Persistent thrombocytopenia is also a manifestation of aGVHD and probably arises from increased consumption rather than marrow suppression.

D) CLINICAL GRADING

Each clinical stage of organ involvement is graded then combined to obtain an overall grade (Table 11.2). Clinically significant aGVHD is usually defined as overall grade II-IV. Grade I is mild, grade II moderate, grade III-IV severe. Patients with moderate to severe GVHD have a significantly higher mortality. In general, relapse rates correlate inversely with severity of aGVHD.

E) HISTOLOGY

Histologically, the essential element of aGVHD is epithelial cell necrosis involving the stem cell regenerating compartment of the skin (basal cell layer), liver (biliary ducts) and gastrointestinal tract (crypts).

1) Skin (Fig. 11.2)

Histologic features of cutaneous aGVHD are vacuolar degeneration, lymphocytic infiltration, dyskeratotic (shrunken/pink) keratinocytes with or without satellite lymphocytes, basal cell necrosis (apoptosis), acantholysis (cell-cell separation) and epidermolysis (dermal-epidermal separation). Unfortunately, the histology is not pathognomonic. The conditioning regimen may cause similar changes, especially if the biopsy is obtained within the first 3 weeks following BMT.

11

2) Liver (Fig. 11.3)

aGVHD of the liver is manifest histologically by portal triad infiltration with lymphocytes, vacuolation of the cytoplasm, loss of nuclei and necrosis of bile duct epithelial cells. In severe cases parenchymal hepatocellular necrosis with acidophilic bodies may be present.

Table 11.2. Grading acute GVHD

Clinical Grading of Individual Organ Systems

Organ	Grade	Description
Skin	+1	Maculo-papular eruption over <25% of body area
	+2	Maculo-papular eruption over 25-50% of body area
	+3	Generalized erythroderma
	+4	Generalized erythroderma with bullous formation and often with desquamation
Liver	+1	Bilirubin 2.0-3.0 mg/dl
	+2	Bilirubin 3.1-6.0 mg/dl
	+3	Bilirubin 6.1-15 mg/dl
	+4	Bilirubin >15 mg/dl
Gut	+1	Diarrhea > 30 ml/kg or > 500 ml/day
	+2	Diarrhea > 60 ml/kg or > 1000 ml/day
	+3	Diarrhea > 90 ml/kg or > 1500 ml/day
	+4	Diarrhea > 90 ml/kg or > 2000 ml/day; or severe abdominal pain ileus

Overall Grade*

Grade	Skin	Liver		Gut	ECOG Performance
I	+1 to +2	0		0	0
II	+1 to +3	+1	and/or	+1	0-1
III	+2 to +3	+2 to +3	and/or	+2 to +3	2-3
IV	+2 to +4	+2 to +4	and/or	+2 to +4	3-4

* If no skin disease the overall grade is the higher single organ grade.

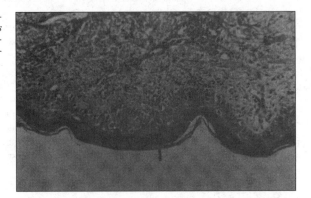

Fig. 11.2. Histology of cutaneous acute GVHD. Arrow points to dyskeratotic body.

11

3) **Gastrointestinal** (Figs. 11.4A and 11.4B)

The histologic picture demonstrates necrosis of crypt cells, accumulation of cellular debris within crypts, and crypt cell drop out leading to loss of crypts and in severe cases total epithelial denudation.

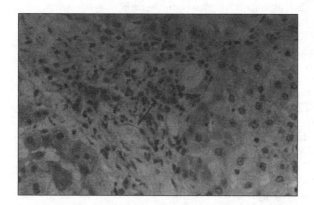

Fig. 11.3. Histology of hepatic acute GVHD. Arrows point to two lymphocytes in biliary ductual of portal triad.

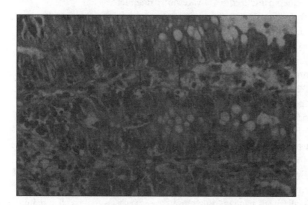

Fig. 11.4A. Histology of acute gastrointestinal GVHD. Arrows point to apoptotic cells in intestinal crypts.

Fig. 11.4B. Histology of acute gastrointestinal GVHD. Marked loss of intestinal crypts.

11

F) PHARMACOLOGIC PREVENTION OF aGVHD

Patients who have undergone transplantation of a solid organ (e.g. heart, liver, kidney) require life-long immunosuppression to prevent rejection. In contrast, after BMT with a healthy allograft (no chronic GVHD), immunosuppression is temporary and usually stopped approximately 6 months after BMT. Some form of prophylaxis is always initiated prior to allogeneic marrow infusion. The mainstay for prevention of aGVHD is cyclosporine (CsA). The three most commonly used CsA regimens are: 1) cyclosporine and methotrexate, 2) cyclosporine and prednisone, 3) cyclosporine, methotrexate, and prednisone.

1) Cyclosporine and Methotrexate (CsA/MTX)

Dose and schedule of cyclosporine and methotrexate as developed at Seattle is shown in Table 11.3. This is the most accepted and widely used regimen. The overall incidence of grade II-IV aGVHD for HLA-matched related bone marrow recipients receiving CsA/MTX is 20-30%.

2) Cyclosporine/Prednisone

No randomized trial has been published comparing CsA and prednisone versus CsA and MTX. However, one study of CsA combined with prednisone has reported only a 12% incidence of grade II-IV aGVHD in sibling matched allogeneic BMT.

3) Cyclosporine/Prednisone/Methotrexate (Table 11.3)

Triple drug prophylaxis in a randomized study had only a 9% incidence of grade II-IV aGVHD. The optimal time of initiating prednisone appears to be day 7. Starting before day 7 did not result in less GVHD while delaying until day 15 increases the incidence of grade II-IV aGVHD.

4) Intravenous Immunoglobulins (IVIG)

IVIG at a dose of 400-500 mg/kg/week from day 0 until day 120 has been reported to significantly decrease the incidence of grade II-IV aGVHD in matched sibling BMT for patients over age 20.

5) Other

Other pharmacologic approaches for the prevention of GVHD, some of which are beginning or are in on-going clinical trials, include new immunosuppressants such as FK 506 (tacrolimus) and possibly rapamycin. Alternatively, immunosuppressive regimens have combined monoclonal antibodies with standard drug prophylaxis. Finally, as the the afferent and efferent pathways leading to GVHD are elucidated, more specific targeting of the selected cytokines involved may lead to more effective approaches.

G) CYCLOSPORINE AND METHOTREXATE TOXICITY

1) CsA Toxicity

The primary toxicity of cyclosporine is its effect on the kidneys. Cyclosporine can cause an elevation of the creatinine and this effect appears to be dose related. Histologically, cyclosporine causes renal tubule vacuolization, interstitial fibrosis and arteriolar hylanization. Cyclosporine-induced liver dysfunction can mimic veno-occlussive disease of the liver resulting in jaundice, hepatorenal syndrome, hepatic encephalopathy and elevated transaminases. There are no unique histologic features to cyclosporine-induced hepatic injury. Cyclosporine associated central nervous system toxicity includes somnolence, headache, anxiety, depression, tremors, parathesias, thrombotic thrombocytopenic purpara (TTP), seizures especially if the concentration of magnesium is low, cortical blindness especially if hypocholesterolemic and a demylinating peripheral neuropathy. Gastrointestinal side effects include nausea, vomiting and anorexia. Another common toxicity is hypertension which is treated with nifedipine, a calcium channel blocker. In severe cases microangiopathic changes are present. Other side effects are hyperkalemia and islet cell damage with hyperglycemia.

a) CsA Assays

Some institutions attempt to maintain blood cyclosporine levels within the therapeutic range of that institution. However, correlation of cyclosporine blood levels and toxicity or prevention of aGVHD

Table 11.3. Two regimens for prophylaxis of acute GVHD

Day	Cyclosporine and Methotrexate	
	Cyclosporine	Methotrexate
-2	5.0 mg/kg IV daily	—
+1	"	15 mg/m^2 IV single dose
+3	"	10 mg/m^2 IV single dose
+4	3.0 mg/kg IV daily	—
+6	"	10 mg/m^2 IV single dose
+11		10 mg/m^2 IV single dose
+15	2.75 mg/kg IV daily	
+36	10 mg/kg PO daily	
+84	8 mg/kg PO daily	
+98	6 mg/kg PO daily	
+120	4 mg/kg PO daily	
+180	off	

Day	Cyclosporine	Prednisone	Methotrexate
-2	5.0 mg/kg IV daily	—	
+1	"	—	15 mg/m^2 IV single dose
+3	"	—	10 mg/m^2 IV single dose
+4	3.0 mg/kg IV daily	—	
+6	"	—	10 mg/m^2 IV single dose
+7	"	0.5 mg/kg IV daily	
	"	"	
+15	2.75 mg/kg IV daily	1.0 mg/kg IV daily	
+29	"	0.8 mg/kg PO daily	
+36	10 mg/kg PO daily	"	
+43	"	0.5 mg/kg PO daily	
+57	"	0.2 mg/kg PO daily	
+84	8 mg/kg PO daily	"	
+98	6 mg/kg PO daily	"	
+120	4 mg/kg PO daily	0.1 mg/kg PO daily	
+180	off	off	

*Divided in two doses 12 hours apart

11

is not always reliable. Different assays are used by different institutions. Cyclosporine in whole blood is distributed in lymphocytes (10%), red blood cells (40%) and plasma (30-40%). Plasma levels of cyclosporine vary by temperature of specimen. Therefore, we recommend that levels be determined on whole blood. Some cyclosporine assays (e.g. monoclonal flourescence polarization immunoassay) are actually measuring cyclosporine and its metabolities, many of which are not immunosuppressive. Other assays such as high performance liquid chromatography (HPLC) are specific for cyclosporine not its metabolites. CsA levels should be troughs drawn 15 minutes before the morning dose. Cyclosporine also binds to catheter side walls. Caution must be exercised and a generous discard must be drawn before a cyclosporine level is drawn from an indwelling catheter. Frequently peripheral blood draws are performed to bypass this problem.

b) CsA Drug Interactions

CsA levels are altered by a multitude of drugs. For example, CsA levels are increased by ketoconazole and cimetidine and decreased by phenytoin, phenobarbital and rifampin.

c) Dose Adjustment of CsA

Institutions vary on cyclosporine dose modification according to renal function (Table 11.4). Institutions such as the Medical College of Wisconsin, which do a large number of unrelated transplants-dose cyclosporine according to levels independent of creatinine. On the other hand, for HLA-matched transplants, dosing CsA by creatinine including withholding CsA for a creatinine > 2.0 mg/dl at Stanford or > 4.0 mg/dl at Johns Hopkins has not been shown to increase GVHD severity.

2) Methotrexate Toxicity

The major toxicities of methotrexate when used for GVHD prophylaxis are renal and hepatic insufficiency as well as the development of severe mucositis. Methotrexate may also contribute to the development of interstitial pneumonitis.

a) Methotrexate Dose Adjustment

The methotrexate dose may be omitted or reduced for severe mucositis, or renal and/or liver insufficiency (Table 11.4).

H) NONPHARMACOLOGIC PREVENTION OF aGVHD

1) Ex-vivo T Cell Depletion (TCD)

One attractive approach to prevent GVHD is to eliminate T lymphocytes from the donor inoculum prior to infusion of the bone marrow. Several techniques have been developed to deplete T lymphocytes from donor bone marrow in humans including a variety of physical separation techniques such as density gradients, selective depletion with lectins, treatment with cytotoxic drugs and the use of anti-T cell serum or monoclonal antibodies, either alone, with complement or conjugated to toxins. The depletion of T lymphocytes from donor bone marrow may also have adverse effects. T cell depletion adversely affects engraftment, the adequacy of immune reconstitution and the incidence of leukemic relapse or infections.

11

Table 11.4. Dose adjustment of cyclosporine and methotrexate

Creatinine (mg/dl)	Stanford CsA Dose Adjustment	Creatinine (mg/dl)	Johns Hopkins CsA Dose Adjustment	Medical College of Wisconsins CsA Dose Adjustment
< 1.5	100%	< 2.2	100%	CsA dose based
1.5-1.7	75%	2.2–3.0	75%	on levels and
1.8- 2.0	50%	3.0-4.0	25%	is not adjusted
> 2.0	hold	> 4.0	hold	by creatinine

Methotrexate % Dose

Bilirubin (mg/dl)	Methotrexate Dose
< 2.0	100%
2.1-3.0	50%
3.1-5.0	25%
> 5.0	hold

a) Nonselective

Several techniques for T cell depletion have been utilized and most studies use ex-vivo treatment of the donor bone marrow with monoclonal antibodies. The most common methods include broadly reactive anti-T cell agents, such as anti-CD2, anti-CD3, anti-CD5. Moreover, a broadly reactive human monoclonal antibody against lymphoid tissues, Campath-1, has also been used. Although these studies show some efficacy, the amount of variability involved between batches of complement and the different antibodies led to the development of newer methods of T cell depletion, such as antibodies bound to ricin A chain, other toxins or antibodies conjugated to magnetic beads, depletion of an antibody bound to target cells, soybean lectin agglutination and E-rosette formation or counterflow elutriation. Most of these techniques generally achieve between a 1.5-4 log reduction of T cells. Unfortunately nonselective T cell depletion methodologies result in increased graft failure and a higher incidence of relapse, and therefore there is no significant overall improvement in disease-free survival.

b) Selective T Cell Depletion

Several other specific T cell targets have been studied. One study which appears to be promising is the selective depletion of CD8+ T lymphocytes for the prevention of GVHD. This monoclonal antibody appears to retain the graft-versus-leukemia effect, as the relapse rate in patients with chronic myelogenous leukemia was not increased, however, the overall incidence of acute GVHD was still high since these patients also received cyclosporine following BMT. The overall incidence of acute GVHD was not different from that observed in small series using cyclosporine alone. Another intriguing study was selective depletion of CD6+ T lymphocytes from donor bone marrow. These patients did not receive any further prophylactic treatment for acute GVHD. Eighteen percent of the patients

11

developed acute GVHD grade II-IV and 2.7% had acute graft failure. These results are encouraging in that depletion of CD6+ cells from the donor marrow reduced the morbidity and mortality associated with BMT secondary to GVHD. This approach using CD6 depletion selectively targeted mature T cells but spared closely related and potentially important cells such as NK cells. This may be one reason for the low incidence of graft failure.

I) THERAPY OF aGVHD

Therapy for aGVHD should be aggressive since completeness of response correlates with survival. The criteria for starting therapy is usually if patients develop grade II or greater GVHD. This is especially true if the change in the grade of GVHD is rapid and progressive.

1) First line Therapy—Glucocorticoids

Since most patients are already on prophylactic CsA, first line therapy should be corticosteroids. Some centers treat with low-dose IV methylprednisolone or oral prednisone, usually at 1-2 mg/kg/day. If after 7-10 days improvement occurs, steroids are tapered. Others institutions treat with high-dose intravenous methylprednisolone at 1.0 g/m^2 times 3 days followed by a rapid consecutive taper of 0.5 g/m^2 for 3 days, 0.25 g/m^2 for 3 days, and 0.125 g/m^2 for 3 days, then 100 mg/day oral prednisone tapered off over 20-30 days. High-dose steroid therapy increases the response rate but does not change overall survival. The overall response rate to steroids is 50-80%. Almost all survivors are destined to develop chronic GVHD.

2) Second Line Therapy—Steroid Resistant GVHD

Patients with acute GVHD resistant to corticosteroids have a poor prognosis. Addition of other immunosuppressive drugs to corticosteroids and cyclosporine is of no proven benefit. However, encouraging but generally transient responses have been obtained with antibody therapy directed against T lymphocytes or their cytokines.

a) Polyclonal Antibody Therapy—Anti-thymocyte Globulin (ATG)

Large trials of ATG for corticosteroid resistant GVHD have not been reported. ATG is commercially available and capable of inducing temporary improvement in steroid resistant GVHD. The dose of ATG has varied. One regimen utilizes 10 mg/kg IV over 6-12 hours daily for 7-10 days. Pretreatment skin testing for hypersensitivity should be done.

b) Monoclonal Antibody Therapy

Since treatment with monoclonal antibody OKT3, specific for the CD3 complex, can reverse or cure rejection of human renal allografts, its use was attractive for the treatment of GVHD. However, these anti-CD3 antibodies were mitogenic, and the resultant T cell activation was dose limiting. Nonmitogenic antibodies against CD3 have also been used. Various other monoclonal antibodies or cytokine receptor antagonists have demonstrated improvement in corticosteroid-resistant GVHD with minimal side effects (Table 11.5). These antibodies are currently in intrainstitutional phase I-II trials and are not commercially available.

11

Table 11.5. *Monoclonal antibody or receptor antagonist therapy for steroid-resistant acute GVHD*

Reference	Other GVHD Therapy	Antibody	Dose/Schedule	Response
Blood 1990; 75(7):1426	Cyclosporine and methylprednisolone	H65-RTA (Zomazyme)—anti-CD5 antibody labeled with ricin A chain	0.05 mg/kg/day to 0.33 mg/kg/day IV for up to 14 consecutive days	16/34 durable complete and partial responses; side effects—fatigue, myalgia, weight loss, hypoalbuminemia
Blood 1992; 79(12):3362	Cyclosporine and methylprednisolone	B-C7, anti-TNF α antibody	0.1-0.4 mg/kg IV daily x 4 days then every other day x 2	15 patients, 74% partial response within 3 days; relapse in most when therapy stopped; no side effects
Blood 1994; 84(4):1342	Cyclosporine and methylprednisolone	IL-1Ra, IL-1 receptor antagonist	400-3200 mg a day continuous IV infusion for 7 days	10/16 improved; side effect-increased transaminase in 2 patients
Transplant Int 1991; 4:3	Cyclosporine and prednisone	25.3, murine anti-LFA-1 (CD11a) antibody	0.1 mg/kg IV over 4 hours daily x 5 days	8/10 (80%) partial response, 7/8 relapse off therapy; no side effects
Blood 1994; 84:(4,):1320	Cyclosporine and prednisone	Humanized anti-Tac antibody, (IL-2 α receptor antibody)	0.5, 1.0, or 1.5 mg/kg IV over 1 hour single dose, repeated once between 11-48 days after first dose in responding patients	4/20 complete response, 4/20 partial response; side effects-chills, diaphoresis in 1 patient
BMT 1994; 13:563	TCD, cyclosporine and methylprednisolone	BT 563 (B-B10), murine anti-human IL-2 α receptor antibody	0.2 mg/kg IV over 30 min daily (mean 27 days, range 12-70 days) until GVHD < grade II for 48 hours	11/15 complete remission, 2/15 partial remission; 6/13 relapsed; no side effects
Blood 1990; 75(4):1017	Cyclosporine and methylprednisolone	BT 563 (B-B10), murine anti-human IL-2 α receptor antibody	5.0 mg IV bolus daily x 10 days then every other day for 10 days	21/32 complete response, 6/32 partial response; 10/27 relapse

11

II) CHRONIC GRAFT-VERSUS-HOST DISEASE

A) INTRODUCTION

Graft-versus-host disease has historically been divided into acute and chronic phases. Traditionally, acute GVHD occurs within the first 100 days and chronic GVHD was defined as GVHD occurring after the first 100 days following allogeneic BMT. It is clear now that cases of acute GVHD generally occur within 30-40 days of BMT and chronic GVHD, both by clinical and histological criteria, can occur as early as 50 days posttransplant. Chronic GVHD is the single major determinant of long-term outcome and quality of life following BMT. Patients who have limited disease (Table 11.6) have a favorable prognosis even without therapy. Patients with extensive, particularly multi-organ disease, have an unfavorable natural history. However, if such patients can be kept alive, the usual course is that the disease will "burn-out" (i.e. tolerance occurs) and the patient improves.

B) INCIDENCE

As many as 50-60% of patients may develop chronic GVHD at some point after the first 100 days, depending on the risk factors present. The greatest risk factor for the development of chronic GVHD is the prior occurrence of acute GVHD. Therefore some of the risk factors are shared between the two diseases. The risk factors for the development of chronic GVHD include degree of HLA matching, age of the recipient, subacute GVHD detected by skin histology and possibly CMV seropositivity. The known predictors for poor outcome are thrombocytopenia on day +100, progressive presentation of GVHD, lichenoid changes in skin histology and elevated serum bilirubin > 1.2 mg/dl.

C) CLINICAL PRESENTATION

The diagnosis of chronic GVHD is generally made on clinical and laboratory parameters. Chronic GVHD presents as an autoimmune phenomena including autoantibody formation. This leads to varied clinical presentation involving many different organ systems. However, CNS involvement is rarely seen in chronic GVHD. Often a biopsy of the affected area will yield the diagnosis.

1) Skin

The most frequent feature of chronic GVHD is involvement of the skin. There are two forms of presentation, lichenoid changes similar to lichen planus and sclerodermatous similar to scleroderma. The onset of skin involvement may be generalized erythema with plaques and waves of desquamation. The end result, without specific treatment, is that of hyperpigmentation or hypopigmentation, hide-like skin and joint contractures similar to scleroderma (Figs. 11.5A and 5B).

Table 11.6. Chronic GVHD grades

Limited	Localized skin involvement and/or hepatic dysfunction
Extensive	Generalized skin involvement or Limited skin involvement or hepatic involvement and any of the following a) liver histology showing chronic progressive hepatits, bridging necrosis or cirrhosis b) eye involvement (Schirmer's test with < 5 mm wetting) c) involvement of minor salivary glands or oral mucosa d) involvement of any other organ

11

a) Differential Diagnosis

Skin changes may be caused by a drug eruption or an infectious agent. However, lichenoid or sclerodermatous changes are chracteristic of chronic GVHD.

2) Oral

The oral mucosa is involved in the majority of patients with extensive GVHD (Fig. 11.6). Patients develop dryness with pain secondary to ulceration. Examination of the mouth usually demonstrates a white lace-like pattern on the buccal mucosa bilaterally.

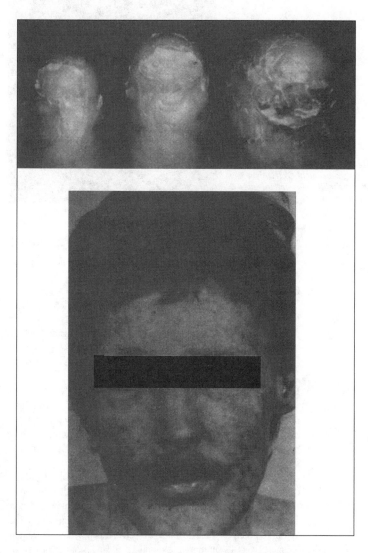

Fig. 11.5A and B. Chronic GVHD-hypo and hyperpigmentation and sclerotic changes.

11

Fig. 11.6. Day 100 lip biopsy revealing subclinical GVHD- lymphocytic infiltrate at dermal epidermal junction.

a) *Differential Diagnosis*

Oral changes may be caused by infectious agents, specifically herpes viruses or candidiasis. Occasionally cultures or microscopic examination is useful.

3) Ocular

Changes in the eyes are another frequent finding. The earliest manifestation is excessive tearing which will frequently lead to dry eyes (sicca syndrome). Conjunctivitis is also seen, usually secondary to the dry eyes.

a) *Differential Diagnosis*

Eye changes may be caused by infectious agents. A culture of the conjunctiva is helpful.

4) Liver/GI

The gastrointestinal tract is another frequent site of GVHD. Hepatic function tests show a predominant cholestatic picture. The degree of hyperbilirubinemia does not reflect the ultimate outcome. Patients may have persistent hyperbilirubinemia for many years followed by improvement or severe hepatic failure. The small bowel and colon may also be involved in chronic GVHD. Radiographic studies of the gastrointestinal tract may include findings such as webs, ring-like narrowing and tapering stricture of the mid and upper esophagus. Diarrhea can also be present but the more common manifestation is a "runting syndrome" (malabsorption, weight loss and a poor performance status) as seen in experimental animal models. Increasing GI symptoms (e.g. dysphagia and indolent weight loss) are associated with progression of GVHD.

11

a) Differential Diagnosis

Persistent GI changes may result from the prior GI damage from the preparatory regimen or GVHD prophylactic regimen. Other possible explanations may be peptic ulcers or infections such as herpes viruses (including CMV) and fungal gastritis.

5) **Autoimmune Manifestations**

Nearly all known autoimmune syndromes have been reported as part of chronic GVHD. Aspects of chronic GVHD may mimic systemic lupus erythematosus, scleroderma, progressive systemic sclerosis, lichen planus, sicca syndrome, eosinophilic fasciitis, rheumatoid arthritis, polymyositis, primary biliary sclerosis, mysthenia gravis, thyroiditis, idiopathic thrombocytopenic purpura and hemolytic anemia. Autoantibodies can also be detected in a similar fashion to those found in connective tissue diseases, such as, antinuclear antibody, anti-smooth muscle, anti-mitochondria, anti-liver kidney microsome and anti-epidermal antibodies. In patients with polymyositis, the creatine phosphokinase (CPK) may be elevated.

a) Differential Diagnosis

Other potential explanations for some of these findings may be related to the underlying disease or concurrent medications. For example, the underlying malignancy may cause an Eaton-Lambert syndrome and cyclosporine may cause a hemolytic anemia.

6) **Pulmonary**

The manifestation of chronic GVHD in the lungs is the development of bronchiolitis obliterans. This complication is manifest by the obliteration of the small bronchioles and best measured by the $FEF_{25-75\%}$ as part of the pulmonary function test. Patients complain of cough and shortness of breath. This complication is frequently fatal although early diagnosis and treatment may result in some responses. The incidence of bronchiolitis obliterans seems to be associated with a decreased serum IgG.

a) Differential Diagnosis

The other possible etiologies of bronchiolitis obliterans are broad and include, pulmonary infections (especially viral or mycoplasma), radiation injury, toxic inhalants, connective tissue diseases and pulmonary alveolar proteinosis. Other causes of pulmonary symptoms are common following BMT including infections such as CMV or other viruses and fungal or bacterial infections. Interstitial pneumonitis, drug related or idiopathic pneumonitis may present in a similar way. Heart failure related to the chemotherapy drugs may also cause pulmonary symptoms.

7) **Other Manifestations**

Gynecological manifestations in GVHD have also been observed. These include vaginal inflammation, sicca syndrome and vaginal stenosis. Membranous nephropathy has also been associated with chronic GVHD.

8) **Immunodeficiency**

Chronic GVHD is associated with marked immunodeficiency since the disease itself is immunosuppressive and the therapy usually involves the use of more immunosuppressive agents. Although this manifestation is listed last because there is no clinical observation of this finding, it is the most important sequelae of chronic GVHD. The recovery of immune

11

function is significantly delayed in comparison to patients without GVHD. These patients remain immunodeficient as long as the disease is active. T and B lymphocyte control remains dysregulated. Many of these patients are functionally asplenic. Recurrent infectious processes may occur in up to 100% of these patients with prolonged observation. These infectious complications largely account for morbidity and mortality associated with chronic GVHD.

D) GRADING

Chronic GVHD can be classified as limited or extensive, depending on the clinical presentation in patients (Table 11.6). Chronic GVHD has further been categorized as "progressive", i.e. evolving directly out of acute GVHD, "quiescent" if acute GVHD had at least transiently responded to therapy, and "de novo" in patients who never showed signs of acute GVHD. There appears to be moderate to high concordance for the diagnosis, grading and treatment of chronic GVHD. However, major disagreements were observed in the diagnosis of uncommon manifestations of chronic GVHD, interpretation of symptoms which occur less than 2 months after transplantation, interpretation of persistent stable symptoms and in deciding whether to treat chronic GVHD limited to skin. One interesting observation was that time of onset is an important clinical feature for some, but not all transplanters in establishing diagnosis of chronic GVHD. Consistency in grading the severity of the symptoms may be improved by using, for example, the Karnofsky performance status to differentiate between limited versus extensive chronic GVHD.

E) HISTOLOGY

1) Skin

A generalized and localized form of this disease has been described. The generalized type has hyperkeratosis, epidermal hypertrophy and a lichenoid reaction at the basal layer. As disease progresses, the dermis becomes atrophic, the inflammatory changes are less striking and the dermal-epidermal junction becomes straightened and obliterated. A second form has localized skin involvement with epidermal atrophy and dense focal dermal fibrosis in the absence of significant inflammation. This type of chronic GVHD resembles morphea and lupus. There is damage to the basal lamina, the basal cell layer and the spinous layer similar to lesions seen in the lichen planus. The epidermis becomes significantly atrophic with prominent poikiloderma.

2) Liver

A biopsy specimen may show lobular hepatitis, chronic persistent hepatitis or chronic active hepatitis. There is a reduction or absence of small bile ducts with cholestasis. The pathophysiology of chronic GVHD is suggestive of primary biliary sclerosis with biliary cell necrosis and thickening of the bile duct basement membranes.

3) Gastrointestinal Tract

GVHD of the portion of the esophagus lined by squamous epithelium is similar to histologic findings in the skin. Chronic GVHD of the stomach, intestine and colon is manifest histologically by fibrosis of the submucosa and sclerosis and hyalinization of small venules.

4) Lung

Bronchiolitis obliterans causes destruction of small airways. Fibrous obliteration of the lumen and the bronchioles is observed in the histology.

11

This granulation tissue often extends into the alveolar ducts. It is unclear, however, whether the lungs are a primary or secondary target in chronic GVHD.

F) PREVENTION OF CHRONIC GVHD

The best prophylaxis of chronic GVHD is effective prevention of acute GVHD. Only 15-20% of patients without acute GVHD will develop "de novo" chronic GVHD, compared to 40-100% of those who suffer from acute GVHD. Thus several attempts have been made to include additional preventive measures in the early posttransplant period.

G) THERAPY OF CHRONIC GVHD

Although debatable, limited stage chronic GVHD is not treated, while extensive stage chronic GHVD is generally treated with alternate day cyclosporine and prednisone. The platelet count is the best predictor of survival for patients with extensive chronic GVHD. Standard risk chronic GVHD, defined by a platelet count greater than or equal to 100,000/µl, has an 80% 5-year survival. High risk chronic GVHD, defined by a platelet count persistently less than 100,000 /µl, has a 30-40% 5-year survival.

1) Prednisone and Cyclosporine

Early treatment with prednisone alone in standard risk chronic GVHD patients was superior to treatment with a combination of prednisone and azathioprine. Azathioprine should not be used unless it is part of a prospective study. The dose of prednisone is usually 1 mg/kg every other day, or 30-60 mg daily depending on the severity of clinical symptoms. However, in high risk patients, defined as persistent thrombocytopenia < 100,000/µl, prednisone alone resulted in only 26% long-term survival. Survival in both standard risk and high risk patients can be improved and transplant-related mortality reduced by treatment with an alternating day regimen of cyclosporine (6 mg/kg orally every 12 hours every other day) and prednisone (1 mg/kg alternating with cyclosporine every other day). These doses are tapered slowly over 3-6 months depending on the response of the patient.

2) Thalidomide

Thalidomide (N-phthalidoglutarimide) originally introduced as a sleeping pill has side effects of neurotoxicity and teratogenicity. It remained in the market due to its efficacy in treating leprosy. Thalidomide also has immunosuppressive activity, possibly due to interference with lymphocyte adhesion molecules. Studies suggested that thalidomide (800-1600 mg per day in divided doses) is effective in treating established chronic GVHD. In a clinical study of salvage therapy for chronic GVHD, overall response rate was 64% and survival was 76%; for the subset of patients with high risk chronic GVHD survival was 48%. Unfortunately, thalidomide is not readily available.

3) Psoralen plus Ultraviolet Irradiation (PUVA)

Ultraviolet irradiation of methoxypsoralen-sensitized patients (PUVA) has been used widely for dermatologic disorders. Experimental models have revealed a profound immunosuppressive effect of PUVA. The proposed mechanism of PUVA is inhibition of DNA transcription and mitosis. Upon activation with ultraviolet A light, photoexcited psoralen covalently binds to one or both strands of DNA. During PUVA treatment, there is a decrease

11

in the number and function of circulating lymphocytes including a decrease in NK activity, T cell proliferation to mitogens and cytokine production, specifically IL-1 and IL-2. In small series of patients, a 60% response rate has been observed. The efficacy of PUVA may be limited by skin complications of chronic GVHD. Also, the rather shallow depth penetration of UVA may limit the application of PUVA to patients without liver or intestinal GVHD. These treatments necessitate the availability of an experienced dermatologist since too high of a dose may result in a severe skin burn and its resultant morbidities.

4) Extracorporeal Photophoresis

Use of methoxypsoralen with extracorporeal photopheresis has also been attempted. The results in a limited number of patients suggest only transient improvement.

5) Ursodeoxycholic Acid

Ursodeoxycholic acid is a relatively nontoxic hydrophilic bile acid with a striking choleresis effect. Patients with chronic GVHD on this drug improved with a 33% decrease in the bilirubin level compared to baseline ($p < 0.005$), 32% decrease in alkaline phosphatase ($p < 0.038$) and a 37% decrease in AST ($p < 0.007$). The levels of these enzymes rose again after discontinuation of the drug. The dose of ursodeoxycholic acid is 300-600 mg PO BID-TID.

H) SUPPORTIVE CARE FOR CHRONIC GVHD

Since chronic GVHD in itself causes immunosuppression, intensive supportive care such as antibiotic prophylaxis, intravenous immunoglobulins if the patient is hypogammaglobulinemic and nutritional support is of paramount importance.

1) Prophylaxtic Antibiotics

Infectious complications are a common occurrence due to the impaired immune system especially in patients with chronic GVHD. The most common pathogens are those found in the upper airways including *Streptococcus* and *Hemophilus* species. One approach is to place patients on oral penicillin (e.g. PenVK 250 mg orally QID) especially if there is evidence of functional asplenia. Patients who have recurrent pulmonary infections may be placed on rotating antibiotics (one antibiotic per week) for 3 weeks with 1 week off. One possible regimen we have utilized consists of using therapeutic doses of a first generation cephlosporin, amoxicillin and erythromycin or trimethoprim-sulfamethoxazole, where each drug is given for 1 week. Prophylaxis for *Pneumocystis carinii* with trimethoprim-sulfamethoxazole is instituted and continued for as long as patients are on corticosteroids. The dose of trimethoprim-sulfamethoxazole is 1 double strength tablet twice daily for 2 days per week. If patients are on ≥ 0.5 mg/kg of prednisone or equivalent steroid dose, the dose of trimethoprim-sulfamethoxazole is increased to 1 double strength tablet daily. Patients who are unable to tolerate trimethoprim-sulfamethoxazole can receive nebulized pentamidine, at a dose of 300 mg each month. Hypogammaglobulinemia is also found in patients who develop chronic GVHD and use of intravenous immunoglobulins may be of help in preventing recurrent infections.

11

2) Physical Therapy

Physical therapy is of critical importance for those patients who present with skin involvment. Aggressive programs for mobility help in preventing some of the sequelae of the sclerodermatous skin changes.

3) Oral Hygiene

Careful attention to oral hygiene is also very important to prevent oral ulcerations or the development of dental caries.

4) Sun Light Exposure

Not infrequently, patients will give a history of photoactivation. UV radiation may damage superficial epidermal cells leading to increased antigen expression and increased cytokine release with the resultant expression of chronic GVHD. Patients should be advised to avoid direct sun exposure and to use sun block lotion with a large hat that shades the face when outdoors.

III) GRAFT-VERSUS-HOST DISEASE IN UNRELATED DONOR BMT

A) INTRODUCTION

The increased incidence and severity of GVHD following unrelated donor marrow transplantation suggests that more aggressive therapies (e.g. triple drug prophylaxis, T cell depletion, anti-Tac monoclonal antibody) are warranted in these patients.

B) ACUTE GVHD FOLLOWING UNRELATED BMT

The reported incidence of acute GVHD for patients who receive matched unrelated donor transplantation is as high as 80% with two drug prophylaxis. Analysis of 462 unmanipulated marrow transplants from unrelated donors by the National Marrow Donor Program revealed that the probability of having grades II-IV acute graft-versus-host disease was 64% whereas 47% had severe grade III-IV acute GVHD. T cell depletion of donor marrow, a more aggressive method of GVHD prophylaxis, results in roughly a 50% incidence of grade II-IV GVHD. One report using T cell depletion for GVHD prophylaxis suggested the actuarial probability of developing grade II-IV graft-versus-host disease was 40% and of developing grades III-IV acute GVHD was only 8.3%.

C) CHRONIC GVHD AFTER UNRELATED BMT

The probability of limited or extensive chronic graft-versus-host disease 1 year after transplantation from an unrelated donor is approximately 55%. In HLA sibling matched transplants, T cell depletion decreases both acute and chronic GVHD but increases the probablity of relapse. In contrast, in unrelated marrow transplants, T cell depletion decreases acute GVHD but appears to have no effect on the incidence of chronic GVHD or relapse. This suggests that the graft-versus-leukemia effect reported with allogeneic BMT may be associated with both chronic and acute GVHD.

SUGGESTED READING

1. Burakoff SJ, Deeg HJ, Ferrara J, Atkinson K eds.Graft vs. Host Disease—Immunology, Pathophysiology, and Treatment. New York: M Dekkar, 1990.
2. Forman SJ, Blume KG, Thomas ED eds. Bone Marrow Transplantation. Boston: Blackwell Scientific Publications, 1994.
3. Ferrara JLM, Deeg HJ. Graft-versus-host disease. N Engl J Med 1991; 324:667-674.

11

12

Graft-Versus-Leukemia Effect of Allogeneic BMT

12

Dimitrios A. Mavroudis

I) INTRODUCTION

Allogeneic BMT cures leukemia by means of myeloablation from the preparative regimen and transfer with the bone marrow graft of immunocompetent donor cells that exert an antileukemic effect called graft-versus-leukemia (GVL). For many years the evidence for this GVL effect was alluded to from experimental animal models and retrospective statistical analysis of leukemia relapse between patients with allogeneic vs syngeneic BMT, presence vs absence of acute or chronic GVHD and unmodified vs T-depleted BMT. More recently direct evidence for this antileukemic effect was demonstrated with the infusion of unmanipulated donor peripheral blood leukocytes into patients who have relapsed after allogeneic BMT (Table 7.4.1 in chapter 7.4: *Relapse After BMT*).

II) GVL EFFECT ASSOCIATED WITH GVHD

The International Bone Marrow Transplant Registry (IBMTR) analyzed data from 2,254 patients who underwent HLA-identical sibling BMT for early leukemia (CML chronic phase, AML and ALL in first remission). After adjusting for other variables that affect relapse, they showed that there was a statistically significant reduction in the relapse risk for patients who developed GVHD. For patients that received unmanipulated BMT and did not develop GVHD the relative risk for relapse was defined as 1.0. Patients who developed only acute GVHD had a relative risk of relapse of 0.68 (p = 0.03); patients with only chronic GVHD had a relative risk of 0.43 (p = 0.01); and patients with both acute and chronic GVHD had a relative risk of 0.33 (p = 0.0001). The relative risk for relapse for patients who developed both acute and chronic GVHD was 0.38 for ALL, 0.34 for AML and 0.24 for CML. This antileukemic effect was conferred predominantly by acute GVHD in ALL patients and chronic GVHD in AML and CML patients. It was also shown that the antileukemic effect parallels the severity of GVHD.

III) LOSS OF GVL WITH T CELL DEPLETION

The removal of mature T cells from the bone marrow graft represents the most effective way to prevent the development of acute and chronic GVHD. This benefit of T cell depletion is offset by increased graft failure and leukemia relapse so that overall survival is not improved. The IBMTR compared HLA-identical sibling transplants in 731 recipients of T cell depleted transplants with 2,480 recipients of non-T cell depleted transplants. The relative risk of relapse after T-depleted transplant was significantly increased compared to non-T-depleted transplant for all types of leukemia (ALL, AML, CML) and all stages (early, intermediate, advanced) of disease with the exception of AML in or beyond second remission and CML in blast crisis. Because T cell depletion decreases GVHD which affects leukemia relapse, the analysis was repeated after adjusting for incidence and severity of acute and chronic GVHD to examine if T-depletion independently affects leukemia relapse. After adjusting for GVHD, the relative relapse risks with T-depletion for AML and ALL in first remission and CML in chronic phase were 1.66 (p = NS), 1.55 (p = NS) and 4.87 (p < .0001) respectively compared to non-T-depleted

Bone Marrow Transplantation, edited by Richard K. Burt, H. Joachim Deeg, Scott Thomas Lothian, George W. Santos. © 1996 R.G. Landes Company.

transplants. Therefore, it appears that T-depletion increases leukemia relapse in AML and ALL because of the loss of GVHD. In the case of CML there is also a GVL effect independent of GVHD which is lost during the process of T cell depletion.

IV) GVL OF ALLOGENEIC VERSUS SYNGENEIC TRANSPLANTATION

BMT from an identical twin donor is associated with minimal or no GVHD but a relatively high leukemia relapse rate. In a matched-pair analysis of IBMTR data, leukemia relapse rates were compared between 103 patients transplanted from identical twin donors and 1,030 patients transplanted from HLA-identical siblings. The 3 year probabilities of relapse for ALL, AML and CML were 36%, 52% and 40% respectively after identical twin transplant, compared with 26%, 16% and 7% after HLA-identical sibling transplant. After adjustment for acute and chronic GVHD the relative risks for relapse of syngeneic compared to allogeneic transplants were 1.4 (CI, 0.6 to 2.8) for ALL, 3.1 (CI, 1.9 to 5.1) for AML and 5.5 (CI, 2.8 to 11.0) for CML. Therefore, it appears that there is an allogeneic GVL effect for AML and CML that is independent of clinically apparent GVHD.

V) GVL ASSOCIATED WITH DONOR LEUKOCYTE INFUSIONS

Kolb et al, (1990) first reported that for patients who relapse with CML after allogeneic BMT, leukocyte infusions from the original donor can reinduce remission. Reports from other groups have confirmed this strong antileukemic effect of peripheral blood mononuclear cells. From the existing reports, it seems that patients with CML who have isolated cytogenetic relapse or chronic phase respond better to this treatment than patients with more advanced disease or acute leukemia. The majority of CML patients (60-80%) will have a complete response usually achieving a polymerase chain reaction negative status for bcr-abl and the responses are durable. A minority of patients with relapsed acute leukemia will also respond to this therapy. Leukocyte infusions are associated with GVHD in most (80%) but not all the responders and this implies that mechanisms for both GVHD-associated GVL and GVL independent of clinically significant GVHD may operate. The cytotoxic effect of leukocyte infusions is so strong that it may lead to marrow aplasia in up to 50% of patients. The optimal dose and interval of donor lymphocyte infusions is unknown. The total dose has varied between $0.3\text{-}16.4 \times 10^8$ cells/kg recipient weight divided into 1-4 fractions, separated by 1-4 week intervals. The time to response is generally 1-3 months. The effector cells for this antileukemic response have not been defined but presumably T cells and/or NK cells are involved.

Table 12.1. GVL effect based on clinical data

	CML	AML	ALL
GVHD data	+++	++	++
T-depletion data	+++	±	±
Syngeneic BMT data	+++	++	+
Leukocyte infusions data	+++	++	±

+++ = Support for GVL effect is strongest
++ = Support for GVL effect is strong
+ = Support for GVL effect is weak
± = Support for GVL effect is questionable

12

VI) GVL EFFECT ASSOCIATED WITH CYCLOSPORINE WITHDRAWAL

There are sporadic case reports of patients with relapsed AML or CML after allogeneic BMT who achieved remission after abrupt discontinuation of cyclosporine. In all these cases the GVL effect was associated with GVHD.

VII) CONCLUSIONS FROM CLINICAL DATA (Table 12.1)

1) Allogeneic BMT is associated with a graft-versus-leukemia (GVL) effect that may be associated with or independent of clinically significant GVHD.

2) The GVHD-dependent GVL is stronger in CML than AML or ALL. It is conferred primarily by acute GVHD in ALL and chronic GVHD in AML and CML. The combination of both acute and chronic GVHD is associated with the strongest antileukemic effect. Severe GVHD usually correlates with strong GVL.

3) The GVL independent of clinically significant GVHD is stronger in CML than AML and weak in ALL.

4) T cell depletion is associated with a loss of the GVL effect especially in CML chronic phase where it results in a 5-fold increase in leukemia relapse.

5) Donor leukocyte infusions posttransplant mediate a strong GVL effect for patients with relapsed CML.

VIII) MECHANISMS OF GVL

Graft-versus-leukemia effect is an immune response of the donor cells against the recipient's leukemia. The effectors and targets for this immune reaction remain undefined although the close association with the GVHD process suggests that common mechanisms may operate. Alloreactivity between HLA-identical siblings is due to differences in minor histocompatibility antigens (mHA) which are peptides derived from cytoplasmic "self" proteins and presented by major histocompatibility molecules (MHC) on the cell surface. Two important characteristics of minor antigens are their tissue restricted distribution and the immunodominance that some of them exhibit. It has been shown recently that some of the known minor antigens are present on the surface of lymphocytic and myeloid leukemic cells. Cytotoxic T lymphocytes (CTLs) can recognize and kill the cells that express "foreign" minor antigens. Because T lymphocytes recognize minor antigens in the context of MHC molecules this is known as MHC-restricted cytotoxicity. Several possible speculative mechanisms could explain the clinical observations.

1) Minor antigens that are present on both leukemic cells and normal host tissues could explain why GVL is closely associated with GVHD. Quantitative differences in the presentation of these antigens by leukemic and normal cells could also create variability in the GVHD-GVL association.

2) Some minor antigens have tissue restricted distribution so that a "myeloid minor antigen" could be present only on normal myeloid cells of the host and on the myeloid leukemia cells. Such a minor antigen could elicit an anti-myeloid effect which would be GVL without GVHD.

3) It is possible that some antigens are present only on the surface of leukemic cells. Such leukemia-specific antigens could also produce GVL without GVHD.

Both CD8+ and CD4+ cells are thought to participate in this MHC-restricted GVL effect. CD8+ and CD4+ T cells can be cytotoxic using either the perforin mechanism that causes direct cell lysis or the Fas mechanism that results in apoptosis through receptor-ligand interaction. CD4+ T cells could enhance the GVL reaction by secreting cytokines such as IL-2 that help CD8+ T cells to clonally expand or other cytokines such as TNFα or IFNγ that could mediate a direct cytotoxic effect or recruit other secondary effector cells, e.g. monocytes.

12

Natural killer (NK) cells are also thought to play a role in the GVL effect by mediating cytotoxicity in a MHC-unrestricted fashion. The mechanism through which NK cells become activated and kill their targets is still unknown. Early after BMT there is a surge of cytokines and NK cells that could mediate a GVL effect independent of alloreactivity as in the case of syngeneic or autologous BMT.

IX) SUMMARY

Patients may have an allogeneic GVL effect that is not associated with clinically evident GVHD. The goal of research in this field is to augment the GVL effect while minimizing or separating it from GVHD. Identification of possible minor antigens that may be responsible for the GVL effect will allow manipulation at the clonal level of the GVL effector cells. More selective T cell depletion techniques may then allow GVHD prevention without significant loss of GVL.

SUGGESTED READING

1. Horowitz MM et al. Graft-versus-leukemia reactions after bone marrow transplantation. Blood 1990; 75:555-562.
2. Marmont AM et al. T cell depletion of HLA-identical transplants in leukemia. Blood 1991; 78:2120-2130.
3. Gale RP et al. Identical-twin bone marrow transplants for leukemia. Ann Intern Med 1994; 120:646-652.
4. Giralt SA, Champlin RE. Leukemia relapse after allogeneic bone marrow transplantation: A review. Blood 1994; 84:3603-3612.

13.0 Toxicity

13

Regimen Related Toxicity— First 30 Days Early Toxicity of High-Dose Therapy

Gwynn D. Long

I) INTRODUCTION

A grading system has been developed for regimen-related toxicity that excludes complications due to infection or graft-versus-host disease (GVHD) and measures toxicity due only to the transplant preparative regimen (Table 13.1.1). Eight organ systems (cardiac, bladder, renal, pulmonary, hepatic, central nervous system, mouth and gastrointestinal tract) are graded for toxicity from 0 (none) to IV (fatal). In general, grade I toxicity is mild and does not necessarily require medical intervention, grade II toxicity is more severe and requires specific medical interventions and grade III toxicity is life-threatening. All patients undergoing bone marrow transplantation are expected to develop grade I toxicity in at least one organ system and most patients will develop grade II toxicity. Toxicity is influenced not only by the intensity of the preparative regimen, but also by factors such as disease status and prior therapy, type of graft (allogeneic versus autologous) and drugs utilized for GVHD prophylaxis.

II) CARDIAC TOXICITY

The incidence of clinically significant cardiac toxicity following high-dose therapy ranges from 2% up to 43% in various reports, with mortality rates of 2%-9%.

A) MARROW INFUSION

Cardiac complications have been reported during the reinfusion of cryopreserved autologous marrow or stem cells. While one case of fatal arrhythmia has been described, most patients only develop mild, self-limited bradycardia and/or hypotension which is treated by temporarily holding or decreasing the infusion rate.

B) CYCLOPHOSPHAMIDE

Cardiac toxicity is generally felt to be due to high-dose cyclophosphamide and is rarely seen at doses less than 120 mg/kg. Clinical events may include arrhythmias, pericarditis and congestive heart failure. The most serious event is cyclophosphamide-induced myocarditis which presents clinically as dyspnea on exertion, weight gain (fluid retention) and orthostatic hypotension with or without evidence of pericardial effusion. The syndrome usually develops within 5-10 days following chemotherapy and is often fatal in spite of hemodynamic support. Surviving patients, however, usually demonstrate complete recovery. Pathologically, the lesions reveal hemorrhagic myocardial necrosis, interstitial edema and fibrin deposition and endothelial damage. In prospective trials, subclinical cardiac toxicity has been shown to be more frequent and is manifested by decreased voltage on electrocardiogram, increased left ventricular mass and decreased fractional shortening on echocardiogram.

Bone Marrow Transplantation, edited by Richard K. Burt, H. Joachim Deeg, Scott Thomas Lothian, George W. Santos. © 1996 R.G. Landes Company.

13.1

Table 13.1.1. Regimen-related toxicity according to organ system

	Grade I	Grade II	Grade III
Cardiac Toxicity	Mild EKG abnormality, not requiring medical intervention; or noted heart enlargement on CXR with no clinical symptoms	Moderate EKG abnormalities requiring and responding to medical intervention; or requiring continuous monitoring without treatment; or congestive heart failure responsive to digitalis or diuretics	Severe EKG abnormalities with no or only partial response to medical intervention; or heart failure with no or only minor response to medical intervention; or decrease in voltage by more than 50%
Bladder Toxicity	Macroscopic hematuria after 2 days from last chemotherapy dose with no subjective symptoms of cystitis and not caused by infection	Macroscopic hematuria after 7 days from last chemotherapy dose not caused by infection; or hematuria after 2 days with subjective symptoms of cystitis not caused by infection	Hemorrhagic cystitis with frank blood, necessitating invasive local intervention with installation of sclerosing agents, nephrostomy or other surgical procedure
Renal Toxicity	Increase in creatinine up to twice the baseline value (usually the last recorded before start of conditioning)	Increase in creatinine above twice baseline but not requiring dialysis	Requirement of dialysis
Pulmonary Toxicity	Dyspnea without CXR changes not caused by infection or congestive heart failure; or CXR showing isolated infiltrate or mild interstitial changes without symptoms not caused by infection or congestive heart failure	CXR with extensive localized infiltrate or moderate interstitial changes combined with dyspnea and not caused by infection or CHF; or decrease of PO_2 ($> 10\%$ from baseline) but not requiring mechanical ventilation or $> 50\%$ O_2 on mask and not caused by infection or CHF	Interstitial changes requiring mechanical ventilatory support or $> 50\%$ oxygen on mask and not caused by infection or CHF

13.1

Table 13.1.1. Regimen-related toxicity according to organ system (cont'd.)

	Grade I	Grade II	Grade III
Hepatic Toxicity	Mild hepatic dysfunction with bilirubin ≥ 2.0 mg % but ≤ 6.0 mg %; or weight gain > 2.5% and < 5% from baseline of noncardiac origin; or SGOT increase more than 2-fold but less than 5-fold from lowest preconditioning	Moderate hepatic dysfunction with bilirubin > 6 mg % < 20 mg %; or SGOT increase > 5-fold from preconditioning; or clinical ascites or image documented ascites > 100 ml; or weight gain > 5% from baseline or noncardiac origin	Severe hepatic dysfunction with bilirubin > 20mg%; or hepatic encephalapathy; or ascites compromising respiratory function
CNS Toxicity	Somnolence but the patient is easily arousable and oriented after arousal	Somnolence with confusion after arousal; or other new objective CNS symptoms with no loss of consciousness not more easily explained by other medication, bleeding, or CNS infection	Seizures or coma not explained (documented) by other medication, CNS infection, or bleeding
Stomatitis	Pain and/or ulceration not requiring a continuous IV narcotic drug	Pain and/or ulceration requiring a continuous IV narcotic drug (morphine drip)	Severe ulceration and/or mucositis requiring preventive intubation; or resulting in documented aspiration pneumonia with or without intubation
GI Toxicity	Watery stools > 500 ml but < 2,000 ml every day not related to infection	Watery stools > 2,000 ml every day not related to infection; or macroscopic hemorrhagic stools with no effect on cardiovascular status not caused by infection; or subileus not related to infection	Ileus requiring nasogastric suction and/or surgery and not related to infection; or hemorrhagic enterocolitis affecting cardiovascular status and requiring transfusion

NOTE: Grade IV regimen-related toxicity is defined as fatal toxicity. Abbreviations: CXR, chest x-ray; IV, intravenous.
Reprinted with permission from: Bearman SI et al. Regimen Related Toxicity in Patients Undergoing Bone Marrow Transplantation. JCO 1988, 6(10):1562-1568.

1) **Risk Factors of Cyclophosphamide Cardiac Toxicity**

   Risk factors for the development of cardiac toxicity following high-dose therapy are not clearly defined. Some studies have suggested that a prior history of anthracycline therapy or mediastinal radiation is associated with a higher risk of cardiac damage, although these results have not been confirmed in prospective studies. Higher daily doses of cyclophosphamide (> 1.55 g/m^2/day) may also be associated with an increased incidence of complications. The majority of transplant centers utilize radionuclide ventriculography (RVG) as a pretransplant screening tool. In a prospective trial, reduced left ventricular ejection fraction (LVEF) was associated with a higher risk of cardiac complications, however, almost all of the complications were minor and not clinically significant. LVEF was not predictive of life-threatening cardiac toxicity and all patients who developed fatal toxicity had normal LVEFs on pretransplant screening. An acceptable lower limit of LVEF cannot be recommended based on the available literature, however, patients with LVEFs less than 40% on repeat testing or with clinical evidence of congestive heart failure should probably be excluded from high-dose therapy trials. Routine pretransplant screening with RVG is not necessary in all patients, but should be utilized in patients with cumulative anthracycline doses above 450 mg/m^2 or in patients with symptoms of congestive heart failure.

2) **Therapy of Cyclophosphamide Cardiac Toxicity**

   No prophylactic measures are known to prevent cardiac toxicity secondary to high-dose cyclophosphamide and routine cardiac monitoring of patients is not recommended. The management of congestive heart failure in this setting as in other clinical situations, includes diuretics, drainage of pericardial effusions if necessary and the use of hemodynamic monitoring and support as required.

III) **HEPATIC TOXICITY-VOD**

   Hepatic toxicity secondary to high-dose chemotherapy and/or radiotherapy is manifest as veno-occlusive disease of the liver (VOD). It may occur in both autologous and allogeneic BMT.

   A) **DIAGNOSIS**

   VOD is characterized by tender hepatomegaly or right upper quadrant tenderness, unexplained weight gain or ascites and hyperbilirubinemia (Table 13.1.2). The clinical diagnosis is based on the presence of at least two out of three of these findings within 21-30 days after marrow transplantation in the absence of other causes of liver damage such as GVHD. Doppler ultrasound may demonstrate alterations in hepatic blood flow in patients with VOD

Table 13.1.2. Clinical criteria for veno-occlusive disease of the liver

McDonald's Criteria -Seattle	Jones Criteria- Johns Hopkins
Before day 30 any two of the following: Bilirubin > 27 µmole/L=1.7 g/dl Hepatomegaly Ascites or weight gain	Before day 21 any two of the following: Bilirubin > 34 µmole/L= 2.0 g/dl Hepatomegaly Ascites Weight gain

and an elevated hepatic vein wedge pressure may be demonstrated at the time of transjugular liver biopsy. Definitive diagnosis is based on liver biopsy which is usually performed through the transjugular approach, if biopsy confirmation is felt necessary.

B) INCIDENCE

The incidence of VOD varies from 5-54% of marrow transplant recipients and is frequently associated with multiorgan failure with mortality rates ranging from 30-50%.

C) ETIOLOGY

The syndrome is thought to be secondary to hepatic sinusoid and venule endothelial cell damage from high-dose chemotherapy and radiation therapy.

D) HISTOLOGY

Early lesions demonstrate subendothelial edema with centrilobular congestion and hepatocyte degeneration, while later lesions reveal fibrous obliteration of the central venules and centrilobular sinusoids. The end result is obstruction of venous blood flow and intrahepatic portal hypertension.

E) RISK FACTORS

The most important risk factor for the development of VOD is pretransplant elevation of transaminase levels. Other risk factors include more intensive cytoreductive regimens (15.75 Gy total body irradiation versus 12.0 Gy), fever requiring antibiotics during cytoreduction, mismatched or unrelated marrow donors, metastatic liver disease, prior abdominal radiotherapy and the use of methotrexate for GVHD prophylaxis, especially after the busulfan/cyclophosphamide preparative regimen. A prior history of viral hepatitis does not appear to be a significant risk factor in patients with normal liver function tests.

F) THERAPY

No effective therapy is currently available for VOD. Management is supportive and symptomatic and includes fluid and sodium restriction, albumin replacement and transfusion support. Preliminary reports suggest that early administration of low-dose tissue plasminogen activator (Table 13.1.3) may alter the natural history of the syndrome in some patients. Prospective, randomized placebo-controlled trials will be necessary to confirm these results. Other attempts at therapy have included porta-caval shunting (surgical and intrahepatic) and liver transplantation.

Successful prophylaxis of VOD with the continuous infusion of low-dose heparin has been demonstrated in a prospective, randomized trial. Heparin infusion at a dose of 100 units/kg/day was started with the beginning of the preparative regimen and continued until day +30 or hospital discharge. The incidence of VOD was reduced from 13.7% in the placebo group to 2.5% in

---

Table 13.1.3. Tissue plasminogen activator and heparin for treatment of veno-occlusive disease of the liver (Fred Hutchinson Cancer Center protocol)

TPA—0.05 mg/kg/hr for 4 hours for 4 consecutive days (maximum dose 2.5 mg/hour). Starting on the same day as TPA, heparin 20 units/kg bolus (maximum 1000 units) then continuous infusion of 150 units/kg/day for 10 days (infusion adjusted to maintain PTT< or = 1.2 times the upper limit of normal).

Contraindication to therapy is: platelet count <15,000/ml despite transfusions, uncontrolled hypertension, evidence of pulmonary or intracranial hemorrhage or surgery within 10 days.

heparin-treated group. The PTT was not prolonged in the treatment group and no increase in major bleeding episodes or transfusion requirements was noted. Patients treated with ursodeoxycholic acid (300-600 mg PO BID-TID) or prostaglandin E1 have had a lower incidence of VOD in some studies and randomized trials are ongoing, however, pentoxifylline does not appear to offer any benefit.

IV) GASTROINTESTINAL TOXICITY

A) Nausea and Vomiting

Nausea and vomiting occur in the vast majority of patients during the administration of high-dose chemotherapy and total body irradiation, due to effects on the midbrain vomiting center and direct toxicity to the gastrointestinal mucosa. Nausea and vomiting is often worse during the first 48 hours of fractionated total body irradiation and declines as the treatment proceeds. Many patients remain mildly nauseated and anorectic for 3-4 weeks. Excellent control of nausea and vomiting during chemoradiotherapy can be obtained with available anti-emetics such as metaclopromide, serotonin antagonists (ondansetron and granisetron), phenothiazines and lorazepam. Nausea and vomiting that develop or continue beyond 3-4 weeks following transplantation is more likely due to GVHD or viral infections such as cytomegalovirus or herpes simplex.

B) DIARRHEA

Most patients also develop diarrhea secondary to gastrointestinal toxicity from high-dose therapy. Chemoradiotherapy-induced mucosal injury causes increased fluid secretion by the intestine. Nonabsorbable antibiotics also cause increased diarrhea. Agents such as diphenoxylate, loperamide or if necessary, tincture of opium are often effective for control of diarrhea. Diarrhea due to the preparative regimen usually resolves by 3 weeks following transplantation. Symptoms that develop or continue beyond that period are more likely due to GVHD or enteric infection than toxicity of the preparative regimen.

C) MUCOSITIS

Oral mucositis (stomatitis) develops to some degree in more than 90% of patients undergoing high-dose therapy. Oral and throat pain secondary to stomatitis is the primary cause of patient discomfort associated with high-dose therapy, and frequently requires narcotic analgesics for control. Increased mucositis is associated with regimens containing total body irradiation, busulfan, etoposide or thiotepa. More severe mucositis also occurs with higher doses of total body irradiation and with the use of posttransplant methotrexate for GVHD prophylaxis. Mucositis usually starts to develop around the time of marrow infusion, peaks 7-14 days later and resolves around the time of neutrophil recovery. The recent use of chemotherapy and/or growth factor "mobilized" peripheral blood progenitor cells for rescue in autologous transplants has significantly shortened the period of neutropenia following high-dose therapy compared to bone marrow rescue, and has correspondingly shortened the duration of mucositis. No measures have been developed to prevent mucositis. Good oral hygiene is essential and most centers use saline or baking soda rinses several times a day along with chlorhexidine rinses and clotriamazole or nystatin. Topical anesthetics provide some pain relief and help patients to perform the mouth care regimen with less discomfort. Reactivation of herpes simplex virus is associated with more severe and more

13.1

prolonged mucositis and the majority of transplant centers utilize prophylactic acyclovir in patients who are serologically positive for the virus.

V) PULMONARY TOXICITY

Pulmonary complications develop in up to 40-60% of bone marrow transplant recipients.

A) INFECTIONS

Infectious causes include bacteria, fungi such as *Aspergillus*, cytomegalovirus (CMV) and *Pneumocystis carinii*. The most likely etiology depends on the time of onset following transplantation. Gram negative and Gram positive bacterial and fungal infections are common during the period of neutropenia. *Pneumocystis carinii* usually presents 2 months after BMT. The risk period for CMV is 30-100 days after BMT, particularly for allogeneic transplants. Patients with chronic GVHD are at risk for bacterial infections with encapsulated organisms and fungi.

B) DIFFUSE ALVEOLAR HEMORRHAGE (DAH)

DAH is a particular form of noninfectious pneumonitis that occurs in 5-30% of recipients of bone marrow transplants, but the incidence varies greatly between centers. The diagnosis is made by the finding of diffuse consolidation on chest x-ray, hypoxemia requiring supplemental oxygen and progressively bloodier fluid on successive bronchoalveolar lavage aliquots with no evidence of pulmonary infection. DAH occurs in both autologous and allogeneic bone marrow transplant patients and the etiology of the syndrome is not known. The median day of onset is approximately 2 weeks after transplant and is associated with white blood cell recovery, fever and renal insufficiency. The mortality rate ranges from 80-100% in some reports, but treatment with high-dose corticosteroids (1 gram of solumedrol per day for 3 days followed by a 50% taper every 3 days) appears to improve survival. Other supportive measures include correction of thrombocytopenia and any coagulation abnormalities.

C) IDIOPATHIC INTERSTITIAL PNEUMONITIS

In up to half of the cases of pneumonitis associated with bone marrow transplantation no infectious agent can be identified. This syndrome has been termed the idiopathic pneumonia syndrome (IPS). Regimen-related IPS has been associated with both total body irradiation and high-dose chemotherapy, especially BCNU, busulfan and cyclophosphamide. These agents are directly toxic to the lungs. Pathology generally reveals acute interstitial pneumonitis or diffuse alveolar damage. A National Heart, Lung and Blood Institute workshop held in 1991 proposed the following definition for IPS.

1) Evidence of widespread alveolar damage such as multilobar infiltrates.

2) Clinical symptoms of pneumonia such as cough, fever and dyspnea.

3) Evidence of abnormal pulmonary physiology such as new or worsening restrictive changes or increased alveolar-arterial oxygen gradient.

4) Absence of lower respiratory tract infection by bronchoalveolar lavage.

The median time of onset of IPS is 42-49 days after bone marrow transplantation with an early peak in the first 14 days and a lower, consistent incidence through 80 days. Most pneumonias occurring during the first 28 days after transplant are idiopathic. Standard management of IPS has not been established. In general, patients are supported with oxygen and mechanical ventilation if necessary. Many centers utilize corticosteroids once infection has been ruled out, usually at a starting prednisone dose of approximately

1-2 mg/kg/day followed by a slow taper once improvement is evident. If treatment is initiated early, the majority of patients recover, however, once the process progresses to the need for ventilatory support, less than 5% of patients survive long-term.

VI) HEMORRHAGIC CYSTITIS

Hemorrhagic cystitis occurs in 70% of patients following high-dose therapy and marrow transplantation without some type of prophylaxis and in 5-35% of patients with prophylaxis.

A) ETIOLOGY

This complication is most common following high-dose ifosfamide or cyclophosphamide, but other agents such as busulfan, etoposide, irradiation and viruses such as cytomegalovirus, BK or adenovirus have also been implicated. The oxazaphosphorine alkylating agents, cyclophosphamide and ifosfamide, are metabolized to acrolein, which is excreted into the urine and is toxic to the bladder epithelium and leads to hemorrhagic cystitis. Hematuria may occur at any time from immediately following the administration of cyclophosphamide to up to 3 months following therapy. An increased incidence of hemorrhagic cystitis has been observed in association with cyclophosphamide-containing preparative regimens versus no cyclophosphamide, with the combination of busulfan and cyclophosphamide, in older patients, in allogeneic compared to autologous transplants and with the presence of adenovirus in urine cultures.

B) PROPHYLAXIS

Prophylactic measures are required to prevent hemorrhagic cystitis following high-dose therapy. The most commonly used approach is hyperhydration with the infusion of at least 3 liters/m²/day of intravenous fluid beginning a few hours before and continuing for at least 24 hours following cyclophosphamide administration. Diuretics are utilized if necessary to maintain urine output and to avoid fluid overload. Continuous bladder irrigation at a rate of 300-1000 ml/hour has also been shown to be effective prophylaxis against hemorrhagic cystitis, but is associated with the infectious risks of bladder catherization and with considerable patient discomfort. MESNA (2-mercaptoethane sodium sulfonate) is a sulfhydral compound which has been shown to be uroprotective in patients treated with cyclophosphamide and ifosfamide. The drug is converted to the inactive compound dimesna in the serum, reactivated to MESNA in the kidney and excreted into the urine. MESNA binds to acrolein, the toxic metabolite of cyclophosphamide, in the urine and prevents damage to the urinary tract epithelium. In several randomized trials, MESNA has not been shown to be clearly superior to either hyperhydration or continuous bladder irrigation in the prevention of hemorrhagic cystitis. The standard methods of prophylaxis vary between centers, but in general include hyperhydration alone, hyperhydration with MESNA or hyperhydration with continuous bladder irrigation. MESNA is generally dosed at 100-160% of the dose of cyclophosphamide and administered in 4 divided daily doses, or by continuous infusion, both beginning at the initiation of cyclophosphamide and continuing for 24 hours after the last dose of cyclophosphamide.

C) THERAPY

The treatment of significant hemorrhagic cystitis (macroscopic hematuria with clots and bladder pain and spasms) includes aggressive hydration and diuresis to maintain urine output, bladder irrigation with water, saline or alum and the correction of thrombocytopenia and clotting abnormalities. Bleeding not responsive to these measures may require cystoscopy for clot removal and the instillation of formalin. Formalin is the most effective local treatment for intractable bladder hemorrhage, but is associated with pain requiring general anesthesia and with bladder scarring and contraction. Other potential therapies include the selective embolization of the anterior branches of the hypogastric arteries or surgical intervention such as cystectomy.

VII) RENAL TOXICITY

Regimen-related renal toxicity is rare, occurring in fewer than 5% of patients following high-dose therapy. The prognosis of bone marrow transplant patients requiring hemodialysis is very poor with a mortality rate greater than 80%.

A) ETIOLOGY

Renal toxicity may be due to the direct effects of nephrotoxic chemotherapeutic agents such as cisplatin, carboplatin or ifosfamide utilized in the preparative regimen. Total body irradiation has been reported to cause nephritis, but the onset is usually several months after transplantation. Renal toxicity following bone marrow transplantation is commonly multifactorial and associated with the administration of nephrotoxic agents such as aminoglycoside antibiotics, amphotericin B and cyclosporine. Severe renal toxicity requiring hemodialysis rarely occurs except in the setting of multi-organ failure, especially hepatic toxicity. Renal failure is also often associated with sepsis and hemodynamic instability. Infusion of cryopreserved bone marrow and/or peripheral blood progenitor cells has also been associated with renal toxicity secondary to cell lysis products and can be prevented with adequate hydration and maintenance of urine output with mannitol or diuretics.

B) THERAPY

Adequate hydration and maintenance of electrolyte and fluid balance is important for the prevention and management of renal insufficiency, as well as the judicious use of nephrotoxic agents following transplantation.

VIII) CENTRAL NERVOUS SYSTEM TOXICITY

Central nervous system toxicity as a direct result of high-dose chemotherapy is rare. Generalized seizures occurred in approximately 10% of patients treated with high-dose busulfan (4 mg/kg/day x 4 days). Dilantin is now routinely used on a prophylactic basis during the administration of busulfan. High-dose cytosine arabinoside may cause cerebellar toxicity and seizures have rarely been reported following high-dose carmustine (BCNU). Mechlorethamine is no longer used in high-dose regimens due to the frequent development of both acute and chronic neurological problems. Cyclosporine may be the most common cause of neurologic complications following bone marrow transplantation. Approximately 20% of patients develop headaches and 20% of patients develop tremors. Cyclosporine can also cause encephalopathy with disorientation, confusion, loss of vision, impaired concentration and memory loss. Cyclosporine may also cause seizures, especially in the setting of hypomagnesemia and hypertension. Focal motor or sensory deficits are unusual and are more likely due to structural lesions. A cyclosporine-associated encephalopathy with microangiopathic hemolytic anemia which resembles thrombotic thrombocytopenic purpura (TTP) may occur post-BMT. Treatment for cyclosporine-associated neurologic dysfunction is immediate discontinuation of the drug and in the case of CsA-associated TTP, plasma exchange. Other causes of neurologic complications following bone marrow transplantation include metabolic encephalopathies secondary to hepatic failure, renal failure or electrolyte disturbances and central nervous system infections such as *Aspergillus* or toxoplasma.

IX) SKIN TOXICITY

Cutaneous toxicity secondary to high-dose chemoradiotherapy occurs in a variety of forms. Alopecia is almost universal. Skin rashes that develop during the first 3 weeks following transplantation may be due to a variety of causes including the preparative regimen, drug allergy or acute GVHD. Skin biopsies obtained during this period generally show nonspecific inflammatory changes and a diagnosis has to be made on clinical grounds. High-dose etoposide, cytosine arabinoside, cyclophosphamide, thiotepa, carmustine (BCNU) and busulfan have been reported to be associated with severe skin toxicity. The syndrome includes the development of painful erythematous macules and papules on the hands and feet and in the axillae and groin. The lesions may progress to bullae formation followed by desquamation. The pathophysiologic basis of the process is not known. Local treatment with application of ice packs helps to relieve discomfort, but narcotic analgesics

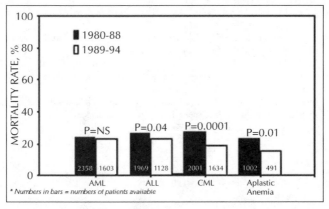

Fig. 13.1.1. 100 day mortality after HLA-identical sibling transplants. Reprinted with permission from IBMTR/ABMTR; 1995.

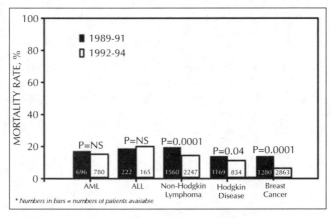

Fig. 13.1.2. 100 day mortality after autotransplants. Reprinted with permission from IBMTR/ABMTR; 1995.

are often required for pain control. Silver sulfadiazine cream is used as a topical antimicrobial agent. The syndrome usually develops in the first 2 weeks following transplantation and resolves over 1-2 weeks. Hyperpigmentation of the skin may occur after busulfan, carmustine (BCNU) or cyclophosphamide conditioning regimens.

13.1

SUGGESTED READING

1. Bearman SI, Appelbaum FR, Buckner CD et al. Regimen-related toxicity in patients undergoing bone marrow transplantation. J Clin Oncol 1988; 6:1562-1568.
2. Hertenstein B, Stefanic M, Schmeiser T et al. Cardiac toxicity of bone marrow transplantation: Predictive value of cardiologic evaluation before transplant. J Clin Oncol 1994; 12:998-1004.
3. McDonald GB, Hinds MS, Fisher LD et al. Veno-occlusive disease of the liver and multiorgan failure after bone marrow transplantation: a cohort study of 355 patients. Ann Intern Med 1993; 118:255-267.
4. Attal M, Hueguet F, Rubie H et al. Prevention of hepatic veno-occlusive disease after bone marrow transplantation by continuous infusion of low-dose heparin: A prospective, randomized trial. Blood 1992; 79:2834-2840.
5. Clark JG, Hansen JA, Hertz MI et al. Idiopathic pneumonia syndrome after bone marrow transplantation. Am Rev Respir Dis 1993; 147:1601-1606.
6. Robbins RA, Linder J, Stahl MG et al. Diffuse alveolar hemorrhage in autologous bone marrow transplant recipients. Am J Med 1989; 87:511-518.
7. Chao NJ, Duncan SR, Long GD et al. Corticosteroid therapy for diffuse alveolar hemorrhage in autologous bone marrow transplant recipients. Ann Intern Med 1991; 114:145-146.
8. Shepherd JD, Pringle LE, Barnett MJ et al. Mesna versus hyperhydration for the prevention of cyclophosphamide-induced hemorrhagic cystitis in bone marrow transplantation. J Clin Oncol 1991; 9:2016-2020.
9. Vose JM, Reed EC, Pippert GC et al. Mesna compared with continuous bladder irrigation as uroprotection during high-dose chemotherapy and transplantation: A randomized trial. J Clin Oncol 1993; 11:1306-1310.

Delayed Complications

H. Joachim Deeg

13.2

I) ETIOLOGY OF DELAYED COMPLICATIONS

Most patients undergoing bone marrow or peripheral blood stem cell transplantation who survive the acute posttransplant period and who do not experience a recurrence of their underlying disease within 18 months to 2 years posttransplant lead normal and productive lives. However, chronic or delayed complications do develop in some. Some delayed complications (e.g. chronic graft-versus-host disease [GVHD] and immunodeficiency) are directly transplant-related; others (e.g. infertility or cataracts) are due to the intensity of the preparative regimen. Others are related to the underlying diagnosis (e.g. disease recurrence) and many are of multifactorial etiology (e.g. chronic pulmonary disease or secondary malignancies). Problems may arise because of direct, therapy-related trauma (e.g. destruction of skeletal growth plates), because of damage related to graft-host interactions (e.g. lymphoproliferative disorders) or because of repair processes that result in scar formation (e.g. bladder dysfunction). Chronic GVHD, disease recurrence and infections are discussed elsewhere in this manual. A list of delayed or chronic problems is given in Table 13.2.1 and Figure 13.2.1.

II) AIRWAY AND PULMONARY DISEASE

Pulmonary tissue is sensitive to various cytotoxic agents and irradiation. The responses to injury are particularly prominent in the interstitium. Repair process may result in scar formation in the interstitial space and interfere with effective gas exchange and respiration. Lungs and airways are also targets for viral, bacterial and fungal infections which, if protracted or associated with incomplete parenchymal repair, further impair pulmonary function. The bronchial tree can furthermore be involved by GVHD. As a result, chronic pulmonary complications affect approximately 15% of patients. Pulmonary dysfunction appears to be an important indicator since patients with abnormal pulmonary function early posttransplant experience significantly higher nonrelapse mortality than do patients with normal lung function.

A) LATE ONSET INTERSTITIAL PNEUMONITIS

Late interstitial pneumonitis occurs almost exclusively in patients with chronic GVHD. Some patients show marked interstitial fibrosis, and improve on immunosuppressive therapy but fail to respond to bronchodilators. Lymphocytic bronchitis, still controversial in regard to its etiology, is often associated with a nonproductive cough, dyspnea, wheezing and occasionally with bronchospasm. Most of these patients also have other typical clinical and pathological findings of chronic GVHD and require therapy with immunosuppressive agents.

B) RESTRICTIVE PULMONARY DISEASE

Restrictive ventilatory defects are common in BMT recipients. In one study, approximately 20% showed a mean loss in total lung capacity of 0.81 liters, a decrease in vital capacity of 0.54 liters and an impairment of diffusing capacity of 4.4 ml/min/mm Hg 1 year after BMT. Lung function improved over the subsequent 3-4 years. Restrictive pulmonary changes are not correlated with the type of conditioning regimen or chronic GVHD. Generally, these changes do not produce severe symptoms and do not require therapeutic intervention. A

Bone Marrow Transplantation, edited by Richard K. Burt, H. Joachim Deeg, Scott Thomas Lothian, George W. Santos. © 1996 R.G. Landes Company.

13.2

Table 13.2.1. Delayed complications

Chronic graft-versus-host disease
Airway and pulmonary disease
Autoimmune dysfunction
Neuroendocrine dysfunction
Impaired growth
Infertility
Ophthalmological problems
Avascular necrosis of the bone
Dental problems
Genitourinary dysfunction
Secondary malignancies
Central and peripheral nervous system
Psychosocial effects and rehabilitation

more recent study comparing results after autologous and allogeneic BMT found marked defects of diffusing capacity, as well as static and dynamic lung volumes, particularly in patients with GVHD. There were no significant differences between patients given autologous BMT and allogeneic BMT recipients without evidence of GVHD.

C) OBSTRUCTIVE PULMONARY DISEASE

Obstructive changes might represent sequelae to extensive restrictive changes in the small airways, or, as in obstructive bronchiolitis, may be due to small airway destruction. Recurrent aspirations associated with esophageal abnormalities or purulent sinus secretions contribute to airway inflammation and obstructive lung disease. Obstructive defects are present in 10-15% of patients with chronic GVHD. Patients present with cough, dyspnea and wheezing at approximately 1-1.5 years. In addition to abnormal pulmonary function tests, immunoglobulin G (IgG) and IgA levels in serum are usually low. There is no response to bronchodilator treatment and only 30-40% of patients improve with immunosuppressive therapy. Aggressive management of infections is indicated.

D) BRONCHIOLITIS OBLITERANS

Progressive bronchiolitis obliterans is seen in approximately 10% of patients with chronic GVHD. The clinical and pathological findings are similar to those seen after lung or heart-lung transplants. It is thought that the histological changes are due to a graft-versus-host reaction and aggravated by infections. Pulmonary infections develop in more than 20% of allogeneic BMT recipients without GVHD and in more than 60% of those with chronic GVHD. The clinical course of bronchiolitis varies. There may be slow deterioration or diffuse necrotizing fatal bronchiolitis. Bronchiolitis obliterans has been observed as early as 3 months and as late as 2 years after marrow grafting. Chest radiographs may show hyperinflation of the lungs and flattening of the diaphragm. Pneumothoraces may develop. Significant morbidity among patients with bronchiolitis obliterans is due to recurrent respiratory tract infections. Characteristically, this disease does not respond well to conventional therapy of chronic GVHD, although immunosuppression with glucocorticoids is helpful in some patients, and intensive infection prophylaxis and therapy should be given. Infections should be managed aggressively.

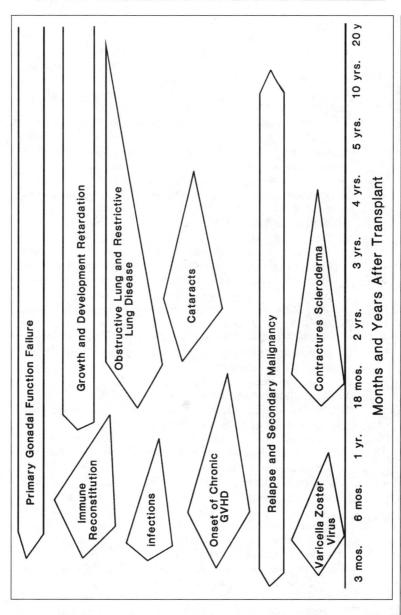

Fig. 13.2.1. Time of onset of late complications following stem cell transplantation. Courtesy of Philip Gold, Mary Flowers and Keith Sullivan.

13.2

III) AUTOIMMUNE DYSFUNCTION

One manifestation of autoimmunity is the presence of autoantibodies which are observed frequently after BMT, particularly in patients with chronic GVHD. Most commonly detected are rheumatoid factor, antinuclear, anti-smooth muscle and antimitochondrial antibodies. Most patients are asymptomatic. In some instances, however, such as in myasthenia gravis, autoantibodies cause severe clinical problems. The antiacetylcholine receptor antibodies mediating this problem are donor-derived. Generally, the disease occurs in patients with chronic GVHD more than 1 or 2 years after BMT. Besides myasthenia gravis, patients with chronic GVHD may develop weakness from polymyositis. Patterns of abnormal immune reactivity of the donor can be transferred from the donor to the patient (e.g. atopic asthma). Several hematological problems, including immune thrombocytopenia, anemia and neutropenia, can be mediated by autoantibodies. All these complications should be treated with modalities as established in nontransplant patients.

IV) NEUROENDOCRINE DYSFUNCTION

A) THYROID FUNCTION

Compensated hypothyroidism is seen in 30-60% of patients conditioned with a regimen including single-dose TBI; the incidence is 15-25% in patients prepared with fractionated TBI. Thyroid function is generally normal in patients prepared with chemotherapy alone. Elevated thyroid-stimulating hormone (TSH) and low T4 thyroxine levels along with overt hypothyroidism may present as early as 1 and as late as 15 years after transplantation. In addition to the immediate metabolic effects, hypothyroidism may also contribute to diminished linear growth. Hyperthyroidism has been described in patients with chronic GVHD. Treatment consists of hormone supplementation and anti-inflammatory therapy.

B) GROWTH HORMONE DEFICIENCY

Forty to sixty percent of children conditioned with high-dose TBI have growth hormone deficiency. The incidence may be 90% in children with previous cranial irradiation, generally developing 2-3 years after treatment. Growth velocity may be normal in children less than 10 years of age at the time of BMT, but is generally reduced in children who are older at the time of BMT. In these children, final adult height is generally markedly reduced to < 10th percentile in patients less than 11 years at transplant and < 50th percentile in those 11-16 years old. Growth velocity is also reduced due to direct effects of TBI on bones, and quite likely by treatment of chronic GVHD with glucocorticoids. Some "catch-up" growth may occur in patients given fractionated rather than single-dose TBI. Administration of growth hormone may improve the growth rate, but less so than seen with growth hormone deficiencies of other etiologies. In part, this is due to the fact that so many other factors, including thyroid dysfunction, steroid administration and irradiation effects, contribute to impaired growth. With the availability of recombinant growth hormone, more comprehensive studies now need to be carried out. After conditioning with cyclophosphamide alone, growth and development is usually normal.

C) SEXUAL DEVELOPMENT

Sexual development often is severely impaired and delayed in both boys and girls, especially after single-dose TBI. The older a patient at the time of transplantation, the lower the probability of achieving sexual maturity and reproductive function. After fractionated TBI about half of the patients show normal pubertal development, and the majority does so after conditioning with chemotherapy only. When evaluating children and adolescents for gonadal function, numerous factors, including gonadal and cranial irradiation (in patients with acute leukemia), steroid therapy and others, need to be taken into consideration. Generally, hypogonadism is primary (albeit iatrogenic) with elevated gonadotropin levels, but exceptions exist. While in adult males testosterone levels are usually normal, they are depressed in adolescents. Workup should be the same as used for other patients with these problems. Therapy may include gonadotrophin administration and, in certain instances, estrogen or testosterone supplementation. This is important also to optimize the milieu for longitudinal growth.

13.2

D) **INFERTILITY**

After TBI, most women experience primary gonadal failure and require cyclic hormone supplementation. However, some women, particularly those < 25 years of age, will recover ovarian function. At least a dozen pregnancies have been reported and several normal children have been born to these women. Most women, however, experience premature menopausal symptoms, and cyclic hormone supplementation should be given to prevent vaginitis, bone loss and other complications. Men usually preserve Leydig cell function and testosterone and luteinizing hormone production, while Sertoli cell function and spermatogenesis are absent or abnormal. Some men recover spermatogenesis usually several years after TBI and several of those patients have fathered children.

After conditioning with cyclophosphamide, many women 25 years of age or younger have a return of ovarian function within months or years; recovery is rare in older patients. Many women have become pregnant, and there have been more than 30 live births reported. Most men recover spermatogenesis and many have fathered normal children. Experience with busulfan conditioning is still too limited for a conclusive assessment.

V) **OPHTHALMOLOGIC PROBLEMS**

The eyes are targets of acute and chronic GVHD and may show changes related to the conditioning regimen, the treatment of GVHD and infections. Irradiation is known to cause cataracts. Posterior capsular cataracts develop in patients conditioned with TBI beginning approximately 1 year after BMT. For patients given single-dose TBI (usually 920-1,000 cGy), the incidence at 5-6 years is 80%. Among patients given fractionated TBI, the incidence of cataracts is lower, approximately 50% at doses greater than 1,200 cGy, and 30-35% at doses of 1,200 cGy or less. Among chemotherapy-conditioned patients, the incidence is 20%. The incidence of cataracts is higher in patients who also received cranial irradiation before BMT and in patients treated with glucocorticoids. Approaches to cataract prevention are experimental. Treatment may require lens extraction and implantation of an artificial lens. Other ocular complications, in particular infections, have generally been related to chronic GVHD. GVHD and infections may also lead to scar formation (e.g. synechiae, ectropion) and even corneal perforation. Long-term antibiotic coverage and artificial tears can prevent some infections and scarring. Obstruction of the nasolacrimal duct, related to GVHD or conditioning-induced fibrosis, has been observed in some patients.

VI) **AVASCULAR NECROSIS OF THE BONE**

Avascular necrosis has long been recognized as a side effect of glucocorticoid therapy. Since many patients receive glucocorticoid therapy before and after transplantation, this complication is common in transplant recipients. It may occur even after short courses of high-dose therapy, as for treatment of acute GVHD. The exact incidence of this complication has not been determined. Some patients show improvement when glucocorticoids are discontinued, but others have required treatment in the form of hip or knee joint replacement.

VII) DENTAL PROBLEMS

The oral mucosa is often affected by chronic GVHD. An oral sicca syndrome may result in poor oral hygiene, recurrent infection and dental decay. Severe dental decay also occurs in nontransplant recipients given head or neck irradiation and, similarly, an oral sicca syndrome, periodontal disease and cavities can occur in patients without GVHD. Many of these problems can be prevented by diligent hygiene, fluoride treatment and other supportive measures. Irradiation may also interfere with the development of teeth and facial bones in children. There may be poor calcification, micrognathia, mandibular hypoplasia, root blunting and apical closure. The changes are most severe in children less than 7 years of age at the time of BMT. Only omission of irradiation from conditioning regimens can prevent these problems.

VIII) GENITOURINARY DYSFUNCTION

Most patients recover quickly from hemorrhagic cystitis. Infrequently, patients will have protracted hematuria, and scarring of the bladder wall with volume loss may present a chronic problem manifested by urinary frequency. Viruria, for example, with adenovirus or polyomavirus, has also been implicated in hematuria.

Radiation nephritis may develop after high-dose irradiation. Doses of TBI used in preparation for BMT are generally thought to be insufficient to induce clinically relevant radiation nephritis. However, patients who have received aggressive antitumor therapy (e.g. with platinum compounds) before BMT may be at risk of suffering renal damage.

Case reports have documented the development of a hemolytic uremic syndrome or microangiopathic hemolytic anemia associated with renal failure, particularly in cyclosporine (CSP)-treated patients, either during therapy or following discontinuation. Similar observations have been made recently in FK 506-treated patients. The mechanism is not clear. Presumably, the intensity of the conditioning regimen itself or chemoradiotherapy given prior to BMT and resulting in endothelial damage sets the stage. The pathways of prostaglandin synthesis, which, in conjunction with endothelial damage, might interfere with coagulation homeostasis may be altered. Plasmapheresis and discontinuation of the presumptive causative agent have been suggested as therapeutic options.

IX) POSTTRANSPLANT MALIGNANCIES

In the 1970s, several case reports described the posttransplant occurrence of leukemia in donor-derived cells. One study suggested that approximately 5% of leukemic recurrences after BMT were in donor-derived cells. Because the number of long-term surviving patients has continued to increase, it has become possible to carry out more detailed analyses. Results in 2,246 patients transplanted between May 1970 and February 1987 at the Fred Hutchinson Cancer Research Center in Seattle indicated that the risk of developing a posttransplant malignancy other than the original disease was approximately 5-6 times higher than in the population at large. Malignancies included posttransplant lymphoproliferative disorders (often occurring in donor cells and showing integration of EBV genomic sequences), new leukemia and solid tumors. The treatment of GVHD with monoclonal anti-CD3 antibody or with polyclonal anti-T cell sera and the use of TBI for conditioning were identified as risk factors for all tumors. HLA nonidentity of donor and recipient and T cell depletion of donor marrow may be additional risk factors for the development of lymphoproliferative disorders. Recent data suggest that the use of azathioprine for the treatment of chronic GVHD is another risk factor, particularly for the development of solid tumors. Therapy includes the infusion of anti-B cell antibodies or donor-derived lymphocytes for the treatment of lymphoproliferative disorders, possibly chemotherapy for leukemia, and aggressive surgical management for solid tumors.

X) CENTRAL AND PERIPHERAL NERVOUS SYSTEM

Peripheral and central nervous system complications more frequently occur in the acute posttransplant period, but some are delayed. In addition to infections, hemorrhage and structural abnormalities, there may also be hypothalamic-pituitary dysfunction, impaired memory, shortened attention span and defects of verbal fluency. Children, particularly those who also received cranial irradiation pretransplant, may score lower than control subjects in visual-motor and processing task and various IQ tests.

XI) PSYCHOSOCIAL EFFECTS AND REHABILITATION

Long-term adjustments and rehabilitation depend strongly on events along the way. Changes in body image and weakness, in addition to medications, especially glucocorticoids and their side effects, may be the most significant factors. Along with the fear of disease recurrence, these factors may lead to depression, changes in partner relationships and family roles. Patients may change their life priorities and perspectives, which, in turn, affects their families.

While patients without chronic problems move toward normal activities within 1 or 2 years after BMT, patients with chronic GVHD or pulmonary disease may be crippled for years. Major adjustments in lifestyle and occupation may be necessary in patients in whom joint contractures or muscle wasting develop. Problems with sexuality arise sooner or later after BMT, both in adolescents and in adults. Healthy partners may put inappropriate demands on BMT recipients which further contributes to intramarital stress, and in addition to social and financial demands, may result in marital dysfunction and divorce.

Problems with employment and insurance occur because many insurance companies are reluctant to provide health or life protection.

Early detection of potential problems and health maintenance are important, possibly for several decades, especially in children and adolescents. The desire to be equal to their peers may cause problems with compliance in this age group. A multidisciplinary approach involving adolescent medicine physicians and endocrinologists along with group therapy appears most promising. Rehabilitation must begin at the time of diagnosis and should involve a long-term treatment plan.

SUGGESTED READING

1. Crawford SW, Pepe M, Lin D, Benedetti F, Deeg HJ. Abnormalities of pulmonary function tests after marrow transplantation predict non-relapse mortality. Am J Respir Crit Care Med (in press).
2. Dahllöf G, Barr M, Bolme P, Modeer T, Lonnqvist B, Ringden O, Heimdahl A. Disturbances in dental development after total body irradiation in bone marrow transplant recipients. Oral Surg Oral Med Oral Pathol 1988; 65:41-44.
3. Deeg HJ. Delayed complications and long-term effects after bone marrow transplantation. Hematol Oncol Clin North Am 1991; 4:641-657.
4. Fischer A, Blanche S, Le Bidois J, Bordigoni P, Garnier JL, Niaudet P, Morinet F, Le Deist F, Fischer AM, Griscelli C, Hirn M. Anti-B-cell monoclonal antibodies in the treatment of severe B-cell lymphoproliferative syndrome following bone marrow and organ transplantation. N Engl J Med 1991; 324:1451-1456.

13.2

13.2

5. Nims JW. Survivorship and rehabilitation. In: Whedon MB, ed. Bone Marrow Transplantation, Principles, Practices and Nursing Insights. Boston: Jones & Bartlett, 1991:334-345.

6. Sanders JE. Growth and development after bone marrow transplantation. In: Forman FJ, Blume KG, Thomas ED, eds: Bone Marrow Transplantation. Boston: Blackwell Scientific Publications, 1994:527-537.

7. Smith CIE, Aarli JA, Biberfeld P, Bolme P, Christensson B, Gahrton G, Hammarstrom L, Lefvert AK, Lonnqvist B, Matell G, Pirskanen R, Ringden O, Svanborg E. Myasthenia gravis after bone-marrow transplantation. Evidence for a donor origin. N Engl J Med 1983; 309:1565-1568.

8. Sullivan KM, Mori M, Sanders J, Siadak M, Witherspoon RP, Anasetti C, Appelbaum FR, Bensinger W, Bowden R, Buckner CD, Clark J, Crawford S, Deeg HJ, Doney K, Flowers M, Hansen J, Loughran T, Martin P, McDonald G, Pepe M, Petersen FB, Schuening F, Stewart P, Storb R. Late complications of allogeneic and autologous marrow transplantation. Bone Marrow Transplant 1992; 10 (Suppl 1):127-134.

9. Tarbell NJ, Guinan EC, Niemeyer C, Mauch P, Sallan SE, Weinstein HJ. Late onset of renal dysfunction in survivors of bone marrow transplantation. Int J Radiat Oncol Biol Phys 1988; 15:99-104.

10. Wingard JR, Curbow B, Baker F, Piantadosi S. Health, functional status, and employment of adult survivors of bone marrow transplantation. Ann Intern Med 1991; 114:113-118.

11. Witherspoon RP, Fisher LD, Schoch G, Martin P, Sullivan KM, Sanders J, Deeg HJ, Doney K, Thomas D, Storb R, Thomas ED. Secondary cancers after bone marrow transplantation for leukemia or aplastic anemia. N Engl J Med 1989; 321:784-789.

14

Outpatient Management of Marrow and Blood Stem Cell Transplant Patients

Philip Gold, Mary E.D. Flowers, Keith M. Sullivan

I) INTRODUCTION TO THE OUTPATIENT DEPARTMENT

With the rising costs of health care, the role of ambulatory care for hematopoietic stem cell transplant patients is expanding and many services traditionally given in the inpatient setting can now be provided in less costly outpatient settings.

The role of the outpatient department is to facilitate comprehensive, multidisciplinary care of the stem cell recipient.

A) GOALS

1) Provide 7 day a week physician and nurse staffing for pretransplant and posttransplant care of stem cell recipients. Provisions for prompt evaluation and treatment should be available 24 hours a day.

2) Evaluate potential transplant recipients and donors.

3) Provide defined areas for essential clinical procedures such as blood sampling, apheresis and stem cell infusions, blood product transfusions, skin biopsies, bone marrow aspirations, intravenous medications and infusions and hyperalimentation.

4) Administer high-dose chemotherapy, as indicated, in an outpatient setting.

5) Provide a multidisciplinary patient care approach with support staff including physicians, nurses, dieticians, dentists, social workers and psychiatrists. The staff should be highly skilled in the care of the transplant recipient including subspeciality components of pulmonary medicine, gastroenterology, pathology, pharmacology, infectious disease, histocompatibility, blood banking, radiology and imaging.

6) Provide surveillance for the complications of transplantation.

  a) Monitor for regimen-related toxicities following high-dose chemotherapy or total body irradiation and provide supportive care with antiemetics and analgesics.

  b) Detect and treat infections in the immunocompromised host.

  c) Diagnose and manage acute and chronic graft-versus-host disease (GVHD).

  d) Monitor for graft failure, relapse of malignancy or late complications.

7) Provide a structure and organization for conduct of clinical investigations and maintain an extensive clinical data base for quality control and protocol analysis.

8) Prepare patients and their referring physicians for discharge home.

9) Perform long-term followup evaluations to monitor the late effects of transplantation.

10) Establish written policies of standard medical practices and procedures.

11) Ensure that physicians are skilled in the practice of high-dose chemotherapy and stem cell transplantation.

Bone Marrow Transplantation, edited by Richard K. Burt, H. Joachim Deeg, Scott Thomas Lothian, George W. Santos. © 1996 R.G. Landes Company.

II) OUTPATIENT MONITORING

A) CRITERIA FOR HOSPITAL DISCHARGE TO THE OUTPATIENT DEPARTMENT

As efforts continue to reduce both the length and cost of hospital care, criteria for hospital discharge continue to evolve. Standard criteria for hospital discharge include: adequate oral intake, absence of fever while off antibiotics, adequate control of nausea, vomiting and diarrhea, granulocyte counts greater than 500/µl, supportable platelet counts greater than 15,000/µland presence of a responsible adult caregiver. Table 14.1 details traditional, interim and experimental discharge criteria currently being investigated.

B) DISCHARGE PLANNING

Patients are typically discharged 10-20 days after autologous and 15-25 days after allogeneic stem cell transplantation. In order to have a smooth transition from inpatient to outpatient services, transition planning begins in the final days of hospitalization. Patients should be seen on the day of discharge by the clinic nurse and by the outpatient physician on the following day. Direct communication between the inpatient and outpatient physicians must take place prior to discharge. Ambulatory monitoring in the outpatient department is conducted during the period of maximum risk of transplant-related complications. Table 14.2 lists complications which may occur within the first few months of transplantation and which are discussed elsewhere in this manual. For unrelated and HLA-mismatched allogeneic stem cell recipients, patients are followed in the outpatient department through day 100 posttransplant. For autologous transplant recipients, patients are typically discharged home within 1 month of transplant. Coordination with the referring physician is crucial to ensure continued effective ambulatory care and to avoid rehospitalization for supportive care or management of complications upon return home.

C) POSTTRANSPLANT OUTPATIENT DEPARTMENT MONITORING

1) Daily rounds are held with the attending physician, primary care physician, primary nurse, dietician, pharmacist and social worker. Results of diagnostic procedures and laboratory studies are presented and further therapies are planned.

2) Comprehensive weekly or twice weekly patient examinations are conducted by the primary care physician. Emergent or drop in visits are triaged by hierarchy of illness. Patients may be monitored in a day bed area for assessment of evolving clinical conditions.

3) Daily or every other day CBC with differential and platelet count and twice or thrice weekly electrolytes, BUN and creatinine determinations are obtained.

4) Weekly cytomegalovirus (CMV) monitoring with CMV antigenemia assays or cultures of blood, urine and throat and weekly cyclosporine levels and liver function tests are reviewed.

5) Bone marrow aspirates are obtained on days 28, 56, 80 and 365 posttransplantation. Cytogenetics and molecular studies of peripheral blood and marrow monitor the status of donor engraftment and markers of hematologic malignancy.

6) Departure studies (see below) are secured to screen for chronic GVHD and prepare the patient for the summary conference before discharge home.

7) Patient updates are given several times per month to the referring physician via written interim summaries and physician telephone calls.

14

Table 14.1. Criteria for hospital discharge to the outpatient department

Parameter	Traditional	Criteria	
		Interim	**Early**
Oral intake as a baseline of requirements	≥ 33%	15-30%	May be 'nil by mouth'
Maximum parenteral fluid required in 24 hours (total fluids to include lipid solutions)	1500 ml/day	50 ml/kg/d < 20 kg: 20-35 kg: ≤ 1500 ml/d > 35-70 kg: ≤ 2000 ml/d > 70 kg: ≤ 2500 ml/d	< 20 kg: < 1000 ml/d 20-35 kg: ≤ 1500 ml/d > 35-70 kg: ≤ 2000 ml/d > 70 kg: ≤ 3000 ml/d
Nausea and vomiting	PO medication	PO medication	IV medication
Diarrhea controlled at	< 500 ml/d	< 500 ml/d	< 500 ml/d
Afebrile (< 38.4)	Yes	Yes	Yes
IV medications	Off for 48 hours	8 or 12 hourly dosing of infusions	4, 6 or 8 hourly or as needed
Platelet count	> 15,000 μl	Supportable > 15,000 μl with no more than 2-3x/week platelet transfusions	Supportable > 15,000 μl with daily platelet transfusions
Granulocytes for 48 hours	> 500/μl	> 500/μl	> 500/μl
Hematocrit: Adult Child	> 30% > 25%	> 26% > 26%	> 30% > 25%
Hours tolerating oral cyclosporine	48	IV CsA	IV CsA
Herpes zoster patients on IV therapy provided lesions have crusted over and patient has < 4 days of therapy remaining	No	Yes	Yes
TPN and pump care teaching completed and demonstrated ability to perform tasks	Yes	Yes	Yes
Responsible adult living in apartment as determined by discharging nurse	Yes	Yes	Yes

Abbreviations: IV, intravenous; TPN, total parenteral nutrition; PO, per os (oral); CsA, cyclosporine

D) HOSPITAL READMISSION

Not uncommonly, patients are readmitted to the hospital from the outpatient department. The primary reasons for readmission include bacteremia, gastrointestinal or nutritional abnormalities and GVHD. Table 14.3 outlines criteria for hospital readmission.

III) DEPARTURE HOME

The outpatient department provides an ideal setting to evaluate and educate patients prior to discharge home. At approximately 80 days after allogeneic transplantation, a departure evaluation is initiated to assess:

1) Hematopoietic engraftment
2) Disease status
3) Organ toxicity
4) Chronic GVHD activity

14

Early detection and treatment of chronic GVHD is essential to prevent long-term complications in allogeneic transplant recipients. Day 80 screening studies performed in the outpatient department are effective, predictive and prognostic tools. As outlined in Table 14.4, biopsies are taken from the skin and oral mucosa to detect subclinical histologic evidence of GVHD. Eyes are examined to detect ocular sicca, and liver function and pulmonary function are examined. A comprehensive physical exam focuses on skin, oral, osteoarticular and cardiopulmonary findings. The functional performance is assessed by standardized criteria (Appendix VI).

Table 14.2. Early complications of stem cell transplantation

Regimen-related toxicity
    Cystitis
    Mucositis
    Pulmonary conditions
    Renal conditions
    Neurologic conditions
Veno-occlusive disease of the liver
Idiopathic pneumonia syndrome
Marrow graft failure
Immunodeficiency
Infection
Acute graft-versus-host disease

Table 14.3. Criteria for hospital readmission

Shaking chills (± fever)
Septicemia
Gram-negative bacteremia or fungemia
Uncontrolled GVHD
Graft failure
Interstitial pneumonia
Varicella zoster
Medical emergencies/acute organ failure
Failure to thrive

- weight loss (% of discharge weight): > 10% loss in adults
- fluid losses exceeding maximal support > 5% loss in children

---

Table 14.4. Screening studies for chronic graft-versus-host disease

Organ/System	Clinical Finding	Screening Study
Dermal	Dyspigmentation, xerosis, erythema, scleroderma, onychodystrophy, alopecia	Skin biopsy—3 mm punch biopsy from posterior iliac crest and forearm areas
Oral	Lichen planus, xerostomia	Oral biopsy from inner lower lip
Ocular	Sicca, keratitis	Schirmer's test
Hepatic	Jaundice	Alkaline phosphatase, SGOT, bilirubin
Pulmonary	Obstructive/restrictive pulmonary disease	Pulmonary function studies and arterial blood gas
Vaginal	Sicca, atrophy	Gynecological evaluation
Nutritional	Protein and caloric deficiency	Weight, muscle/fat store measurement
Clinical performance	Contractures, debility	Karnofsky score or Lansky play index

Abbreviations: SGOT, serum glutamic oxaloacetic transaminase

Table 14.5. Late complications of stem cell transplantation

Regimen-related toxicity
    Cataracts
    Neurologic conditions
    Gonadal conditions
    Endocrine conditions
    Growth and development
Immunodeficiency
Infection
Chronic graft-versus-host disease
Relapse of malignancy
Secondary malignancy

Prior to discharge home the data from the departure evaluation are reviewed to determine the need for additional intervention. Table 14.5 summarizes late complications that may develop after day 100. Figures 13.2.1 and 14.1 show the etiologies and time of onset of these late complications. Before departure home, patients may be entered on additional treatment protocols for:

1) Prevention and treatment of chronic GVHD
2) Posttransplant immune modifier or cellular therapy for prevention or treatment of recurrent malignancy
3) Pediatric growth and development studies
4) Psychologic or neuropsychometric studies
5) Gonadal hormone replacement therapy

IV) **LONG-TERM FOLLOWUP**

Approximately 100 days after transplant most allogeneic stem cell recipients return home from the transplant center. After the departure evaluation is completed, a discharge conference is held with the attending physician, the primary care

14

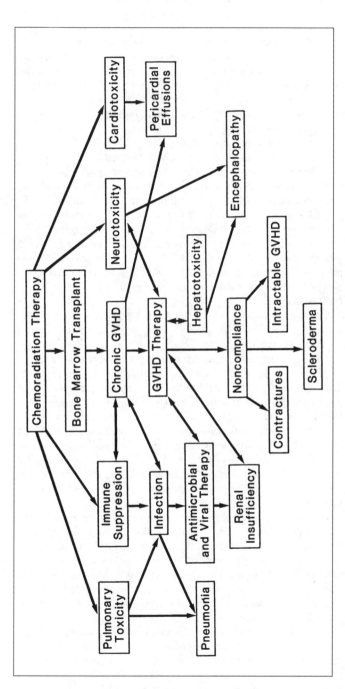

Fig. 14.1. Interrelationships of late complications following stem cell transplantation. Reprinted with permission from Nims JW. In: Kasprisin CA, Snyder EL, eds. Bone Marrow Transplantation: A Nursing Perspective. Arlington: American Associational Blood Banks 1990:45-57.

physician and nurse. Results of screening studies and evaluations are presented and patients enter chronic GVHD or other research protocols as indicated. The referring physician is contacted by telephone and the patient hand carries medical summaries and guidelines for continuing care to the referring physician.

Upon return home, the outpatient department physicians and the referring physician remain in close contact. In addition, nurses remain in contact with the patient. Information and questions obtained from the physician's office and the patient are presented in daily rounds to an attending physician for analysis and recommendations to the referring doctor. This ongoing dialogue between the outpatient department and referring physicians ensures:

1) Continued appropriate long-term care of the transplant recipient
2) A large clinical data base which can be updated regularly
3) Long-term monitoring of the late results of transplant regimens.

Life-long monitoring is provided through telephone updates, review of referring physicians' office notes and yearly questionnaires sent to the patient (two page self assessment mailer) and to the primary physician (three page survey of medical status, complications and medications). Patients are encouraged to return to the outpatient department for comprehensive evaluations at 1, 5 and 10 years after transplantation. Chronic GVHD status, immunologic recovery, hematopoietic function, pediatric growth and development, and psychosocial states are examined with the goal of maintaining disability-free survival.

V) IMMUNIZATIONS

Despite prompt return of normal neutrophil and lymphocyte counts, humoral and cellular immunity may take as long as 1 year to fully recover after stem cell transplantation. The tempo of immunologic reconstitution is dependent upon the type of marrow graft (autologous versus allogeneic), the type of immunosuppressive therapy and, importantly, whether or not chronic GVHD has developed. Patients with chronic GVHD are at risk to develop bacterial and opportunistic infections, and recipients of unrelated stem cell grafts appear to have the highest risk of late infection. During the first year posttransplant, patients may be unable to develop antibody response to immunizations. Antibody titers and immunoglobulin levels are checked to determine the quantitative adequacy of humoral function. Most stem cell recipients surviving beyond 1 year posttransplant, who are free of chronic GVHD, develop specific IgG antibody titers to antigens such as tetanus toxoid, diphtheria and measles. This suggests transfer of specific antibody producing B cells from the donor to the recipient.

At the first year posttransplant, patients free of chronic GVHD are likely to respond to booster immunizations with pneumococcal, inactivated polio, influenzae, diphtheria, pertussis, tetanus toxoid, hepatitis B and hemophilus influenza type B vaccines (Table 14.6). Antibody titers can be tested 2 and 4 weeks postvaccination to evaluate immune responsiveness. Individuals should be given the current influenza vaccine each November. Live virus biologicals such as Sabin oral polio, mumps, measles, rubella (MMR), BCG and oral typhoid vaccines carry risk in the immunocompromised host. Current studies suggest that MMR and probably the varicella vaccine can be safely given after the second year posttransplant in patients free of chronic GVHD and immunosuppressive drug treatment. Contacts and family members should not receive oral polio vaccine during the first posttransplant year since the patient may be at risk from live virus shedding which may occur for up to 12 weeks after immunization of family members.

14

Table 14.6. Recommended immunizations after hematopoietic transplantation

Year One*	Diphtheria, pertussis, tetanus (DPT)
	H. influenzae (Hib) conjugate
	Hepatitis B
	Influenza (repeat every November)
	Salk poliovirus (inactivated vaccine)
Year Two**	Measles, mumps, rubella (MMR)
	Varicella vaccine (Varivax)
Family Members	
	No Sabin poliovirus (OPV) during year one
	MMR does not pass to others

*Patients with chronic GVHD may not benefit
**ONLY in patients free of chronic GVHD and immunosuppressive treatment

14

ACKNOWLEDGMENTS

This work was supported in part by grants CA 18029, CA 15704, CA 47748, CA 18221 and HL 36444 from the National Institutes of Health, DHHS.

SUGGESTED READING

1. Flowers MED, Sullivan KM. Preadmission procedures, marrow transplant hospitalization, and posttransplant outpatient monitoring. In: Atkinson K, ed.. Textbook on Bone Marrow Transplantation. London: Cambridge University Press, 1992; 75-86.
2. Peters WP, Ross M, Vredenburgh JJ et al. The use of intensive clinic support to permit outpatient autologous bone marrow transplantation for breast cancer. Semin Oncol 1994; 21:25-31.
3. Corcoran-Buchsel P, Parchem C. Ambulatory care of the bone marrow transplant patient. Semin Oncol Nursing 1988; 6(1):41-46.
4. Rowe JM, Ciobanun C et al. Recommended guidelines for the management of autologous and allogeneic bone marrow transplantation. Ann Intern Med 1994; 120:143-158.
5. Phillips G, Armitage J, Bearman S et al. American Society for Blood and Marrow Transplantation Guidelines for Clinical Centers. Biol Blood Marrow Transplantation 1995, 1:54-55.
6. Sullivan KM, Wade JC, Bowden RA, Reed EC. Management of the immunocompromised host. In: Schrier SL, McArthur JR, eds. Hematology. American Society of Hematology, 1993; 163-174.
7. Ljungman P, Fridell E, Lönnquist B et al. Efficacy and safety of vaccination of marrow transplant recipients with a live attenuated measles, mumps, and Rubella vaccine. J Infect Dis 1989; 159:620-615.
8. Loughran TP, Sullivan KM, Morton T et al. Value of day 100 screening studies for predicting the development of chronic graft-versus-host disease after allogeneic bone marrow transplantation. Blood 1990; 76(1):228-234.
9. Moinpour CM, Chapko MK, Sullivan KM, Kirk S. Cost issues for BMT. In: Buchsel PC, Whedon MD, eds. Bone Marrow Transplantation: Administrative and Clinical Strategies. Boston: Jones and Bartlett, Inc., 1995; 427-442.
10. Sullivan KM, Mori M, Sanders J et al. Late complications of allogeneic and autologous transplantation. Bone Marrow Transplantation 1992; 10:127-134.
11. Nims JW. Late effects of bone marrow transplantation: A nursing perspective. In: Kasprisin CA, Snyder EL, eds. Bone Marrow Transplantation: A Nursing Perspective. Arlington: American Associational Blood Banks, 1990; 45-57.
12. Bush NE, Haberman M, Donaldson G, Sullivan KM. Quality of life of 125 adults surviving 6-18 years after bone marrow transplantation. Soc Sci Med 1995; 40:479-490.
13. Sullivan KM, Siadak MF. Long-term followup after hematopoietic stem cell transplantation. In: Johnson FI, Virgo KS, eds. Cancer Patient Followup. St Louis: Mosby Inc., (in press).

15

Gene Therapy

Christopher E. Walsh

I) INTRODUCTION

Gene therapy involves the insertion of a functional gene into a target cell population that subsequently corrects a genetic defect or adds a desirable function to those cells. Clinical application of gene transfer involves: 1) isolation, characterization, and expansion of target cells and 2) efficient and safe methods of gene transfer resulting in stable expression of the inserted gene.

Bone marrow and peripheral blood are sources of either mature cells or pluripotent stem cells that are target cells for gene transfer. Easy access to target cells has simplified the application of basic gene transfer techniques to hematological diseases. Both congenital and acquired hematological disorders are potentially amenable to gene therapy (Table 15.1).

II) HEMATOPOIETIC TARGET CELLS

A) CHARACTERIZATION

The pluripotent hematopoietic stem cell (PHSC) is a self-renewing cell with long-term repopulating capacity. These biological characteristics of PHSCs have been defined by bone marrow transplantation experiments in animals (mice). Transplanted mouse donor cells can be easily identified from the recipient by a variety of biochemical (isozymes, hemoglobin) or genetic markers (Y-chromosome, translocations). Stable reconstitution of all lineages after a period of several months indicates that donor pluripotent cells have engrafted; thus PHSCs are defined on the basis of bone marrow repopulating capability. No such in vivo system exists to assay human PHSCs. In vitro systems, such as long-term culture-initiating cell (LTC-IC) or high proliferative potential-colony forming cell (HPP-CFC), assay the proliferative capacity of human cells but cannot measure long-term reconstitution (see Fig. 15.1).

It is thought that the most primitive stem cells are quiescent, with only 1-5% of cells actively cycling and contributing to hematopoiesis. These cells appear to reenter the cell cycle randomly. Individual stem cells may vary in their contribution to hematopoiesis, over time giving rise to a pattern of clonal succession. If successfully targeted for gene transfer, stem cells would ensure the continuous production of genetically-modified cells over the lifetime of the patient. Mature hematopoietic cells are inappropriate targets due to their lack of self-renewal and long-term survival. Lymphocytes are an important exception due to their long life span. Human PHSCs are found in the bone marrow, peripheral blood and umbilical cord blood.

B) ISOLATION

Current techniques to purify PHSCs employ cell surface markers and physical cell separation methods. Immunoaffinity techniques using monoclonal antibodies directed to cell surface antigens such as CD34, c-kit or HLA antigens are commonly used in positive selection. Lineage markers are used for negative selection (removal) of differentiated cells. PHSCs actively expel certain drugs via a cell membrane P-glycoprotein pump. This pump function can be measured by the expulsion of fluorescent dyes such as rhodamine 123.

15

Bone Marrow Transplantation, edited by Richard K. Burt, H. Joachim Deeg, Scott Thomas Lothian, George W. Santos. © 1996 R.G. Landes Company.

Table 15.1. Gene therapy

Inherited Genetic Disorders	Target Cell

Severe combined immunodeficiency (SCID)
   adenosine deaminase deficiency
   X-linked SCID
	Peripheral blood lymphocyte

Hemogloblinopathies
   β-thalassemia
   sickle cell disease
	Hematopoietic stem cell

Lysosomal storage disorders Gaucher disease	Hematopoietic stem cell

Leukocyte disorders
   chronic granulomatous disease
   leukocyte adhesion deficiency
	Hematopoietic stem cell

Bone marrow failure Fanconi's anemia	Hematopoietic stem cell
Acquired Genetic Disorders	
Acquired immunodeficiency syndrome	Peripheral blood lymphocyte
Neoplasia	Hematopoietic stem cell
Bone Marrow Transplantation	
Modulation of graft-vs-host disease	Donor lymphocytes

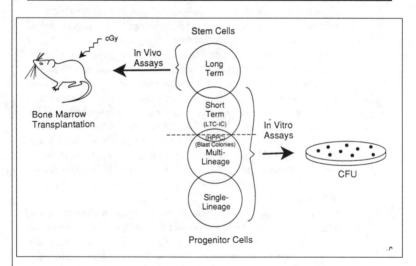

Fig. 15.1. Assays used to detect populations of hematopoietic progenitors and stem cells. In vitro assays are suitable, for cells lack self-renewal and have limited repopulating growth potential. Cells are grown in a suspension culture or on a stromal feeder layer and scored for the number and type of colonies they produce in semisolid media. Progenitor cell growth is restricted to 2-3 weeks in culture and generate single or mixed lineage colonies. More immature progenitors or short-term stem cells are cultured for periods of 2-3 months and produce multi-lineage colonies. In vivo assays define long-term repopulating stem cells (PHSCs) by their ability to reconstitute myeloablated animals.

When analyzed for fluorescence intensity, rhodamine dull cells (decreased fluorescence intensity) are enriched for repopulating and proliferative cells carrying the membrane pump. Cell separation methods based on cell size and density have been used successfully for the isolation of murine PHSCs.

C) TRANSDUCTION

Genetic material (RNA or DNA) may be transferred by a variety of viral or nonviral vectors. Transfer of genetic material into a target cell is termed transduction. Due to hematopoietic stem cell clonal succession, gene transduction of a significant proportion of stem cells may be necessary. Thus, efficient stem cell transduction is required if gene therapy is to be clinically useful. The transfer efficiency may be dependent on the type of vector used.

III) METHODS OF GENE TRANSFER (TABLE 15.2)

Genetic material may integrate into the chromosomal DNA of an appropriate target cell resulting in the stable passage of the integrated gene to daughter cells with each mitosis. Alternatively, genetic material may remain as an episome or nonintegrated extrachromosomal DNA that is not replicated with each cell division. Episomal forms are transient requiring repeated transduction to sustain the gene in the target cell population. The most clinically important aspect of gene therapy is the efficiency of gene integration and expression. Virus-mediated transfer of genetic material is currently the most efficient method for gene transfer.

15

A) NONVIRAL METHODS OF GENE TRANSFER

Nonviral methods of gene transfer rely on the adsorption and entry of DNA into cells by physical interaction such as calcium-phosphate co-precipitation, liposomal encapsulation, electroporation or direct injection into target cells. These methods are limited due to their toxicity, poor transduction efficiency and lack of stable integrated DNA. Gene transfer has also been accomplished by using ligands linked to DNA which bind to specific cell surface receptors.

1) Advantages
-avoids use of virus

2) Disadvantages
-nonintegrating
-gene expression transient
-toxic to primary cells

B) VIRAL METHODS FOR GENE TRANSFER

The use of recombinant viral vectors to target hematopoietic cells capitalizes on the inherent efficiency of viruses to infect cells and deposit their genetic material. A variety of viral vectors include recombinant murine retroviruses, recombinant adenoviruses, vaccinia virus, herpes viruses and recombinant adeno-associated virus (rAAV). Currently, most clinical applications of gene transfer for investigation and treatment of hematologic disease are based on the use of retroviral vectors.

1) Retrovirus

Retroviruses encapsidate a single-strand RNA genome which is converted into double-stranded DNA in the infected cell; the DNA then inserts (integrates) randomly into the host cell chromosomal genome and is termed a provirus. The provirus functions as a template for the generation of new RNA molecules which code for the necessary structural proteins and enzymes required for packaging and assembly of new virus. Progeny virus are then released from the cell, wrapped in an envelope composed of portions of the cell membrane.

15

Recombinant retroviruses used for therapeutic gene transfer purposes are derived from murine leukemia virus (MuLV) whose genome is shown in Figure 15.2. The coding sequences of the genome are flanked by long terminal repeats (LTR) that contain transcriptional control elements, poly-adenylation signals and sequences required for replication and integration. The encoded genes represent structural proteins (gag), enzymes such as reverse transcriptase (pol) and envelope proteins (env). Incorporation of retroviral RNA into capsids is facilitated by the packaging signal (ψ, psi).

The ability to use retroviruses as vectors has been enhanced by the development of packaging cell lines engineered to produce recombinant retroviruses. The cell lines contain the coding sequences for the viral structural proteins. Bacterial plasmids are constructed that link the foreign gene to be transferred to the elements necessary for viral RNA encapsulation, replication and integration. Transfection of the packaging cell line with recombinant plasmid results in the production of recombinant virus. To prevent the production of replication-competent virus, the helper genome within the packaging cell line and the viral genome within the plasmid are designed to minimize overlapping sequences, thus making the chances of making replication-competent virus unlikely (Fig. 15.3). This strategy for retrovirus production is necessary to prevent the production of replication-competent retroviruses (wild-type or helper virus) which are known to cause tumors. Clinical-grade virus must be certified to be free of replication-competent virus. Newer packaging cell lines have been developed to eliminate helper virus generation.

Retroviruses may be either amphotropic (capable of infecting cells from a wide range of species) or ecotropic (only capable of infecting rodent cells). This cell restriction of retrovirus infection is due to the envelope protein that surrounds the retrovirus. Therefore, the producing cell retrovirus particles may be ecotropic or amphotropic, depending on the envelope protein provided by the packaging cell line.

Induction of stem cell cycling is necessary for retroviral transduction ex vivo. Cell cycling is induced by recombinant growth factors such as

Table 15.2. Methods of gene transfer

	Vector			
	Retrovirus	**AAV**	**Adenovirus**	**Nonviral**
Genetic material inserted	RNA	DNA	DNA	DNA
Integration	Yes	Yes	No	No
Replication-defective	Yes	Yes	Yes	Yes
Cell cycle dependent	Yes	No	No	No
Infect hematopoietic progenitor cells	Yes	Yes	No	?
Virus purification	No	Yes	Yes	–
Clinical trials	Yes	No	Yes	No

interleukin-3 (IL-3), interleukin-6 (IL-6), granulocyte macrophage-colony stimulating factor (GM-CSF) and stem cell factor (SCF). PHSCs are incubated in media containing growth factors and admixed with a media suspension containing the transducing recombinant virions. Bone marrow stromal cells that produce multiple hematopoietic growth factors can also serve as a substitute for defined recombinant factors. In this case, PHSCs are overlaid on a bed of autologous stromal cells and incubated with recombinant retrovirus. The cells are collected, washed and injected into an appropriate recipient.

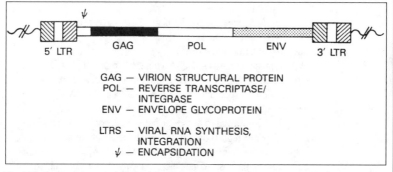

Fig. 15.2. Integrated proviral form of the Moloney murine leukemia virus (MoMuLV).

15

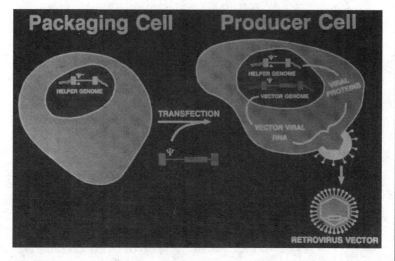

Fig. 15.3. Recombinant retrovirus generation from a producer cell line. The strategy used to generate recombinant virus containing a gene of interest is shown. Plasmid DNA containing the gene of interest and the retroviral psi (ψ)region is transfected into a packaging cell containing a retroviral wild-type helper genome lacking psi (-ψ). Integration of the vector genome allows for stable generation of recombinant virions. Clones are screened for the highest production of recombinant virus particles and used for gene transfer studies.

a) *Advantages of Retrovirus*

    i) Production of high-titer replication defective virus (free of wild-type virus) are generated from genetically engineered "packaging cell lines".

    ii) Retroviral gene transfer of hematopoietic cells has been demonstrated in human clinical trials.

b) *Disadvantages of Retroviruses*

    i) Mitosis or cell-cycling of the target cell is necessary for retroviral integration.

    ii) Retrovirus preparations are difficult to purify.

    iii) Contaminating wild-type virus is capable of producing lymphoma in immunosuppressed animals

    iv) Low gene transfer efficiency (1-5%) of target hematopoietic cells has occurred in human gene marking trials.

2) Adeno-associated Virus (AAV)

AAV is among the smallest and structurally simplest of DNA viruses. AAV propagation depends upon coinfection with a second virus for a productive infection to occur. Infection of permissive cells with AAV in the presence of helper virus allows productive AAV generation and causes host cell lysis. Adenovirus and herpes helper virus encode ancillary proteins required for AAV replication and virion formation. Without an appropriate helper virus, productive infection does not occur and the AAV genome integrates into the host cell genome, producing a latent state. Subsequent helper virus infection of latent cells allows for rescue of the AAV genome and a productive infection to ensue. AAV infects a broad range of host cell types but has never been associated with human disease. Unique among viruses, AAV integrates specifically at a locus on chromosome 19q.

The AAV genome is a single-stranded DNA molecule consisting of 4700 bases. The coding sequences are flanked by two inverted terminal repeat sequences (ITR) which are themselves sufficient for viral replication, packaging and integration. The genome is organized with one intron separating replication *(rep)* and capsid *(cap)* genes.

Recombinant AAV (rAAV) particles are generated by cotransfection of a recombinant plasmid and helper plasmid. The recombinant plasmid contains the gene of interest placed between two ITRs, and the helper plasmid contains the remainder of the AAV genome. A simultaneous infection of the producer cell with adenovirus is required. Due to the lack of sequence homology between the helper and vector plasmid, recombination and generation of wild type AAV is precluded. Producer cells are subsequently lysed to release the virus from the nucleus, heated to destroy contaminating adenovirus and concentrated using isopycnic density centrifugation (Fig. 15.4). Initial crude cell lysate titers range from 10^3-10^5 infectious particles per ml but can be concentrated to 10^7-10^8 particles per ml.

Unlike most retroviruses which integrate in an apparent random fashion, wild-type AAV integrants are restricted to specific chromosomal regions. However, rAAV integrates randomly in the genome. This implies that the AAV ITRs allow for integration but that other viral elements are required for site-specific integration. Viral targeting to an integration locus might eliminate the possibility of gene disruption or proto-oncogene activation. Recent data indicate that rAAV infect nondividing cells and integrate when

15

cell cycling is initiated. Thus strategies to induce PHSCs to cycle for viral transduction to occur may not be necessary for rAAV.

　　　a) *Advantages of AAV*
　　　　　-integrating viral vector
　　　　　-wild-type virus integrates site-specifically (chromosome 19q)
　　　　　-stable virus, purified by isopycnic density centrifugation
　　　　　-wild-type AAV nonpathogenic
　　　　　- infect both dividing and nondividing cells
　　　b) *Disadvantages of AAV*
　　　　　-recombinant virus lack site-specific integration
　　　　　-no "packaging cell line"

3) Other Viruses (Adenovirus, Herpes Virus)

　　Adenovirus infects a variety of cell types (liver, respiratory, epithelial, muscle) but exists in an episomal form. Herpes simplex virus maintains a specific tropism for central nervous tissue and also exists as an episome. Both viral genomes are quite large, 30 Kb (adenovirus)-150 Kb (herpes virus), and express several viral-specific gene products in addition to the transferred gene of interest. In vivo animal and human clinical studies indicate that the host immune system may either prevent viral infection or

15

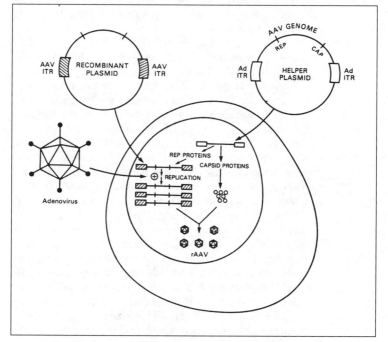

Fig.15.4. Recombinant adeno-associated virus production. A cell line permissive for AAV generation is cotransfected with helper and recombinant plasmids and infected with adenovirus. The helper plasmid contains coding sequences for viral capsid proteins (cap) and replication proteins (rep) necessary for recombinant (rAAV) AAV replication and packaging. Recombinant virion formation occurs in the cell nucleus, and both rAAV and adenovirus are released following cell lysis. Adenovirus is removed by either heat treatment or density gradient centrifugation.

initiate an inflammatory reaction because of virus-specific protein expression. These processes may be harmful to the patient and limit the use of these vectors.

IV) CLINICAL APPLICATIONS

A) GENE REPLACEMENT

1) Adenosine Deaminase Deficiency

The first gene transfer trial with clinical intent involved patients with adenosine deaminase deficiency (ADA). Deficiency of this enzyme leads to accumulation of toxic nucleoside metabolites which inhibit normal lymphocyte development. In this trial children were selected who lacked a suitable sibling bone marrow donor and did not respond to PEG-ADA enzyme administration. Patients' peripheral blood T lymphocytes were transduced with a retrovirus carrying the normal ADA gene, multiple times over a period of several months. In two patients, marked improvement of lymphocyte function and number were documented.

The introduction of the ADA gene into bone marrow PHSCs may eliminate the requirement for repeated correction of lymphocytes. In one study, three patients received retroviral-transduced PHSCs without prior myeloablation, however the gene was not detected after 6 months. Thus, either committed cells were initially transduced or prior marrow conditioning may be important for the transduced cells to engraft and proliferate.

2) Fanconi's Anemia

Fanconi's anemia (FA) is an inherited bone marrow failure disorder characterized by pancytopenia, physical anomalies and predisposition to malignancy. FA cells exhibit hypersensitivity to DNA-damaging agents leading to chromosomal instability and cell death. The DNA instability suggests that the defect in FA resides in the inability to repair damaged DNA. Four genetic loci have been defined to be defective in this disorder. One gene, FA complementing gene C (*facc*) has been cloned.

The cDNA of the *facc* gene has been engineered into both retroviral and rAAV vectors. Phenotypic correction of mutant FA cell lines was demonstrated by normalized cell growth in the presence of DNA-damaging agents. Both vectors were used to transduce hematopoietic progenitors from patients carrying mutant *facc* alleles. Gene transduction resulted in enhanced hematopoietic colony formation in the absence and presence of DNA-damaging agents. Transduction and expression of *facc* in murine hematopoietic cells using the retroviral vector has been demonstrated. Long-term reconstitution assays using rAAV transduction of human PHSCs engrafted in a SCID mouse xenograft model resulted in positively marked cells at 3-4 months following transplantation. Gene corrected PHSCs may have a selective growth advantage when transplanted into patients, mitigating the requirement for highly efficient gene transduction. Clinical protocols designed to test each vector are now being developed.

3) Hemoglobinopathies

The introduction of a normal β-globin gene into PHSCs of a patient with homozygous β-thalassemia should be useful in correcting defective globin production. Alternatively, increased fetal hemoglobin (HbF) production due to γ-globin synthesis ameliorates the severity of sickle cell and thalassemia. HbF levels of > 20% should reduce the clinical sequelae associated in both these disorders. Therefore, transfer and high-level production of a γ-globin gene leading to accumulation of HbF may be therapeutic.

Retroviral vectors designed to transfer the β-globin gene were successfully introduced into murine hematopoietic progenitor cells. However, the level of globin expression was too low to be of any therapeutic value. Regulatory sequences termed the locus control region (LCR) are located several kilobases upstream from the β-globin gene and induce high level globin expression. Unfortunately, globin genes linked to LCR elements produced either low titer retrovirus or proviral rearrangement. Retroviral vectors using smaller LCR elements may overcome these problems.

AAV vectors containing γ- and β- genes linked to LCR elements generate high-level globin expression in erythroid cell lines and in human hematopoietic progenitor cells. To determine potential clinical efficacy, in vivo testing in murine and nonhuman primates using rAAV-globin vectors are in progress.

4) Gaucher's Disease

Gaucher's disease is one of several lysosomal storage disorders amenable to gene therapy. Accumulation of complex carbohydrates in macrophages throughout the body due to an inherited deficiency of glucocerebrosidase results in liver/spleen enlargement, cytopenias and central nervous system involvement. Retroviral-mediated transfer of the glucocerebrosidase gene corrects the defect in patient's fibroblasts and lymphocytes. Transduction of murine and primate PHSCs has been achieved resulting in long-term expression in the macrophages of transplanted animals. Clinical gene therapy trials for this disorder have been approved.

B) GENE MARKING

Using vectors carrying a benign bacterial gene (such as the neomycin selectable gene *neor*) the efficiency of transduction can be easily determined by polymerase chain reaction (PCR) based methods. Gene marking studies have been performed using retroviral-transduced bone marrow from patients undergoing autologous bone marrow transplantation for acute myelogenous leukemia, chronic myelogenous leukemia and neuroblastoma. PCR analysis of patient cells following engraftment and marrow reconstitution determined that 1-5% of peripheral blood cells contained the transferred gene. In patients that relapsed following marrow reinfusion containing marked cells, the resurgent blast cells contained the transferred gene, thus, the remission marrow contributed to disease recurrence.

C) MODULATION OF GRAFT-VS-HOST DISEASE

Infusion of donor lymphocytes following allogeneic BMT may prevent multiple complications resulting from profound immunosuppression. However, severe graft-vs-host disease (GVHD) may result from this therapeutic intervention. To circumvent the complications of GVHD, investigators transduced donor lymphocytes with a retroviral "suicide" vector carrying the herpes simplex thymidylate kinase (*hsv-tk*) gene. Activation of the suicide gene by the antiviral agent ganciclovir destroys only cells carrying the gene. Transduced lymphocytes (carrying the *tk* gene) were enriched by immunoaffinity selection for a cotransduced foreign membrane protein. Marked donor lymphocytes were infused into patients suffering from either CML and AML in relapse or EBV-lymphoproliferative disease. Two patients with clinical signs of GVHD received multiple doses of ganciclovir resulting in the elimination of marked lymphocytes and resolution of all clinical signs of GVHD.

D) AIDS

The concept for a gene therapy approach to acquired immunodeficiency syndrome is based upon the introduction of genes that would either prevent HIV infection or destroy previously infected HIV cells. Multiple strategies (see reference #2) using vectors targeted to peripheral lymphocytes and PHSCs have already been approved for clinical trials.

V) FDA/RAC APPROVAL

NIH guidelines for research involving recombinant DNA molecules have been established for human clinical trials. As of 1995, over 100 gene therapy protocols have been approved by the Recombinant DNA Advisory Committee (RAC). Guidelines have been published in the Federal Register July 5, 1994. RAC review requires that each protocol already have IRB and biosafety approval and requires preclinical "proof of concept" data. Food and Drug Administration (FDA) approval requires data demonstrating that the vectors to be used are effective and safe. Worldwide, over 100 patients have received retrovirally modified cells without a report of any adverse side effects.

GLOSSARY OF TERMS

Amphotropic retrovirus- recombinant retrovirus capable of infecting both murine and nonmurine cells.

Ecotropic retrovirus- recombinant retrovirus capable of infecting murine cells.

Integration- the process that exogenously added genetic material is incorporated into the target cell genome. Stable integration refers to a latent proviral state of viruses.

Episome- refers to a transient state where the transferred genetic material is not incorporated into the host genome, but exists as an extrachromosomal element which is eventually lost following cell division.

Packaging cell- a murine fibroblast cell line containing integrated retroviral genome lacking the packaging function (ψ) and LTRs.

Producer cell- a packaging cell line transduced with the appropriate retroviral vector carrying LTR elements, packaging function and gene(s) of interest. Introduction of the vector allows for recombinant retroviral virus production.

Transduction - the transfer of genetic material into target cells. The transferred material is either permanently or transiently expressed due to the method of transduction.

Transfection - the uptake of foreign genetic material into a cell. Transfection is accomplished by physical methods (DNA-calcium phosphate complexes, electroporation).

Vector - vehicle designed to introduce RNA and DNA molecules into cells.

Suggested Reading

1. Nienhuis AW, Walsh CE, Liu J. Viruses as therapeutic gene transfer vectors. In: Young, NS ed. Viruses and Bone Marrow. New York: Marcel Dekker, 1993:353-414.
2. Mulligan RC. The basic science of gene therapy. Science 1993; 260:926.
3. Karlsson S. Treatment of genetic defects in hematopoietic cell function by gene transfer. Blood 1991; 78:2481.
4. Blaese RM. Development of gene therapy for immunodeficiency: Adenosine deaminase deficiency. Pediatr Res 1993; 33(suppl):S49.
5. Brenner MK, Rill Moen RC, Krance RA, Mirro JM, Anderson WF, Ihle JN. Gene marking to trace the origin of relapse after autologous bone marrow transplantation. Lancet 1993; 341:85.

16

Medications

Dominic A. Solimando

I) ANTIDIARRHEAL

Diarrhea frequently accompanies chemotherapy as the rapidly dividing cells of the gastric mucosa are damaged or killed. In these patients it is important to consider other causes such as medications, infection or GVHD.

A) ALUMINUM HYDROXIDE (AMPHOGEL®)

Class—antacid with constipating effect.
Dose—960-1920 mg (15-30 ml) PO q 2-4 hours.
Side effects—hypophosphatemia, constipation.

B) ATROPINE (ANTROCOL® ELIXIR (COMBINATION WITH PHENOBARBITAL); SAL-TROPINE® TABLETS)

Class—antimuscarinic.
Mechanism of action—inhibits GI motility and propulsion by acting on intestinal smooth muscle.
Dose—0.4-0.6 mg PO, IM, IV or SQ q 4-6 hours.
Side effects—xerostoma, blurred vision, mydriasis, urinary retention, tachycardia, lightheadedness.
Contraindications—GI infection, intestinal obstruction, paralytic ileus, toxic megacolon, glaucoma, obstructive uropathy, myasthenia gravis.

C) CHOLESTYRAMINE RESIN (QUESTRAN®)

Class—bile salt sequestrant.
Dose—4 grams PO TID (doses up to 24 grams/day in 3 divided doses have been used).
Indications—used successfully for pseudomembranous colitis diarrhea, relief of pruritus due to cholestasis.

D) COLESTIPOL RESIN (COLESTID®)

Class—bile salt sequestrant.
Dose—5 grams PO TID (doses up to 30 grams/day in 3 divided doses have been used).
Indications—used successfully for pseudomembranous colitis diarrhea, relief of pruritus due to cholestasis.

E) KAOLIN AND PECTIN (KAOPECTATE®)

Class—kaolin is aluminum silicate. Pectin is an extract from the rind of citrus fruits.
Mechanism of action—adsorbents of bacteria and toxins.
Dose—usual adult dose is 45-90 ml of the concentrated, or 60-120 ml of the regular strength suspension orally after each loose bowel movement.
Indications—should be used in the symptomatic treatment as a temporary measure until the cause of the diarrhea is known.
Side effects—may impair oral absorption of other drugs (e.g. digoxin) when given concomitantly. Therefore, kaolin/pectin should be given several hours after the dose of digoxin to minimize this potential interaction.

16

Bone Marrow Transplantation, edited by Richard K. Burt, H. Joachim Deeg, Scott Thomas Lothian, George W. Santos. © 1996 R.G. Landes Company.

F) **DIPHENOXYLATE HYDROCHLORIDE/ATROPINE SULFATE (LOMOTIL®)**

Class—opiate agonist.

Mechanism of action—inhibits GI motility and propulsion by acting on intestinal smooth muscle. The atropine content is subtherapeutic and present only to inhibit abuse of the drug.

Dose—Adults: diphenoxylate 5 mg (2 tablets) PO QID.

Side effects—nausea, ileus, toxic megacolon, sedation, urinary retention, xerostomia, dry skin, tachycardia, flushing.

Contraindications—use may not be indicated with infectious diarrhea or pseudomembranous colitis. May precipitate hepatic coma in patients with cirrhosis or advanced hepato-renal disease, and should be used with extreme caution in those situations.

G) **LOPERAMIDE (IMODIUM®)**

Class—piperidine derivative.

Mechanism of action—slows intestinal motility through an effect on nerve endings, decreasing transit time of intestinal contents; thus decreasing loss of fluid and electrolytes.

Dose—Adults: 4 mg (2 capsules) orally, followed by 2 mg after each loose stool; not to exceed 16 mg/day.

Contraindications—should not be used with infectious diarrhea.

Side effects—nausea, dry mouth, fatigue, dizziness.

H) **OCTREOTIDE (SANDOSTATIN®)**

Class—synthetic polypeptide related to somatostatin.

Mechanism of action—acts similarly to somatostatin, inhibiting gastric acid, GI hormone secretion, GI motility, secretin, motilin, VIP and pancreatic polypeptide among others.

Dose—Adults: initiate at 50-100 µg subcutaneously or IV 2-3 times a day, gradually increasing to 600 µg 2-3 times a day. The total daily dose may be given by continuous infusion when added to the TPN bag.

Indications—controlling diarrhea due to GVHD.

Side effects—may inhibit motility of the gallbladder and biliary tree, causing cholestasis and jaundice. Other effects are nausea, cramping, diarrhea, alteration in glucose metabolism and absorption, flushing and edema.

I) **PSYLLIUM HYDROPHILIC MUCILLOID (METAMUCIL®)**

Class—methylcellulose from the seeds of Plantago plant (bulk forming laxative).

Dose—Adults: 1-2 Tsp (3-4 g/Tsp) PO 1-3 times daily.

Indications—chronic watery diarrhea.

Contraindications—phenylketonuria (contains aspartame which is metabolized to phenylalanine).

J) **TINCTURE OF OPIUM OR PAREGORIC**

Class—opiate narcotic.

Mechanism of action—inhibits GI motility and propulsion, and decreases digestive secretions due to morphine content.

Dose—Opium tincture—0.6 ml PO QID.

Paregoric—5-10 ml PO QID.

NOTE: Opium tincture contains ~25 times MORE morphine than paregoric.

Side effects—nausea, prolonged use may produce opiate dependence. Use with caution in patients with hepatic disease.

16

II) ANTIEMETICS

The symptoms of acute nausea and vomiting usually occur within 24 hours of chemotherapy administration. These symptoms may lead to dehydration, aspiration and electrolyte imbalances. Delayed nausea and vomiting usually occur 48-72 hours after chemotherapy is administered, and may last for up to 7 days. Anticipatory nausea and vomiting develop prior to the actual administration of the chemotherapy, and may occur in 30-60% of patients. Antiemetics are much more effective if given prophylactically on a fixed schedule. Prevention of nausea or vomiting is more easily accomplished than stopping the conditions once they occur. Antiemetics should continue on a fixed schedule throughout the chemotherapy administration instead of on an "as needed" basis. Antiemetics can be switched to an "as needed" schedule after the treatment has ended. Chronic GVHD and viral infections of the upper GI tract (e.g. CMV) also cause nausea and vomiting.

A) DEXAMETHASONE (DECADRON®)

Class—fluorinated corticosteroid.

Mechanism of action—unclear. Dexamethasone combinations are superior to the dopamine (e.g. phenothiazines, butyrophenones, metoclopramide) or serotonin (e.g. granisetron, ondansetron) antagonists alone. Also useful for treating delayed nausea or vomiting.

Dose—10-20 mg PO or IV over 3-5 minutes.

Side effects—insomnia, hyperglycemia, euphoria and perineal irritation (with rapid IV infusion).

B) DIPHENHYDRAMINE (BENADRYL®)

Class—antihistamine.

Mechanism of action—H_1 receptor antagonist. Useful in preventing/treating extrapyramidal reactions from dopamine antagonists.

Dose—Adults: 25-50 mg IVP or PO q4-6 hours; do not exceed 300 mg/24 hours.

Side effects—CNS depression sedation, lassitude, confusion, restlessness, delirium, tremor, nervousness; anticholinergic effects-dry mouth, urinary retention, blurred vision.

C) DIMENHYDRINATE (DRAMAMINE®)

Class—antihistamine.

Mechanism of action—H_1 receptor antagonist.

Dose—Adults: 50-100 mg IVP, IM or PO q4 hours.

Indications—prevention of nausea and vomiting due to movement or vestibular or inner ear damage from high-dose chemoradiation therapy or medications.

Side effects—CNS depression sedation, lassitude, confusion, restlessness, delirium, tremor, nervousness; anticholinergic effects-dry mouth, urinary retention, blurred vision, tinnitus.

D) DRONABINOL (MARINOL®, δ-9-TETRAHYDROCANNABINOL)

Class—cannabinoids.

Mechanism of action—acts on the cortex, inhibiting the vomiting center.

Dose—Adults: 5-10 mg/m² PO q4 hours.

Side effects—euphoria, increased appetite, hallucinations, drowsiness.

E) DROPERIDOL (INAPSINE®)

Class—butyrophenone.

Mechanism of action—dopaminergic antagonist.

Dose—Adults: 2.5 mg IM or IVP q2-3 hours. High-dose, continuous infusion regimens (5-7.5 mg rapid IVP followed by 0.5-2.5 mg/hour as a continuous infusion); may also be used in refractory patients.

Side effects—transient hypotension, sedation, extrapyramidal effects (may be prevented or treated with diphenhydramine or benztropine (Cogentin®)).

Table 16.1. Emetic potential of chemotherapy agents

Incidence	Agent	Dose (mg/m^2)
Very high (> 90%)	Carmustine	
	Cisplatin	> 20/day
	Cyclophosphamide	2000
	Cytarabine	> 1000
	Dacarbazine	
	Mechlorethamine	
	Melphalan (IV)	
	Streptozocin	
High (60-90%)	Actinomycin	
	Carboplatin	
	Cyclophosphamide	1000-2000
	Daunorubicin	60-75
	Doxorubicin	60-75
	Ifosfamide	
	Lomustine	
	Procarbazine	
Moderate (30-60%)	Asparaginase	
	Azacytidine	
	Busulfan	
	Cisplatin	20/day
	Cyclophosphamide	< 1000
	Cytarabine	< 1000
	Daunorubicin	< 60
	Doxorubicin	< 60
	Etoposide	
	Hexamethylmelamine	
	Methotrexate	> 1000
	Mitoxantrone	
	Mitomycin	
	Teniposide	
Low (10-30%)	Bleomycin	
	Fluorouracil	
	Hydroxyurea	
	Melphalan (oral)	
	Mercaptopurine	
	Methotrexate	< 1000
	Steroids	
	Tamoxifen	
	Thioguanine	
	Thiotepa	
	Vinblastine	
Very Low (< 10%)	Chlorambucil	
	Vincristine	
	Vinorelbine	

16

F) GRANISETRON (KYTRIL®)
Class—serotonin antagonist.
Mechanism of action—selectively binds to type 3 serotonin receptors, blocking emetic impulses at central and peripheral sites.
Dose—10 µg/kg (or 1 mg) IV or 2 mg PO q 24 hours, or 1 mg PO q12 hours
Side effects—headache, constipation, diarrhea, elevated SGOT/SGPT.

G) HALOPERIDOL (HALDOL®)
Class—butyrophenone.
Mechanism of action—dopaminergic antagonist.
Dose—Adults: 1-5 mg PO or IVP q4-6 hours.
Side effects—sedation, extrapyramidal effects (may be prevented or treated with diphenhydramine or benztropine (Cogentin®)).

H) HYDROXYZINE (VISTARIL®, ATARAX®)
Class—antihistamine.
Dose—25-100 mg PO or IM QID.
Indications—anxiolytic effect; useful for anticipatory nausea.
Side effects—sedation and anticholinergic effects.

I) LORAZEPAM (ATIVAN®)
Class—benzodiazepine.
Dose—0.5-2 mg IVP or PO q6 hours.
Indications—anxiolytic effect; useful for anticipatory nausea.
Side effects—sedation, confusion, anterograde amnesia. May cause respiratory arrest in patients with compromised pulmonary function.

J) METOCLOPRAMIDE (REGLAN®)
Class—benzamide.
Mechanism of action—dopaminergic/serotonin antagonist.
Dose—1-2 mg/kg IV; given 30 minutes prior to chemotherapy, then q2 hours for 2 doses and q3 hours for 3 doses. For delayed nausea 10-40 mg PO or IV q6hr.

K) ONDANSETRON (ZOFRAN®)
Class—serotonin antagonist.
Mechanism of action—selectively binds to type 3 serotonin receptors; blocking emetic impulses at central and peripheral sites.
Dose—0.45 mg/kg (maximum dose 32 mg) IV q24 hours; or 0.15 mg/kg IV 30 minutes before and 4 and 8 hours after chemotherapy.
Side effects—headache, blurred vision, diplopia, constipation, diarrhea, elevated SGOT/SGPT.

L) PROCHLORPERAZINE (COMPAZINE®)
Class—phenothiazine.
Mechanism of action—dopaminergic antagonist.
Dose—Adults: 10 mg PO, IM, IVPB or 25 mg rectally q4-6 hours.
     Children: 2.5-5 mg PO, IM, IVPB or 2.5 mg rectally q4-6 hours.
Side effects—sedation, extrapyramidal effects (may be prevented or treated with diphenhydramine or benztropine (Cogentin®)).

M) PROMETHAZINE, (PHENERGAN®)
Class—phenothiazine/antihistamine.
Dose—12.5-25 mg PO, IVPB, IM or rectally q4-6 hours.
Side effects—dry mouth, dry eyes, blurred vision, confusion, jaundice, extrapyramidal effects (may be prevented or treated with diphenhydramine or benztropine (Cogentin®)).

16

N) **SCOPOLAMINE, (TRANSDERM-SCOP®)**

Class—antimuscarinic.

Indications—prevention of nausea or vomiting due to movement or vestibular or inner ear damage from high-dose chemoradiation therapy or medications. Also useful for prolonged low-grade nausea, without vomiting, occasionally seen following high-dose chemotherapy.

Dose—Adults: 0.3-0.65 mg IM, IV, SQ q6-8 hours. Transdermal patch 0.5 mg q72 hours.

O) **THIETHYLPERAZINE, (TORECAN®)**

Class—phenothiazine.

Mechanism of action—dopaminergic antagonist.

Dose—Adults: 10 mg PO or IM q4-6 hours.

Side effects—sedation, extrapyramidal effects (may be prevented or treated with diphenhydramine or benztropine (Cogentin®)).

III) **ANTIFUNGAL**

A) **AMPHOTERICIN B**

Class—polyene antibiotic.

Mechanism of action—binds to sterols in the cell membrane, causing the membrane to lose its ability to function as a barrier. Amphotericin is not active against organisms that do not contain sterols in their cell membrane.

Dose—0.4-1.5 mg/kg/day by IV infusion. Guidelines for test doses and prophylaxis against acute drug reactions and renal failure are empiric and varied. One approach is a 1 mg test dose, infused over 20 minutes, before the first 1 or 2 doses. Many institutions routinely use diphenhydramine, meperidine, hydrocortisone or/and acetaminophen to reduce febrile reactions, chills and rigors during the drug infusion.

Side effects—fever, chills, rigors, nausea and vomiting, and nephrotoxicity (renal tubular acidosis with excessive bicarbonate excretion, hypokalemia, hypocalcemia and hypomagnesemia); and hypersensitivity reactions (bronchospasm, tachypnea, dyspnea) are common. Concomitant use of other nephrotoxic agents (e.g. aminoglycoside antibiotics, furosemide, cisplatin) may increase the risk of nephrotoxicity. The initial dose is sometimes given without premedication to assess the patient's tolerance of the drug. Subsequent doses are accompanied by a premedication regimen targeted to the symptoms and severity, seen with the initial dose.

B) **AMPHOTERICIN LIPID FORMULATION, (AMPHOTERICIN B COLLOIDAL DISPERSION (ABCD), AMPHOTERICIN B LIPID COMPLEX (ABCL), LIPOSOMAL AMPHOTERICIN, AMPHOCIL®)**

Class—lipid formulations of amphotericin. Are at least as efficacious as, but significantly less toxic (especially nephrotoxicity), than amphotericin.

Dose—4-6 mg/kg/day IV.

C) **CLOTRIMAZOLE (MYCELEX® TROCHE; LOTRIMIN® CREAM OR LOTION)**

Class—imidazole.

Mechanism off action—binds to phospholipids in the cell membrane, altering membrane permeability.

Dose—For oral candiasis:

    1) Prophylaxis
    one 10 mg troche dissolved orally TID.
    2) Treatment
    one 10 mg troche dissolved orally 5 times a day (q4 hours while awake).

16

For cutaneous candiasis: topical application of 1-2% cream or lotion BID.
Indications—Prophylaxis against, and treatment of, oral and cutaneous candidiasis. Not useful in treatment of systemic infections.

D) FLUCONAZOLE (DIFLUCAN®)

Class—triazole.

Mechanism of action—alters cell membrane with increased permeability.

Dose—200-400 mg PO or IVPB daily. The drug is >90% bioavailable, making
oral and IV dosing essentially identical. A 50% dose reduction is recommended
for CrCl of 30-60 ml/min/70kg; a 75% reduction (give 25% of the dose) is recommended for CrCl <30 ml/min/70 kg.

Side effects—nausea, vomiting, headache, skin rash, diarrhea, abdominal pain.
May cause an increase in cyclosporine levels.

E) ITRACONAZOLE (SPORANOX®)

Class—triazole.

Mechanism of action—alters cell membrane with increased permeability.

Dose—200 mg PO BID. Only the oral formulation is currently available.

Indications—prophylaxis against most species of *Candida* and, in vitro against
Aspergillus sp. Is considered inadequate in vivo therapy for *Aspergillus* in neutropenic patients.

Side effects—concomitant use of histamine receptor 2 blockers may decrease
bioavailability up to 20%. The drug may also cause elevations in liver function
tests. Elimination is by hepatic microsomal enzyme metabolism. A variety of
drug interactions are reported with itraconazole, including: cardiovascular
events (ventricular tachycardia, torsade de pointes and death) with terfenidine;
increased levels and/or effect of cyclosporine, phenytoin, sulfonylureas and
warfarin; and decreased levels and/or effect of H2 antagonists, isoniazid,
omeprazole and rifampin.

F) KETOCONAZOLE (NIZORAL®)

Class—imidazole.

Mechanism of action—alters cell membrane with increased permeability.

Dose—200 mg PO daily.

Side effects—Absorption may be incomplete or impaired due to decreased acid
secretion. Gynecomastia may occur due to displacement of testosterone from
binding proteins and suppression of gonadal testosterone synthesis. May increase cyclosporine levels.

G) MICONAZOLE

Class—imidazole.

Mechanism of action—alters cell membranes.

Dose—Candidiasis: 600-1800 mg IV daily.
       Coccidiomycosis: 1.8-3.6 g IV daily.
       Cryptococcosis: 1.2-2.4 g IV daily.

H) NYSTATIN

Class—polyene.

Mechanism of action—binds to sterols in cell wall.

Dose—500,000-1,000,000 units as an oral suspension—swish and swallow q4-6
hours.

Indications—management of oral thrush; not useful for treatment of systemic
infections.

Side effects—nausea, vomiting.

IV) ANTIHISTAMINES (H2 ANTAGONISTS)

Patients undergoing hematopoietic stem cell transplant are treated with high-dose chemoradiation therapy; in the case of allogeneic transplants, there is the additional risk of GVHD. Both situations place the patient at high risk for disruption of GI mucosal integrity, requiring therapy with H_2 blockers.

Drug		Dose
Cimetidine (Tagamet®)	Maintenance:	400 mg PO QHS 800 mg PO QHS
	Treatment of active ulcers:	300 mg IV q6-8h 900-1200 mg IV QD as a continuous infusion
Famotidine (Pepcid®)	Maintenance:	20 mg PO QHS 40 mg PO QHS
	Treatment of active ulcers:	20 mg IV q12h 40 mg IV QD as a continuous infusion
Ranitidine (Zantac®)	Maintenance:	150 mg PO QHS 300 mg PO QHS
	Treatment of active ulcers:	50 mg IV q8h 150-200 mg IV QD as a continuous infusion
Omeprazole (Prilosec®)*		20-40 mg PO QD

* A proton pump inhibitor, not an H_2 antagonist per se.

V) ANTIMICROBIALS

Patients with fever and neutropenia should receive parenteral, not oral, therapy at maximum tolerated doses. Appropriate combination antibiotic regimens for fever and neutropenia are discussed in the chapter on infections.

Drug	Adult Dose (Normal renal and hepatic function)	Pediatric Dose (> 6 months old; normal renal and hepatic function)
Aminoclycosides		
Amikacin	7.5 mg/kg IV q8h	15 mg/kg/d IV divided q8-12h
Gentamycin	1-2.5 mg/kg IV q8h	2-2.5 mg/kg IV q8h
Kanamycin	7.5 mg/kg IV q12h	15 mg/kg/d IV divided q8h
Neomycin	12.5 mg/kg PO BID	50-100 mg/kg/d PO divided q6-8h
Tobramycin	2-2.5 mg/kg IV q8h	2-2.5 mg/kg IV q8h
Cephalosporins		
Cefaclor	250-500 mg PO q8h	20-40 mg/kg/d PO divided TID
Cefadroxil	500-1000 mg PO q12h	15 mg/kg PO q12h
Cefamandole	500-2000 mg IV q4-8h	100-150 mg/kg/d IV divided q4-6h
Cefazolin	1000-2000 mg IV q8h	150-150 mg/kg/d IV dvd q8h
Cefixime	200 mg PO q12h 400 mg PO QD	4 mg/kg PO q12h 8 mg/kg PO QD

16

16

Cefonicid	1000 mg IV QD	–
Cefoperazone	1-6 g IV q12h	50-75 mg/kg IV q12h
Cefotaxime	1-2 g IV q6-8h	100-150 mg/kg/d IV divided q6-8h
Cefoxitin	1-2 g IV q6-8h	100-160 mg/kg/d IV divided q6-8h
Cefprozil	250-500 mg PO q12h	7.5-15 mg/kg PO q12h
Ceftazidime	1-2 g IV q8-12h	30-50 mg/kg IV q8h
Ceftizoxime	1-4 g IV q8h	150-200 mg/kg/d IV divided q8h
Ceftriaxone	1-2 g IV q12-24h	25-50 mg/kg IV q12h
Cefuroxime	750-1500 mg IV q8h	35-50 mg/kg IV q8h
	250-500 mg PO q12h	10 mg/kg PO q12h
Cephalexin	250-1000 mg PO q6h	8-15 mg/kg PO q6h
Cephalothin	500-2000 mg IV q4-6h	80-150 mg/kg/d IV divided q4-6h
Cephapirin	500-1000 mg IV q6h	10-20 mg/kg IV q6h
Cephradine	500-1000 mg IV q6h	12-25 mg/kg IV q6h
	250-500 mg PO q6-12h	6-12 mg/kg PO q6h

Penicillins

Amoxicillin	250-500 mg PO q8h	8-12 mg/kg PO q8h
Amoxicillin/	250-500 mg PO q8h	8-12 mg/kg PO q8h
clavulanate	(dose by amoxicillin)	(dose by amoxicillin)
Ampicillin	500-3000 mg IV q4-6h	100-400 mg/kg/d IV divided q4-6h
	250-1000 mg PO q6h	6-25 mg/kg PO q6h
Ampicillin/sulbactam	1.5-3 g IV q6h	15-30 mg/kg IV q6h
Carbenicillin	382-764 mg PO q6h	7.5-12 mg/kg PO q6h
Dicloxacillin	125-500 mg PO q6h	3-6 mg/kg PO q6h
Methicillin	1-2 g IV q4-6h	150-400 mg/kg/d IV divided q4-6h
Mezlocillin	3-4 g IV q4-6h	200-300 mg/kg/d IV divided q4-6h
Nafcillin	500-2000 mg IV q4-6h	50-200 mg/kg/d IV divided q4-6h
Oxacillin	250-2000 mg IV q4-6h	35-50 mg/kg IV q6h
Penicillin G	350,000-6,000,000 units	15,000-40,000 units IV q4h
	IV q4h	
Penicillin V	250-500 mg PO q6h	25-50 mg/kg/d PO divided q6-8h
Piperacillin	3-4 g IV q4-6h	200-300 mg/kg/d IV divided q4-6h
Piperacillin/ tazobactam	3.375-4.5 g IV q4-6h	–
Ticarcillin	1-4 g IV q4-6h	200-300 mg/kg/d IV divided q4-6h
Ticarcillin/	3.1 g IV q4-6h	200-300 mg/kg/d IV divided q4-6h
clavulanate		

Macrolides

Clarithromycin	250-500 mg PO q12h	7.5 mg/kg PO q12h
Erythromycin	2.5-4.5 mg/kg IV q6h	3-7 mg/kg IV q6h

VI) ANTIMUCOSITIS DRUGS

At non-bone marrow ablative doses, the antineoplastic drugs most commonly associated with mucositis are: actinomycin, bleomycin, doxorubicin, fluorouracil,

Fluoroquinolones

Ciprofloxacin	250-750 mg PO q12h	–
	200-400 mg IV q12h	–
Norfloxacin	400 mg PO q12h	–
Ofloxacin	200-400 mg PO/IV q12h	–

Other

Aztreonam	1-2 g IV q8-12h	90-120 mg/kg/d IV divided q6-8h
Clindamycin	300-900 mg IV q6-8h	25-40 mg/kg/d IV divided q6-8h
Co-trimoxazole		
PCP Prophylaxis	1 DS Tablet PO TIW	2.5 mg TMP/kg PO TIW
PCP Treatment	5 mg TMP/kg IV q6h	5 mg TMP/kg IV q6h
Imipenem/cilastatin	500-1000 mg IV q6-8h	15-25 mg/kg IV q6h
Metronidazole	250-750 mg IV q8h	10 mg/kg IV q8h
Pentamidine		
PCP prophylaxis	300 mg IH q month	300 mg IH q month
PCP treatment	4 mg/kg IV QD	4 mg/kg IV QD
Teicoplanin	6-12 mg/kg IV QD	–
Vancomycin	15 mg/kg IV q12h	20 mg/kg IV q12h

16

methotrexate and vinblastine. At the high-doses required for bone marrow conditioning regimens, mucositis may be a dose-limiting toxicity with the following agents: busulfan, cytarabine, etoposide and thiotepa. Treatment of mucositis if primarily palliative, and prevention of complications such as infection. Chronic GVHD is also associated with xerostoma and painful oral lesions.

A) ARTIFICIAL SALIVA (SALIVART®)
Mechanism of action—an electrolyte solution that mimics the composition of saliva.
Dose— as needed.
Indications—mouth moisturizer in chronic GVHD.

B) BENZOCAINE (ORAGEL® AEROSOL, CETACAINE® AEROSOL, GEL, SOLUTION, CEPACOL® ANESTHETIC LOZENGES, VICKS® THROAT LOZENGES)
Mechanism of action—local anesthetic.
Dose—topically, or dissolved in the mouth, as needed.

C) CHLORHEXIDINE (PERIDEX®)
Mechanism of action—antimicrobial.
Dose—15 ml mouth rinse—swish and expectorate BID; starting with conditioning regimen and continued until resolution of neutropenia.
Side effects—alcohol content may cause burning or irritation of inflamed tissue, bitter taste, tooth discoloration.

D) COCAINE
Mechanism of action—local anesthetic.
Dose—1-10% solution, applied with a cotton applicator, or as a mouth rinse.
Indications—rarely used in transplant patients.
Side effects—systemic absorption may cause confusion, restlessness, hallucinations, delirium, nausea, vomiting, headache, hypertension, cardiac arrhythmias, physical dependence.

E) DYCLONINE (DYCLONE®, SUCRETS®)

Mechanism of action—topical anesthetic.

Dose—5-10 mg of 0.5-1% solution, swish and expectorate, or 1 lozenge dissolved in the mouth, as needed.

Indications—severe, painful mucositis.

Side effects—inhibition of the gag reflex, loss of sensation resulting in the patient biting his/her tongue; aspiration. Systemic absorption may result in arrhythmias.

F) HYDROMORPHONE (DILAUDID®)

Class—opiate analgesic.

Dose—titrate to the patient's need/tolerance; there is no ceiling on total dose. Doses should be administered on a fixed schedule. If given parenterally, a patient controlled analgesia (PCA) program is preferred.

Indications—severe mucositis.

Side effects—sedation, dysphoria, euphoria, hypotension, nausea, vomiting, constipation, dizziness, visual hallucinations, pruritus, urticaria, urinary retention, oliguria, respiratory depression. Prolonged use results in physical dependence.

G) LIDOCAINE

Class—local anesthetic.

Mechanism of action—local anesthesia.

Dose—5-10 ml of a 2% solution or gel as a mouth rinse (swish and swallow or expectorate) as needed.

Indications—painful mucositis.

Side effects—inhibition of the gag reflex, loss of sensation resulting in the patient biting his/her tongue; aspiration. Systemic absorption of large amounts may result in arrhythmias.

H) MORPHINE

Class—opiate analgesic.

Dose—titrate to the patient's need/tolerance; there is no ceiling on total dose. Doses should be administered on a fixed schedule. If given parenterally, a patient controlled analgesia (PCA) program is preferred.

Indications—severe mucositis.

Side effects—sedation, dysphoria, euphoria, hypotension, nausea, vomiting, constipation, dizziness, visual hallucinations, pruritus, urticaria, urinary retention, oliguria, respiratory depression. Prolonged use results in physical dependence.

I) SALT AND SODA MOUTH RINSE

Mechanism of action—promotes oral hygiene and keeps mouth moist and alkaline.

Composition—Sodium bicarbonate (baking soda) 1.8%, sodium chloride (salt) 0.9% (or 1 tsp each to a liter of H_2O).

Dose—10-15 ml swish and expectorate at least 5 times a day, starting with the conditioning regimen and continuing until resolution of neutropenia.

J) SULCRALFATE SLURRY

Mechanism of action—forms a protective coating on GI mucosa that serves as a barrier.

Dose—1 gm in 10-15 ml, swish and swallow q6h. Doses extemporaneously prepared from tablets must be freshly prepared at the time of administration.

Side effects—constipation, interference with absorption of other drugs.

K) LOCAL MOUTH RINSES

Many institutions prepare combinations of two or more of the agents listed above. Such products frequently contain a local anesthetic and antacid; some institutions also include diphenhydramine in their recipe. Dosage, indications and side effects of such combinations are similar to those described for each component ingredient.

VII) ANTITUBERCULOSIS

Ideally, patients with a positive PPD test, or anergy and granulomas on chest radiograph, are placed on tuberculosis suppressive therapy prior to the transplant. These patients should remain on suppressive therapy until recovery from neutropenia and discontinuation of immunosuppressive medications. Unfortunately, toxicity (predominately hepatitis) frequently requires interruption of tuberculosis prophylaxis.

Drug	Adult Dose (Normal renal and hepatic function)	Comment
Aminosalicylic acid	150 mg/kg/d PO divided q8-12h	Bacteriostatic
Capreomycin	15-30 mg/kg IM QD (maximum: 1 g/d)	Bacteriostatic, nephrotoxic, ototoxic
Cycloserine	250-500 mg PO q12h	Bacteriocidal
Ethambutol	15-25 mg/kg PO QD (maximum 2.5 g/d)	Bacteriostatic, retrobulbar neuritis
Ethionamide	250-500 mg PO q12h	Bacteriocidal
Isoniazid	5-10 mg/kg PO/IV QD (maximum: 300 mg/day)	Bacteriocidal, peripheral neuropathy, hepatitis
Pyrazinamide	15-30 mg/kg/d PO QD	Bacteriocidal
Rifampin	10 mg/kg PO/IV QD (maximum: 600 mg/day)	Bacteriocidal, hepatitis, orange-red discoloration of urine/ tears/sweat

16

VIII) ANTIVIRAL

Timing of viral infections correlates with the interval after transplant. Herpes simplex is seen in the first 30 days; CMV between days 30-90; and herpes zoster after day 100. Prophylaxis against herpes simplex mucositis with acyclovir from the beginning of the conditioning regimen to the recovery from neutropenia is standard therapy in both autologous and allogenic transplants. Prevention of CMV disease by ganciclovir therapy at the first evidence of CMV infection (i.e. + shell vial from blood or bronchioalveolar lavage) is indicated after allogeneic, but not autologous, transplants (see viral chapter). Herpes zoster infection requires hospitalization, with isolation and intravenous acyclovir therapy. Due to toxicity, foscarnet is reserved for therapy of resistant viruses.

A) ACYCLOVIR (ZOVIRAX®)

Class—purine nucleoside.

Mechanism of action—interferes with DNA synthesis and inhibits viral replication.

Dose (adults)—herpes simplex: 250 mg/m^2 IVPB over 1 hour every 8 hours; varicella zoster: 500 mg/m^2 IVPB over 1 hour every 8 hours.

Side effects—dose adjustments must be made for impaired renal function (see chapter on viral diseases). Renal toxicity may occur, especially in patients who are dehydrated or on other nephrotoxic medications. Adequate hydration and urine output should be maintained to prevent precipitation of the drug in renal tubules, leading to acute renal failure. A minimum urine output of 500 ml/24 hours/gram of acyclovir is recommended. Headache, tremor, delirium and dizziness occur rarely; however, the drug should be used cautiously in patients with underlying neurologic abnormalities, renal, hepatic or electrolyte abnormalities. Concurrent probenecid may decrease the renal clearance and excretion of acyclovir.

B) **GANCYILOVIR (CYTOVENE®)**

Class—purine nucleoside

Mechanism of action—interferes with DNA synthesis. The drug's activity is dependent on intracellular conversion to ganciclovir triphosphate by thymidine kinase.

Dose—5 mg/kg IVPB every 12 hours. Dose adjustments must be made for impaired renal function (see chapter on viral disease).

Side effects—adverse effects of ganciclovir are usually reversible. The most common toxicities are hematologic (neutropenia, thrombocytopenia, anemia and eosinophilia). Elevations in liver function tests occur in about 2-3% of patients. Nausea, vomiting, CNS manifestations, renal dysfunction and phlebitis at the infusion site may also occur. Concurrent use of other renal toxic drugs (e.g. cyclosporine) increases the risk of renal toxicity.

C) **FOSCARNET (FOSCAVIR®)**

Class—phosphonoformic acid, organic analog of pyrophosphate.

Dose—60-90 mg/kg IVPB over 1 hour every 12 hours.

Side effects—dose adjustments must be made for impaired renal function. Common side effects include: renal insufficiency, hypocalcemia, hypomagnesemia, hypokalemia, headache, confusion, tremor, delirium, ataxia, seizures, myelosuppression (less than with ganciclovir), nausea, vomiting, hypo- or hypertension and arrhythmias.

IX) **HORMONAL THERAPY**

Women undergoing transplants are at risk of menstrual bleeding, and are placed on prophylactic therapy with oral contraceptives.

Drug	Category	Dose
Norethindrone	Progesterone	1 PO QD
Norethindrone and mestranol	Combination progesterone/estrogen	1/50 or 1/35 PO QD
Leuprolide	LHRH agonist	3.75 mg IM q month

X) **CHEMOTHERAPY**

For the toxicity of high-dose chemoradiation therapy, see the chapters on conditioning regimens, regimen-related toxicity, and TBI.

Drug	Standard Doses	Transplant Doses
Amsacrine	60-160 mg/m² IV day 1 q3-4w	–
(AMSA, m-AMSA)	40-120 mg/m² IV days 1-5 q3-4w	

Asparaginase (*E. Coli* strain)	6000 units/m² IM TIW for 2-3 weeks 10,000 units/m² IV q2-3 weeks 200 units/kg IV daily for 28 days 500 units/kg IV daily for 15 days	–
Azacytidine	50-200 mg/m² IV for 5-7 days q14-28d 2.4 mg/kg IV daily for 10 days 75 mg/m² SQ daily for 7 days	–
Bleomycin	4-15 units/m² IM/SQ/IV for 7 days q2-4w 10-20 units/m² IM/SQ/IV weekly or twice weekly	–
Busulfan (Myleran®)	1-3 mg PO daily for several days 1-12 mg PO daily for several weeks 1.8 mg/m² PO daily for several days	**1 mg/kg PO q6h (total dose = 14-16 mg/kg)**
Carboplatinum (CBDCA)	400 mg/m² IV over 1-2 days q35d 60-100 mg/m² IV days 1-5 q4-5w 175 mg/m² IV weekly for 4 weeks 125 mg/m² IV days 1-3 q4-6 weeks	**450-1800 mg/m² IV (total dose)**—given over 3-4 days
Carmustine (BCNU)	75-200 mg/m² IV q6-8 weeks	**250-600 mg/m² IV** (total dose)–given as a single dose or over 3 days
Chlorambucil (Leukeran®)	0.1-0.2 mg/kg PO daily for 3-6 weeks 2-4 mg PO daily 16 mg/m² PO days 1-5 q4 weeks 0.4 mg/kg PO q4 weeks	–
Cisplatin (DDP, CDDP)	50-120 mg/m² IV q3-4 weeks 10-20 mg/m² IV days 1-5 q3-4 weeks	**150-180 mg/m² IV (total dose)**—usually given over 3 days (50-60 mg/m²/day)
Cladribine (2-CDA)	0.09 mg/kg/day IV for 7 days	
Cyclophosphamide (Cytoxan®)	60-120 mg/m² IV or PO daily for 14 days q4-6 weeks 30-40 mg/kg IV or PO q2-4 weeks 500-1500 mg/m² IV or PO q2-4 weeks	**120-200 mg/kg IV (total dose)**—given as 50-60 mg/kg/day for 2-4 days; or 5.25-7.5 g/m² IV divided over 3 days
Cytarabine	100-200 mg/m² IV for 5-10 days	3 g/m² IV q12h for

16

16

(Ara-C)	0.5-3 g/m² IV q12h for 2-6 days	6 days (total dose 36 g/m²)
Dacarbazine (DTIC)	2-4.5 mg/kg IV days 1-10 q4 weeks 650-1450 mg/m² IV q4-6 weeks	–
Dactinomycin (Actinomycin-D)	15 µg/kg IV days 1-5 q3-4 weeks 400-600 µg/m² IV days 1-5 q3-4 weeks	–
Daunomycin (Daunorubicin)	25-60 mg/m² IV daily for 3-5 days 30-60 mg/m² IV q3-4 weeks	–
Doxorubicin (Adriamycin®)	60-90 mg/m² IV q3 weeks 20-30 mg/m² IV days 1-3 q3 weeks 10-20 mg/m² IV weekly	–
Etoposide (VP-16)	50-150 mg/m² IV for 3-7 days q2-5 weeks 120 mg/m² IV q4 weeks 100-300 mg/m² PO days 1-5 q2-4 weeks	60 mg/kg or 450-2400 mg/m² IV (total dose)—given over 1-3 days
Fludarabine	18-30 mg/m² IV days 1-5 q4 weeks 125 mg/m² IV q4 weeks	–
Fluorouracil (5FU)	200-600 mg/m² IV q week; or days 1&8 q4w 300-450 mg/m² IV days 1-5 q4 weeks 6-12 mg/kg IV days 1-4 q4 weeks 10-15 mg/kg IV weekly	–
Hexamethyl-melamine (Altretamine)	10-12 mg/kg PO qd for 21-28 days 6 mg/kg PO for 15 days q8 weeks 8 mg/kg PO days 1-21 q6 weeks	–
Hydroxyurea (Hydrea®)	80 mg/kg PO q 3d 20-30 mg/kg/d PO divided BID-TID	–
Idarubicin	12-13 mg/m² IV daily for 3 days	–
Ifosfamide	0.6-1.2 g/m² IV days 1-5 q2-4 weeks 35 mg/kg IV days 1-5 q2-4 weeks 4 g/m² IV q2-4 weeks	12-16 g (total dose)– usually divided over 3 days
Lomustine (CCNU)	100-130 mg/m² PO q6 weeks 60-75 mg/m² PO q4 weeks 30 mg/m² PO q3 weeks	200-500 mg/m² IV— given as a single dose

Mechlorethamine (Nitrogen mustard)	0.4 mg/kg IV q4-6 weeks 6 mg/m^2 IV days 1&8 q4 weeks	–
Melphalan	10 mg/m^2 PO q4-6 weeks 6 mg/m^2 PO days 1-5 q4-6 weeks 16 mg/m^2 IV q2 weeks	140-200 mg/m^2 IV **(total dose)**—given as a single dose
Mitomycin (Mitomycin-C)	5-20 mg/m^2 IV q6-8 weeks 2 mg/m^2 IV days 1-5 & 8-12 q6-8 weeks	–
Mitoxantrone (Novantrone®)	8-33 mg/m^2 IV q3-4 weeks 2-10 mg/m^2 IV days 1-5 q4 weeks 5-8 mg/m^2 IV weekly	–
Paclitaxel (Taxol®)	135-175 mg/m^2 IV q21-28 days	625 mg/m^2 IV —as a single dose
Pentostatin (Deoxycoformycin)	4-5 mg/m^2 IV q2 weeks	–
Procarbazine	100 mg/m^2 PO daily for 30 days 100 mg/m^2 PO days 1-14 q28 days	–
Taxotere® (Docetaxel)	55-100 mg/m^2 IV q21 days 8-14 mg/m^2 IV days 1-5 q21 days 50 mg/m^2 IV days 1&8 q21 days	–
Teniposide (VM-26)	100-130 mg/m^2 IV weekly for 4-8 weeks 30-60 mg/m^2 IV days 1-5 q2-4 weeks 40-80 mg/m^2 CIVI for 5 days	–
Thiotepa	0.4 mg/kg IV q1-4 weeks 6 mg/m^2 IV days 1-4 q2-4 weeks 1-10 mg/m^2 IT 1-2 times/week	10 mg/kg or 500-800 mg/m^2 IV **(total dose)**—usually given over 3 days
Topotecan	0.5-2.0 mg/m^2 IV days 1-5 q3-4 weeks	–
Vinblastine (Velban®)	4-20 mg/m^2 IV weekly 1.4-2 mg/m^2/day CIVI days 1-5 q3-4 weeks	–
Vincristine	0.4-2 mg/m^2 IV weekly 1.4 mg/m^2 IV days 1&5 q5-6 weeks 1.4 mg/m^2 IV days 1&8 q4 weeks 1-2 mg/m^2 IV weekly for 4-8 weeks	–

16

16

XI) DIURETICS

During the early transplant period (first 30 days), fluid overload and weight gain occur in all patients undergoing hematopoietic stem cell transplant. This situation arises from aggressive hydration to prevent hemorrhagic cystitis, cytokine release, regimen-related capillary leakage secondary to endothelial cell damage and veno-occlusive disease of the liver. Diuretics are administered to maintain the baseline dry weight.

Drug	Adult Dose	Site of Action
Albumin	6.25-25 g IV	Given concomitantly with diuretics to increase renal perfusion pressure
Acetazolamide	250 mg IV or PO daily	Carbonic anhydrase inhibitor
Amiloride	5-10 mg PO QD-BID	Distal tubule
Bumetanide (Bumex®)	0.5-1 mg IV q2-3h 0.5-2 mg PO q12-24 h (maximum: 10 mg/day)	Loop of Henle
Chlorothiazide (Diuril®)	250-1000 mg PO q12h 100-500 mg IV QD	Distal tubule
Chlorthalidone	50-100 mg PO QD	Distal tubule
Dopamine	1-5 μg/kg/minute CIVI	Increases renal perfusion
Ethacrynic acid	0.5-1 mg/kg IV QD (maximum: 100 mg) 25-100 mg PO QD	Loop of Henle
Furosemide (Lasix®)	20-40 mg IV q2h 10-80 mg PO q6-8h	Loop of Henle
Hydrochlorothiazide (HCTZ, HydroDIURIL®, Esidrix®)	25-100 mg PO q8-24h	Distal tubule
Hydrochlorothiazide/ triampterene (Dyazide®, Maxide®)	25-100 mg (HCTZ) PO QD	Distal tubule
Spironolactone (Aldactone®)	25-100 mg PO q12-24h	Distal tubule
Triampterene	50-150 mg PO BID	Distal tubule

XII) ELECTROLYTES

Patients undergoing BMT often require supplemental electrolytes from diarrhea or renal wasting secondary to chemotherapy (e.g. ifosfamide), cyclosporine or antibiotics (e.g. amphotercin).

A) BICARBONATE

Oral—Shohl's solution 15-30 ml PO as needed or sodium bicarbonate tablets 325 or 650 mg PO 1-4 PO as needed.

Intravenous—44 meq or 50 meq amps IV push over more than 5 minutes. May be added as supplement to IV bag, 1 amp (44 meq) has equivalent osmolality to $\frac{1}{4}$ normal saline, 2 amps (44 meq amp) are equivalent to $\frac{1}{2}$ normal saline.

B) CALCIUM

Oral—calcium carbonate tablets (200 mg or 500 mg of elemental calcium per tablet). Usual dose 1-2 tablets PO BID to QID.

Intravenous—1 amp = 5 meq (10 ml) calcium gluconate = 90 mg elemental calcium; IVP only for emergency as 1 amp over 10 minutes, may be added to IV fluids.

C) MAGNESIUM

Oral—magnesium oxide tablet 400 mg 1-2 tablets PO BID or TID.

Intravenous— magnesium sulfate 1-4 grams IVPB in D5W infused no faster than 2 g per hour.

D) PHOSPHORUS

Oral—NeutroPhos® contains 250 mg elemental phosphorus per packet or dissolving capsule; usual dose 1-2 packets/capsules BID to QID.

Intravenous—range 0.01-0.24 mmoles per kg over 6 hours in 500 ml NS, may be given as sodium or potassium salt.

E) POTASSIUM

Oral—10 meq or 20 meq tablets/capsules or 20 meq powder or liquid preparation 10, 15, 20, 25, 30, 40 or 45 meq/15 ml. Do not administer more than 40meq PO per dose due to esophagitis. Usual schedule QD, BID, TID, QID.

Intravenous—10-60 meq diluted in D5W or normal saline IV, not more than 10-20 meq per hour via peripheral vein, nor more than 20 meq per hour via central vein.

XIII) GROWTH FACTORS/BIOLOGIC RESPONSE MODIFIERS

Growth factors are commonly used for peripheral blood stem cell mobilization, and recovery from neutropenia after high-dose chemoradiation therapy (see chapter on growth factors). Biologic response modifiers (e.g. interferons, interleukins) are being evaluated as immunotherapy after transplant in attempts to decrease relapse or modulate GVHD.

Drug	Representative Doses
Erythropoietin (Epoetin, Epogen®, Procrit®)	50-150 units/kg SQ or IV 3 times a week
Filgrastim (G-CSF, Filgrastim®)	5-10 µg/kg SQ or IV daily
Interferon α (Intron-A®; Roferon®)	$2\text{-}3 \times 10^6$ units SQ 3 times a week 30×10^6 units/m^2 IM days 1-5 q3 weeks 50×10^6 units/m^2 IM 3 times a week
Interferon γ	50 µg/m^2 3 times a week
Sargramostim (GM-CSF, Leukine®, Prokine®)	250 µg/m^2 SQ or IV daily

XIV) UROPROTECTANTS

Hydration with diuresis is the standard for prevention of hemorrhagic cystitis (HC) from high-dose chemotherapeutic agents such as cyclophosphamide, ifosfamide or etoposide. If HC occurs, placement of a Foley catheter, and bladder irrigation with 250-300 ml/hour of normal saline solution is used to prevent obstructive uropathy from blood clots.

Drug	Dose	Comment
Alum	10% solution in normal saline at 250-300 ml/hr	
Flavoxate	100-200 mg PO q6-8 hours	Antispasmodic for (Urispas®) relief of urgency and bladder discomfort
Mesna (2-mercaptoethane sulfonate)	**Total dose = 60-80% of the daily ifosfamide/cyclophosphamide dose** Loading dose = 10% of cytotoxic drug, the remainder as a continuous infusion over 24 hours.	May be given in 3-5 divided doses starting 20-30 minutes before chemotherapy.
Oxybutynin	5 mg PO q6-12 hours	Antispasmodic for (Ditropan®) relief of urgency and bladder discomfort
Phenoazo pyridine (Pyridium®)	100-200 mg PO TID	Urinary analgesic

16

XV) IMMUNIZATIONS

Ideally, serologic response should be checked after immunization. If there is no response, immunization should be delayed for 6 months, then repeated. Immunizations should be given only if the patient is **off** immunosuppression and has had **no** GVHD for 6 months. The following are recommended immunizations following allogeneic bone marrow transplantation.

A) INFLUENZA
Schedule—given 1 year after BMT, then annually in early autumn.
Dose—0.5 ml IM.

B) HEPATITIS B
Schedule—given 1 year after BMT.
Dose—1 ml IM monthly for 2 doses, then at 6 months.

C) MEASLES/MUMPS/RUBELLA (MMR)
Schedule—given 2 years after BMT.
Dose—0.5 ml SQ.

D) POLIO (SALK) VACCINE
(DO NOT USE SABIN VACCINE)
Schedule—given 1 year after BMT.
Dose—0.5 ml SQ q8 weeks for 2 doses, then a third dose 6-12 months after the second dose.

E) TETANUS AND DIPHTHERIA TOXOID
(PEDIATRIC DOSE)
Schedule—given 1 year after BMT.
Dose—0.5 ml IM q8 weeks for 3 doses, then a fourth dose 1 year after the third dose.

XVI) INTRAVENOUS IMMUNOGLOBULIN (IVIG)

IVIG is obtained from pooled plasma of over 1000 donors. Preparation inactivates hepatitis B and HIV, and removes most IgA and vasoregulatory peptides (kinins, kalikrein, etc.). All IgG subclasses and complement activity remain. Several brands are commercially available, including: Sandoglobulin®, Gammagard®, Gammar-IV®, Venoglobulin-I®, Gamimune N®, Polygam® and Iveegam®.

A) INDICATIONS

1) Prophylaxis

No proven role in autologous BMT. In allogeneic BMT, prophylactic IVIG decreases bacterial sepsis, CMV disease, interstitial pneumonitis and acute GVHD in recipients of HLA-matched sibling BMT over age 20. It may also decrease platelet transfusion requirements. In hypogammaglobulinemic patients with chronic GVHD, IVIG decreases bacterial infections.

2) CMV pneumoniti

Decreases mortality when given in combination with ganciclovir.

B) CONTRAINDICATIONS

Patients with a history of hypersensitivity or anaphylaxis, or individuals with selective IgA deficiency.

C) ADVERSE REACTIONS

Chills occur in less than 10% of patients; fever, nausea or headache in less than 2-4%; hypo- or hypertension in less than 1%. Side effects may be ameliorated by slowing the infusion rate or premedicating the patient with diphenhydramine, acetaminophen or corticosteroids. Urticaria, angioedema, nephrotic syndrome and anaphylaxis are rare but do occur. These reactions may be due to low concentrations of vasoactive agents (e.g. kinins, kalikren, plasmin), or IgA infused into a patient with selective IgA deficiency. A hypersensitivity syndrome, consisting of flushing, chest tightness, chills, hypotension, diaphoresis and urticaria, may occur in hypogammaglobulinemic patients who have sufficient quantities of circulating antibodies to induce immune complex formation and activation of complement.

16

D) CAUTION

IVIG preparations contain antibodies to viruses (e.g. hepatitis A, B, C) and red blood cell antigens (Rh antigen). These may cause false positive viral antibody screens and difficulties in cross-matching blood.

E) DOSE

1) Prophylaxis

400-500 mg/kg IV weekly from day 7-100, or as long as chronic GVHD persists.

2) CMV pneumonitis

500 mg/kg IV every other day for 7 doses, combined with ganciclovir therapy. (There is no proof that CMV hyperimmune globulin is more effective than IVIG.)

F) ADMINISTRATION

IV infusion over 3-8 hours. Most commonly 3, 5 or 6% solutions are used, although 9 and 12% solutions may be used if given through a central line. The initial infusion rate is 0.5-1 ml/kg/hour, increasing in 0.5-1 ml/kg/hour, at 15 minute intervals, to a maximum of 5 ml/kg/hour.

XVII) IMMUNOSUPPRESSIVE AGENTS

A) AZATHIOPRINE (IMURAN®)

Mechanism of action—azathioprine inhibits the proliferation of T and B lymphocytes by interfering with nucleotide synthesis.

Indications—treatment of GVHD refractory to cyclosporine and steroids. Since the introduction of cyclosporine, use of azathioprine has decreased considerably.

Contraindications—hypersensitivity to the drug.

Adverse reactions—bone marrow suppression, nausea, vomiting, anorexia, diarrhea, increased liver function tests, rash, fever, hypersensitivity reactions.

Dose—1-2.5 mg/kg PO daily. Dose reduction is required for patients with renal impairment or who are receiving concurrent allopurinol therapy.

B) **CYCLOSPORINE (CYCLOSPORINE A)**

Class—macrolide antibiotic.

Mechanism of action—inhibits activated T lymphocytes by preventing the transcription of interleukin-2 (IL-2). Immunosuppression is most effective when cyclosporine is present prior to antigenic stimulation.

Dose—Initial: 5-6 mg/kg/day IV or 10-18 mg/kg/day PO.

Maintenance—3-15 mg/kg/day PO.

Guidelines for cyclosporine dosage, therapeutic monitoring and dosage adjustment vary; local institutional guidelines and protocols should be consulted for specific recommendations. Infusions should be given over 8-24 hours, and the patient switched to oral therapy at the earliest opportunity.

Drugs that inhibit the cytochrome P-450 system (e.g. corticosteroids, calcium channel blockers, cimetidine, erythromycin, itraconazole and ketoconazole) may increase cyclosporine levels.

Drugs that stimulate the cytochrome P-450 system (e.g. rifampin, phenobarbital, phenytoin) may decrease cyclosporine levels.

Side effects—hypertension, renal toxicity, hirsuitism, flushing, pain in palmar and plantar surfaces, hyper-sensitivity reactions (most likely with the Cremophor EL® diluent of the parenteral formulation), parathesias, tremor, thrombotic thrombocytopenic purpura.

C) **DEXAMETHASONE MOUTH WASH**

Class—corticosteroid.

Indications—treatment of chronic GVHD of the mouth.

Dose—10-15 ml of a 1% solution, swish and expectorate QID.

D) **TACROLIMUS (PROGRAF®, FK 506)**

Class—macrolide antibiotic.

Mechanism of action—inhibits interleukin-2 (IL-2). Similar to, but more potent than, cyclosporine.

Dose—Initial: 0.03-0.05 mg/kg/day CI or 0.075-0.15 mg/kg PO q12h.

Maintenance—adjust dose to maintain a trough plasma concentration of 0.5-2 ug/L in plasma or 15-20 ug/L in whole blood.

Side effects—nephrotoxicity, neurotoxicity, hyperkalemia, hypomagnesemia, hypertension, hirsuitism, hypercholesterolemia, gingival hyperplasia.

E) **METHOTREXATE**

Class—antimetabolite.

Mechanism of action—irreversible inhibition of dihydrofolate reductase. Methotrexate's immunosuppressive activity may be due to inhibition of lymphocyte multiplication.

Dose—Methotrexate prophylaxis for GVHD varies by institution. A common regimen is MTX 15 mg/m^2 IV on day 1, then 10 mg/m^2 IV on day 3, 6 and 11. Dose may be adjusted or held for mucositis and renal or liver insufficiency.

Side effects—myelosuppression, mucositis, photosensitivity.

16

F) METHOXSALEN (METHOXYPSORALEN)

Class—psoralen.

Mechanism of action—suppresses DNA synthesis and cell division by covalent binding to pyrimidine bases in DNA.

Dose—0.6 mg/kg PO 1.5-2.0 hours before exposure to UV light. Treatment schedules vary; local institutional guidelines and protocols should be consulted for specific recommendations.

Side effects—nausea, pruritus, edema, hypotension. Prolonged exposure to UV light can result in severe burning, pealing and blistering of the skin.

G) METHYLPREDNISOLONE

Class—corticosteroid.

Indications—prevention and treatment of acute and chronic GVHD. Usually used in combination with cyclosporine.

Mechanism of action—apoptosis; stabilization of leukocyte lysosomal membrane, inhibition of chemotaxis and cytokine release.

Dose—2 mg/kg/day IV for 10-14 days, then taper every 4 days as tolerated; or, 1 g/m^2/day IV for 3 days, then taper by 50% every 3 days.

Side effects—hyperglycemia, hypertension, insomnia, euphoria, depression, gastritis, peptic ulcers, fluid and electrolyte alterations, headache, pseudotumor cerebri, skin atrophy, aseptic necrosis, myopathy, impaired wound healing, Cushingoid state, adrenal insufficiency.

H) PREDNISONE

Class—corticosteroid.

Indications—prevention and treatment of acute and chronic GVHD. Usually used in combination with cyclosporine.

Mechanism of action—apoptosis; stabilization of leukocyte lysosomal membrane, inhibition of chemotaxis and cytokine release.

Dose—Acute GVHD: 0.5-1 mg/kg PO q12h taper over 3-5 weeks according to response.

—Chronic GVHD: 0.5-1 mg/kg PO daily or every other day for 6 months or longer.

Side effects—hyperglycemia, hypertension, insomnia, euphoria, depression, gastritis, peptic ulcers, fluid and electrolyte alterations, headache, pseudotumor cerebri, skin atrophy, aseptic necrosis, myopathy, impaired wound healing, Cushingoid state, adrenal insufficiency.

I) PUVA

Photochemotherapy with a psoralen and ultraviolet-A (UV-A) radiation. The psoralen, most commonly used methoxsalen (see above), is given orally, or applied topically, 1-2 hours prior to radiation. UV-A exposure times vary, but 20-30 minute exposure times appear most common. Treatments are given 1-2 times weekly for several weeks. Treatment schedules vary; local institutional guidelines and protocols should be consulted for specific recommendations.

Side effects—skin dryness, hyperpigmentation and actinic keratoses.

J) THALIDOMIDE

Mechanism of action—uncertain, but may interfere with adhesion molecules.

Dose—100 mg PO QID, increasing to a total dose of 800-1600 mg/day.

Indications—treatment or prophylaxis of chronic GVHD.

Side effects—**Thalidomide is a potent teratogen and is contraindicated in patients who are, or are likely to become, pregnant.** Other side effects

16

include: sedation, constipation, abdominal distension, peripheral neuropathy, skin rash (which may mimic GVHD) and neutropenia.

XVIII) VENO-OCCLUSIVE DISEASE MEDICATIONS

Drug	Dose
Heparin	100 units/kg/day CIVI, starting with the initiation of the preparative regimen and continuing to day 30
Tissue plasminogen activator (TPA)	Dosage varies; see Table 13.1.3 in the chapter on regimen related toxicity for one example. Local institutional guidelines and protocols should also be consulted for specific recommendations.
Ursodeoxycholic acid (Ursodiol)	300 mg PO BID or TID

16

Appendix I:
Organizations Relevant to
Bone Marrow Transplantation

I) RELATED ALLOGENEIC DATA REGISTRY
International Bone Marrow Transplant Registry (IBMTR)
8701 Watertown Plank Road
Milwaukee, Wisconsin 53226
phone: 414-456-8325; fax: 414-266-8471

II) AUTOLOGOUS DATA REGISTRY
North American Autologous Bone Marrow Transplantation Registry (ABMTR)
8701 Watertown Plank Road
Milwaukee, Wisconsin 53226
phone: 414-456-8325; fax: 414-266-8471

III) UNRELATED DONOR REGISTRY
1) AMERICA (UNITED STATES)
a) National Marrow Donor Program (NMDP) (US)
3433 Broadway St. N.E. Suite 400
Minneapolis, Minnesota 55413
phone: 800-526-7809; fax: 1-612-627-5899
b) The American Bone Marrow Donor Registry (ABMDR)
The Caitlin Raymond International Registry
The University of Massachusetts Medical Center
55 Lake Ave. North
Worcester, Massachusetts 01655
phone: 508-756-6444; fax 508-752-1496
c) New York Blood Center Cord Blood Registry
The Fred H. Allen Laboratory of Immunogenetics
The New York Blood Center
310 East 67th Street
New York, New York 10021
phone: 212-570-3230

2) AUSTRIA
Austrian Bone Marrow Donors
Institute for Blood Group Serology
University of Vienna
Florianigasse 38/12, A-1080
Vienna, Austria
phone: 43-1-403-7193; fax: 43-1-408-2321

3) AUSTRALIA
Australian Bone Marrow Donor Registry
Bone Marrow Transplant Unit
153 Clarence Street
Sydney, NSW 2000 Australia
phone: 61-2-229-4361; fax: 61-2-229-4474

4) **BELGIUM**
National Marrow Donors Program-Belgium
1616 Edmond Picard Street
B-1060 Brussels, Belgium
phone: 32-2-347-28-04; fax: 32-2-347-03-01

5) **CANADA**
Canadian Unrelated Bone Marrow Donor Registry
Canadian National Coordinating Centre
401-555 West Eighth Avenue
Vancouver, British Columbia, Canada V5Z 1C6
phone: 604-879-5269; fax: 604-879-4255

6) **CZECHOSLOVAKIA**
a) The Czech Bone Marrow Donor Registry
HLA Lab, Department of Immunology
IKEM, Videnska 800
140 00 Prague 4, Czech Republic
phone: 42-2-472-2242; fax: 42-2-472-1603
b) Central Bone Marrow Donor Registry
1st Clinic of Internal Medicine
alej Svobody 80, 323 18 Plzen
Czech Republic
phone: 42-19-533-300; fax: 42-19-533-896

7) **CYPRUS**
Cyprus Paraskevaidio Bone Marrow Donor Registry
4A Char. Mouskos St.
PO Box 4307
Nicosia, Cyprus
phone: 357-2-455382; fax: 357-2-474397

8) **DENMARK**
Skejby Hospital Bone Marrow Donor Registry
Department of Clinical Immunology
University Hospital
Skejby Sygehus, Brend Strupgards
Vej. DK-8200 Arhus N, Denmark
phone: 45-89-49-53-10 ext. 5300; fax: 45-89-49-60-07

9) **FRANCE**
France Greffe de Moelle
Hospital Saint Louis
1, Avenue Claude Vellefaux
75010 Paris Cedex, France
phone: 33-1-4-803-1233; fax: 33-1-4-803-0202

10) **GERMANY**
a) ZKRD-German Registry of Bone Marrow Donors
Helmholtzstrasse 10
D-89081 Ulm, Germany
phone: 49-731-954300; fax -49-731-9543050

b) German Branch of the European Cord Blood Bank
Heinrich-Heine University
Moorenstr. 5, Building 14.80
Postfach 101007, 40001 Dusseldorf, Germany
phone: 49-211-311-8686/8684; fax: 49-211-934-8435

11) **FINLAND**
Finnish Bone Marrow Donor Registry
Finnish Red Cross Blood Transfusion Service
Kivihaantie 7, FIN-00310, Helsinki, Finland
phone: 358-0-580-1227; fax: 358-0-580-1429

12) **GREECE**
a) Athens Bone Marrow Donor Registry
Department of Immunology and Natl. Tissue Typing Lab
General Hospital of Athens
154, Mesogion Ave.
GR-11527 Athens, Greece
phone: 30-1-771-1914; fax: 30-1-777-4395
b) "Macedonian" Bone Marrow Donors Association Thessaloniki
Dept. of Immunol. and Regional Tissue Typing Lab
Hippokration Gen. Hospital of Thessaloniki
50, Papanastassiou Street
54642 Thessaloniki, Greece
phone: 30-3-181-2957; fax: 30-3-181-2957

13) **HUNGARY**
Hungarian Bone Marrow Registry
Natl. Inst. of Haematol. Blood Transfusion and Immunol.
24, Daroczi Street
Budapest, Hungary H-1113
phone: 36-1-166-5822; fax: 36-1-166-7020

14) **IRELAND**
The Irish Unrelated Bone Marrow Panel
The Blood Transfusion Service Board
Pelican House, PO Box 97
40 Mespil Road
Dublin 4, Ireland
phone: 353-1-660-3333; fax: 353-1-660-3419

15) **ISRAEL**
a) Israeli Donor Bank Registry
Tissue Typing Unit
Hadassah University Hospital
PO Box 12000
il-91-120 Jerusalem, Israel
phone: 972-241-5074; fax: 972-243-3165
b) Israeli Cord Blood Registry, same as above

16) ITALY

a) Italian Bone Marrow Donor Registry
Laboratorio di Istocompatibilita
E.O. Ospedale Galliera
Via A. Volta 19/2
16128 Genova, Italy
phone: 39-10-563-2340; fax: 39-10-563-2544
b) Milan Cord Blood Registry
Centro Transfusionale e di Immunologia dei Trapianti
Ospedale Maggiore Policlinico
Via Francesco Sforza 35
20122 Milan, Italy
phone: 39-2-5503-4053/5; fax: 39-2-545-8129

17) NETHERLANDS

Europdonor Foundation
Bldg 1, E3-Q, University Hospital of Leiden
Rijnsburgerweg 10
2333 AA Leiden, Netherlands
phone: 31-7-126-8002; fax: 31-7-121-0457

18) NORWAY

The Norwegian Bone Marrow Donor Registry
Institute of Transplant Immunology
Rikshospitalet (I.T.I.), 0027
Oslo 1, Norway
phone: 47-2-286-8555; fax: 47-2-220-3693

19) POLAND

National Polish Bone Marrow Donor Registry
K. Diuski Hospital/ L. Hirsfeld Inst. of Immunology and Experimental Therapy
Grabiszynska 105, 53-439 Wroclaw, Poland
phone: 48-7-167-9081; fax: 48-7-167-9111

20) PORTUGAL

Portuguese Bone Marrow Donors Registry
Centro de Histocompatibilidade do Sul
Campo de Santana
130 1100 Lisbon, Portugal
phone: 351-1-8850485; fax: 351-1-8850118

21) RUSSIA

a) Russian Bone Marrow Donor Registry
Research Center for Hematology
Novozykovsky Lane 4a
125167 Moscow, Russia
phone: 7-095-213-2476; fax: 7-095-212-4302
b) RICEI Bone Marrow Donor Registry
Research Inst. Clin. Exp. Immunol.; Bldg. 16
1st Leonovskaya Street
129226 Moscow, Russia
phone/fax: 7-095-187-6111

22) SAUDIA ARABIA
Section of Hematology and Bone Marrow Transplantation
Department of Oncology
King Faisal Specialist Hospital and Research Centre
Riyadh 11211
Kingdom of Saudi Arabia
phone: 966-1-464-7272; fax: 966-1-441-4839

23) SLOVENIA
Slovenija Bone Marrow Donor Registry
Tissue Typing Center
Blood Transfusion Centre of Slovenia
SLO-61000 Ljubljana
Slajmerjeva 6, Slovenia
phone: 386-61-302-313; fax: 386-61-1312-304

24) SOUTH AFRICA
South African Bone Marrow Registry
Provincial Lab. for Tissue Immunology
Private Bag 4
Observatory 7935
Cape Province, South Africa
phone: 27-2-147-3080; fax: 27-21-448-6107

25) SPAIN
REDMO, Fundacio International Josep Carreras
Muntaner 383-2o2a
08021 Barcelona, Spain
phone: 34-3-414-5566; fax: 34-3-201-5588

26) SWEDEN
Tobias Registry of Swedish Bone Marrow Donors
Department of Clinical Immunology
F79 Huddinge Hospital
S-141 86 Huddinge, Sweden
phone: 46-8-746-8020; fax: 46-8-746-6869

27) SWITZERLAND
Foundation Swiss Bone Marrow Donor Registry
Wankdorfstrasse 10
3000 Bern 22, Switzerland
phone: 41-31-330-0440; fax: 41-31-332-5991

28) UNITED KINGDOM
Anthony Nolan BM Trust
The Royal Free Hospital Unit 2
Hethcot Towers
75287 Agin Court Road
London, NW3 2NT, United Kingdom
phone: 44-1-71-284-1234; fax: 44-1-71-284-8226

IV) PHYSICIAN BMT ORGANIZATIONS

The American Society for Blood and Marrow Transplantation (ASBMT)
The Emory Clinic
1365 Clifton Road NE
Atlanta, Georgia 30322
phone: 404-778-3774; fax: 404-778-5020

> To become a member contact:
> ASBMT c/o KLUGE Carden Jennings Publishing
> 1224 West Main Street, Suite 200
> Charlottesville, Virginia, 22903
> phone: 804-979-4913

International Society for Hematotherapy
and Graft Engineering (ISHAGE)
Department of Pediatrics
University of Florida
Box 100296, JHMHC
Gainsville, Florida 32610
phone: 904-392-4472; fax: 904-392-8725

European Group for Bone Marrow Transplantation (EBMT)
EBMT Secretariat, Congress House
65 West Drive
Cheam, Sutton, Surrey, SM2 7NB
United Kingdom
phone: 44-81-661-0877; fax: 44-81-661 9036

V) PATIENT BMT INFORMATION

BMT Newsletter
1985 Spruce Ave
Highland Park, Illinois 60035
phone: 708-831-1913; fax: 708-831-1943
Bi-monthly newsletter for patients, family, friends, medical professionals. List of attorneys with successful track record in obtaining insurance payment for BMT. List of nationwide support groups.

BMT Link
29209 Northwestern Highway #624
Southfield, Michigan 48034
phone: 313-932-8483

National Coalition of Cancer Survivorship
1010 Wayne Ave., 7th floor
Silver Spring, Maryland 20910
phone: 301-650-8868; fax: 301-565-9670

Candlelighters Childhood Cancer Foundation
7910 Woodmount Ave., Suite 460
Bethesda, Maryland 20814
phone: 800-366-2223; fax: 301-718-2686

VI) BMT FINANCIAL SUPPORT

National Children's Cancer Society
1015 Locust, #1040
St. Louis, Missouri 63101
phone: 800-882-6227; fax: 314-241-6229

Children's Organ Transplantation Fund
2501 Cota Dr.
Bloomington, Indiana 47403
phone: 812-336-8872; 800-366-2682; fax: 812-336-8885

Organ Transplant Fund
1027 S. Yates
Memphis, Tennessee 38119
phone: 901-684-1697; 800-489-3863; fax: 901-684-1128

Cancer Fund of America
2901 Breezewood Lane
Knoxville, Tennessee 37921
phone: 423-938-5282; fax: 423-938-2968

VII) CANCER INFORMATION

PDQ- Physician Data Query
Cancer Information Service
National Cancer Institute
9000 Rockville Pike; Bldg. 31; Room 10A07
Bethesda, Maryland 20814
phone: 800-422-6237 (800-4-cancer)
Cancer Fax: 301-402-0555 (obtaining cancer data on your fax)

American Cancer Society
1599 Clifton Road NE
Atlanta, Georgia 30329
phone: 404-320-3333; fax: 404-325-2217

Cancer Federation
PO Box 1298
Banning, California 92220
phone: 909-849-4325; fax: 909-849-0156

National Brain Tumor Foundation
323 Geary St.; Suite 510
San Francisco, California 94102
phone: 415-284-0208; fax: 415-284-0209

National Alliance of Breast Cancer Organizations (NABCO)
9 East 37th Street, 10th floor
New York, New York 10016
phone: 212-719-0154; fax: 212-689-1213

Y-ME National Organization for Breast Cancer
Information and Support
212 W Van Buren
Chicago, Illinois 60607
phone: 800-221-2141; fax: 312-986-0020

VIII) LEUKEMIA INFORMATION
Leukemia Society of America
600 3rd Ave., 4th floor
New York, New York 10016
phone: 212-573-8484; fax: 212-338-0323

National Leukemia Society
585 Stewart Ave, #536
Garden City, New York 11530
phone: 516-222-1944; fax: 516-222-0457

IX) APLASTIC ANEMIA / FANCONI'S ANEMIA INFORMATION
Aplastic Anemia Foundation of America
PO Box 22689
Baltimore, Maryland 21203
phone: 410-955-2803

Fanconi Anemia Research Fund
1902 Jefferson Street #2
Eugene, Oregon 97405
phone: 541-687-4658; fax: 541-687-0548

X) IMMUNODEFICIENCY INFORMATION
Immune Deficiency Foundation
25 W Chesapeake Ave., Suite 206
Towson, Maryland
phone: 410-321-6647; fax: 410-321-9165

XI) INSURANCE INFORMATION
Health Insurance Association of America
555 13th Street NW Suite 600 East
Washington, DC 20004
phone: 202-824-1600; fax: 202-824-1722

Appendix II: Allogeneic BMT Patient Pre-admission Evaluation

Appendix

Patient name/sex/age/Diagnosis	Donor name/age/sex
DOB/social security #	DOB/social security #
phone (H)/(W)	phone (H)/(W)
address	address
insurance	insurance
referring Dr	referring Dr

Patient Test/Evaluation	Scheduled Date	Result
history, ROS, PE		
chest x-ray		
CT scan restaging, if indicated, i.e. lymphoma		
lumbar puncture, if indicated		
EKG		
CBC, diff, plt		
electrolytes, Cr, Mg		
LFT		
PT, PTT		
HIV		
hepatitis A, B, C		
CMV		
HSV		
pregnancy test		
marrow aspirate		
cytogenetics		
Patient Test/Evaluation	**Scheduled Date**	**Result**
2-D echo or Mugga		
PFTs, FEV1, DLCO		

ABG		
sinus x-ray, if indicated		
ABO type		
HLA typing		
radiation onc. consult		
dental consult		
central line placement		
oocyte/sperm banking		
family conference		
consent		
admission date		
start conditioning		
HSC infusion (day 0)		

Appendix III: Leukocyte Antigens

Cluster of Differentiation (CD) Designation	Cell Distribution	Comment
CD1a, b, c, d	cortical thymocytes– immature T cells dendritic cells, B cells (CD 1c)	similar structure to MHC class I molecules; IgSF
CD2	T, NK	receptor for sheep red blood cell rosette formation, binds CD58 and CD16; IgSF
CD3	T	T cell receptor complex- γ, δ, ϵ, η, ζ chains, IgSF
CD4	T cell subset "helper"	accessory molecule for TCR-MHC-antigen recognition; IgSF
CD5	T cell, some B cells (eg. B cell CLL)	binds to CD72
CD6	T cell, some B cells (eg. B cell CLL)	possibly involved in signal transduction
CD7	T cell, hematopoietic progenitors	unknown function; IgSF
CD8	T cell subset "suppressor" "cytotoxic"	accessory molecule for TCR-MHC-antigen recognition; IgSF
CD9	platelets, early B, eosinophils, basophils, some T cells	possible role in platelet aggregation
CD10	early B and T cells	CALLA- common acute lymphocytic leukemia antigen
CD11a	lymphocytes, granulocytes, monocytes, macrophages	integrin heterodimer associated with CD18- the CD11/CD18 heterodimer involved in cell-cell and cell-matrix adhesion
CD11b	myeloid, NK	integrin heterodimer associated with CD18- the CD11/CD18 heterodimer involved in cell-cell and cell-matrix adhesion

CD11c	myeloid, macrophages, some B cells, eg. hairy cell leukemia, some B-CLL	integrin heterodimer associated with CD18- the the CD11/CD18 heterodimer involved in cell-cell and cell-matrix adhesion
CDw12	monocytes, granulocytes, platelets	
CD13	myeloid, basophils, eosinophils	aminopeptidase N—a zinc binding metalloproteinase
CD14	myeloid, monocytes, macrophages	lipopolysaccharide (LPS) receptor
CD15	granulocytes, eosinophils, monocytes	Lewis X (Le^x)
CD16	NK, macrophages, neutrophils, some cytotoxic T cells	IgSF; low affinity IgG Fc receptor
CDw17	neutrophils, monocytes, platelets	glycosphingolipid lactosylceramide
CD18	lymphocytes, granulocytes, monocytes, macrophages, NK	integrin heterodimer associated with CD11a, b, c,
CD19	B	involved in B cell proliferation; IgSF
CD20	B	involved in B cell activation and proliferation
CD21	B cells	receptor for Epstein-Barr virus and receptor for complement fragment C3d
CD22	B	IgSF
CD23	B	IgE receptor
CD24	B	
CD25	T, B, monocytes	a subunit of interleukin-2 receptor–α chain (Tac)
CD26	T, B, macrophages	dipeptidyl peptidase- a cell surface protease
CD27	T	NGF receptor superfamily; binds CD 70
CD28	T, activated B cell	costimulatory signal - CD28 binds B7 (CD80); IgSF

Cluster of Differentiation (CD) Designation	Cell Distribution	Comment
CD29	some leukocytes, some T helper subsets (Helper-inducer or "memory" T cell)	integrin-forms a heterodimer with CD49 to mediate cell-cell and cell-matrix adhesion
CD30	activated B and T cells, Reed-Sternberg cell	Ki-1; NGF receptor superfamily
CD31	monocytes, platelets, granulocytes, B, endothelial	platelet endothelial cell adhesion molecule-1 (PECAM-1); IgSF
CD32	B, monocytes, neutrophils	receptor for aggregated IgG; IgSF
CD33	granulocytes, macrophages, expressed on myeloid cells after CD34	IgSF
CD34	early hematopoietic pro-genitor cells, endothelial cells	ligand for CD62L
CD35	monocytes, B, some T, erythrocytes, monocytes, macrophages	complement receptor 1– binds complement fragments C3b and C4b
CD36	platelets, monocytes	platelet glycoprotein IV
CD37	B, plasma cell, some T	
CD38	early B and T lineage, activated B and T, not on resting B or T	
CD39	activated NK and B and some T, not on resting B or T	
CD40	B, monocyte, dendritic	receptor for B cell co-stimulatory signal; NGF receptor superfamily
CD41	platelets, megakaryocytes	GPIIb - involved in platelet adhesion and aggregation, part of GPIIb/GPIIIa complex (CD41/CD61)
CD42a, b, c, d	platelets, megakaryo-cytes	GPIb - binds to von Willibrand factor
CD43	T, granulocytes, monocytes, macrophages, NK, platelets, plasma cell, hematopoietic progenitor cells, activated B	leukosialin, binds ICAM-1 (CD54)

CD44	T, B, monocytes, granulocytes, erythrocytes	hyaluronate receptor, part of RBC Lutheran blood group
CD45	leukocytes	leukocyte common antigen (LCA), isoforms that occur by differential exon splicing are CD45RA, CD45RB, CD45RC, CD45RO
CD46	T, B, monocytes, granulocytes, NK, platelets, endothelial	binds complement C3b and C4b, protecting cells from lysis
CD47	all hematopoietic cells	IgSF
CD48	T,B, eosinophils, neutrophils	IgSF
CD49a, b, c, d, e, f	T, B, monocytes, platelets	very late antigen (VLA), an integrin associated with CD29; binds collagen, laminin
CD50	T, B, monocytes, granulocytes	ICAM-3; binds CD11a/CD18
CD51	platelets, megakaryocytes	vitronectin receptor–an integrin associated with CD61 that binds to RGD amino acid sequences in vitronectin, von Willibrand factor, fibrinogen
CD52	T, B, granulocytes, eosinophils, monocytes	used as a target for complement mediated killing of lymphocytes via the antibody campath -1
CD53	T, B, monocytes, granulocytes, platelets	
CD54	T, B, endothelial cells	ICAM-1, binds to CD11/CD18 causing adhesion; IgSF
CD55	all hematopoietic cells	DAF (decay acclerating factor)– prevents membrane lysis by complement by blocking C3 convertase
CD56	NK	neural cell adhesion molecule (NCAM); IgSF
CD57	NK, some T and B	
CD58	macrophages, B, T	LFA-3; binds to CD2; IgSF
CD59	all hematopoietic cells	prevents membrane lysis by complement by binding C8 and C9

Cluster of Differentiation (CD) Designation	Cell Distribution	Comment
CD60	T, platelets, monocytes	
CD61	platelets, megakaryocytes, macrophages	GPIIIa - involved in platelet adhesion and aggregation, part of GPIIb/GPIIIa complex (CD41/CD61)
CD62 E, L, P	megakaryocyte, platelet, endothelium	adhesion molecule, E, L or P-selectin
CD63	activated platelets, monocytes, macrophages	platelet activation antigen
CD64	macrophages, monocytes	IgG receptor; IgSF
CDw65	myeloid	ceramide dodecasaccharide
CD66a, b, c, d, e	neutophils	IgSF; CD66e is carcinoembryonic antigen (CEA)
CD67	granulocytes	CD67 is the same as CD 66b
CD68	macrophages, mono-cytes, neutrophils, basophils	
CD69	platelets, some activated lymphocytes	
CD70	some T and B, RBC, platelets	Ligand for CD27
CD71	leukocytes, expression increases with activation	transferrin receptor
CD72	B	ligand for CD5
CD73	some lymphocytes and endothelial cells	ecto-5'-nucleotidase, dephospho-cylates nucleotides to allow cellular uptake
CD74	B, monocytes, macrophages	MHC class II invariant chain
CD75	B some T	
CD80	B	B7.1 costimulatory signal -the ligand for CD28 and CTLA-4; IgSF
CD88	granulocytes, macrophages, mast cells	C5a complement receptor
CD89	monocytes, granulocytes, macrophages, some T and B cells	IgA receptor; IgSF

CD95	complete distribution unknown, present on T cells	Fas - important in apoptosis and negative selection of lymphocytes; NGF receptor superfamily
CD97	myeloid cells, epithelial cells	tumor necrosis factor I and II receptor
CD98	monocytes, B, T, NK, granulocytes	4F2
CD102	lymphocytes, monocytes, endothelial cells	ICAM-2—a ligand for the integrin LFA-1(CD11a/CD18); IgSF
CD115	monocytes, macrophages	macrophage colony stimulating (M-CSF) receptor; IgSF
CDw116	monocytes, neutrophils, eosinophils, endothelial cells	GM-CSF receptor
CD117	mast cells and hemato-poietic progenitor cells except B lineage cells	c-Kit - the ligand for "steel factor"; stem cell factor (SCF) receptor
CDw119	macrophages, monocytes, B cells, endothelium	interferon γ receptor
CDw121a, b	T, B, macrophages, monocytes, endothelial cells	interleukin-1 receptor; IgSF
CD123	hematopoietic progenitor cells, granulocytes, basophils, eosinophils, monocytes, megakaryocytes, erthyroid cells	interleukin-3 receptor α chain
CD124	B, T, endothelial cells, hemato-poietic progenitor cells	interleukin-4 receptor
CD125	eosinophils, basophils	interleukin-5 receptor α chain
CD126	B, plasma cells	interleukin-6 receptor α chain
CD127	pro-B cells, T	interleukin-7 receptor
CD128	granulocytes, basophils, monocytes, some T cells	interleukin-8 receptor

B = B cells   NK = natural killer cells   T = T cells
IgSF = immunoglobulin superfamily   NGF - nerve growth factor

Appendix IV:
NCI Common Toxicity Criteria

NCI COMMON TOXICITY CRITERIA

		0	1	2	3	4
Leukopenia	WBC x 10^3	≥ 4.0	3.0–3.9	2.0–2.9	1.0-1.9	< 1.0
	granulocytes/bands	≥ 2.0	1.5–1.9	1.0–1.4	0.5-0.9	< 0.5
	lymphocytes	≥ 2.0	1.5–1.9	1.0–1.4	0.5-0.9	< 0.5
Thrombocytopenia	plt x 10^3	WNL	75.0–normal	50.0–74.9	25.0–49.9	< 25.0
Anemia	Hgb	WNL	9.5–10.9	8.0–9.4	6.5–7.9	< 6.5
Hemorrhage (clinical)	******	none	mild, no transfusion	gross, 1-2 units transfusion/episode	gross, 3-4 units transfusion/episode	massive > 4 units transfusion/episode
Infection	******	none	mild, no active Rx	moderate, localized infection requires active Rx	severe, systemic infection requires active Rx, specify site	life-threatening, sepsis, specify site
Fever in absence of infection	******	none	37.1-38.0 C 98.7-100.4 F	38-40C 100.4-104F	> 40.0 C (> 104.0 F) despite anti-pyretics	> 40.0 C (>104.0 F) for > 24 hrs or fever with hypotension

Fever felt to be caused by drug allergy should be coded as allergy.
Fever due to infection is coded under infection only.

		0	1	2	3	4
GU	creatinine	WNL	< 1.5 x N	1.5-3.0 x N	3.1-6.0 x N	> 6.0 x N
	proteinuria	no change	1+ or < 0.3g% or < 3g/L	2-3 and/or 0.3-1.0g% or 3-10g/L	4+ or > 1.0g% or > 10g/L	nephrotic syndrome
	hematuria	neg	micro only	gross, no clots	gross + clots	requries transfusion
	*BUN	< 1.5 x N	1.5-2.5 x N	2.8-5 x N	5.1-10 x N	> 10 x N

Urinary tract infection should be coded under infection, not GU.
Hemarturia resulting from thrombocytopenia should be coded under hemorrhage, not GU.

		0	1	2	3	4
GI	nausea	none	able to eat reasonable intake	intake significantly decreased but can eat	no significant intake	******
	vomiting	none	1 episode in 24 hours	2-5 episodes in 24 hours	5-10 episodes in 24 hours	> 10 episodes in 24 hrs or requiring parenteral support

		0	1	2	3	4
	diarrhea	none	increase of 2-3 stools/day over pre-Rx	increase of 4-6 stools/day, or nocturnal stools, or moderate cramping	increase of 7-9 stools/day or incontinence, or severe cramping	increase of > 10 stools/day or grossly bloody diarrhea, or need for parenteral support
	stomatitis	none	painless ulcers, erythema, or mild soreness	painful erythema, edema, or ulcers, but can eat	painful erythema, edema or ulcers, and cannot eat	requires parenteral or enteral support
Liver	bilirubin	WNL	1.5-2.5 x N	2.6-5.0 x N	5.1-20.0 x N	> 20.0 x N
	transaminase (SGOT, SGPT)	WNL	1.5-2.5 x N	2.6-5.0 x N	5.1-20.0 x N	> 20.0 x N
	alk phos or 5'nucleotidase	WNL	1.5-2.5 x N	2.6-5.0 x N	5.1-20.0 x N	> 20.0 x N
	liver–clinical	no change from baseline	******	*******	precoma	hepatic coma

Viral hepatitis should be coded as infection rather than liver toxicity.

		0	1	2	3	4
Pulmonary		****	none or no change	asymptomatic with abnormality in PFTs	dyspnea on significant exertion	dyspnea at normal level of activity	dyspnea at rest

Pneumonia is considered infection and not graded as pulmonary toxicity unless felt to be resultant from pulmonary changes directly involved by treatment.

		0	1	2	3	4
Cardiac	cardiac dysrhythmias	none	asymptomatic transient, requiring no therapy	recurrent or persistent, no therapy required	requires treatment	requires monitoring, or hypotension or ventricular tachycardia or fibrillation
	cardiac function	none	asymptomatic, decline of resting ejection fraction by less than 20% of baseline value	asymptomatic, decline of resting ejection fraction by more than 20% of baseline value	mild CHF, responsive to therapy	severe or refractory CHF
	cardiac-ischemia	none	non-specific T wave flattening	asymptomatic ST and T wave changes suggesting ischemia	angina without evidence for infarction	acute myocardial infarction
	cardiac-pericardial	none	asymptomatic effusion, no intervention required	pericarditis (rub, chest pain, ECG changes)	symptomatic effusion; drainage required	tamponade; drainage urgently required

		0	1	2	3	4
Blood Pressure	hypertension	none or no change	asyptomatic, transient increase by > 20 mm Hg (D) or to > 150/100 if previously WNL; no treatment required	recurrent or persistent increase by > 20 mm Hg (D) or to > 150/100 if previously WNL; no treatment required	requires therapy	hypertensive crisis
	hypotension	none or no change	changes requiring no therapy (including transient orthostatic hypotension)	requires fluid replacement or other therapy, but not hospi-talization	requires thera-py and hospi-talization; re-solves within 48 hours or stopping the agent	requires therapy and hospitaliza-tion for > 48 hours after stopping the agent
Skin	******	none or no change	scattered macular or papular eruption or erythema that is asymp-tomatic	scattered macular or papular eruption or erythema with pruritus or other associated symptoms	generalized symptomatic macular, papular or vesicular eruption	exfollative dermatitis or ulcerating dermatitis
Allergy	*******	none	transient rash, drug fever < 38C, 100.4 F	unicaria, drug fever > 38 C, 100.4 F, mild broncho-spasm	serum sickness, bron-chospasm, requires parenteral meds	anaphylaxis
Local		none	pain	pain and swelling, with inflam-mation or phlebitis	ulceration	plastic surgery indicated
Alopecia	*******	no loss	mild hair loss	pronounced or total hair loss	*******	********
Weight gain/loss	*******	< 5.0%	5.0-9.9%	10.0-19.9%	> 20.0%	********
Neurologic						
Sensory	neuro-sensory	none or no change	mild paresthesias: loss of deep tendon reflexes	mild or moderate objective sensory loss: moderate paresthesias	severe objective sen-sory loss or paresthesias that interfere with function	complete loss of sensation

		0	1	2	3	4
	neuro-vision	none or no change	decreased activity	Finger counting only	light perception only	total loss of vision-blind
	neuro-hearing	none or no change	asymptomatic hearing loss on audio-metry only	tinnitus	hearing loss interfering w/ function but correctable with hearing aid	deafness, not correctable
Motor	neuro-motor	none or no change	subjective weakness: no objective findings	mild objec-tive weakness without significant impairment of function	objective weakness with impairment of function	paralysis
	neuro-constipation	none or no change	mild	moderate	severe	ileus > 99 hours
Psych	neuro-mood	no change	mild anxiety or depression	moderate anxiety or depression	severe anxiety or depression	suicidal ideation
Clinical	neuro-cortical	none	mild somnolence or agitation	moderate somnolence or agitation	severe somno-lence or agita-tion, confusion, disorientation or hallucinations	coma, seizures, toxic psychosis
	neuro-cerebellar	none	slight incoordination, dysdiado-kinesis	intention tremor, dys-metria, slurred speech, nystagmus	locomotor alexia	cerebellar necrosis
	neuro-headache	none	mild	moderate or severe but transient	unrelenting and severe	*******
Metabolic	hyperglycemia	< 116	116-160	161-250	251-500	> 500 or ketoacidosis
	hypoglycemia	> 64	55-64	40-54	30-39	< 30
	amylase	WNL	< 1.5 x N	1.5-2.0 x N	2.1-5.0 x N	> 5.1 x N
	hypocalcemia	> 8.4	8.4-7.8	7.7-7.0	6.9-6.1	< 6.1
	hypomagne-semia	> 1.4	1.4-1.2	1.1-0.9	0.8-0.6	< 0.5
Coagulation	fibrinogen	WNL	0.99-0.75x N	0.74-0.50x N	0.49-0.25 x N	< 0.25 x N
	prothrombin time	WNL	1.01-1.25x N	1.26-1.50 x N	1.51-2.00 x N	> 2.00 x N
	partial thrombo-plastin time	WNL	1.01-1.66 xN	1.67-2.33 x N	2.34-3.00 x N	> 3.00 x N

Appendix V: Nomogram for Adult Body Surface Area

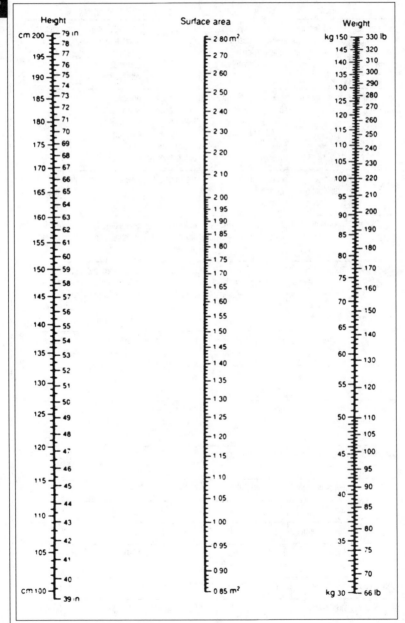

Appendix VI: Karnofsky and Lansky Play Performance Scales

Karnofsky Performance Scale

(For use with persons ages ≥ 17 years)

100%	=	Normal; no complaints, asymptomatic
90%	=	Able to carry on normal activity; minor signs or symptoms of disease
80%	=	Normal activity with effort, some signs or symptoms of disease
70%	=	Cares for self; unable to carry on normal activity or to do active work
60%	=	Requires occasional assistance, but is able to care for most of own needs
50%	=	Requires considerable assistance and frequent medical care
40%	=	Disabled, requires special care and assistance
30%	=	Severely disabled; hospitalization is indicated although death is not imminent
20%	=	Hospitalization necessary, very sick; active supportive treatment necessary
10%	=	Moribund, fatal processes progressing rapidly
0%	=	Dead

Modified Lansky Play Performance Scale

(For use with persons ages 1 through 16 years)

100%	=	Fully active, normal
90%	=	Minor restrictions in physically strenuous activity
80%	=	Active, but tires more quickly
70%	=	Both greater restriction of, and less time spent in, play activities
60%	=	Up and around, but minimal active play; keeps busy with quieter activities
50%	=	Gets dressed but lies around much of the day; no active play; able to participate in all quiet play and activities
40%	=	Mostly in bed; participates in quiet activities
30%	=	Often sleeping; play entirely limited to very passive activities
20%	=	No play; does not get out of bed
10%	=	Unresponsive
0%	=	Dead

Appendix VII: American Heart Association Life Support Algorithms

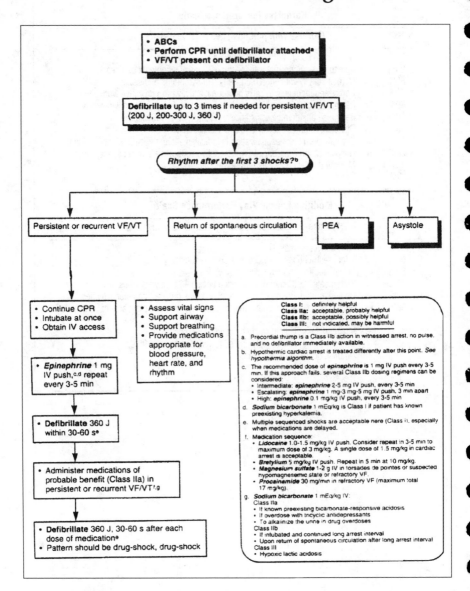

- **ABCs**
- **Perform CPR until defibrillator attached[a]**
- **VF/VT present on defibrillator**

Defibrillate up to 3 times if needed for persistent VF/VT (200 J, 200-300 J, 360 J)

Rhythm after the first 3 shocks?[b]

Persistent or recurrent VF/VT | Return of spontaneous circulation | PEA | Asystole

- Continue CPR
- Intubate at once
- Obtain IV access

- **Epinephrine** 1 mg IV push,[c,d] repeat every 3-5 min

- **Defibrillate** 360 J within 30-60 s[e]

- Administer medications of probable benefit (Class IIa) in persistent or recurrent VF/VT[f,g]

- **Defibrillate** 360 J, 30-60 s after each dose of medication[e]
- Pattern should be drug-shock, drug-shock

- Assess vital signs
- Support airway
- Support breathing
- Provide medications appropriate for blood pressure, heart rate, and rhythm

Class I: definitely helpful
Class IIa: acceptable, probably helpful
Class IIb: acceptable, possibly helpful
Class III: not indicated, may be harmful

a. Precordial thump is a Class IIb action in witnessed arrest, no pulse, and no defibrillator immediately available.

b. Hypothermic cardiac arrest is treated differently after this point. *See hypothermia algorithm.*

c. The recommended dose of *epinephrine* is 1 mg IV push every 3-5 min. If this approach fails, several Class IIb dosing regimens can be considered:
   - Intermediate: *epinephrine* 2-5 mg IV push, every 3-5 min
   - Escalating: *epinephrine* 1 mg-3 mg-5 mg IV push, 3 min apart
   - High: *epinephrine* 0.1 mg/kg IV push, every 3-5 min

d. *Sodium bicarbonate* 1 mEq/kg is Class I if patient has known preexisting hyperkalemia.

e. Multiple sequenced shocks are acceptable here (Class I), especially when medications are delayed.

f. Medication sequence:
   - *Lidocaine* 1.0-1.5 mg/kg IV push. Consider repeat in 3-5 min to maximum dose of 3 mg/kg. A single dose of 1.5 mg/kg in cardiac arrest is acceptable.
   - *Bretylium* 5 mg/kg IV push. Repeat in 5 min at 10 mg/kg.
   - *Magnesium sulfate* 1-2 g IV in torsades de pointes or suspected hypomagnesemic state or refractory VF.
   - *Procainamide* 30 mg/min in refractory VF (maximum total 17 mg/kg).

g. *Sodium bicarbonate* 1 mEq/kg IV:
   Class IIa
   - If known preexisting bicarbonate-responsive acidosis
   - If overdose with tricyclic antidepressants
   - To alkalinize the urine in drug overdoses
   Class IIb
   - If intubated and continued long arrest interval
   - Upon return of spontaneous circulation after long arrest interval
   Class III
   - Hypoxic lactic acidosis

Electrical cardioversion algorithm. Reprinted with permission from JAMA 1992; 268:219902235. Copyright 1992, American Medical Association.

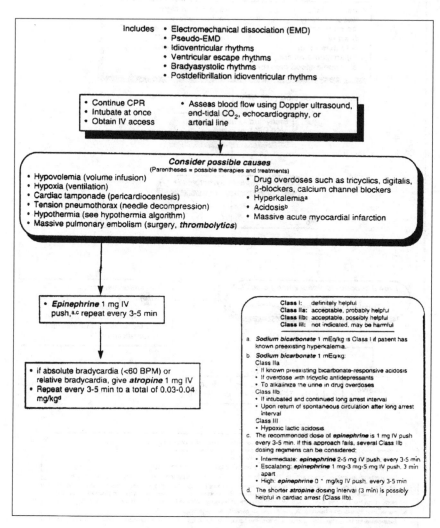

Includes • Electromechanical dissociation (EMD)
          • Pseudo-EMD
          • Idioventricular rhythms
          • Ventricular escape rhythms
          • Bradyasystolic rhythms
          • Postdefibrillation idioventricular rhythms

- Continue CPR
- Intubate at once
- Obtain IV access

- Assess blood flow using Doppler ultrasound, end-tidal CO_2, echocardiography, or arterial line

Consider possible causes
(Parentheses = possible therapies and treatments)

- Hypovolemia (volume infusion)
- Hypoxia (ventilation)
- Cardiac tamponade (pericardiocentesis)
- Tension pneumothorax (needle decompression)
- Hypothermia (see hypothermia algorithm)
- Massive pulmonary embolism (surgery, *thrombolytics*)

- Drug overdoses such as tricyclics, digitalis, β-blockers, calcium channel blockers
- Hyperkalemia[a]
- Acidosis[b]
- Massive acute myocardial infarction

- *Epinephrine* 1 mg IV push,[a,c] repeat every 3-5 min

- if absolute bradycardia (<60 BPM) or relative bradycardia, give *atropine* 1 mg IV
- Repeat every 3-5 min to a total of 0.03-0.04 mg/kg[d]

Class I: definitely helpful
Class IIa: acceptable, probably helpful
Class IIb: acceptable, possibly helpful
Class III: not indicated, may be harmful

a *Sodium bicarbonate* 1 mEq/kg is Class I if patient has known preexisting hyperkalemia.
b *Sodium bicarbonate* 1 mEq/kg:
   Class IIa
   • If known preexisting bicarbonate-responsive acidosis
   • If overdose with tricyclic antidepressants
   • To alkalinize the urine in drug overdoses
   Class IIb
   • If intubated and continued long arrest interval
   • Upon return of spontaneous circulation after long arrest interval
   Class III
   • Hypoxic lactic acidosis
c The recommended dose of *epinephrine* is 1 mg IV push every 3-5 min. If this approach fails, several Class IIb dosing regimens can be considered:
   • Intermediate: *epinephrine* 2-5 mg IV push, every 3-5 min
   • Escalating: *epinephrine* 1 mg-3 mg-5 mg IV push, 3 min apart
   • High: *epinephrine* 0 · mg/kg IV push, every 3-5 min
d The shorter *atropine* dosing interval (3 min) is possibly helpful in cardiac arrest (Class IIb).

Pulseless electrical activity (PEA) algorithm (electromechanical dissociation [EMD]). Reprinted with permission from JAMA 1992; 268:2199-2235. Copyright 1992, American Medical Association.

Appendix

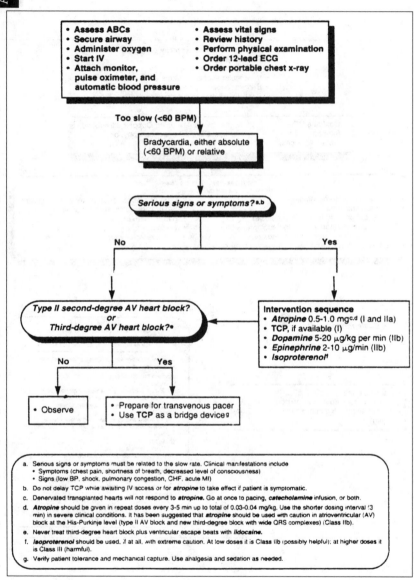

Bradycardia algorithm. Reprinted with permission from JAMA 1992; 268:2199-2235. Copyright 1992, American Medical Association.

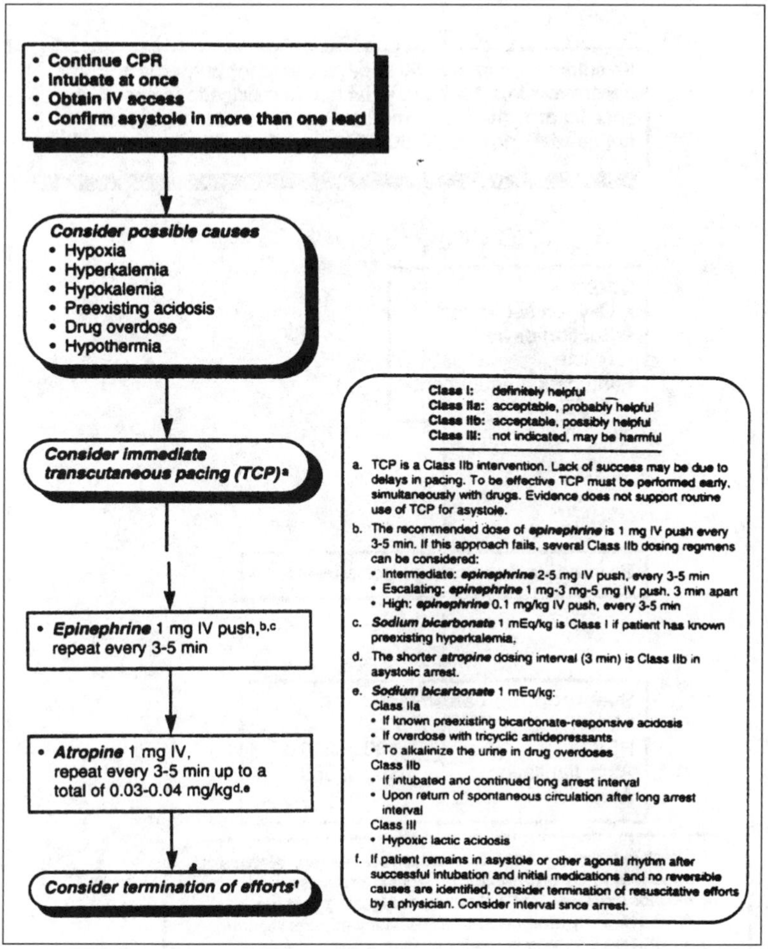
• Continue CPR
• Intubate at once
• Obtain IV access
• Confirm asystole in more than one lead
Consider possible causes
• Hypoxia
• Hyperkalemia
• Hypokalemia
• Preexisting acidosis
• Drug overdose
• Hypothermia
Consider immediate transcutaneous pacing (TCP)a
• Epinephrine 1 mg IV push,b,c repeat every 3-5 min
• Atropine 1 mg IV, repeat every 3-5 min up to a total of 0.03-0.04 mg/kgd,e
Consider termination of effortsf
Class I: definitely helpful
Class IIa: acceptable, probably helpful
Class IIb: acceptable, possibly helpful
Class III: not indicated, may be harmful
a. TCP is a Class IIb intervention. Lack of success may be due to delays in pacing. To be effective TCP must be performed early, simultaneously with drugs. Evidence does not support routine use of TCP for asystole.
b. The recommended dose of epinephrine is 1 mg IV push every 3-5 min. If this approach fails, several Class IIb dosing regimens can be considered:
• Intermediate: epinephrine 2-5 mg IV push, every 3-5 min
• Escalating: epinephrine 1 mg-3 mg-5 mg IV push, 3 min apart
• High: epinephrine 0.1 mg/kg IV push, every 3-5 min
c. Sodium bicarbonate 1 mEq/kg is Class I if patient has known preexisting hyperkalemia.
d. The shorter atropine dosing interval (3 min) is Class IIb in asystolic arrest.
e. Sodium bicarbonate 1 mEq/kg:
Class IIa
• If known preexisting bicarbonate-responsive acidosis
• If overdose with tricyclic antidepressants
• To alkalinize the urine in drug overdoses
Class IIb
• If intubated and continued long arrest interval
• Upon return of spontaneous circulation after long arrest interval
Class III
• Hypoxic lactic acidosis
f. If patient remains in asystole or other agonal rhythm after successful intubation and initial medications and no reversible causes are identified, consider termination of resuscitative efforts by a physician. Consider interval since arrest.

Asystole algorithm (patient is not in cardiac arrest). Reprinted with permission from JAMA 1992; 268:2199-2235. Copyright 1992, American Medical Association.

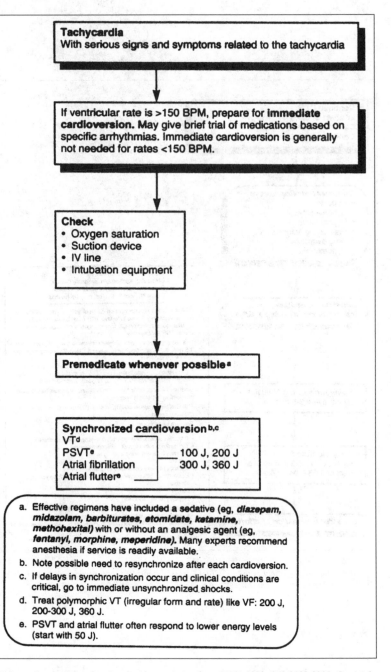

Tachycardia
With serious signs and symptoms related to the tachycardia

⬇

If ventricular rate is >150 BPM, prepare for **immediate cardioversion.** May give brief trial of medications based on specific arrhythmias. Immediate cardioversion is generally not needed for rates <150 BPM.

⬇

Check
- Oxygen saturation
- Suction device
- IV line
- Intubation equipment

⬇

Premedicate whenever possible[a]

⬇

Synchronized cardioversion[b,c]
VT[d]
PSVT[e]   100 J, 200 J
Atrial fibrillation   300 J, 360 J
Atrial flutter[e]

a. Effective regimens have included a sedative (eg, *diazepam, midazolam, barbiturates, etomidate, ketamine, methohexital*) with or without an analgesic agent (eg, *fentanyl, morphine, meperidine*). Many experts recommend anesthesia if service is readily available.
b. Note possible need to resynchronize after each cardioversion.
c. If delays in synchronization occur and clinical conditions are critical, go to immediate unsynchronized shocks.
d. Treat polymorphic VT (irregular form and rate) like VF: 200 J, 200-300 J, 360 J.
e. PSVT and atrial flutter often respond to lower energy levels (start with 50 J).

Tachycardia algorithm. Reprinted with permission from JAMA 1992; 268:2199-2235. Copyright 1992, American Medical Association.

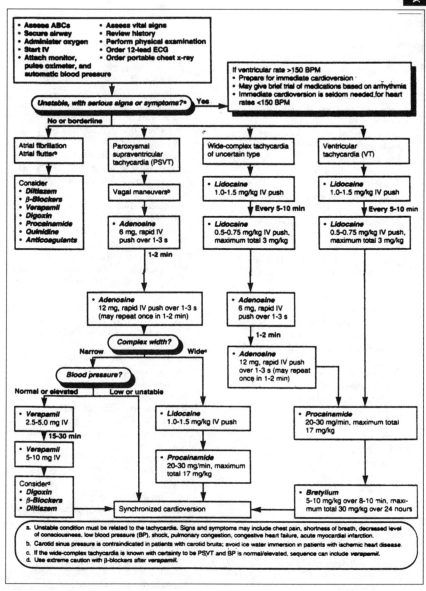

Emergency cardiac care. Reprinted with permission from JAMA 1992; 268:2199-2235. Copyright 1992, American Medical Association.

Index

Index

Notes

Notes

Notes

Notes

Notes

Notes

Notes